Springer Undergraduate Texts in Mathematics and Technology

Springer Undergraduate Texts in Mathematics and Technology (SUMAT) publishes textbooks aimed primarily at the undergraduate. Each text is designed principally for students who are considering careers either in the mathematical sciences or in technology-based areas such as engineering, finance, information technology and computer science, bioscience and medicine, optimization or industry. Texts aim to be accessible introductions to a wide range of core mathematical disciplines and their practical, real-world applications; and are fashioned both for course use and for independent study.

More information about this series at https://www.springer.com/series/7438

Wolfram Koepf

Computer Algebra

An Algorithm-Oriented Introduction

 Springer

Wolfram Koepf
Institut für Mathematik
Universität Kassel
Kassel, Germany

ISSN 1867-5506   ISSN 1867-5514   (electronic)
Springer Undergraduate Texts in Mathematics and Technology
ISBN 978-3-030-78019-7   ISBN 978-3-030-78017-3   (eBook)
https://doi.org/10.1007/978-3-030-78017-3

Mathematics Subject Classification: 68W30, 11Y05, 11Y11, 11Y16, 11R04, 12D05, 12-08, 13F20, 13F25, 33F10

Translation and update from the German language edition: *Computeralgebra - Eine algorithmisch orientierte Einführung* by Wolfram Koepf, © Springer-Verlag Berlin Heidelberg 2006. Published by Springer-Verlag GmbH Germany. All Rights Reserved.

This Springer imprint is published by the registered company Springer Nature Switzerland AG
The registered company address is: Gewerbestrasse 11, 6330 Cham, Switzerland

Preface

This textbook about computer algebra gives an introduction to this modern field of Mathematics. It was published in German language in 2006, based on lectures given at the University of Kassel and containing the material of two lectures of four hours per week each, with additional two hours of exercise classes. When the German edition appeared, I had already taught this topic several times, and in the meantime I have taught it a dozen times. My students were bachelor and master students of Mathematics and Computer Science, as well as high school teacher students with Math as a major. The requirements for this course are low, namely just some knowledge on linear algebra and calculus is required. Hence the book can also be used by high school teachers—who might use computer algebra equipment in their classrooms—as a source for better understanding the underlying computer algebra algorithms. I tried hard to include the most important concepts and algorithms of computer algebra, whose proofs are demonstrated as elementarily as possible. Since I did not want to require a course in algebra, I refrained from using deep algebraic arguments. Rather, the presentation is quite implementation-oriented, and therefore more algorithmic than algebraic. For example, the Chinese remainder theorem is presented as an algorithm and not as an isomorphism theorem. For these reasons, this book does not replace a higher algebra course, but advertises an algebraic consolidation.

In my introductory computer algebra course, I normally lecture the contents of the first eight chapters. These topics seem to be an indispensable foundation of computer algebra. Of course all lecturers can set their own priorities. For example, if the lecturing time is too short, I tend to leave out some theorems on finite fields and point the students to the proofs in the book instead. In contrast, another lecturer might omit the chapter about coding theory and cryptography, although my experience is that the students like these kinds of applications. Another option is to teach the algorithms for integers and for polynomials (for example, the extended Euclidean algorithm) in parallel instead of sequentially. Besides, if knowledge about higher algebra can be assumed, some proofs get much shorter. In any case, all lecturers have enough possibilities to keep their own style.

In the first two chapters, we present the abilities of a general-purpose computer algebra system, and we show how mathematical algorithms can be programmed in such a system. Next, integer arithmetic is introduced, which defines the core of every computer algebra system. After discussing modular arithmetic with the Chinese remainder theorem, the Little Fermat theorem and primality testing, a chapter about coding theory and cryptography follows in which these number-theoretic foundations are applied. Here, the RSA cryptographic scheme is treated, among others.

Next, we consider polynomial arithmetic, which inherits many algorithms from integer arithmetic. Kronecker's algorithm to factorize an integer-coefficient polynomial comes next. In the chapter about algebraic numbers, modular arithmetic is considered from a new point of view, as a result of which finite fields and resultants can be introduced. In the following chapters about polynomial factorization, modern (and much more efficient) algorithms are treated and implemented.

Chapters 9 to 12 build the basis of a more advanced lecture on computer algebra and are influenced by my own research interests. The selected content is very well suited for such an advanced course, since the algorithms treated in the first part are now applied to topics which should be known to the students from their previous studies and are also of interest in applications: simplification and normal forms, power series, summation formulas, and integration. Whereas power series are already treated in higher calculus and turn out to be necessary in Physics, Computer Science, and probability theory, summation formulas appear in many parts of Mathematics and applications. Definite integration is considered a difficult part of analysis, and it is not immediately clear that this problem can be treated with algebraic methods. As a model for algebraic integration, we shall treat the case of rational functions in detail.

Of course I could not include all relevant topics of modern computer algebra. For example, I decided to omit the theory of Gröbner bases although they play a very important role in modern computer algebra systems, in particular for the solution of non-linear systems of equations. In the meantime, Gröbner bases have surpassed the use of resultants for this purpose. However, in order to treat this topic thoroughly, one needs either higher algebra or the space of a book on its own. Therefore, I preferred to refer to existing literature, for example to the very well suited book by Cox, Little, and O'Shea [CLO1997]. Also, I would have liked to include the LLL algorithm by Lenstra, Lenstra, and Lovász [LLL1982] which has very important and interesting applications, but it had to be omitted due to time and space restrictions. Still, treating these two topics seems to be mandatory for a deep understanding of modern computer algebra.

All algorithms considered in this book are implemented and presented in the general-purpose computer algebra system *Mathematica*. Nevertheless, lecturers are absolutely not bound to this system, since the book's definitions and theorems are completely independent of a chosen computer algebra system. Consequently, all computer algebra sessions can be downloaded as worksheets in the three systems *Mathematica*, *Maple*, and *Maxima* from the book's internet page www.computer-algebra.org. *Maxima* is freely available and was used by the author when teaching the same course several times at AIMS Cameroon.[1] The three mentioned systems are general-purpose systems that treat computer algebra in a wide range and are therefore perfectly suitable for this book. They all have the feature that computations can be saved as sessions that can subsequently be reloaded. All instructions which are saved in such a notebook (or worksheet) can be easily computed by the system again after loading the worksheet, which allows the lecturer to use the worksheets for demonstration purposes. In my lectures, I always load the worksheets and let the system process one command line after the other in real-time with the help of a projector, so that the students can see how the system works and generates the computational results. While demonstrating the system's behavior, the lecturer should always present the background information that is needed for the correct and smart use of the system. This procedure gives the students the chance to directly follow the algorithms, to observe whether they are fast or slow, and to suggest their own examples. Furthermore, this scheme gives the students the opportunity to use the algorithms for testing purposes in the sense of experimental Mathematics.

For this reason, I found it important not to describe the algorithms—as in most other books—by so-called pseudo code, but in contrast as executable programs. All sessions considered in the

---

[1] The *African Institute of Mathematical Sciences*, founded in South Africa, now has branches in South Africa, Senegal, Ghana, Cameroon, Tanzania, and Rwanda, and it is still expanding.

framework of this book can be downloaded on www.computer-algebra.org as *Mathematica*, *Maple*, or *Maxima* code, which you can then run on your own computer with the respective computer algebra system. My feeling is that this should be quite useful for many readers. Concerning the exercise classes, in all three systems the students are able to write down their exercise solutions, containing both computations and texts, in respective notebooks (worksheets), which they can hand in electronically for grading. Afterwards, the tutor can return the corrected worksheets to the students.

In the first decade of the current century, I was the chairman of the German Fachgruppe Computeralgebra (www.fachgruppe-computeralgebra.de). In this position, I organized several meetings about *Computer Algebra in Education*, where I was often asked by high school teachers about literature on the essentials of computer algebra for the non-specialist. The current book gives an elementary introduction to the algorithmic foundation of computer algebra systems that is suitable for high school teachers. Since the book is rather elementary and contains a detailed index, it can also be used as a reference manual on algorithms of computer algebra.

As a reference for the first 9 chapters of the present book, I could resort to the books [Chi2000], [GCL1992] and [GG1999]. The book by Lindsay Childs, already written in 1979, is surely one of the first books about computer algebra algorithms, although its title *A Concrete Introduction to Higher Algebra* does not suggest this. Another very important reference is the book by Geddes, Czapor, and Labahn from 1992, a very careful and readable description of the capabilities of *Maple*. Finally, in 1999 the book by Joachim von zur Gathen and Jürgen Gerhard was published, which could well be called the computer algebra bible. However, because of its opulence and the requirement to be familiar with higher algebra, this book seems not directly suitable as an introductory lecture.

The German version of this book would not have been possible without the help of numerous colleagues, staff, and students who helped me with words and deeds, smoothed my text on several occasions, and corrected many small and large errors in my first draft. My special thanks goes to my research colleague Dr. Dieter Schmersau—who unfortunately passed away in 2012—with whom I had endless discussions about the whole material. Further suggestions for improvement have been given by Imran Hafeez, Dr. Peter Horn, Dr. Detlef Müller, Prof. Dr. Reinhard Oldenburg, Dr. Torsten Sprenger, and Jonas Wolf.

One of the typical problems when working with software is that every new release gives you a headache, since some things do not work as they did before. Therefore, when I decided in 2016 to translate my text into English, it was also clear that I had to adapt the computer algebra sessions to the current releases. For example, *Mathematica* had meanwhile decided to incorporate many packages into the core language, so that loading of packages is no longer necessary. Even worse, loading a package as described might even result in errors! Also, some language changes had to be taken care of. Consequently, please be aware that I cannot guarantee that all the sessions will work as described in future releases.

Last but not least, I would like to thank the University of Kassel for granting me three sabbaticals that I could use to write the book and its translation. Also, I would like to thank Clemens Heine from Springer Heidelberg, who was already my project partner for the German version and who now made the English edition possible, and Dr. Loretta Bartolini from Springer New York for providing excellent support during the publication process. Finally, I want to thank Tobias Schwaibold for his thorough copy editing prior to the publication of the English edition.

I should mention that I did not foresee how much work is involved in such a translation and how much time it would occupy when I signed my contract in 2017. Unfortunately, it turned out that the progress was very slow, since I had to manage many different activities in parallel, in particular teaching, research, hosting guests from all over the world (mostly from Africa), and being active in the executive committee of the *Deutsche Mathematiker-Vereinigung* (www.mathematik.de). Among other things, I organized the two workshops *Introduction to Computer Algebra and Applications* and *Introduction to Orthogonal Polynomials and Applications*, together with my Cameroonian colleague Mama Foupouagnigni, which took place in 2017 and 2018 and were generously funded by the *VolkswagenStiftung* (www.aims-volkswagen-workshops.org).

Now that the English translation is ready to be published, I would like to thank Alta Jooste from Pretoria, South Africa, very much, who read the English version extremely carefully and gave many valuable suggestions for improvement. This was of great help to me regarding the completion of the manuscript. Furthermore, I would like to thank one of my Master students, Thomas Izgin, who provided an interesting argument to simplify the proof of Theorem 12.9. My sincere thanks also go to my former PhD students Torsten Sprenger, Henning Seidler, and Kornelia Fischer who had found mistakes in the German Edition that could now be eliminated.

Kassel, April 20th, 2021   Wolfram Koepf

E-Mail: koepf@mathematik.uni-kassel.de
www.mathematik.uni-kassel.de/~koepf

Contents

Chapter 1

Introduction to Computer Algebra

1.1 Capabilities of Computer Algebra Systems

Before we discuss mathematical algorithms and their programming, we want to show the capabilities of a general computer algebra system such as *Mathematica*.[1]

To start with, one can use a computer algebra system like a calculator. Entering as input[2]

$$In[1]:= \frac{1.23+2.25}{3.67}$$

Mathematica computes the output

$Out[1]=$ 0.948229

as expected.

However, computations with decimal numbers go far beyond the capabilities of a calculator. Using the function N, numbers are converted to decimal form. With an (optional) second argument, one can specify an arbitrary precision, bounded only by memory and time restrictions. For example, the command[3]

$In[2]:=$ **N[π, 500]**

computes and subsequently displays the first 500 decimal places of the circular number π:

$Out[2]=$ 3.1415926535897932384626433832795028841971693993751058209749445923078640628620899862803482534211706798214808651328230664709384460955058223172535940812848111745028410270193852110555964462294895493038196442881097566593344612847564823378678316527120190914564856692346034861045432664821339360726024914127372458700660631558817488152092096282925409171536436789259036001133053054882046652138414695194151160943305727036575959195309218611738193261179310511854807446237996274956735188575272489122793818301194 91

---
[1] The same questions can be also treated with the systems *Maple* and *Maxima*, and the corresponding worksheets can be downloaded from `www.computer-algebra.org`.

[2] This text is entered in an empty line by using the key combination <SHIFT> <RETURN>. For editing purposes, one can use the palettes—in particular the palette `Basic Math Input`—which can be loaded via |Palettes, Other, Basic Math Input| if the palette is not visible on the screen. With the help of the palette, one can enter ratios, square roots, etc. A ratio can also be entered using /.

[3] Please note the *square brackets*. Besides, entering π can be done either by `Pi`, through <ESC> p <ESC> or via one of the palettes.

© Springer Nature Switzerland AG 2021
W. Koepf, *Computer Algebra*, Springer Undergraduate Texts in Mathematics and Technology,
https://doi.org/10.1007/978-3-030-78017-3_1

But the characteristic property of a computer algebra system is not the computation with decimal numbers, but *rationally exact arithmetic*. In particular, *Mathematica* can work with arbitrary large integers. For example, the request

```
In[3]:= 100!
Out[3]= 93326215443944152681699238856266700490715968264381621468592963895
           217599993229915608941463976156518286253697920827223758251185210
           91686400000000000000000000000000
```

computes the *factorial* $100! = 100 \cdot 99 \cdots 1$.

Similarly, we get[4]

```
In[4]:= 2^64 - 1
Out[4]= 18446744073709551615
```

```
In[5]:= N[%]
Out[5]= 1.84467 × 10^19
```

where the symbol % refers to the last output.

Note that in numerically oriented programming languages like Pascal or C (on a computer with 64-bit word size), the computed number $2^{64} - 1$ might be the largest integer (`maxint`). In such programming languages, every variable occupies a *fixed* amount of memory, and for this reason all variables have to be declared at the beginning of the program. In computer algebra systems, this is completely different, here the memory amount of every variable is *dynamically* determined at run time. Additionally, the type of a variable is completely free and can be changed arbitrarily.

The example

```
In[6]:= 100!
        ────
        2^100
Out[6]= 58897122236768765137162784634680788828847238288331257425324980425
           6440585603406374176100610302040933304083276457607746124267578125/8
```

shows that *Mathematica* has the ratio *put in lowest terms* automatically.

The sum

$$In[7]:= \sum_{k=1}^{100} \frac{1}{k}$$

$$Out[7]= \frac{14466636279520351160221518043104131447711}{2788815009188490865813523574124921422272}$$

also results in lowest terms, computing the sum of the first 100 positive integer reciprocals. Note that the function Sum represented by the $\sum$ sign is a top-level programming procedure which saves the programming of a loop. *Mathematica* possesses many such top-level programming procedures. The numerical value of the computed sum is obtained as[5]

```
In[8]:= % //N
Out[8]= 5.18738
```

The computed result shows how slowly this series is growing.

Next, let us ask what the divisors of 100! are. This is computed using the command

---

[4] Powers can be entered via the palettes or using the power operator ^.

[5] All *Mathematica* functions can be entered either in *standard form* (N[%]) with square brackets or in postfix form (% // N) with the help of the operator //.

$In[9] :=$ **FactorInteger[100!]**

with the result[6]

$$Out[9] = \begin{pmatrix} 2 & 97 \\ 3 & 48 \\ 5 & 24 \\ 7 & 16 \\ 11 & 9 \\ 13 & 7 \\ 17 & 5 \\ 19 & 5 \\ 23 & 4 \\ 29 & 3 \\ 31 & 3 \\ 37 & 2 \\ 41 & 2 \\ 43 & 2 \\ 47 & 2 \\ 53 & 1 \\ 59 & 1 \\ 61 & 1 \\ 67 & 1 \\ 71 & 1 \\ 73 & 1 \\ 79 & 1 \\ 83 & 1 \\ 89 & 1 \\ 97 & 1 \end{pmatrix}$$

From this output, we can read off that 100! contains 97 times the divisor 2, 48 times the divisor 3, etc. Obviously all prime numbers up to 100 occur as divisors of 100!.

The request

$In[10] :=$ **PrimeQ[1234567]**
$Out[10] =$ False

has the value `False`, hence 1234567 is no prime number.[7] Here are its divisors:

$In[11] :=$ **FactorInteger[1234567]**
$$Out[11] = \begin{pmatrix} 127 & 1 \\ 9721 & 1 \end{pmatrix}$$

Now we *implement* a *Mathematica* function `Nextprime[n]`,[8] which outputs $n+1$, if $n+1$ is a prime, and computes the next larger prime number otherwise.[9]

---

[6] The result is a list of lists which is shown in the given *matrix form* if the setting `TradionalForm` is selected via Format, Option Inspector, Cell Options, New Cell Defaults, Common Default Format Types, Output, TraditionalForm . In this book, we generally use this output form. Lists are entered in curly braces, for example {a,b}.

[7] If `PrimeQ[n]` has the value `False`, then n is provably composite; however, if `PrimeQ[n]` is `True`, then n is a *pseudoprime number*, i.e., it is a prime with a very high probability, see also Section 4.6. *Mathematica's* help pages state under *Implementation notes* that `PrimeQ` first tests for divisibility using small primes, then uses the Miller-Rabin strong pseudoprime test base 2 and base 3, and then uses a Lucas test.

[8] *Mathematica* has a built-in function `NextPrime` for the same purpose.

[9] Functions are usually declared using the *delayed assignment* `:=`. Such assigments are executed once the functions are called (and not immediately), therefore the definition does not generate an output line. For the definition of one function, multiple assignments are admissible. This generates simple and well-organized case distinctions which we will frequently use. Note that the character `_` signals that n is a variable, hence `n_` stands for "a variable named n".

In[12]:= **Nextprime[n_] := n + 1/; PrimeQ[n + 1]**
         Nextprime[n_] := Nextprime[n + 1]

This definition reads as follows: Nextprime[n] equals $n + 1$ if $n + 1$ is a prime; otherwise the next prime of n is declared as the next prime of $n + 1$. Hence, if $n + 1$ is not a prime, then n is increased (using the second condition) until a prime is found and the first condition applies. In such a *recursive program*, a defined function calls itself until a *break condition* is satisfied. In our example, the break condition is given by the first line. More about this and other programming techniques in *Mathematica* can be found in Chapter 2.

Now we can compute

In[13]:= **Nextprime[1234567]**
Out[13]= 1234577

Hence the next prime larger than 1234567 is 1234577. We check this by

In[14]:= **PrimeQ[%]**
Out[14]= True

Mathematica can deal symbolically with square roots and other algebraic numbers, see Chapter 7. The input

In[15]:= **x = $\sqrt{2} + \sqrt{3}$**
Out[15]= $\sqrt{2} + \sqrt{3}$

provides the sum of two square roots and assigns the result to the variable x. An automatic simplification is not executed even if possible. Next, we compute the reciprocal

In[16]:= $\dfrac{1}{x}$
Out[16]= $\dfrac{1}{\sqrt{2} + \sqrt{3}}$

and simplify

In[17]:= $\dfrac{1}{x}$**//Simplify**
Out[17]= $\dfrac{1}{\sqrt{2} + \sqrt{3}}$

In the current situation, *Mathematica's* simplification command Simplify is not successful. However, the command FullSimplify, which takes care of quite many simplifications and therefore often needs long computation times, is able to simplify our number in a certain way:

In[18]:= $\dfrac{1}{x}$**//FullSimplify**
Out[18]= $\sqrt{5 - 2\sqrt{6}}$

In Section 7.3, we will discuss in detail how such a simplification is possible.

We will find yet another simplification by declaring

In[19]:= **y = $\sqrt{3} - \sqrt{2}$**
Out[19]= $-\sqrt{2} + \sqrt{3}$

Then the following simplification is indeed executed[10]

In[20]:= **x * y//Simplify**
Out[20]= 1

---

[10] which is easy by the binomial formula $(a + b)(a - b) = a^2 - b^2$

showing that the reciprocal is given by $\frac{1}{x} = y$. Such representations are also introduced in Section 7.3.

Mathematica also knows some special algebraic values of certain *transcendental functions*:

$In[21]:=$ **Sin[$\frac{\pi}{5}$]**

$Out[21]=$ $\sqrt{\frac{5}{8} - \frac{\sqrt{5}}{8}}$

$In[22]:=$ **Sin[$\frac{\pi}{5}$] Cos[$\frac{\pi}{5}$]//Simplify**

$Out[22]=$ $\frac{1}{8} \sqrt{\frac{1}{2} (5 - \sqrt{5}) (1 + \sqrt{5})}$

For this example, **FullSimplify** is more successful:

$In[23]:=$ **Sin[$\frac{\pi}{5}$] Cos[$\frac{\pi}{5}$]//FullSimplify**

$Out[23]=$ $\frac{1}{4} \sqrt{\frac{1}{2} (5 + \sqrt{5})}$

In certain cases, an automatic simplification occurs:

$In[24]:=$ **Sin[π]**
$Out[24]=$ 0

Mathematica can also deal with *complex numbers and functions*:

$In[25]:=$ **$\frac{1 + i}{1 - i}$**
$Out[25]=$ i

$In[26]:=$ **Re[2 Exp[3x + i y]]**
$Out[26]=$ $2 e^{3(\sqrt{2} + \sqrt{3})} \cos(\sqrt{2} - \sqrt{3})$

$In[27]:=$ **Clear[x, y]**

$In[28]:=$ **ComplexExpand[Re[2 Exp[3x + i y]]]**
$Out[28]=$ $2 e^{3x} \cos(y)$

With **Clear[x,y]** we deleted the previously assigned values of the variables x and y. The special symbols i and e represent the imaginary unit i and the Eulerian number e, respectively.[11] The fact that computer algebra systems can do computations with symbols, like in the above examples, distinguishes them most remarkably from numeric programming languages. Note that **ComplexExpand** simplifies under the condition that the occurring variables are real.

Mathematica can deal with *polynomials and rational functions*:

$In[29]:=$ **pol = (x + y)^{10} - (x - y)^{10}**
$Out[29]=$ $(x + y)^{10} - (x - y)^{10}$

$In[30]:=$ **Expand[pol]**
$Out[30]=$ $20 x^9 y + 240 x^7 y^3 + 504 x^5 y^5 + 240 x^3 y^7 + 20 x y^9$

With **Expand** polynomials are expanded, whereas **Factor** carries out polynomial factorizations:

$In[31]:=$ **Factor[pol]**

---

[11] The input of i or e can be entered either by using I or E, or by using <ESC> ii <ESC> or <ESC> ee <ESC>, or by using the palettes.

$Out[31] = 4xy\left(5x^4 + 10y^2x^2 + y^4\right)\left(x^4 + 10y^2x^2 + 5y^4\right)$

Polynomial factorization is one of the real highlights of computer algebra systems, since such factorizations are practically impossible by manual computation. Many algebraic concepts are utilized for this purpose, some of which we will deal with in Section 6.7 and in Chapter 8.

When entering a rational function,

$In[32] := \mathbf{rat} = \dfrac{\mathbf{1 - x^{10}}}{\mathbf{1 - x^4}}$

$Out[32] = \dfrac{1 - x^{10}}{1 - x^4}$

the expression is not automatically simplified, i.e., reduced to lowest terms. This reduction is handled by the simplification command `Together`:

$In[33] := \mathbf{Together[rat]}$

$Out[33] = \dfrac{x^8 + x^6 + x^4 + x^2 + 1}{x^2 + 1}$

Rational functions can also be factorized:

$In[34] := \mathbf{Factor[rat]}$

$Out[34] = \dfrac{(x^4 - x^3 + x^2 - x + 1)\,(x^4 + x^3 + x^2 + x + 1)}{x^2 + 1}$

Note that factorizations depend on the considered ring. E.g., the computation

$In[35] := \mathbf{Factor[x^6 + x^2 + 1]}$

$Out[35] = x^6 + x^2 + 1$

shows that the polynomial $x^6 + x^2 + 1$, considered as element of the polynomial ring $\mathbb{Z}[x]$, i.e., the polynomials with integer coefficients, is not reducible. However, if the "same polynomial" is considered modulo 13, then nontrivial factors exist:[12]

$In[36] := \mathbf{pol = Factor[x^6 + x^2 + 1, Modulus \rightarrow 13]}$

$Out[36] = \left(x^2 + 6\right)\left(x^4 + 7x^2 + 11\right)$

By the way, the fact that the resulting polynomial agrees with the original one modulo 13 can be seen by considering the difference

$In[37] := \mathbf{Expand[pol - (x^6 + x^2 + 1)]}$

$Out[37] = 13x^4 + 52x^2 + 65$

is a polynomial whose coefficients are divisible by 13.

The polynomial $x^4 + 1 \in \mathbb{Z}[x]$ is irreducible:

$In[38] := \mathbf{Factor[x^4 + 1]}$

$Out[38] = x^4 + 1$

In the algebraic extension ring $\mathbb{Z}[i]$ of $\mathbb{Z}$, however, a factorization exists:

$In[39] := \mathbf{Factor[x^4 + 1, GaussianIntegers \rightarrow True]}$

$Out[39] = \left(x^2 - i\right)\left(x^2 + i\right)$

Another factorization can be found in the extension field $\mathbb{Q}(\sqrt{2})$:

$In[40] := \mathbf{Factor[x^4 + 1, Extension \rightarrow \{\sqrt{2}\}]}$

$Out[40] = -\left(-x^2 + \sqrt{2}\,x - 1\right)\left(x^2 + \sqrt{2}\,x + 1\right)$

---

[12] The sign $\rightarrow$ can be entered via the palette or by typing `->` on the keyboard.

Unfortunately, in general, *Mathematica* cannot help us to find an appropriate extension field over $\mathbb{Q}$ for a successful factorization.

Computations in residue class rings like $\mathbb{Z}_{13}$ will be treated in Section 4.1, $\mathbb{Z}[i]$ in Example 6.1, and algebraic extension fields are treated in Section 7.1.

Mathematica also has many capabilities in *linear algebra*. As a fist example, we compute the *Hilbert matrix* for $n = 7$:

$In[41]:=$ **H = HilbertMatrix[7]**

$$Out[41]=\begin{pmatrix} 1 & \frac{1}{2} & \frac{1}{3} & \frac{1}{4} & \frac{1}{5} & \frac{1}{6} & \frac{1}{7} \\ \frac{1}{2} & \frac{1}{3} & \frac{1}{4} & \frac{1}{5} & \frac{1}{6} & \frac{1}{7} & \frac{1}{8} \\ \frac{1}{3} & \frac{1}{4} & \frac{1}{5} & \frac{1}{6} & \frac{1}{7} & \frac{1}{8} & \frac{1}{9} \\ \frac{1}{4} & \frac{1}{5} & \frac{1}{6} & \frac{1}{7} & \frac{1}{8} & \frac{1}{9} & \frac{1}{10} \\ \frac{1}{5} & \frac{1}{6} & \frac{1}{7} & \frac{1}{8} & \frac{1}{9} & \frac{1}{10} & \frac{1}{11} \\ \frac{1}{6} & \frac{1}{7} & \frac{1}{8} & \frac{1}{9} & \frac{1}{10} & \frac{1}{11} & \frac{1}{12} \\ \frac{1}{7} & \frac{1}{8} & \frac{1}{9} & \frac{1}{10} & \frac{1}{11} & \frac{1}{12} & \frac{1}{13} \end{pmatrix}$$

The Hilbert matrix is notorious for its weak *condition* for large n, i.e., its *inverse* and *determinant* cannot be computed accurately using decimal arithmetic. The inverse of the Hilbert matrix has integer coefficients,

$In[42]:=$ **Inverse[H]**

$$Out[42]=\begin{pmatrix} 49 & -1176 & 8820 & -29400 & 48510 & -38808 & 12012 \\ -1176 & 37632 & -317520 & 1128960 & -1940400 & 1596672 & -504504 \\ 8820 & -317520 & 2857680 & -10584000 & 18711000 & -15717240 & 5045040 \\ -29400 & 1128960 & -10584000 & 40320000 & -72765000 & 62092800 & -20180160 \\ 48510 & -1940400 & 18711000 & -72765000 & 133402500 & -115259760 & 37837800 \\ -38808 & 1596672 & -15717240 & 62092800 & -115259760 & 100590336 & -33297264 \\ 12012 & -504504 & 5045040 & -20180160 & 37837800 & -33297264 & 11099088 \end{pmatrix}$$

and its determinant is a very small rational number which can be determined exactly by using rational arithmetic:

$In[43]:=$ **Det[H]**

$$Out[43]= \frac{1}{2067909047925770649600000}$$

The following function defines the *Vandermonde matrix*:

$In[44]:=$ **Vandermonde[n_] := Table[x_j^{k-1}, {j,n}, {k,n}]**

Then the Vandermonde matrix with 5 variables $x_1, ..., x_5$ is given by

$In[45]:=$ **V = Vandermonde[5]**

$$Out[45]=\begin{pmatrix} 1 & x_1 & x_1^2 & x_1^3 & x_1^4 \\ 1 & x_2 & x_2^2 & x_2^3 & x_2^4 \\ 1 & x_3 & x_3^2 & x_3^3 & x_3^4 \\ 1 & x_4 & x_4^2 & x_4^3 & x_4^4 \\ 1 & x_5 & x_5^2 & x_5^3 & x_5^4 \end{pmatrix}$$

and its determinant is the very complicated multivariate polynomial[13]

$In[46]:=$ **Expand[Det[V]]**

$Out[46]=$ $-x_2\,x_3^2\,x_4^3\,x_1^4 + x_2^2\,x_3\,x_4^3\,x_1^4 + x_2\,x_3^2\,x_5^3\,x_1^4 - x_2\,x_4^2\,x_5^3\,x_1^4 + x_3\,x_4^2\,x_5^3\,x_1^4 - x_2^2\,x_3\,x_5^3\,x_1^4 +$
$x_2^2\,x_4\,x_5^3\,x_1^4 - x_3^2\,x_4\,x_5^3\,x_1^4 + x_2\,x_3^3\,x_4^2\,x_1^4 - x_3^3\,x_4^2\,x_1^4 - x_2\,x_3^3\,x_5^2\,x_1^4 + x_2\,x_4^3\,x_5^2\,x_1^4 -$
$x_3\,x_4^3\,x_5^2\,x_1^4 + x_2^3\,x_3\,x_5^2\,x_1^4 - x_2^3\,x_4\,x_5^2\,x_1^4 + x_3^3\,x_4\,x_5^2\,x_1^4 - x_2^3\,x_3\,x_4\,x_1^4 + x_2^3\,x_3^2\,x_4\,x_1^4 +$
$x_2^2\,x_3^3\,x_5\,x_1^4 - x_2^2\,x_4^3\,x_5\,x_1^4 + x_3^2\,x_4^3\,x_5\,x_1^4 - x_2^3\,x_3^2\,x_5\,x_1^4 + x_2^3\,x_4^2\,x_5\,x_1^4 - x_3^3\,x_4^2\,x_5\,x_1^4 +$
$x_2\,x_3^3\,x_4^4\,x_1^3 - x_2^2\,x_3\,x_4^4\,x_1^3 - x_2\,x_3^3\,x_5^4\,x_1^3 + x_2\,x_4^3\,x_5^4\,x_1^3 - x_3\,x_4^3\,x_5^4\,x_1^3 + x_2^2\,x_3\,x_5^4\,x_1^3 -$
$x_2^2\,x_4\,x_5^4\,x_1^3 + x_3^2\,x_4\,x_5^4\,x_1^3 - x_2\,x_3^4\,x_4^2\,x_1^3 + x_4^2\,x_3\,x_4^4\,x_1^3 + x_2\,x_4^3\,x_5^4\,x_1^3 -$
$x_3\,x_4^4\,x_5^2\,x_1^3 - x_2^4\,x_3\,x_5^2\,x_1^3 + x_2^4\,x_4\,x_5^2\,x_1^3 - x_3^4\,x_4\,x_5^2\,x_1^3 + x_2^4\,x_3\,x_4\,x_1^3 -$
$x_2^4\,x_3^2\,x_4\,x_1^3 - x_2^2\,x_3^4\,x_5\,x_1^3 + x_2^2\,x_4^4\,x_5\,x_1^3 - x_3^2\,x_4^4\,x_5\,x_1^3 + x_2^4\,x_3^2\,x_5\,x_1^3 - x_2^4\,x_4^2\,x_5\,x_1^3 +$
$x_3^4\,x_4^2\,x_5\,x_1^3 - x_2\,x_3^3\,x_4^4\,x_1^2 + x_2^2\,x_3\,x_4^4\,x_1^2 + x_2\,x_3^3\,x_5^4\,x_1^2 - x_2\,x_4^3\,x_5^4\,x_1^2 + x_3\,x_4^3\,x_5^4\,x_1^2 -$
$x_2^3\,x_3\,x_5^4\,x_1^2 + x_2^3\,x_4\,x_5^4\,x_1^2 - x_3^3\,x_4\,x_5^4\,x_1^2 + x_2\,x_3^4\,x_4^3\,x_1^2 - x_2\,x_3\,x_4^3\,x_1^2 - x_2\,x_3^4\,x_5^3\,x_1^2 +$
$x_2\,x_4^4\,x_5^3\,x_1^2 - x_3\,x_4^4\,x_5^3\,x_1^2 + x_2^4\,x_3\,x_5^3\,x_1^2 - x_2^4\,x_4\,x_5^3\,x_1^2 + x_3^4\,x_4\,x_5^3\,x_1^2 - x_2^3\,x_3^4\,x_4\,x_1^2 +$
$x_2^4\,x_3^3\,x_4\,x_1^2 + x_2^3\,x_3^4\,x_5\,x_1^2 - x_2^3\,x_4^4\,x_5\,x_1^2 + x_3^3\,x_4^4\,x_5\,x_1^2 - x_2^4\,x_3^3\,x_5\,x_1^2 + x_2^4\,x_4^3\,x_5\,x_1^2 -$
$x_3^4\,x_4^3\,x_5\,x_1^2 + x_2^2\,x_3^3\,x_4^4\,x_1 - x_2^3\,x_3^2\,x_4^4\,x_1 - x_2^2\,x_3^3\,x_5^4\,x_1 + x_2^2\,x_4^3\,x_5^4\,x_1 -$
$x_3^2\,x_4^3\,x_5^4\,x_1 + x_2^3\,x_3^2\,x_5^4\,x_1 - x_2^3\,x_4^2\,x_5^4\,x_1 + x_3^3\,x_4^2\,x_5^4\,x_1 - x_2^2\,x_3^4\,x_4^3\,x_1 + x_2^3\,x_3^4\,x_5^3\,x_1 -$
$x_2^2\,x_4^4\,x_5^3\,x_1 + x_3^2\,x_4^4\,x_5^3\,x_1 - x_2^4\,x_3^2\,x_5^3\,x_1 + x_2^4\,x_4^2\,x_5^3\,x_1 - x_3^4\,x_4^2\,x_5^3\,x_1 + x_2^3\,x_3^4\,x_4^2\,x_1 -$
$x_2^4\,x_3^3\,x_4^2\,x_1 - x_2^3\,x_3^2\,x_4^4\,x_1 + x_2^2\,x_3^4\,x_4^3\,x_5 - x_2^2\,x_3^3\,x_4^4\,x_5 +$
$x_2\,x_3^3\,x_4^2\,x_5^4 + x_2^3\,x_3\,x_4^2\,x_5^4 + x_2^2\,x_3\,x_4^3\,x_5^4 - x_2^3\,x_3^2\,x_4\,x_5^4 - x_2\,x_2^2\,x_4^4\,x_5^3 + x_2^2\,x_3\,x_4^4\,x_5^3 +$
$x_2\,x_3^4\,x_4^2\,x_5^3 - x_2^2\,x_3\,x_4^2\,x_5^3 - x_2^2\,x_3^4\,x_4\,x_5^3 + x_2^4\,x_3^2\,x_4\,x_5^3 + x_2\,x_3^4\,x_4^4\,x_5^2 -$
$x_2^3\,x_3\,x_4^4\,x_5^2 - x_2\,x_4^4\,x_3^2\,x_5^2 + x_2^4\,x_3\,x_3^2\,x_5^2 + x_2^3\,x_3^4\,x_4\,x_5^2 - x_2^4\,x_3^3\,x_4\,x_5^2 -$
$x_2^2\,x_3^3\,x_4^4\,x_5 + x_2^3\,x_3^2\,x_4^4\,x_5 + x_2^2\,x_3^4\,x_4^3\,x_5 - x_2^4\,x_3^2\,x_4^3\,x_5 - x_2^3\,x_3^4\,x_4^2\,x_5 + x_2^4\,x_3^3\,x_4^2\,x_5$

which is very simple in factorized form:[14]

$In[47]:=$ **Factor[Det[V]]**

$Out[47]=$ $(x_1-x_2)\,(x_1-x_3)\,(x_2-x_3)\,(x_1-x_4)\,(x_2-x_4)\,(x_3-x_4)\,(x_1-x_5)\,(x_2-x_5)\,(x_3-x_5)\,(x_4-x_5)$

With *Mathematica* one can also solve *equations* and *systems of equations*. As an example, here are the solutions of a quadratic equation:[15]

$In[48]:=$ **s = Solve[x^2 - 3x - 1 == 0, x]**

$Out[48]=$ $\left\{\left\{x \to \frac{1}{2}\left(3-\sqrt{13}\right)\right\}, \left\{x \to \frac{1}{2}\left(3+\sqrt{13}\right)\right\}\right\}$

The solutions of an equation are given in the form of a list. In our case, there are two solutions, therefore our solution list contains two elements, each of them being lists of the form $\{x \to a\}$, where x denotes the variable and a the corresponding solution. Such solution lists can be used—with the substitution command / .—for substitution purposes:

$In[49]:=$ **x/.s**

$Out[49]=$ $\left\{\frac{1}{2}\left(3-\sqrt{13}\right), \frac{1}{2}\left(3+\sqrt{13}\right)\right\}$

---

[13] The order in which these summands are displayed depends on the choice of **StandardForm** or **TraditionalForm**, but also on the used *Mathematica* version. Since the outputs in different *Mathematica* releases vary, we did not change the presentations of rather long, but equivalent outputs given in the German edition of this book. Some of the *Mathematica* outputs in this book might therefore differ from what the reader obtains when using another version of *Mathematica*.

[14] Can you see why the computed factors *must* be divisors of the determinant?

[15] An equation is entered by using the sign ==.

Next, we solve an equation of degree three:

```
In[50]:= s = Solve[x³ - 3x - 1 == 0, x, Cubics → True]
```

$$Out[50]= \left\{\left\{x \to \frac{1}{\sqrt[3]{\frac{1}{2}\left(1+i\sqrt{3}\right)}} + \sqrt[3]{\frac{1}{2}\left(1+i\sqrt{3}\right)}\right\},\right.$$

$$\left\{x \to -\frac{1}{2}\left(1-i\sqrt{3}\right)\sqrt[3]{\frac{1}{2}\left(1+i\sqrt{3}\right)} - \left(\frac{1}{2}\left(1+i\sqrt{3}\right)\right)^{2/3}\right\},$$

$$\left.\left\{x \to -\frac{1-i\sqrt{3}}{2^{2/3}\sqrt[3]{1+i\sqrt{3}}} - \frac{\left(1+i\sqrt{3}\right)^{4/3}}{2\sqrt[3]{2}}\right\}\right\}$$

Its solution contains rather complicated roots. Their decimal values are given by

```
In[51]:= s//N
```

$$Out[51]= \left\{\{x \to 1.87939 + 0.\, i\}, \{x \to -1.53209 + 2.22045 \times 10^{-16}\, i\},\right.$$
$$\left.\{x \to -0.347296 + 1.11022 \times 10^{-16}\, i\}\right\}$$

All three zeros are potentially real which can be confirmed using numerical methods with
NSolve:

```
In[52]:= NSolve[x³ - 3x - 1 == 0, x]
```

$$Out[52]= \{\{x \to -1.53209\}, \{x \to -0.347296\}, \{x \to 1.87939\}\}$$

Note that the above algebraic representation by *Cardano's formula* explicitly uses complex
numbers.[16] The fact that the polynomial has three real zeros, can also be seen from the graphical
representation:

```
In[53]:= Plot[x³ - 3x - 1, {x, -3, 3}]
```

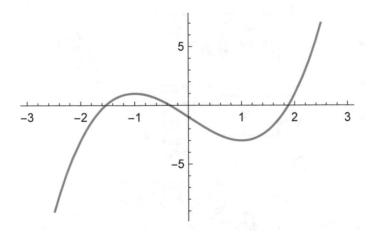

As is well known, the solutions of polynomial equations up to degree four can be represented
by (nested) roots. These solutions are, however, often quite useless because of their complexity:

```
In[54]:= s = Solve[x⁴ - 3x - 1 == 0, x, Quartics → True]
```

---

[16] The current case is called the *casus irreducibilis*, where the solutions also have a real representation using
trigonometric functions. This representation, however, is not supported by *Mathematica*.

$$Out[54]= \left\{\left\{x \to -\frac{1}{2}\sqrt{-4\sqrt[3]{\frac{2}{3\left(81+\sqrt{7329}\right)}}+\frac{\sqrt[3]{\frac{1}{2}\left(81+\sqrt{7329}\right)}}{3^{2/3}}-}\right.\right.$$

$$\frac{1}{2}\sqrt{\left(4\sqrt[3]{\frac{2}{3\left(81+\sqrt{7329}\right)}}-\frac{\sqrt[3]{\frac{1}{2}\left(81+\sqrt{7329}\right)}}{3^{2/3}}-\right.}$$

$$\left.\left.\left.\frac{6}{\sqrt{-4\sqrt[3]{\frac{2}{3\left(81+\sqrt{7329}\right)}}+\frac{\sqrt[3]{\frac{1}{2}\left(81+\sqrt{7329}\right)}}{3^{2/3}}}}\right.\right)\right\},$$

$$\left\{x \to -\frac{1}{2}\sqrt{-4\sqrt[3]{\frac{2}{3\left(81+\sqrt{7329}\right)}}+\frac{\sqrt[3]{\frac{1}{2}\left(81+\sqrt{7329}\right)}}{3^{2/3}}+}\right.$$

$$\frac{1}{2}\sqrt{\left(4\sqrt[3]{\frac{2}{3\left(81+\sqrt{7329}\right)}}-\frac{\sqrt[3]{\frac{1}{2}\left(81+\sqrt{7329}\right)}}{3^{2/3}}-\right.}$$

$$\left.\left.\frac{6}{\sqrt{-4\sqrt[3]{\frac{2}{3\left(81+\sqrt{7329}\right)}}+\frac{\sqrt[3]{\frac{1}{2}\left(81+\sqrt{7329}\right)}}{3^{2/3}}}}\right)\right\},$$

$$\left\{x \to \frac{1}{2}\sqrt{-4\sqrt[3]{\frac{2}{3\left(81+\sqrt{7329}\right)}}+\frac{\sqrt[3]{\frac{1}{2}\left(81+\sqrt{7329}\right)}}{3^{2/3}}-}\right.$$

$$\frac{1}{2}\sqrt{\left(4\sqrt[3]{\frac{2}{3\left(81+\sqrt{7329}\right)}}-\frac{\sqrt[3]{\frac{1}{2}\left(81+\sqrt{7329}\right)}}{3^{2/3}}+\right.}$$

$$\left.\left.\frac{6}{\sqrt{-4\sqrt[3]{\frac{2}{3\left(81+\sqrt{7329}\right)}}+\frac{\sqrt[3]{\frac{1}{2}\left(81+\sqrt{7329}\right)}}{3^{2/3}}}}\right)\right\},$$

$$\left\{x \to \frac{1}{2}\sqrt{-4\sqrt[3]{\frac{2}{3\left(81+\sqrt{7329}\right)}}+\frac{\sqrt[3]{\frac{1}{2}\left(81+\sqrt{7329}\right)}}{3^{2/3}}+}\right.$$

$$\frac{1}{2}\sqrt{\left(4\sqrt[3]{\frac{2}{3\left(81+\sqrt{7329}\right)}}-\frac{\sqrt[3]{\frac{1}{2}\left(81+\sqrt{7329}\right)}}{3^{2/3}}+\right.}$$

$$\left.\left.\left.\frac{6}{\sqrt{-4\sqrt[3]{\frac{2}{3\left(81+\sqrt{7329}\right)}}+\frac{\sqrt[3]{\frac{1}{2}\left(81+\sqrt{7329}\right)}}{3^{2/3}}}}\right)\right\}\right\}$$

The solution has the numerical representation

$In[55]:=$ **s//N**
$Out[55]=$ $\{\{x \rightarrow -0.605102 - 1.26713\, i\},$
$\{x \rightarrow -0.605102 + 1.26713\, i\}, \{x \rightarrow -0.329409\}, \{x \rightarrow 1.53961\}\}$

with two real solutions, which can also be seen from the following plot:

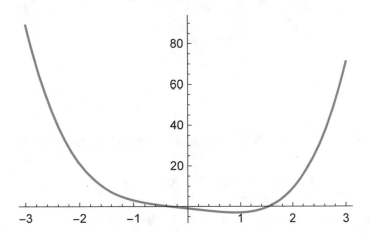

Solutions of polynomial equations of degree larger than four are generally represented as algebraic objects:

$In[56]:=$ **s = Solve[x^5 - 3x - 1 == 0, x]**
$Out[56]=$ $\{\{x \rightarrow \text{Root}[\#1^5 - 3\,\#1 - 1\,\&, 1]\},$
$\{x \rightarrow \text{Root}[\#1^5 - 3\,\#1 - 1\,\&, 2]\}, \{x \rightarrow \text{Root}[\#1^5 - 3\,\#1 - 1\,\&, 3]\},$
$\{x \rightarrow \text{Root}[\#1^5 - 3\,\#1 - 1\,\&, 4]\}, \{x \rightarrow \text{Root}[\#1^5 - 3\,\#1 - 1\,\&, 5]\}\}$

This answer seems to just repeat the question.[17] However, the outputs are `Root` objects that enable computations both numerically,[18]

$In[57]:=$ **s//N**
$Out[57]=$ $\{\{x \rightarrow -1.21465\}, \{x \rightarrow -0.334734\}, \{x \rightarrow 1.38879\},$
$\{x \rightarrow 0.0802951 - 1.32836\, i\}, \{x \rightarrow 0.0802951 + 1.32836\, i\}\}$

and symbolically:

$In[58]:=$ $\displaystyle\prod_{k=1}^{5}$ **(x/.s[[k]])//RootReduce**
$Out[58]=$ 1

The last example computed the product of the 5 different zeros, three of which are real, as the following plot confirms:

---

[17] The solution can be interpreted as follows: first zero, second zero, ..., fifth zero of the polynomial $x^5 - x - 1$.
[18] The input `s[[k]]` yields the k-th element of the list s.

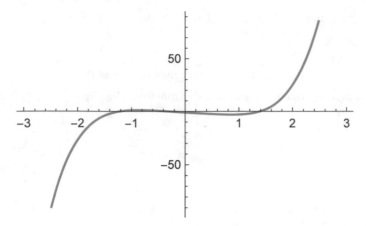

As another highlight of modern computer algebra, Solve can also solve systems of polynomial equations:

$In[59]:=$ **Solve[{x² + y² == 1, -x² + 2y² + 1 == 0}, {x, y}]**

$Out[59]=$ $\{\{x \to -1, y \to 0\}, \{x \to 1, y \to 0\}\}$

For the graphical representation of the two implicit equations, we use the command ContourPlot:

$In[60]:=$ **ContourPlot[{x² + y² == 1, -x² + 2y² + 1 == 0},**
        {x, -3, 3}, {y, -3, 3}]

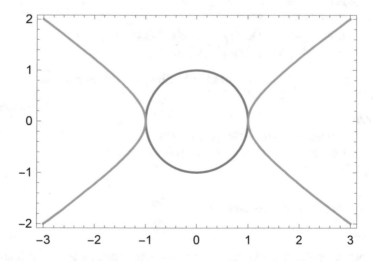

In accordance with the computation, the circle and the hyperbola have two (double) intersection points, whereas the following ellipse and hyperbola have four different intersections:

$In[61]:=$ **Solve$\left[\left\{\dfrac{x^2}{4} + y² == 1, -x² + 2y² + 1 == 0\right\}, \{x, y\}\right]$**

$Out[61]=$ $\left\{\left\{x \to -\sqrt{2}, y \to -\dfrac{1}{\sqrt{2}}\right\}, \left\{x \to -\sqrt{2}, y \to \dfrac{1}{\sqrt{2}}\right\},\right.$

$\left.\left\{x \to \sqrt{2}, y \to -\dfrac{1}{\sqrt{2}}\right\}, \left\{x \to \sqrt{2}, y \to \dfrac{1}{\sqrt{2}}\right\}\right\}$

$In[62]:=$ **ContourPlot** $\left[\left\{\frac{\mathbf{x}^2}{4}+\mathbf{y}^2 == 1, -\mathbf{x}^2 + 2\mathbf{y}^2 + 1 == 0\right\},\right.$
$\left.\{\mathbf{x}, -3, 3\}, \{\mathbf{y}, -3, 3\}\right]$

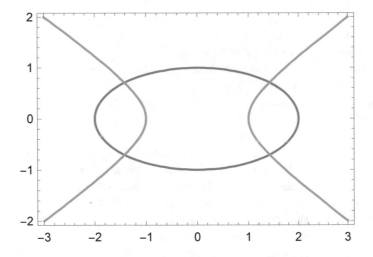

The solution of polynomial systems of equations will be treated in Section 7.6.

Another highlight of *Mathematica* are graphical representations, some of which we have already seen. Here are the graphs of several trigonometric functions:

$In[63]:=$ **Plot[{Sin[x], Cos[x], Tan[x]}, {x, -5, 5},**
PlotStyle → {RGBColor[1, 0, 0], RGBColor[0, 1, 0],
RGBColor[0, 0, 1]}, PlotRange → {-2, 2}]

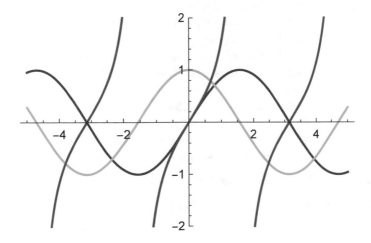

The `Plot` function has many options with the help of which the representations can be adapted:

In[64]:= **Options[Plot]**

Out[64]= $\{$AlignmentPoint $\rightarrow$ Center, AspectRatio $\rightarrow \dfrac{1}{\phi}$,

Axes $\rightarrow$ True, AxesLabel $\rightarrow$ None,

AxesOrigin $\rightarrow$ Automatic, AxesStyle $\rightarrow \{\}$,

Background $\rightarrow$ None, BaselinePosition $\rightarrow$ Automatic,

BaseStyle $\rightarrow \{\}$, ClippingStyle $\rightarrow$ None,

ColorFunction $\rightarrow$ Automatic, ColorFunctionScaling $\rightarrow$ True,

ColorOutput $\rightarrow$ Automatic, ContentSelectable $\rightarrow$ Automatic,

CoordinatesToolOptions $\rightarrow$ Automatic,

DisplayFunction $:\rightarrow$ $DisplayFunction$, Epilog $\rightarrow \{\}$,

Evaluated $\rightarrow$ Automatic, EvaluationMonitor $\rightarrow$ None,

Exclusions $\rightarrow$ Automatic, ExclusionsStyle $\rightarrow$ None,

Filling $\rightarrow$ None, FillingStyle $\rightarrow$ Automatic,

FormatType $:\rightarrow$ TraditionalForm,

Frame $\rightarrow$ False, FrameLabel $\rightarrow$ None,

FrameStyle $\rightarrow \{\}$, FrameTicks $\rightarrow$ Automatic,

FrameTicksStyle $\rightarrow \{\}$, GridLines $\rightarrow$ None,

GridLinesStyle $\rightarrow \{\}$, ImageMargins $\rightarrow$ 0.,

ImagePadding $\rightarrow$ All, ImageSize $\rightarrow$ Automatic,

ImageSizeRaw $\rightarrow$ Automatic, LabelStyle $\rightarrow \{\}$,

MaxRecursion $\rightarrow$ Automatic, Mesh $\rightarrow$ None,

MeshFunctions $\rightarrow \{\#1\&\}$, MeshShading $\rightarrow$ None,

MeshStyle $\rightarrow$ Automatic, Method $\rightarrow$ Automatic,

PerformanceGoal $:\rightarrow$ $PerformanceGoal$,

PlotLabel $\rightarrow$ None, PlotLabels $\rightarrow$ None,

PlotLegends $\rightarrow$ None, PlotPoints $\rightarrow$ Automatic,

PlotRange $\rightarrow \{$Full, Automatic$\}$, PlotRangeClipping $\rightarrow$ True,

PlotRangePadding $\rightarrow$ Automatic, PlotRegion $\rightarrow$ Automatic,

PlotStyle $\rightarrow$ Automatic, PlotTheme $:\rightarrow$ $PlotTheme$,

PreserveImageOptions $\rightarrow$ Automatic, Prolog $\rightarrow \{\}$,

RegionFunction $\rightarrow$ (True&), RotateLabel $\rightarrow$ True,

ScalingFunctions $\rightarrow$ None, TargetUnits $\rightarrow$ Automatic,

Ticks $\rightarrow$ Automatic, TicksStyle $\rightarrow \{\}$,

WorkingPrecision $\rightarrow$ MachinePrecision$\}$

Finally, let us view some three-dimensional graphics. First, a *saddle point*:

In[65]:= **Plot3D[$x^2 - y^2$, {x, -2, 2}, {y, -2, 2}]**

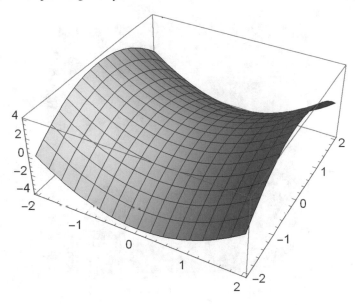

Next, a function which is partially differentiable but not continuous at zero, see e.g. [Koe1994]:

$In[66]:=$ **Plot3D$\left[\dfrac{x\,y}{x^2+y^2},\{x,-2,2\},\{y,-2,2\},\text{PlotPoints}\to 50\right]$**

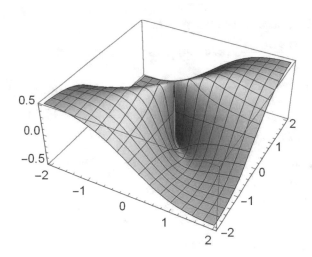

And finally, a *contour diagram*:

$In[67]:=$ **ContourPlot[Sin[x] Sin[y], {x,-3,3},**
          {y,-3,3}, ColorFunction → Hue, PlotPoints → 100]

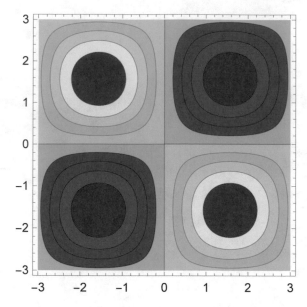

Mathematica can also solve problems from calculus and analysis, for example *limits*:

$In[68]:=$ **Limit $\left[\dfrac{\text{Exp}[x] - 1}{x}, x \to 0\right]$**

$Out[68]=$ 1

Taylor polynomials can also be handled:

$In[69]:=$ **Series $\left[\dfrac{\text{Exp}[x] - 1}{x}, \{x, 0, 10\}\right]$**

$Out[69]=$ $1 + \dfrac{x}{2} + \dfrac{x^2}{6} + \dfrac{x^3}{24} + \dfrac{x^4}{120} + \dfrac{x^5}{720} + \dfrac{x^6}{5040} +$

$\dfrac{x^7}{40320} + \dfrac{x^8}{362880} + \dfrac{x^9}{3628800} + \dfrac{x^{10}}{39916800} + O\left(x^{11}\right)$

Derivatives are also included,[19]

$In[70]:=$ **derivative $= \partial_x \dfrac{\text{Exp}[x] - 1}{x}$**

$Out[70]=$ $\dfrac{e^x}{x} - \dfrac{-1 + e^x}{x^2}$

just as *integrals* are:

$In[71]:=$ $\displaystyle\int$ **derivative dx**

$Out[71]=$ $\dfrac{e^x - 1}{x}$

Note, however, that it is not always so simple to reconstruct a function by integration and subsequent differentiation. As an example, we consider the rational function

$In[72]:=$ **input $= \dfrac{x^3 + x^2 + x - 1}{x^4 + x^2 + 1}$**

$Out[72]=$ $\dfrac{x^3 + x^2 + x - 1}{x^4 + x^2 + 1}$

First, let us view the graph:

---

[19] The derivative ∂_x and the integral $\int dx$ can be entered as the functions D and Integrate, respectively, or via the palettes.

$In[73]:=$ **Plot[input, {x, -5, 5}]**

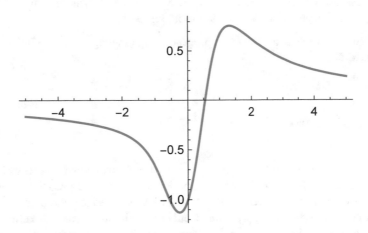

Next, we integrate the function,[20]

$In[74]:=$ **integral = $\int$ input dx**

$Out[74]=$ $\dfrac{1}{4\sqrt{3}}\left(\sqrt{3}\,\log[1+x^2+x^4]-2\,\arctan\left[\dfrac{\sqrt{3}}{1+2x^2}\right]+\right.$

$\sqrt{2-2i\sqrt{3}}\,(-3i+\sqrt{3})\,\arctan\left[\dfrac{1}{2}\left(-i+\sqrt{3}\right)x\right]+$

$\left.\sqrt{2+2i\sqrt{3}}\,(3i+\sqrt{3})\,\arctan\left[\dfrac{1}{2}\left(i+\sqrt{3}\right)x\right]\right)$

and take its derivative:

$In[75]:=$ **result = ∂_x integral**

$Out[75]=$ $\dfrac{1}{4\sqrt{3}}\left(\dfrac{8\sqrt{3}\,x}{(1+2x^2)^2\left(1+\dfrac{3}{(1+2x^2)^2}\right)}+\right.$

$\dfrac{\sqrt{2-2i\sqrt{3}}\,(-i+\sqrt{3})\,(-3i+\sqrt{3})}{2\left(1+\frac{1}{4}\left(-i+\sqrt{3}\right)^2 x^2\right)}+$

$\left.\dfrac{\sqrt{2+2i\sqrt{3}}\,(i+\sqrt{3})\,(3i+\sqrt{3})}{2\left(1+\frac{1}{4}\left(i+\sqrt{3}\right)^2 x^2\right)}+\dfrac{\sqrt{3}\,(2x+4x^3)}{1+x^2+x^4}\right)$

The result does *not* look like the input function at all, although the fundamental theorem of calculus tells that the two functions agree.[21] In order to see this, we have to simplify the result with the help of FullSimplify:

$In[76]:=$ **result//FullSimplify**

$Out[76]=$ $\dfrac{x^3+x^2+x-1}{x^4+x^2+1}$

---

[20] In the $\boxed{\text{TraditionalForm}}$ mode, all inverse functions of the trigonometric and hyperbolic functions are denoted by the $^{-1}$-notation, for example $\tan^{-1} = \arctan$. Please note that although the integrand is real, *Mathematica* produces a *complex* version of the antiderivative!

[21] In Chapter 12, we will show how unnecessary roots can be avoided in the integration result. However, it will turn out, that in our case the use of $\sqrt{3}$ is *unavoidable*.

The simplification of rational functions is considered in Section 6.9. Taylor polynomials and series are discussed in Chapter 10. Furthermore, we will develop an algorithm for differentiation in Section 2.7, and Chapter 12 is devoted to rational integration.

Of course, *Mathematica* is also a very rich programming language, which will be introduced in detail in the next chapter.

1.2 Additional Remarks

Mathematica came into the market in 1988 and is currently available as Release 12. It was the first computer algebra system combining symbolics, numerics and graphics in one platform, and therefore became a market leader quite fast. The user interface is rather advanced and is much better than in the competing systems. Concerning the mathematical kernel, however, it heavily depends on the asked question which computer algebra system performs best, see, e.g., [Wes1999]. Therefore, if *Mathematica* cannot handle your question, then you may try *Axiom*, *Maple*, *Maxima*, *Reduce*, *Sage* or one of the many special systems.

On the other hand, I do not want to conceal an important disadvantage of *Mathematica*: While some other computer algebra systems, in particular *Maple*, provide the opportunity to uncover the process behind the used algorithms, this is generally not possible with *Mathematica*. There, the techniques behind the computations are treated as trade secrets. This seems to be the price for its commercialization.

Please remember that this book can be used equally well with *Mathematica*, *Maple* and *Maxima*, since the computer algebra sessions can be downloaded for each of these three systems from `www.computer-algebra.org`. Hence, if one of these systems is available at your department or if you have a certain personal preference, you can go ahead with this particular system. If you don't have a computer algebra system at your disposal, you can download the system *Maxima* for free.

In the next chapter, we will discuss *Mathematica's* programming capabilities, which often enable the direct use of a mathematical definition instead of complicated programming structures.

1.3 Exercises

Exercise 1.1. (Mersenne Numbers)
If p is a prime number, then the numbers

$$M_p := 2^p - 1 \qquad (p \text{ prime})$$

are called the Mersenne numbers.

The French mathematician Marin Mersenne conjectured that the numbers M_p are only prime for $p = 2, 3, 5, 7, 13, 17, 19, 31, 67, 127$ and 257. This conjecture is false.

Currently 51 Mersenne primes are known. The largest known Mersenne prime is the number $M_{82\,589\,933}$, which has $24\,862\,048$ decimal digits and is the largest known prime.[22]

(a) *Confirm the number of decimal places of $M_{82\,589\,933}$. Hint: Utilize* N.

About Mersenne's conjecture:

(b) *M_{61}, M_{89} and M_{107} are primes that are missing in Mersenne's list.*

(c) *M_{67} is composite, in particular*

$$M_{67} = 147\,573\,952\,589\,676\,412\,927 = 193\,707\,721 \cdot 761\,838\,257\,287\,.$$

(d) *M_{257} is composite, in particular*

$$M_{257} = 535\,006\,138\,814\,359 \cdot 1\,155\,685\,395\,246\,619\,182\,673\,033$$
$$\cdot\, 374\,550\,598\,501\,810\,936\,581\,776\,630\,096\,313\,181\,393\,.$$

Prove (b), (c) and (d) with Mathematica, using PrimeQ *and* FactorInteger. *Please note the short time it takes Mathematica for the factorization of M_{67}.[23] With current releases of Mathematica, even the factorization in (d) is feasible.*

Exercise 1.2. (Twin Primes)
Two primes p_1 and $p_2 = p_1 + 2$ are called twin primes. Find the first twin primes that are larger than 1.000, 10^{10} and 10^{100}, respectively.

Exercise 1.3. (Robertson Conjecture)

(a) *The mathematician Robertson conjectured in 1989 [Rob1989][24] that the coefficients a_k of the Taylor (Maclaurin) series of the function*

$$f(x) = \sqrt{\frac{e^x - 1}{x}} = 1 + a_1 x + a_2 x^2 + \cdots = \sum_{k=0}^{\infty} a_k x^k$$

are all positive. Check this conjecture!

(b) *The coefficients $B_k(x)$ of the Taylor series (at $z_0 = 0$)*

$$F(z) = \sqrt{\frac{\left(\frac{1+z}{1-z}\right)^x - 1}{2xz}} = \sum_{k=0}^{\infty} B_k(x) z^k$$

are polynomials of degree k in x (you don't have to prove this). Compute $B_k(x)$ for $k = 0, \ldots, 5$.

It turns out that the coefficients of the polynomial $B_k(x)$ are all nonnegative for every k, such that $a_k > 0$ in (a). However, if $a_k < 0$ in (a) for some $k \in \mathbb{N} := \{1, 2, 3, \ldots\}$, then at least

---

[22] Please note that in the age of computer algebra systems such records typically do not last long. Therefore you should check whether the record is still valid. News about Mersenne primes are given in detail on the web page www.mersenne.org. For the history of Mersenne primes, see for example www.utm.edu/research/primes/mersenne.

[23] This prime was first factorized in 1903 by F. N. Cole. He answered the question of how long it took him to find the factorization of M_{67} by "Three years of Sundays". By using *Mathematica* or another computer algebra system, he would had saved three years ...

[24] Note that there is another famous Robertson conjecture [Rob1936] which was *proved* in 1984 by de Branges [DeB1984], together with Bieberbach's and Milin's conjectures.

one coefficient of the polynomial $B_k(x)$ must be negative (see [Rob1978]–[Rob1989]). Check whether such a case exists! If yes, what is the smallest such k? Compute $B_k(x)$ in this case.

Exercise 1.4. (Kinetic Energy)
In classical mechanics, a body with rest mass m_0 and speed v has the kinetic energy

$$E_{\mathrm{kin}} = \frac{1}{2} m_0 v^2 \, .$$

In special relativity it is shown that for large v a mass change occurs. The mass can then be computed by the formula

$$m = m_0 \, \frac{1}{\sqrt{1 - \frac{v^2}{c^2}}} \, ,$$

where c denotes the speed of light.

Einstein's celebrated formula for energy

$$E = m c^2 = E_{\mathrm{rest}} + E_{\mathrm{kin}} = m_0 c^2 + E_{\mathrm{kin}}$$

hence yields the relativistic formula

$$E_{\mathrm{kin}} = m_0 c^2 \left(\frac{1}{\sqrt{1 - \frac{v^2}{c^2}}} - 1 \right)$$

for kinetic energy.

Prove—using a Taylor expansion of order five—that the limit $v \to 0$ yields the classical case. How big is the difference up to order five?

Exercise 1.5. (Factorization)
Find out for which $a \in \mathbb{N}$, $a \leq 1000$, the polynomial $x^4 + a \in \mathbb{Z}[x]$ has a proper integer factorization.

Chapter 2

Programming in Computer Algebra Systems

2.1 Internal Representation of Expressions

In *Mathematica*[1], the internal representation of every expression is a list structure of the form
Head[*Argument1*, *Argument2*, ..., *Argument n*] with finitely many arguments. The following
table gives some simple examples for internal representations.[2]

InputForm	FullForm (internal representation)	Head
2	Integer[2]	Integer
2.0	Real[2.0]	Real
4/6	Rational[2,3]	Rational
2+I/3	Complex[2,Rational[1,3]]	Complex
x	Symbol[x]	Symbol
x+y	Plus[x,y]	Plus
2*x*y	Times[2,x,y]	Times
{1,x,z}	List[1,x,z]	List
Sin[x]	Sin[x]	Sin
E^x	Power[E,x]	Power
Sqrt[x]	Power[x,Rational[1,2]]	Power

Of course, by nesting it is possible to compose arbitrarily complicated expressions. For example, the following is a simple composite expression:[3]

InputForm	FullForm (internal representation)
{x,x y,3+2 z}	List[x,Times[x,y],Plus[3,Times[2,z]]]

Hence every expression *expr* has a head which can be accessed by Head[*expr*]. With
FullForm[*expr*], we get the internal representation, and InputForm[*expr*] shows every
expression in a form suitable for input. In a *Mathematica* notebook, all expressions can also be
entered by the palettes.

Please note that internally the exponential function Exp[x] is represented by the expression
Power[E,x], exactly like E^x. Therefore we realize that some simplifications are executed

---

[1] In principle, this is similar in *Maple* and *Maxima*.

[2] For numbers *Mathematica* hides the FullForm, since otherwise every mathematical expression would look
complicated. Nevertheless, the head—and therefore implicitly the internal form—can be found by Head.

[3] Multiplication signs can be replaced by space characters.

© Springer Nature Switzerland AG 2021

W. Koepf, *Computer Algebra*, Springer Undergraduate Texts in Mathematics and Technology,
https://doi.org/10.1007/978-3-030-78017-3_2

automatically and cannot be avoided. The same applies to rational numbers which are always written in lowest terms. The latter requires the computation of the greatest common divisor of the numerator and denominator; we will discuss this algorithm later. Automatic simplification is considered in more detail in the next section.

2.2 Pattern Matching

The working principle behind *Mathematica* is mainly based on *pattern matching*. Once you enter an expression, it is transformed with the aid of built-in transformation rules until no change occurs any more. For example, the expression `Sin[Pi]` is simplified towards 0 since this replacement rule is built in, whereas for the generic expression `Sin[x]` no transformation rules are available. Therefore the latter expression remains unchanged. The following table gives examples how *Mathematica*'s automatic simplification works.

Input	*Mathematica*'s output	Head
`12/4`	3	Integer
`5!^3/10!`	$\frac{10}{21}$	Rational
`Exp[x]`	e^x	Power
`Exp[0]`	1	Integer
`Exp[1]`	e	Symbol
`Sqrt[9]`	3	Integer
`Sqrt[10]`	$\sqrt{10}$	Sqrt
`Sqrt[10]^2`	10	Integer
`(1+I)*(1-I)`	2	Integer
`(x+y)*(x-y)`	$(x-y)(x+y)$	Times
`Product[k,{k,1,5}]`	120	Integer
`Product[k,{k,1,n}]`	$n!$	Factorial
`Sum[k,{k,1,5}]`	15	Integer
`Sum[k,{k,1,n}]`	$\frac{n(1+n)}{2}$	Times
`Sum[a[k],{k,1,n}]`	$\sum_{k=1}^{n} a_k$	Sum
`Integrate[Sin[x],x]`	$-\cos x$	Times
`Integrate[Sin[x]/Log[x],x]`	$\int \frac{\sin x}{\log x}\, dx$	Integrate

It is also possible to access pattern matching directly, and for this purpose you should know the internal formats. For example, the pattern `Exp[x]` with head `Exp` cannot be accessed in *Mathematica* by the user, as was shown above!

An arbitrary pattern is accessed by *Mathematica*'s variable symbol _, i.e., the underscore. For example, the expression `Sin[_]` denotes every expression of the form "sine of anything" where "anything" (hence the underscore) can represent any arbitrary subexpression. If a name is needed for our "anything", then the name is written before the underscore. Hence `Sin[x_]` represents "sine of anything, with name *x*". This approach is important, for example, if we want to replace *x* by something else, like in a function definition `f[x_]:=`...

The *Mathematica* procedure `Position` computes the position of a certain pattern in a given expression. However, the output {{2}, {3, 2}} of the command

```
Position[1+Sin[3]-Sin[x]+Sin[Pi],Sin[_]]
```

can only be understood if the automatic simplification of `1+Sin[3]-Sin[x]+Sin[Pi]` towards $1 + \sin 3 - \sin x$, hence to `Plus[1,Sin[3],Times[-1,Sin[x]]]`, is taken under consideration.

Further important *Mathematica* procedures in connection with pattern matching are `Cases` and `Select`.

As for the `Select` procedure, assume we wish to select the squares that are contained in a list of integers. First we generate a list of 100 random integers between 1 and 100:

```
In[1]:= list = Table[Random[Integer, {1, 100}], {100}]
```
Out[1]= {8, 50, 25, 14, 35, 77, 65, 35, 74, 10, 15, 48, 54, 44, 44, 86, 100, 39, 59, 27, 34, 26,
          66, 58, 7, 55, 94, 64, 61, 63, 73, 52, 83, 80, 51, 26, 9, 12, 1, 66, 32, 40, 31, 38,
          42, 70, 18, 52, 89, 10, 89, 67, 53, 7, 35, 8, 84, 67, 83, 51, 81, 71, 54, 91, 97, 75,
          18, 37, 100, 54, 15, 93, 67, 32, 15, 37, 20, 19, 36, 27, 19, 45, 91, 64, 59, 1, 37,
          59, 4, 83, 81, 47, 95, 25, 89, 86, 95, 82, 32, 39}

Next, we apply the square root function to each element of our list and select those elements that are integers. After squaring again, we receive the list of squares of our list:

$$In[2]:= \texttt{Select}\left[\sqrt{\texttt{list}}, \texttt{IntegerQ}\right]^2$$
Out[2]= {25, 100, 64, 9, 1, 81, 100, 36, 64, 1, 4, 81, 25}

However, now we have "forgotten" at which positions of our original list the squares appeared. These can be computed by

$$In[3]:= \texttt{Flatten}\left[\texttt{Position}\left[\texttt{list}, \texttt{x\_/; IntegerQ}\left[\sqrt{\texttt{x}}\right]\right]\right]$$
Out[3]= {3, 17, 28, 37, 39, 61, 69, 79, 84, 86, 89, 91, 94}

The same effect can be achieved without using a name (here *x*) for the variable subexpression with the aid of a *pure function*:

$$In[4]:= \texttt{Flatten}\left[\texttt{Position}\left[\texttt{list}, \_?\left(\texttt{IntegerQ}\left[\sqrt{\texttt{\#}}\right]\&\right)\right]\right]$$
Out[4]= {3, 17, 28, 37, 39, 61, 69, 79, 84, 86, 89, 91, 94}

A *pure function* is a function whose formal definition (using `:=`) is suppressed and which therefore does not need a function name. Such functions are constructed using `Function`, mostly abbreviated by the postfix notation `&`, and the variable of a pure function is generally indicated by `#`. If a pure function has several arguments, then they are called `#1`, `#2`, etc.

Analogously the square numbers of our list can be selected using `Cases`:

$$In[5]:= \texttt{Cases}\left[\texttt{list}, \_?\left(\texttt{IntegerQ}\left[\sqrt{\texttt{\#}}\right]\&\right)\right]$$
Out[5]= {25, 100, 64, 9, 1, 81, 100, 36, 64, 1, 4, 81, 25}

Mathematica's pattern matching capabilities are very rich. You should study the syntax of the commands `Cases`, `Count`, `DeleteCases`, `Position`, `Select` in detail.

2.3 Control Structures

In *Mathematica*, all Pascal- or C-like constructions are available, although for many situations high-level programming constructs can be used, which make the former obsolete.

Of course there is an if-then-else clause: The request If [*condition*, *term1*, *term2*, *term3*] works as follows: if *condition* is satisfied, then *term1* is evaluated, if *condition* is false, then *term2* is evaluated, otherwise *term3* is evaluated, whereupon the third and the fourth argument are optional. The If clause evaluates towards *termj* for the suitable $j \in \{1, 2, 3\}$.

For example, the function

In[1]:= **f[x_] := If[x > 0, "positive", "not positive", "undecidable"]**

has the values

In[2]:= **f[2]**
Out[2]= positive

In[3]:= **f[-3]**
Out[3]= not positive

and

In[4]:= **f[x]**
Out[4]= undecidable

Such a *trivalent logic* for the If clause is common because of the symbolic options in a computer algebra system. The last example shows why such a third option is helpful in a symbolic programming language. A function which enables more complicated case distinctions without nesting is Which.

Next we consider *loops* in *Mathematica*. The usual *counting loop* (in Pascal: for loop) is called Do [*expr*, *iterator*],[4] where *expr* is a command or a sequence of commands which are separated by semicola. Generally, one should observe that the different arguments of a function are always separated by *commas* whereas *semicola* separate several commands. This is a typical source for input errors!

Please note further that in *Mathematica* loop constructs do *not* possess a value![5] Hence, for the effect of a loop, you must work with assignments.

We consider the example

In[5]:= **x = 1; Do[x = k*x, {k, 100}]; x**

with the result

Out[5]= 93326215443944152681699238856266700490715968264381621468592963895217599993229915608941463976156518286253697920827223758251185210916864000000000000000000000000000

The above loop has computed 100! by iteratively allocating the values 1, ..., 100 to the temporary variable k and recomputing the value of the temporary variable x in each iteration step by multiplying by k. For this purpose the initial value of x is 1.

Often such constructions can be realized in simpler forms using high-level iteration constructs such as Table, Sum or Product. Table generates a list of numbers to be multiplied, and Apply replaces the head of the list by Times:

In[6]:= **Apply[Times, Table[k, {k, 100}]]**

---

[4] An iterator is a list of the form {k, k1, k2} with a variable k which runs from k_1 to k_2. If $k_1 = 1$, then this number can be omitted. If the name of the variable is not needed, then the iterator can be reduced to {k2}. In the long form {k, k1, k2, k3} k_3 denotes the step size.

[5] Every loop has the empty result Null.

Out[6]= 933262154439441526816992388562667004907159682643816214685929638952175999932299156089414639761565182862536979208272237582511852109168640000000000000000000000000000

Meanwhile, `Product` generates the result directly:

$$In[7]:= \prod_{k=1}^{100} k$$

Out[7]= 933262154439441526816992388562667004907159682643816214685929638952175999932299156089414639761565182862536979208272237582511852109168640000000000000000000000000000

In our case we can even utilize the built-in function `Factorial`:

In[8]:= **100!**

Out[8]= 933262154439441526816992388562667004907159682643816214685929638952175999932299156089414639761565182862536979208272237582511852109168640000000000000000000000000000

It is left to the reader to check the help pages for the loop types `While` and `For`.

More interesting are the iteration constructs `Nest`, `NestList`, `NestWhile`, `NestWhileList`, `Fold`, `FoldList`, `ComposeList`, `FixedPoint` and `Fixed-PointList`. Some of these constructs are considered in the next section.

2.4 Recursion and Iteration

In the last section, we defined the factorial function *iteratively*. This loop can be composed towards a complete program:

```
In[1]:= Fact1[n_] := Module[{x, k},
            x = 1;
            Do[x = k * x, {k, n}];
            x]
```

The `Module` construct has two arguments: the first argument (in our case `{x, k}`) is a list of *local variables*, and the second argument is generally a sequence of assignments which are separated by semicola. The *result* of the module is the last statement of the module. In our case, the result of the call `Fact1[n]` is the number x at the end of the call, so that `Fact1[100]` again computes 100!.[6]

In the above program, defining the local variables is important since otherwise there could be a conflict with some already used (global) variables of the same name in your notebook. As a side effect the global variable x would be changed by every function call, if x is not declared as local. In general, we recommend to declare all variables used in your program modules as local.

Of course, all the considered high-level language versions can also be written as programs:

```
In[2]:= Fact2[n_] := Apply[Times, Table[k, {k, n}]]
```

---

[6] At the end of your module you should *not* finish the sequence of statements by a semicolon (like in Pascal). In this case the result of the module is always `Null`.

$$In[3] := \textbf{Fact3[n\_]} := \prod_{k=1}^{n} \textbf{k}$$

Since in these examples, k is an iterator in a `Table` or `Product`, and iterators are automatically considered local by *Mathematica*, modules are not necessary.

All previous examples are *iterative* programs for the computation of the factorial function. In an iteration the underlying recurrent instruction—in our case

$$n! = n \cdot (n-1)! \tag{2.1}$$

—is translated by the programmer as a loop, namely $n! = 1 \cdot 2 \cdots (n-1) \cdot n$. The computation starts with $k = 1$ and ends with $k = n$, which is why such an approach is also called *bottom-up*. This iterative method is completely different from *recursive programming* which is a *top-down* approach. In a recursive program, the recurrent instruction is directly used for action. The characteristics of a recursive program is *that it calls itself*. Every recursive program needs a stopping rule (or initial value) as *break condition* since otherwise it creates an *infinite loop*.

For the factorial function, we can easily program this approach as follows:

```
In[4] := Fact4[0] = 1;
         Fact4[n_] := n * Fact4[n-1]
```

The first definition implements the break condition and the second definition is a direct translation of the recurrence equation (2.1). Whereas in an iterative program the program process and the administration of the local variables is left to the programmer, the recursive program is mainly executed in the computer memory at run time. Until the break condition is reached, all intermediate results cannot be (finally) evaluated and must therefore reside in the computer memory for further evaluation.

Please note that principally our function `Fact4` can be called with an argument $n < 0$ or with symbolic n. In this case, the computation will turn into an infinite loop since the stopping rule cannot be reached. In order to avoid such an event, we can set a condition:

```
In[5] := Fact4[0] = 1;
         Fact4[n_] := n * Fact4[n-1] /; IntegerQ[n] && n > 0
```

The `Trace` command shows the internal evaluation process:

```
In[6] := Trace[Fact1[5],x]
```

$$Out[6] = \begin{pmatrix} \begin{pmatrix} (x\$218 & 1) \\ (x\$218 & 1) \\ (x\$218 & 2) \\ (x\$218 & 6) \\ (x\$218 & 24) \end{pmatrix} & \{x\$218, 120\} \end{pmatrix}$$

Note that the local variables of name x are numbered by $x\$n$.

```
In[7] := Trace[Fact4[5],Fact4]
```

$Out[7] =$ {Fact4(5), 5 Fact4(5 − 1), {Fact4(4), 4 Fact4(4 − 1),

　　　　　　{Fact4(3), 3 Fact4(3 − 1), {Fact4(2), 2 Fact4(2 − 1), {Fact4(1), 1 Fact4(1 − 1),

　　　　　　{Fact4(0), 1}}}}}}

The last command informs about all recursive calls of `Fact4`.

Finally, we want to program the factorial function using the high-level language constructs `Fold` and `Nest`. The command `Nest[f,x,n]` applies the function f n times, starting with the initial value x. Please note that f must be a *function*, not an *expression*! For example, we get

In[8]:= **Nest[Sin, x, 5]**
Out[8]= sin(sin(sin(sin(sin(*x*)))))

If you want to apply the same function *f* a fixed number of times, say *n* times, then Nest is exactly what you need. However, this is not the case for the factorial function: Although there is a multiplication in each iteration step, the multiplier changes every time. *Mathematica's* procedure Fold is devoted to such situations. In the call Fold[*f*, *x*, *list*] the function *f*, which must be a bivariate function, is applied iteratively such that the second argument of *f* runs through the list *list*. The argument *x* is the starting value of the iteration. Therefore the program

In[9]:= **Fact5[n_] := Fold[#1 * #2&, 1, Table[k, {k, n}]]**

or shorter

In[10]:= **Fact6[n_] := Fold[Times, 1, Range[n]]**

yields another option to compute the factorial function. Note that Fold expects as first argument a function of two variables, which we realized (in the first definition) by using a pure function.

If we want to use Nest for the computation of the factorial, it gets more complicated. For this purpose—besides the intermediate results—we need the iteration variable. The idea is to memorize the pair {*k*, *k*!} in the *k*th step. This can be easily programmed using Nest, and after *n* steps we return the second element of the pair, hence *n*!:

In[11]:= **Fact7[n_] :=**
            Nest[{#[[1]] + 1, #[[2]] * #[[1]]}&, {1, 1}, n][[2]]

By the way, the command NestList returns the whole iteration list.

The routines FixedPoint and FixedPointList are the same as Nest and NestList. The only difference is that you are allowed to omit the third argument (number of iterations) in FixedPoint and FixedPointList. In this case, these procedures stop when the last element is a repetition of a former element. Using FixedPoint, we can, e.g., compute the fixed point of the cosine function,

In[12]:= **FixedPoint[Cos, 1.]**
Out[12]= 0.739085

i.e., a solution of the equation $x = \cos x$, which corresponds to the intersection of the graph of the cosine function and the graph of the identity,[7]

In[13]:= **Plot[{x, Cos[x]}, {x, 0, 1}, AspectRatio → Automatic]**

---

[7] Using the option AspectRatio → Automatic we ensure that both axes have the same scaling.

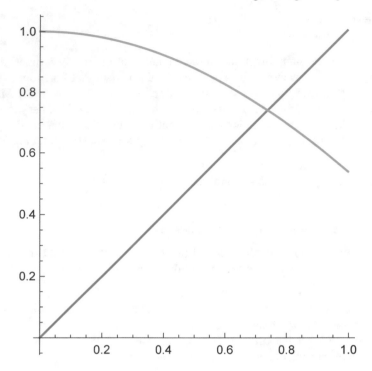

FixedPoint can easily be used to program the *Newton-Raphson method* to (numerically) compute a zero of a real function. A short version of such a program is given by[8]

$In[14]:=$ **NewtonMethod[f_, init_] := FixedPoint** $\left[\# - \frac{\texttt{f[\#]}}{\texttt{f'[\#]}} \&, \texttt{init} \right]$

Using this procedure, it is easy to compute (in decimal arithmetic) the first positive zero of the sine function,

$In[15]:=$ **NewtonMethod[Sin, 3.]**
$Out[15]=$ 3.14159

hence π. However, FixedPoint is also very useful for non-numerical problems, as we shall see later.

2.5 Remember Programming

In this section, we will consider the *Fibonacci numbers* and show how *remember programming* can strongly reduce the *complexity* of a recursive program.[9]

The Fibonacci numbers F_n ($n \in \mathbb{N}_{\geq 0} := \{0, 1, 2, \ldots\}$) are declared by the recurrence equation

$$F_n = F_{n-1} + F_{n-2}$$

---

[8] Please note that the function NewtonMethod—according to our definition—needs a *function* as first argument, and not an expression!

[9] In the current setting the overall number of recursive calls of a recursive program is called its complexity. For more about complexity and run time of algorithms, see Section 3.2.

with the initial values $F_0 = 0$ and $F_1 = 1$. The first computations lead to $F_2 = 1$, $F_3 = 2$, $F_4 = 3$, $F_5 = 5$.

In *Mathematica*, the Fibonacci recurrence can be programmed as

```
In[1]:= Fib1[0] = 0;
        Fib1[1] = 1;
        Fib1[n_] := Fib1[n-1] + Fib1[n-2]
```

We easily get

```
In[2]:= Timing[Fib1[10]]
Out[2]= {0. Second, 55}
```

but it already takes much longer to compute F_{35}:

```
In[3]:= Timing[Fib1[35]]
Out[3]= {22.2769 Second, 9227465}
```

What is happening here? For the computation of F_{35} we need F_{34} and F_{33}. However, for the computation of F_{34} we need F_{33} a second time. It is easy to see that F_{33} has to be computed twice, F_{32} three times, F_{31} five times and finally the stopping rule F_1 is called F_{35} times. Since the Fibonacci numbers grow exponentially, see Exercise 2.8, exponentially many function calls[10] are necessary for the computation of F_n.

However, linearly many function calls $O(n)$ turn out to be sufficient, if we *remember* each computed result, keeping these values in the memory.[11] This is exactly how *remember programming* works:

```
In[4]:= $RecursionLimit = ∞;
        Fib2[0] = 0;
        Fib2[1] = 1;
        Fib2[n_] := Fib2[n] = Fib2[n-1] + Fib2[n-2]
```

First, we reset the global variable $RecursionLimit. Since high recursion depths often indicate missing initial values or other programming bugs, the variable $RecursionLimit has the default value 1024, and if a recurrence needs more than 1024 recursive calls, an error message is displayed. Our definition avoids such an error message.

The above definition of Fib2 uses an additional *undelayed asignment* by = with which the computed value is archived in memory for immediate recall. Therefore the next computation is very fast:

```
In[5]:= Timing[Fib2[35]]
Out[5]= {0. Second, 9227465}
```

However, one should keep in mind that this method might be very memory consuming. In our example *Mathematica* has remembered the following values:

```
In[6]:= ?Fib2
Global`Fib2
```

---

[10] With $f(n) = O(g(n))$ we denote—as usual—the Landau notation with the property $\left|\frac{f(n)}{g(n)}\right| < \infty$ $(n \to \infty)$. From Exercise 2.12 it follows that for each $k = 0, \ldots, n$ we need $O((\frac{1+\sqrt{5}}{2})^k)$ operations, and $\sum_{k=0}^{n}(\frac{1+\sqrt{5}}{2})^k = O((\frac{1+\sqrt{5}}{2})^n)$.

[11] In our case we then need exactly $2n$ function calls. Linear complexity can be also reached if the program is written iteratively, see Exercise 2.7. In this approach every intermediate result is kept in a local variable. However, if you call F_{1001} after the computation of F_{1000}, then the full computing time is needed again, whereas in remember programming the second computation needs only a single recursive step.

```
Fib2(0) = 0
Fib2(1) = 1
Fib2(2) = 1
Fib2(3) = 2
Fib2(4) = 3
Fib2(5) = 5
Fib2(6) = 8
Fib2(7) = 13
Fib2(8) = 21
Fib2(9) = 34
Fib2(10) = 55
Fib2(11) = 89
Fib2(12) = 144
Fib2(13) = 233
Fib2(14) = 377
Fib2(15) = 610
Fib2(16) = 987
Fib2(17) = 1597
Fib2(18) = 2584
Fib2(19) = 4181
Fib2(20) = 6765
Fib2(21) = 10946
Fib2(22) = 17711
Fib2(23) = 28657
Fib2(24) = 46368
Fib2(25) = 75025
Fib2(26) = 121393
Fib2(27) = 196418
Fib2(28) = 317811
Fib2(29) = 514229
Fib2(30) = 832040
Fib2(31) = 1346269
Fib2(32) = 2178309
Fib2(33) = 3524578
Fib2(34) = 5702887
Fib2(35) = 9227465
Fib2(n_) := Fib2(n) = Fib2(n-2) + Fib2(n-1)
```

Now we can compute the Fibonacci numbers in a reasonable time for larger values of n as well:

In[7]:= **Timing[res1 = Fib2[10000];]**
Out[7]= {0.0312002 Second, Null}

In[8]:= **Timing[res2 = Fibonacci[10000];]**
Out[8]= {0. Second, Null}

In[9]:= **res2 - res1**
Out[9]= 0

Note that the built-in function Fibonacci is much faster, which becomes even more evident for larger n. The reason is that a *more efficient algorithm* is used. However, with the defining recurrence the timings cannot be accelerated. In the next section, you will learn how a faster computation is possible nevertheless.

2.6 Divide-and-Conquer Programming

To compute the Fibonacci numbers more efficiently, we can also use the *Divide-and-Conquer paradigm*, with the help of which a problem of order n is reduced to problems of order $n/2$. This often leads to quite efficient algorithms. Hence we must try to compute the nth Fibonacci number F_n with the aid of Fibonacci numbers whose index is around $n/2$. For this purpose, however, another type of recurrence equation is needed.

The Fibonacci numbers indeed satisfy the following recurrence equations (that will be proved in Exercise 2.5):

$$F_{2n} = F_n \cdot (F_n + 2\, F_{n-1}) \qquad \text{and} \qquad F_{2n+1} = F_{n+1}^2 + F_n^2 \qquad (n \in \mathbb{N}_{\geq 0}). \tag{2.2}$$

Through these equations, they can be computed rather efficiently:[12]

```
In[1]:= $RecursionLimit = ∞;
        Fib3[0] = 0;
        Fib3[1] = 1;
        Fib3[n_] :=
          (Fib3[n] = Module[{fib = Fib3[n/2]},
          fib * (fib + 2 * Fib3[n/2 - 1])]) /; EvenQ[n]

        Fib3[n_] := (Fib3[n] = Fib3[(n+1)/2]^2 + Fib3[(n-1)/2]^2) /; OddQ[n]
```

Note that both recurrences in (2.2) have to be rewritten in such a way that F_n occurs on the left-hand side. The first definition is used for even n, the second one for odd n.

Since most preceding Fibonacci numbers do not have to be computed, this implementation has much better run times:

```
In[2]:= Timing[res1 = Fib3[10000000];]
Out[2]= {0.0624004 Second, Null}

In[3]:= Timing[res2 = Fibonacci[10000000];]
Out[3]= {0.0468003 Second, Null}

In[4]:= res2 - res1
Out[4]= 0
```

The number of function calls to compute F_{2^m} is now bounded by $2m$ so that this algorithm has a complexity $O(\log_2 n)$, whereas the defining recurrence led to linear complexity $O(n)$.[13]

As a rule of thumb one should remember that the same problem can often be solved using different methods which can be quite different in their efficiency.

---

[12] Please observe the parenthesizing!

[13] The computation times suggest that *Mathematica* also uses this type of algorithm for the built-in function Fibonacci.

2.7 Programming through Pattern Matching

In this section, we will show that differentiation can be treated as a purely algebraic process, and we will realize how easily it can be programmed in few lines of code using *Mathematica's* pattern matching capabilities.

We would like to implement a function $\texttt{Diff}[f,x]$ which differentiates an expression f with respect to the variable x. (Of course this is not really necessary since this capability is available through the *Mathematica* function D and works very well.)

For the computation of derivatives we do not need limits if we do it exactly in such a way as we differentiate in practice, namely successively using the *differentiation rules*.

We start with the derivative of the powers (*power rule*):

```
In[1]:= Diff[c_,x_] := 0/; FreeQ[c,x]
        Diff[x_^n_.,x_] := n x^n-1 /; FreeQ[n,x]
```

With these two definitions we declare how powers are differentiated. Note that a symbol c is recognized as a constant, if it does not contain the symbol x (FreeQ).

Also note that the pattern x_^n_ . ending with a period includes the power $x = x^1$ although the pattern "anything to the power anything" is not visible in this case.

Now, we can differentiate powers:

```
In[2]:= Diff[x^3,x]
Out[2]= 3 x^2
```

Even rational powers are allowed:

```
In[3]:= Diff[√x,x]
Out[3]= 1/(2 √x)
```

However, *Mathematica* does not yet know how to differentiate $3x^2$:

```
In[4]:= Diff[3 x^2,x]
Out[4]= Diff(3 x^2, x)
```

So let us teach *Mathematica* to do so! Differentiation is linear, hence we generally have (*linearity*)

```
In[5]:= Diff[c_ * f_,x_] := c * Diff[f,x] /; FreeQ[c,x]
        Diff[f_ +g_,x_] := Diff[f,x] + Diff[g,x]
```

Now we can derive $3x^2$:

```
In[6]:= Diff[3 x^2,x]
Out[6]= 6 x
```

The derivative of every linear combination can also be computed, e.g. as

$$In[7]:= \texttt{Diff}\Big[\sum_{k=0}^{10} k\, x^k, x \Big]$$

$$Out[7]= 100\, x^9 + 81\, x^8 + 64\, x^7 + 49\, x^6 + 36\, x^5 + 25\, x^4 + 16\, x^3 + 9\, x^2 + 4\, x + 1$$

since *Mathematica* knows that addition is associative (Flat) and commutative (Orderless):

In[8]:= **??Plus**

x + y + z represents a sum of terms.

Attributes [Plus] =

    {Flat, Listable, NumericFunction, OneIdentity, Orderless, Protected}

Default [Plus] := 0

However, the following polynomial is not a sum, but a product:

In[9]:= **Diff[(x² + 3) (3x + 2) , x]**
Out[9]= $\text{Diff}((3x+2)(x^2+3), x)$

Hence, we have to declare the product rule:

In[10]:= **Diff[f_ * g_, x_] := g * Diff[f, x] + f * Diff[g, x]**

Now we can solve the above problem:

In[11]:= **Diff[(x² + 3) (3x + 2) , x]**
Out[11]= $2x(3x+2)+3(x^2+3)$

If we also want to differentiate transcendental functions, *Mathematica* must know the derivatives of the elementary functions:

In[12]:= **Diff[Log[x_], x_] :=** $\dfrac{1}{x}$

      Diff[Sin[x_], x_] := Cos[x]

      Diff[Cos[x_], x_] := - Sin[x]

      Diff[Tan[x_], x_] := 1 + Tan[x]²

      Diff[ArcSin[x_], x_] := $\dfrac{1}{\sqrt{1-x^2}}$

      Diff[ArcTan[x_], x_] := $\dfrac{1}{1+x^2}$

Note that you can always extend this list by defining the derivatives of further functions.

We are not yet able to differentiate composite functions:

In[13]:= **Diff[Sin[x²], x]**
Out[13]= $\text{Diff}(\sin(x^2), x)$

For this purpose, we need to implement the chain rule:[14]

In[14]:= **Diff[g_[f_], x_] := Diff[f, x] * Diff[g[f], f]**

Now we can treat the above example:

In[15]:= **Diff[Sin[x²], x]**
Out[15]= $2x \cos(x^2)$

However, if the composition is disguised as a power, then the given pattern $g(f)$ does not apply because the power function has two arguments:

In[16]:= **Diff[(1 + x²)^Sin[x] , x]**
Out[16]= $\text{Diff}\big((x^2+1)^{\sin(x)}, x\big)$

---

[14] This definition is not an easy one and might be difficult to understand at first sight. Hence you should check this definition carefully in order to understand the mode of operation of the pattern matching here!

Hence, we finally need to implement the chain rule for powers:

$In[17]:=$ **Diff[f_$^{g-}$,x_] := f^g * Diff[g * Log[f],x]**

Then we get the desired result indeed:

$In[18]:=$ **Diff[(1+x^2)$^{Sin[x]}$,x]**

$Out[18]=$ $(x^2+1)^{\sin(x)}\left(\dfrac{2\,x\,\sin(x)}{x^2+1} + \cos(x)\,\log(x^2+1)\right)$

Now we can also differentiate the exponential function (why?),

$In[19]:=$ **Diff[Exp[x],x]**

$Out[19]=$ e^x

and quotients can be treated as well (without declaring the quotient rule, why?):

$In[20]:=$ **Diff$\left[\dfrac{x^2+3}{3x+2},x\right]$**

$Out[20]=$ $\dfrac{2\,x}{3\,x+2} - \dfrac{3\,(x^2+3)}{(3\,x+2)^2}$

We conclude by differentiating a fantasy function:

$In[21]:=$ **Diff$\left[\dfrac{Log[Sin[x]]}{ArcSin[x]} + \dfrac{ArcTan\left[\frac{1}{x}\right] * Tan[x]}{Log[x]},x\right]$**

$Out[21]=$ $\dfrac{\cot(x)}{\sin^{-1}(x)} - \dfrac{\log(\sin(x))}{\sqrt{1-x^2}\,\sin^{-1}(x)^2} -$

$\dfrac{\tan(x)}{\left(1+\frac{1}{x^2}\right)x^2\,\log(x)} + \tan^{-1}\left(\dfrac{1}{x}\right)\left(\dfrac{\tan^2(x)+1}{\log(x)} - \dfrac{\tan(x)}{x\,\log^2(x)}\right)$

Now you should test the procedure `Diff` with some of your own complicated input functions, see also Exercises 2.14–2.15!

2.8 Additional Remarks

Recommended books for programming with *Mathematica* are [Mae1997], [Tro2004b] and of course the handbook [Wol2003]. Michael Trott, one of the best insiders of *Mathematica*, has written some further comprehensive books about *Mathematica* [Tro2004a]–[Tro2005].

The divide-and-conquer formula (2.2) for the Fibonacci numbers can be found in [GKP1994], formula (6.109). The introduction of differentiation (Section 2.7) was given in this form in [Koe1993b].

2.9 Exercises

Exercise 2.1. (Internal Representation and Simplification)

(a) *What are the internal representations of the expressions* 10, 10/3, 10.5, x, 1+x, 2*x, x^2, {1,2,3}, f[x], a->b, a:>b, a==b, a=b, a:=b? *What are the corresponding heads? Hint: To get the full form of the last two expressions, you must use the* Hold *procedure. Why?*

(b) *What are the heads of the expressions* `1*y`, `Sqrt[x]`, `x+x`, `Sum[x^k,{k,0,5}]`, `Sum[x^k,{k,0,n}]`, `Product[k,{k,1,3}]`, `Product[k,{k,1,n}]`, `Nest[f,x,5]`,`1/Cos[x]`,`1/Tan[x]`? *Explain!*

(c) *Define* `x=1; y=x; x=2; y`. *What is the value of* `y`? *Delete* `x` *and* `y` *using* `Clear[x,y]`.
Define next `x:=1; y:=x; x:=2; y`. *What is the value of* `y` *now? Why? Why are the input lines in Mathematica numbered by* `In[n]:=...` *and the output lines are numbered by* `Out[n]=...?*

Exercise 2.2. *Declare the matrix* `M={{1,2,3},{4,5,6},{7,8,9}}` *and note that* $2^{15} = 32\,768$. *Compute the power* $M^{32\,768}$ *of M using* `Nest` *and compare the result and computation time with* `MatrixPower`. *Explain! How can you speed up your computation?*

Exercise 2.3. *Define a list of replacement rules corresponding to the logarithm rules*

$$\ln x + \ln y \to \ln(xy), \quad n \ln x \to \ln(x^n).$$

Apply these rules to the expression

$$2\ln x + 3\ln y - 4\ln z.$$

Hint: The logarithm function ln *is called* `Log` *in Mathematica. Use* `//.` *for the substitution. Why?*

Exercise 2.4. *Measure the computation times (using* `Timing`*) for the built-in function* `n!` *in comparison to our functions* `Fact1` *to* `Fact7`, *and explain the timings. Use* $n = 100\,000$ *and suppress the output using* `;` *to get the pure computation time. Set* `$RecursionLimit` *in a suitable way.*

Exercise 2.5. *Show the divide-and-conquer recurrences (2.2) using mathematical induction.*

Exercise 2.6. *Implement the Fibonacci numbers* F_n *iteratively, using the defining recurrence equation and* `Nest`.

Exercise 2.7. *Implement the Fibonacci numbers* F_n *iteratively, using the divide-and-conquer recurrence equations. Hint: The use of the binary representation of n is helpful, see Section 3.1.*

Exercise 2.8. *Show, with the help of mathematical induction, that the Fibonacci numbers have the representation*

$$F_n = \frac{1}{\sqrt{5}}\left(\left(\frac{1+\sqrt{5}}{2}\right)^n - \left(\frac{1-\sqrt{5}}{2}\right)^n\right).$$

From this, we get in particular

$$F_n = \text{round}\left(\frac{1}{\sqrt{5}}\left(\frac{1+\sqrt{5}}{2}\right)^n\right),$$

where round(x) (`Round(x)`) *rounds the real number* $x \in \mathbb{R}$ *to the next integer.*

When computing the Fibonacci numbers numerically (as decimal numbers), the latter representation is especially efficient for large n where rounding can be neglected.

Measure the timing of this approach for the computation of $F_{10^{10}}$ and compare it with the timings of a numerically modified version of Section 2.6 and the built-in function Fibonacci. Note that the built-in function needs more time, but gives the most accurate result. Check this using a higher accuracy for the other two methods.

Exercise 2.9. (Fibonacci Polynomials)
The Fibonacci polynomials are declared by

$$F_n(x) = x\, F_{n-1}(x) + F_{n-2}(x), \qquad F_0(x) = 0, F_1(x) = 1.$$

(a) *Extend the programs for the computation of Fibonacci numbers from Section 2.5 and compute $F_{100}(x)$ using the different methods. Hint: The use of the Expand function should be avoided if this accelerates the computation. You may have to increase n slowly to realize some effects.*
(b) *Using the first 10 coefficients, verify that the Fibonacci polynomials have the generating function*

$$\sum_{k=0}^{\infty} F_n(x)t^n = \frac{t}{1 - xt - t^2}.$$

Exercise 2.10. (Map and Apply)
Assume that the variable list *is a list of integers. Explain the following commands and test them using* list=Table[Random[Integer,{1,100}],{k,1,100}].

(a) Map[PrimeQ,list]
(b) Count[Map[PrimeQ,list],True]
(c) Count[Map[EvenQ,list],True]
(d) Map[(#^2)&,list]
(e) Apply[Plus,list]/Length[list]
(f) Sqrt[Apply[Plus,Map[(#^2)&,list]]]
(g) N[Sqrt[Apply[Plus,Map[(#^2)&,list]]]]

Try to find two more interesting examples for the use of Map *and* Apply *and carefully describe their effect.*

Exercise 2.11. Nested Expressions
With the help of pure functions and Nest, *nested expressions can be generated easily.*

(a) *Generate the continued fraction*

$$\cfrac{1}{1 + \cfrac{1}{1 + \cfrac{1}{1 + \cfrac{1}{1 + \cfrac{1}{1 + \frac{1}{1+x}}}}}}.$$

(b) *Generate the nested square root*

$$\sqrt{1 + \sqrt{1 + \sqrt{1 + \sqrt{1 + \sqrt{1 + \sqrt{1 + \sqrt{2}}}}}}}$$

Exercise 2.12. *Show that, for* $x \in (0, 1)$ *and* $n \to \infty$,

$$\sum_{k=0}^{n} x^k = O(x^n) \,.$$

Exercise 2.13. Number Conversion

(a) *Test the procedure*

```
digitstonumber[digits_]:=Fold[(10 #1+#2)&,0,digits]
```

   for the example `digitstonumber[1,2,3,4,5,6,7,8,9]` *and explain how the conversion works.*

(b) *Implement the inverse function* `numbertodigits` *of the procedure* `digitstonumber` *recursively. For this purpose, you will need some of the functions* `Append`, `Prepend`, `First`, `Rest` *and* `Mod`.

(c) *Convert* 100! *to its corresponding number sequence.*

(d) *Which built-in function correlates to* `numbertodigits`?

Exercise 2.14. *Check the computation of the derivative of the tangent function of Section* 2.7 *using the product rule*

```
Diff[Tan[x_],x_]:=Diff[Sin[x]/Cos[x],x]
```

instead of the one given there. This rule does not work. Why not?

Exercise 2.15. *Mathematica considers the trigonometric functions* $\sin x$, $\cos x$, $\tan x$, $\cot x$, $\sec x :=$ $\frac{1}{\cos x}$ *and* $\csc x := \frac{1}{\sin x}$ *as independent functions and "simplifies" them automatically, for example* $\frac{1}{\cos x}$ *towards* $\sec x$. *Therefore you must complete the differentiation rules given in Section* 2.7. *Execute this completion, test your code, and compare it with the built-in function* D.

Exercise 2.16. *Determine the degree of a polynomial* p *in* x *using the function* `Cases`. *Write a function* `degree[p,x]`.

Exercise 2.17. *Use pattern matching to write a Mathematica function* `DEOrder[DE, f[x]]` *which computes the order of the ordinary differential equation DE with respect to the function* $f(x)$. *Compute* `FullForm[f'[x]]` *to find out how Mathematica represents derivatives internally!*

Test your procedure for the following differential equations:

(a) $f'(x) = 1 + f(x)^2$;

(b) $y''(x) + x^2 y''(x) - y(x) = \sin x$;

(c) $(1 - t^2) f'''(t) + 2t f''(t) - f'(t) = 0$;

(d) $\sin y''(t) + e^{y'(t)^2} = \cos \sqrt{y(t)}$.

Hint: If necessary, use the Mathematica function `Max`.

Exercise 2.18. *Implement a recursive or iterative function* `reverse`, *which reverses a list (like the built-in function* `Reverse`).

Exercise 2.19. *Implement a (probably recursive) procedure* `Powerset[list]`, *which computes and returns the power set of a set* `list`. *Apply your procedure to* $\{1, 2, 3, 4, 5\}$. *What is the computation time for the call* `PowerSet[Range[10]]`?

Hint: Construct the power set of a set `list` *with* n *elements from the power set of the set* `Rest[list]` *which has* $n-1$ *elements. If applicable, use the Mathematica functions* `First`, `Rest`, `Append`, `Prepend`, `Map`, `Union`.

Exercise 2.20. *Implement a (probably recursive) procedure* `Subsets[list,k]`, *which computes and returns the subsets with* k *elements of the set* `list`. *Compute the sets with two elements of the set* $\{1, 2, 3, 4, 5\}$.

Hint: Take the union of those subsets with k *elements of the set* `list`, *which contain the first element* `First[list]`, *with those that do not contain the first element.*

Chapter 3
Number Systems and Integer Arithmetic

3.1 Number Systems

In this chapter we consider the arithmetics of arbitrary large integers. Integer arithmetic is the backbone of every computer algebra system, hence we would like to discuss this topic in some detail. In order to be able to compute with integers, they must be suitably represented. In the history of Mathematics, several representations of integers have been introduced, two of which are still in use: the roman and the arabic number systems.[1]

Although the roman number system has a certain aesthetic attraction,[2] it turns out to be unfavorable with respect to *algorithmic computations*, hence the computation according to given computational schemes. On the other hand, the arabic position system is very well suited for this purpose.

In the arabic position system, positive integers are represented as linear combinations of powers of a given *basis* $B \in \{2, 3, \ldots\}$. A positive integer z therefore has the B-adic representation

$$z = z_{n-1} B^{n-1} + z_{n-2} B^{n-2} + \cdots + z_1 B + z_0 = \sum_{k=0}^{n-1} z_k B^k \qquad (z_k \in \{0, 1, \ldots, B-1\}) \qquad (3.1)$$

and is mostly denoted by $z = z_{n-1} z_{n-2} \ldots z_1 z_0$ or—if the basis is unclear—by $z = z_{n-1} z_{n-2} \cdots z_1 z_{0_B}$. The *digits* z_k $(k = 0, \ldots, n-1)$ are elements of the digit set $\{0, 1, \ldots, B-1\}$.[3] The number of digits n of the positive integer z is also called the *length of the number*. The *binary number system* $(B = 2)$ is particularly simple. This system has only two digits 0 and 1, and the representations can be quite long. For example, the roman number MM has the binary representation MM $= 11111010000_2$. In computers the octal system with $B = 8$ and the *hexadecimal system* with $B = 16$ are also used, where for example MM $= 3720_8 = 7D0_{16}$. Note that the hexadecimal system has the digit set $\{0, 1, 2, 3, 4, 5, 6, 7, 8, 9, A, B, C, D, E, F\}$, where A, B, C, D, E and F are used for $10, \ldots, 15$, respectively.

When realizing integer arithmetic on a computer, then as a basis one can also use a memory unit which is called *word*, which in a 64-bit processor consists of 2^{64} different states. In this case, we get the basis $B = 2^{64} = 18\,446\,744\,073\,709\,551\,616$.

---

[1] The simplest representation of a positive integer consists of tally marks. This is of course too cumbersome to be suitable for *efficient arithmetic*.

[2] The mystic year MM has a very simple number representation, in contrast to its pre-predecessor MCMXCVIII.

[3] If more than ten digits are necessary, different digit symbols can be used.

© Springer Nature Switzerland AG 2021
W. Koepf, *Computer Algebra*, Springer Undergraduate Texts in Mathematics and Technology,
https://doi.org/10.1007/978-3-030-78017-3_3

We are most familiar with the *decimal system* where $B = 10$. Of course MM $= 2000_{10}$.

Session 3.1. *Mathematica can work with different bases. The conversion of a decimal number to another basis is done by the function* BaseForm:

In[1]:= **BaseForm[2000,2]**
Out[1]= 11111010000_2

In[2]:= **BaseForm[2000,8]**
Out[2]= 3720_8

In[3]:= **BaseForm[2000,16]**
Out[3]= $7d0_{16}$

One can enter an integer with respect to an arbitrary basis $2 \leqq B \leqq 36$ *using* ^^:[4]

In[4]:= **16^^7D0**
Out[4]= 2000

Let us convert this number to the base 7:

In[5]:= **BaseForm[16^^7D0,7]**
Out[5]= 5555_7

A list of digits is generated by IntegerDigits:

In[6]:= **IntegerDigits[16^^7D0,7]**
Out[6]= $\{5,5,5,5\}$

The following command computes the number of digits:

In[7]:= **Length[IntegerDigits[16^^7D0,7]]**
Out[7]= 4

Note that this result equals n in equation (3.1).

Negative numbers contain an appropriate mark (sign). Their addition and subtraction is reduced to addition and subtraction of positive integers ($x + (-y) = x - y$ and $(-x) + (-y) = -(x + y)$ for $x, y > 0$), and multiplication can also be reduced to multiplication of positive integers ($x \cdot (-y) = -(x \cdot y)$ and $(-x) \cdot (-y) = x \cdot y$ for $x, y > 0$). Hence it is enough to consider addition, subtraction and multiplication of positive integers. This is the content of the next section.

3.2 Integer Arithmetic: Addition and Multiplication

How does integer arithmetic work? The algorithms that we treat in the sequel can be used with respect to every base B and are maybe the simplest with respect to the base $B = 2$. We will, however, present the algorithms in the decimal system that we are all used to.

In high school, we learn algorithms for addition, subtraction and multiplication of (positive) integers. When adding two positive integers, we have to add the same positions of the two summands according to the following *digit addition table*, starting with the rightmost position:

---

[4] What is the mystery of the upper bound $B = 36$?

```
+|0  1  2  3  4  5  6  7  8  9
0|0  1  2  3  4  5  6  7  8  9
1|1  2  3  4  5  6  7  8  9 10
2|2  3  4  5  6  7  8  9 10 11
3|3  4  5  6  7  8  9 10 11 12
4|4  5  6  7  8  9 10 11 12 13
5|5  6  7  8  9 10 11 12 13 14
6|6  7  8  9 10 11 12 13 14 15
7|7  8  9 10 11 12 13 14 15 16
8|8  9 10 11 12 13 14 15 16 17
9|9 10 11 12 13 14 15 16 17 18
```

Shorter summands are completed by leading zeros, and possible *carries*.

Obviously the addition of two n-digit integers needs exactly n elementary additions (of the digit addition table). Additionally we have at most n carry additions. In total the addition algorithm needs $\leq 2n$ or $O(n)$ elementary operations. (In this lecture, we are generally satisfied with an asymptotic analysis and therefore with such an order term.)

Multiplication of two positive integers can also be done using the high school algorithm. This method yields an iterative scheme using the following *digit multiplication table*, with suitable carries:

```
·|0  1  2  3  4  5  6  7  8  9
0|0  0  0  0  0  0  0  0  0  0
1|0  1  2  3  4  5  6  7  8  9
2|0  2  4  6  8 10 12 14 16 18
3|0  3  6  9 12 15 18 21 24 27
4|0  4  8 12 16 20 24 28 32 36
5|0  5 10 15 20 25 30 35 40 45
6|0  6 12 18 24 30 36 42 48 54
7|0  7 14 21 28 35 42 49 56 63
8|0  8 16 24 32 40 48 56 64 72
9|0  9 18 27 36 45 54 63 72 81
```

Example 3.2. (**High School Multiplication I**) We consider an exemplary multiplication $x \cdot y$ for $x = 1234$ and $y = 5678$ using the high school algorithm:

$$
\begin{array}{r}
1234 \cdot 5678 \\
\hline
6170 \\
7404 \\
8638 \\
9872 \\
\hline
7006652
\end{array}
$$

The algorithm divides the full problem (using the distributive law) into the subproblems $1234 \cdot 5$, $1234 \cdot 6$, $1234 \cdot 7$ and $1234 \cdot 8$, and these subproblems are eventually divided into problems of the digit multiplication table.

A closer examination shows that for the full computation we need $4 \cdot 4 = 16$ elementary multiplications (of the digit multiplication table) plus some additions. If we count only the multiplications for the moment, then we see that for the multiplication of two n-digit numbers we need $O(n^2)$ elementary multiplications. One can show that there are also at most $O(n^2)$ elementary additions.[5] Overall, high school multiplication has $O(n^2)$ elementary operations, where we count every operation of the digit addition and multiplication tables.[6] $\triangle$

These demonstrations show that *algorithms exist* for the execution of the elementary operations $+$ and $\cdot$. Whereas high school addition is essentially the best possible approach, it will turn out that high school multiplication can be made more efficient.

In order to ro reach a more efficient approach to multiplication, we first give a different, namely *recursive*, presentation of high school multiplication. For this purpose, let us assume that we want to multiply two positive integers, $x, y \in \mathbb{N} := \{1, 2, 3, \ldots\}$, both of length n and hence with n decimal digits. If one of the factors is smaller, we can complete it by leading zeros. Without loss of generality we can furthermore assume that n is a power of 2.[7]

Next, we partition the representations of both x and y,

$$x = a \cdot 10^{n/2} + b \qquad \text{and} \qquad y = c \cdot 10^{n/2} + d,$$

in two equally long digit parts and compute the product $x \cdot y$ according to the formula

$$x \cdot y = a \cdot c \cdot 10^n + (a \cdot d + b \cdot c) \cdot 10^{n/2} + b \cdot d. \tag{3.2}$$

This procedure can be carried on recursively, thus leading to a typical *divide-and-conquer algorithm*.

Example 3.3. (**High School Multiplication II**) Let us try out this method with the above example. For the computation of $1234 \cdot 5678$ we use the decompositions

$$1234 = 12 \cdot 100 + 34 \qquad \text{and} \qquad 5678 = 56 \cdot 100 + 78.$$

Hence $a = 12$, $b = 34$, $c = 56$ and $d = 78$. According to (3.2), for the computation of the product $x \cdot y$ the four multiplications $a \cdot c = 12 \cdot 56$, $a \cdot d = 12 \cdot 78$, $b \cdot c = 34 \cdot 56$ and $b \cdot d = 34 \cdot 78$ are needed. The application of (3.2) to any of these four multiplications requests four elementary multiplications. Obviously, the whole procedure finally yields *the same* 16 elementary multiplications as before, but this time they are evaluated in a different order. $\triangle$

Now we will show that the above recursive method needs $O(n^2)$ elementary operations for two factors of length n.

For this, $K(n)$ denotes the number of necessary elementary operations. $K(n)$ is called the *complexity* or the *run time* of the algorithm. Due to the recursive structure of the method, we can deduce a recurrence equation for $K(n)$. Formula (3.2) shows that the computation of a product of two n-digit numbers can be realized by the computation of 4 multiplications of integers with $n/2$ digits plus some additions of n-digit numbers.[8] Hence we have $K(1) = 1$ (for the multiplication of two one-digit numbers we need exactly one elementary operation) and

---

[5] We will prove this soon using another approach.

[6] In more detail, an elementary operation is any memory access to the stored digit addition and multiplication tables. Every such access needs a certain amount of time, whether or not the operation is called one or several times.

[7] Otherwise we complete once more with leading zeros.

[8] Please observe that only the middle addition in (3.2) is a real addition. The two other additions actually represent shifts in the position system and need only carries. However, for an asymptotic analysis using order terms $O(f(n))$ the actual number of additions is not relevant.

$$K(n) = 4 \cdot K(n/2) + C \cdot n$$

for a constant C which denotes the number of necessary additions in each recursive step. For $n = 2^m$ this yields

$$
\begin{aligned}
K(n) = K(2^m) &= 4 \cdot K(2^{m-1}) + C \cdot 2^m \\
&= 4 \cdot (4 \cdot K(2^{m-2}) + C \cdot 2^{m-1}) + C \cdot 2^m \\
&= 4^2 \cdot K(2^{m-2}) + C \cdot 2^m(1+2) = \cdots \\
&= 4^m \cdot K(1) + C \cdot 2^m \cdot (1 + 2 + \cdots + 2^{m-1}) \\
&\leqq 4^m \cdot K(1) + C \cdot 4^m \\
&= D \cdot 4^m = D \cdot n^2
\end{aligned}
$$

for a constant D. As usual, the dots $\cdots$ can be verified by mathematical induction: The induction basis is clear and the inductive step follows from the computation

$$
\begin{aligned}
&4^j \cdot K(2^{m-j}) + C \cdot 2^m(1 + 2 + \cdots + 2^{j-1}) \\
&= 4^j \cdot (4 \cdot K(2^{m-j-1}) + C \cdot 2^{m-j}) + C \cdot 2^m(1 + 2 + \cdots + 2^{j-1}) \\
&= 4^{j+1} \cdot K(2^{m-j-1}) + C \cdot 2^{m+j} + C \cdot 2^m(1 + 2 + \cdots + 2^{j-1}).
\end{aligned}
$$

Hence we have shown that this method—similar to the high school method—needs $O(n^2)$ operations indeed.

Session 3.4. *Mathematica includes very efficient and fast integer arithmetics. Nevertheless, we would like to check how the above algorithm works. For this, we declare*[9]

```
In[1]:= Clear[Multiply]

        Multiply[x_,y_] :=
          Module[{a,b,c,d,n,m},
              n = Length[IntegerDigits[x]];
              m = Length[IntegerDigits[y]];
              m = Max[n,m];
              n = 1;
              While[n < m, n = 2*n];
              b = Mod[x, 10^(n/2)];
              a = (x-b)/10^(n/2);
              d = Mod[y, 10^(n/2)];
              c = (y-d)/10^(n/2);
              Multiply[a,c]*10^n +
                 (Multiply[a,d] + Multiply[b,c])*10^(n/2)
                 +Multiply[b,d]]/;
            Min[Length[IntegerDigits[x]],
                Length[IntegerDigits[y]]] > 1

        Multiply[x_,y_] := x*y
```

---

[9] It is good programming style in *Mathematica* to write all definitions of a function declaration in *one input cell* which starts with a Clear command. If you have to change your definition—for example to delete a programming mistake—the prior function definitions are deleted so that the new definitions can be effective and are not superimposed by the old ones.

In this code, the second definition `Multiply[x_,y_]:=x*y`, *for which the built-in mul-tiplication is used, applies only if the condition* `Min[Length[IntegerDigits[x]],` `Length[IntegerDigits[y]]]>1` *is invalid, hence if both numbers x and y are digits and therefore the digit multiplication table comes into effect.*

The call

```
In[2]:= Multiply[1234,5678]
Out[2]= 7006652
```

provides the wanted result 7006652. *However, how was this result obtained? With other words: Which intermediate results were computed? This question is answered by* Mathematica *with the aid of the* `Trace` *command:*

```
In[3]:= Flatten[
           Trace[Multiply[1234,5678],Multiply[n_,m_] → {n,m}]
        ]//MatrixForm
```

$$
Out[3]= \begin{pmatrix}
\{1234,5678\} \\
\{12,56\} \\
\{1,5\} \\
\{1,6\} \\
\{2,5\} \\
\{2,6\} \\
\{12,78\} \\
\{1,7\} \\
\{1,8\} \\
\{2,7\} \\
\{2,8\} \\
\{34,56\} \\
\{3,5\} \\
\{3,6\} \\
\{4,5\} \\
\{4,6\} \\
\{34,78\} \\
\{3,7\} \\
\{3,8\} \\
\{4,7\} \\
\{4,8\}
\end{pmatrix}
$$

The output provides a list of argument pairs of the function `Multiply` *in the order in which they were called. You can see that in total* $4 \cdot 4 = 16$ *elementary multiplications were computed. These are exactly the multiplications that also occur in the high school algorithm.*

This recursive scheme gives us the chance to find a more efficient method, the *Karatsuba algorithm* [Kar1962]. This algorithm uses the identity

$$a \cdot d + b \cdot c = a \cdot c + b \cdot d + (a-b) \cdot (d-c)$$

(or a similar one). Using this identity the recursive step of the multiplication algorithm can be rewritten in the form

$$x \cdot y = a \cdot c \cdot 10^n + \left(a \cdot c + b \cdot d + (a-b) \cdot (d-c) \right) \cdot 10^{n/2} + b \cdot d . \tag{3.3}$$

The obvious advantage of this method (which is not immediately apparent) is the fact that in each recursive step only three instead of four multiplications are needed, namely $a \cdot c$, $b \cdot d$ (each used twice) and $(a - b) \cdot (d - c)$.[10]

Hence we get for the complexity of the Karatsuba recurrence

$$K(1) = 1 \qquad \text{with} \qquad K(n) = 3 \cdot K(n/2) + C \cdot n,$$

for which we conclude

$$
\begin{aligned}
K(n) = K(2^m) &= 3 \cdot K(2^{m-1}) + C \cdot 2^m \\
&= 3 \cdot (3 \cdot K(2^{m-2}) + C \cdot 2^{m-1}) + C \cdot 2^m \\
&= 3^2 \cdot K(2^{m-2}) + C \cdot 2^m \left(1 + \frac{3}{2}\right) = \cdots \\
&= 3^m \cdot K(1) + C \cdot 2^m \cdot \left(1 + \frac{3}{2} + \cdots + \left(\frac{3}{2}\right)^{m-1}\right) \\
&= 3^m + C \cdot 2^m \frac{\left(\frac{3}{2}\right)^m - 1}{\frac{3}{2} - 1} \\
&\leq D \cdot 3^m = D \cdot n^{\log_2 3} \approx D \cdot n^{1.585}.
\end{aligned}
$$

Clearly the constant in Karatsuba's algorithm is much larger than the one in high school multiplication, so that this method is better for relative large input numbers only. However, our computation shows that asymptotically (hence for *very large* input) the new method is *much better*!

Let us collect the previous results about the complexity of integer arithmetic.

Theorem 3.5. (Complexity of Basic Arithmetic Operations) High school addition and multiplication need $O(n)$ and $O(n^2)$ elementary operations, respectively. However, Karatsuba's multiplication algorithm has a complexity $O\left(n^{\log_2 3}\right)$. The complexity of these algorithms is independent of the used basis B.

Proof. For $B = 10$ we have shown these results already. Now, we prove that the complexity does not depend on the choice of the basis. Assume that $z \in \mathbb{N}$ has a representation with respect to the basis B, then its number of digits n is given by the formula[11]

$$n \approx \log_B z = \frac{\log z}{\log B}.$$

For the number of digits n' with respect to another basis B', we therefore get

$$n' \approx \log_{B'} z = \frac{\log z}{\log B'} = \frac{\log z}{\log B} \cdot \frac{\log B}{\log B'} = \frac{\log B}{\log B'} n = \log_{B'} B \cdot n.$$

Hence the complexity results are valid with respect to every basis, although with different constants. □

---

[10] In contrast to the high school algorithm this method needs a larger "digit multiplication table" since negative numbers also come into the play. Moreover four additions of length n are needed in the recursive step instead of one addition.

[11] The exact integer value is $n = \lfloor \log_B z + 1 \rfloor$. The *floor function* $\lfloor x \rfloor$ is treated on page 50. If we use the log function without basis, then the basis is assumed to be arbitrary. In Discrete Mathematics mostly $B = 2$ is used.

In the above computations n is the number of digits of an integer z. Hence, if we would like to express the complexity by z itself, we get for the Karatsuba algorithm $K(z) = O((\log z)^{\log_2 3})$. However, the usual reference for complexity measures is the representation length n of an integer.

Session 3.6. *The Karatsuba algorithm can be implemented by*

```
In[1]:= Clear[Karatsuba]

        Karatsuba[x_, y_] :=
          Module[{a, b, c, d, n, m, atimesc, btimesd},
              n = Length[IntegerDigits[x]];
              m = Length[IntegerDigits[y]];
              m = Max[n, m];
              n = 1;
              While[n < m, n = 2 * n];
              b = Mod[x, 10^(n/2)];
              a = (x - b)/10^(n/2);
              d = Mod[y, 10^(n/2)];
              c = (y - d)/10^(n/2);
              atimesc = Karatsuba[a, c];
              btimesd = Karatsuba[b, d];
              atimesc * 10^n +
                (atimesc + btimesd + Karatsuba[a - b, d - c]) * 10^(n/2)
                + btimesd]/;
            Min[Length[IntegerDigits[x]],
                Length[IntegerDigits[y]]] > 1

        Karatsuba[x_, y_] := x * y
```

Again, we visualize the intermediate results.:

```
In[2]:= Flatten[
            Trace[Karatsuba[1234, 5678], Karatsuba[n_, m_] -> {n, m}]
            ]//MatrixForm
```

$$
Out[2]= \begin{pmatrix} \{1234, 5678\} \\ \{12, 56\} \\ \{1, 5\} \\ \{2, 6\} \\ \{-1, 1\} \\ \{34, 78\} \\ \{3, 7\} \\ \{4, 8\} \\ \{-1, 1\} \\ \{-22, 22\} \\ \{-3, 2\} \\ \{8, 2\} \\ \{-11, 0\} \end{pmatrix}
$$

Note that this recursive algorithm has negative intermediate results indeed, but needs only 9 instead of 16 elementary multiplications.

In the given form the program is not yet efficient since every call of Karatsuba *causes three further calls of the function, which all—without knowledge of the results of the parallel calls—have to be evaluated. Therefore every elementary addition and multiplication has to be multiply evaluated. This situation was considered in Section 2.5.*

The slightly modified versions of the high school and Karatsuba algorithms are given by

```
In[3]:= $RecursionLimit = ∞;
        Clear[Karatsuba]
        Karatsuba[x_, y_] :=
          (Karatsuba[x, y] = Module[{a, b, c, d, n, m, atimesc, btimesd},
                n = Length[IntegerDigits[x]];
                m = Length[IntegerDigits[y]];
                m = Max[n, m];
                n = 1; While[n < m, n = 2 * n];
                b = Mod[x, 10^(n/2)];
                a = (x - b) / 10^(n/2);
                d = Mod[y, 10^(n/2)];
                c = (y - d) / 10^(n/2);
                atimesc = Karatsuba[a, c];
                btimesd = Karatsuba[b, d];
                atimesc * 10^n + (atimesc + btimesd
                  + Karatsuba[a - b, d - c]) * 10^(n/2) + btimesd]) /;
                 Min[Length[IntegerDigits[x]],
                 Length[IntegerDigits[y]]] > 1
        Karatsuba[x_, y_] := x * y
```

```
In[4]:= Clear[Multiply]
        Multiply[x_, y_] :=
          (Multiply[x, y] =
               Module[{a, b, c, d, n, m},
                n = Length[IntegerDigits[x]];
                m = Length[IntegerDigits[y]];
                m = Max[n, m];
                n = 1; While[n < m, n = 2 * n];
                b = Mod[x, 10^(n/2)];
                a = (x - b) / 10^(n/2);
                d = Mod[y, 10^(n/2)];
                c = (y - d) / 10^(n/2);
                Multiply[a, c] * 10^n +
                  (Multiply[a, d] + Multiply[b, c]) *
                    10^(n/2) + Multiply[b, d]]) /;
                 Min[Length[IntegerDigits[x]],
                 Length[IntegerDigits[y]]] > 1
        Multiply[x_, y_] := x * y
```

Now, we are able to compare the efficiency of both algorithms under consideration. For this, we multiply two random integers of 100 digits (using Random*),*

```
In[5]:= x = Random[Integer, {10^99, 10^100}];
        y = Random[Integer, {10^99, 10^100}];
```

first using the built-in multiplication routine,

```
In[6]:= Timing[z1 = x * y; ]
Out[6]= {0. Second, Null}
```

and then with our Karatsuba *implementation:*[12]

$In[7]:=$ **Timing[z2 = Karatsuba[x,y];]**
$Out[7]=$ {0.0780005 Second, Null}

$In[8]:=$ **z2 - z1**
$Out[8]=$ 0

Of course the built-in function is considerably better since the integer arithmetic, as the most important module of computer algebra, is completely implemented in the programming language C and resides in the kernel of Mathematica. Programs that are written in Mathematica's high-level language cannot have the same efficiency as kernel functions.

Next, we can compare our implementation with the high school algorithm:

$In[9]:=$ **Timing[z3 = Multiply[x,y];]**
$Out[9]=$ {0.124801 Second, Null}

We find that in Mathematica's high-level language the Karatsuba algorithm is already faster for 100-digit integers.

Eventually, we would like to show the extraordinary efficiency of the built-in multiplication. Mathematica multiplies two integers with one million digits each without any difficulties:

$In[10]:=$ **x = Random[Integer, {10^{999999}, $10^{1000000}$}];**
            y = Random[Integer, {10^{999999}, $10^{1000000}$}];

$In[11]:=$ **Timing[x*y;]**
$Out[11]=$ {0.0468003 Second, Null}

Such an efficiency cannot be reached by any program in the Mathematica high-level language.

Example 3.7. **(Karatsuba Algorithm)** We consider the application of Karatsuba's algorithm to the example $1234 \cdot 5678$ in more detail. We have $x = 1234$, $y = 5678$ and therefore $n = 4$, $a_0 = 12$, $b_0 = 34$, $c_0 = 56$ and $d_0 = 78$. Hence the computation of $x \cdot y$ is given by

$$x \cdot y = a_0 \cdot c_0 \cdot 10^4 + \left(a_0 \cdot c_0 + b_0 \cdot d_0 + (a_0 - b_0) \cdot (d_0 - c_0) \right) \cdot 10^2 + b_0 \cdot d_0, \qquad (3.4)$$

and therefore the subproblems $a_0 \cdot c_0$, $b_0 \cdot d_0$ and $(a_0 - b_0) \cdot (d_0 - c_0)$ must be solved. These can be treated using the formula (3.3) again by elementary operations, which leads to $a_0 \cdot c_0 = 12 \cdot 56 = 672$, $b_0 \cdot d_0 = 34 \cdot 78 = 2652$ as well as $(a_0 - b_0) \cdot (d_0 - c_0) = (12 - 34) \cdot (56 - 78) = -484$. What remains is the computation

$$x \cdot y = 672 \cdot 10000 + \left(672 + 2652 - 484 \right) \cdot 100 + 2652.$$

Observe that the middle additions and subtractions are *not* elementary, but operations of numbers of length n. Once they are performed, the final result can be computed using elementary additions and carries:

		00	00
	672	00	00
+	28	40	00
+		26	52
	700	66	52

---

[12] Unfortunately, the *Mathematica* function Karatsuba does not tolerate much larger input, since this leads to a collapse of *Mathematica's* kernel. In some instances this can already happen for the data given here. Apparently there are problems with the recursive structure. This could be prevented by an iterative implementation, see Section 2.4. Previous *Mathematica* releases were more stable with this respect.

This completes the example calculations. $\triangle$

3.3 Integer Arithmetic: Division with Remainder

After having algorithms for addition, subtraction and multiplication at hand, we are still missing an algorithm for division.

Let two positive integers x and y, that we want to divide, be given. Of course, in general, this does not yield an integer, but the rational number $\frac{x}{y} \in \mathbb{Q}$. However, in the environment of integer arithmetic, another question is important, namely to find the integer part of the fraction $\frac{x}{y}$, therefore not leaving the integers $\mathbb{Z} := \{\ldots, -2, -1, 0, 1, 2, \ldots\}$.

This problem is solved by the *long division algorithm* that you also know from high school. This algorithm gives an *integer division with remainder*. The following theorem can be established.

Theorem 3.8. (Division with Remainder) Let $x \in \mathbb{N}_{\geq 0}$ and $y \in \mathbb{N}$. Then there exists *exactly one pair* (q, r) with $q \in \mathbb{N}_{\geq 0}$, $r \in \mathbb{N}_{\geq 0}$ and $0 \leq r < y$ such that

$$x = qy + r \qquad \text{or} \qquad \frac{x}{y} = q + \frac{r}{y}$$

is valid. q is called the *integer quotient* of x and y or the *integer part* of $\frac{x}{y}$ and is denoted by $\lfloor \frac{x}{y} \rfloor$. The number r is called the *integer remainder* of the division of x by y.

Proof. Of course everybody knows how to generate the sought solution, namely via using the long division algorithm from high school. Nevertheless, we will give an independent mathematical proof for both the existence of a solution and its uniqueness. The latter cannot be determined from the known algorithm since there could be a method to produce a different solution.

First we prove the *existence* of such a solution. Let $y \in \mathbb{N}$ be given. For $0 \leq x < y$ we set $(q, r) = (0, x)$, and we are done, therefore we can assume in the sequel that $x \geq y$.

Now we show, using mathematical induction with respect to x, that for every $x \geq y$ there are integers q and r as claimed. Induction basis is the statement for $x = y$: $(q, r) = (1, 0)$. Now we assume that a pair (q, r) exists for all $z < x$. Then we choose $z = x - y \in \mathbb{N}_{\geq 0}$. Obviously it follows that $z < x$, and by the induction hypothesis there are (q_0, r_0) with $z = q_0 y + r_0$. Then we get

$$x = z + y = (q_0 + 1)y + r_0$$

which is a representation of the claimed type.

In terms of the uniqueness, let us assume that there are two representations

$$x = q_1 y + r_1 = q_2 y + r_2 .$$

Then we get

$$(q_2 - q_1)y = r_1 - r_2 .$$

Suppose $r_1 \geqq r_2$ (without loss of generality).[13] Then it follows that $q_2 - q_1 \geqq 0$ since $y > 0$. Furthermore, due to $r_1 < y$, we get

$$(q_2 - q_1)y = r_1 - r_2 < y,$$

and hence $q_2 - q_1 < 1$.

Now we have found out that $q_2 - q_1$ is an integer with the property $0 \leqq q_2 - q_1 < 1$. The only integer in this interval is $q_2 - q_1 = 0$. From this result it easily follows that $r_1 - r_2 = (q_2 - q_1)y = 0$ is valid. This completes the proof of uniqueness.   □

Note again that in contrast to "full division" the division with remainder does not leave the integer set $\mathbb{Z}$.

In the sequel, we use the following notations:

$$q = \text{quotient}(x, y) \quad \text{and} \quad r = \text{rem}(x, y) \quad \text{or} \quad r = \text{mod}(x, y).$$

Session 3.9. *In Mathematica the integer quotient q of two integers x and y is computed by the function* Quotient *and the remainder by* Mod:

In[1]:= **x = 1000; y = 9;**

In[2]:= **q = Quotient[x, y]**
Out[2]= 111

In[3]:= **r = Mod[x, y]**
Out[3]= 1

Hence we have

In[4]:= **q * y + r**
Out[4]= 1000

The function Floor[14] *computes the integer part of a rational number:*

In[5]:= **Floor$\left[\dfrac{x}{y}\right]$**
Out[5]= 111

Note that the functions Quotient, Mod *and* Floor *are also defined for negative arguments:*

In[6]:= **q = Quotient[x = -1000, y = 9]**
Out[6]= −112

In[7]:= **r = Mod[x, y]**
Out[7]= 8

In[8]:= **q * y + r**
Out[8]= −1000

In[9]:= **Floor$\left[\dfrac{x}{y}\right]$**
Out[9]= −112

In this connection the relation $0 \leqq r < y$ *holds for* $r = $ Mod$[x, y]$ *and therefore the relation* $\lfloor \frac{x}{y} \rfloor \leqq \frac{x}{y}$ *is always valid.*

---

[13] Otherwise, just renumber the two representations.

[14] The Floor function rounds down.

To examine the complexity of division with remainder, we apply long division to an example:

$$123456 : 789 = 156 \text{ remainder } 372$$

$$
\begin{array}{l}
\underline{789} \\
4455 \\
\underline{3945} \\
5106 \\
\underline{4734} \\
372
\end{array}
$$

If we divide an integer of length n by an integer of length m, then apparently we need $m(n-m)$ elementary multiplications,[15] since every digit of y must be multiplied by every digit of the result. Additionally we have $O(m)$ additions. Hence the division algorithm has a complexity of $O(m(n-m))$. In particular, if x has twice as many digits as y, then the complexity is $O(n^2) = O(m^2)$.

As we saw, the division algorithm can be extended to $x, y \in \mathbb{Z}$, $y \neq 0$, see also Exercise 3.4. The division algorithm has very interesting consequences which we will consider in the next section.

Next, we would like to examine some questions around divisibility. For this purpose, we first give some general definitions.

Definition 3.10. If after division of $x \in \mathbb{Z}$ by $y \in \mathbb{Z}$, the remainder is 0, then we call x *divisible* by y or a *multiple* of y, and we call y a *divisor* of x: $\frac{x}{y} = q \in \mathbb{Z}$. If x is divisible by y, then we write $y \mid x$ (in words: y divides x).

Now let R be a commutative ring with unity 1. Then $a \in R$ is called a divisor of $b \in R$, denoted by $a \mid b$, if there is a $c \in R$ such that $b = c \cdot a$. In particular, every element of R is a divisor of 0. We also call $b = c \cdot a$ *composite*.

An element $u \in R$, which has an inverse $v \in R$, i.e. $u \cdot v = 1$, is called a *unit*. The only units in $\mathbb{Z}$ are ± 1. Units are automatically divisors of every element of F.[16] Therefore a theory of divisibility only makes sense if there are not too many units in a ring. In a *field* like $\mathbb{Q}$, all elements besides 0 are units, therefore $\mathbb{Q}$ has no theory of divisors.

If $a \in R$ and $b \in R$, then $c \in R$ is called a *common divisor* of a and b, if $c \mid a$ and $c \mid b$. If the only common divisors of a and b are units, then we call a and b *relatively prime* or *coprime*.

A number $c \in R$ is called *greatest common divisor* of a and b if c is a common divisor of a and b, and if the relation $d \mid c$ is valid for all common divisors d of a and b. We write $c = \gcd(a, b)$. In particular, we have $\gcd(a, 0) = a$. Please note that, in general, greatest common divisors do not have to exist, see Exercise 3.5. The greatest common divisor of three numbers is given by $\gcd(a, b, c) = \gcd(\gcd(a, b), c)$. Similarly, $\gcd(a_1, a_2, ..., a_n)$ can be defined recursively.

If for $a, b \in R$ both conditions $a \mid b$ and $b \mid a$ are satisfied, then a and b are called *associated*, and we write $a \sim b$. The relation $\sim$ is an equivalence relation. Two numbers a and b are associated if and only if there is a unit $u \in R$ with $b = u \cdot a$, see Exercise 3.7.

---

[15] Before one can execute these divisions, one must find the correct divisor. In any case $Bm(n-m)$ elementary multiplications are enough for this purpose.

[16] $a = a \cdot 1 = a \cdot (u \cdot v) = u \cdot (a \cdot v)$

Hence, if it exists, the greatest common divisor is not uniquely determined, but all associated members are greatest common divisors.[17] In $\mathbb{Z}$ we can make the greatest common divisor unique by the condition $\gcd(a, b) > 0$.

In a similar way we call $c \in R$ the *least common multiple* of $a \in R$ and $b \in R$, if $a \mid c$ and $b \mid c$, and if, furthermore, for all $d \in R$ the relations $a \mid d$ and $b \mid d$ imply that $c \mid d$. We write $c = \operatorname{lcm}(a, b)$ and make the least common multiple in $\mathbb{Z}$ unique by the condition $\operatorname{lcm}(a, b) > 0$. $\triangle$

3.4 The Extended Euclidean Algorithm

Division with remainder leads to a method for computing the greatest common divisor of two integers. This algorithm is called the *Euclidean algorithm*.

Example 3.11. We compute the greatest common divisor of $x = 123$ and $y = 27$ by successive division with remainder using the following scheme:

$$
\begin{aligned}
123 &= 4 \cdot 27 + 15 \\
27 &= 1 \cdot 15 + 12 \\
15 &= 1 \cdot 12 + \mathbf{3} \\
12 &= 4 \cdot 3 + 0
\end{aligned}
$$

This scheme is called the Euclidean algorithm. In the first line we have computed $4 = q = \operatorname{quotient}(123, 27)$ and $15 = r = \operatorname{rem}(123, 27)$ using long division. Next, long division is applied to q and r, and so on. The algorithm terminates when $r = 0$ appears, and the last nonvanishing remainder which was computed—shown in bold—yields the greatest common divisor $\gcd(x, y)$ of $x \in \mathbb{N}$ and $y \in \mathbb{N}$. The number of iterations (= lines) is called the *Euclidean length* of x and y, denoted by $\operatorname{eucl}(x, y)$. In our example we find $\operatorname{eucl}(x, y) = 4$.

The iteration scheme can be written in the form

$$
x_{k-1} = q_k x_k + r_k,
$$

where the initial values $x_0 = x$ and $x_1 = y$ are given, and in every step the numbers

$$
q_k = \operatorname{quotient}(x_{k-1}, x_k), \qquad r_k = \operatorname{rem}(x_{k-1}, x_k) \qquad \text{as well as} \qquad x_{k+1} = r_k
$$

are computed. $\triangle$

Now let us prove this algorithm.

Theorem 3.12. (Euclidean Algorithm) The Euclidean algorithm, applied to the pair $x, y \in \mathbb{Z}$, computes the greatest common divisor $\gcd(x, y)$.

Proof. The fact that the Euclidean algorithm indeed computes a greatest common divisor, is essentially due to the relation[18]

---

[17] Hence, $\gcd(a, b)$ ist an equivalence class.

[18] This equal sign refers to $\sim$.

$$\gcd(x, y) = \gcd(y, r), \tag{3.5}$$

which is valid for the identity $x = qy + r$ of division with remainder.

To prove this relation, we set $g := \gcd(x, y)$ and $h := \gcd(y, r)$. The relation $r = x - qy$ implies that g is a divisor of r. Hence g is a common divisor of y and r, and it follows that $g \mid h$. On the other hand, the relation $x = qy + r$ implies that h is a divisor of x, and therefore a common divisor of x and y, so that $h \mid g$. Therefore g and h are associated, and (3.5) is valid.

The Euclidean algorithm leads to the computation of $\gcd(x, y)$ by iterating the fundamental identity (3.5). This iteration terminates since the relation $0 \leqq r < y$ shows that the pairs get smaller and smaller (which is an induction with respect to $\mathrm{eucl}(x, y)$, i.e., the number of steps). $\qquad\square$

Sometimes additional information is necessary about the greatest common divisor, which can be found using the following extension of the Euclidean algorithm. Using successive back substitution of the intermediate results in the computed value for $\gcd(x, y)$, we can write the result $\gcd(x, y)$ as a linear combination (with integer coefficients) of the two starting values x and y:

$$
\begin{aligned}
3 &= 15 - 1 \cdot 12 \\
&= 15 - 1 \cdot (27 - 1 \cdot 15) \\
&= -27 + 2 \cdot 15 \\
&= -27 + 2 \cdot (123 - 4 \cdot 27) \\
&= -9 \cdot 27 + 2 \cdot 123
\end{aligned}
$$

Obviously, this method always leads to a representation of the form

$$\gcd(x, y) = s \cdot x + t \cdot y \qquad \text{with } s, t \in \mathbb{Z}. \tag{3.6}$$

Theorem 3.13. (Extended Euclidean Algorithm) Let $x, y \in \mathbb{Z}$. Then there is a representation of the form (3.6) with *Bézout coefficients* $s, t \in \mathbb{Z}$, which can be determined by the described algorithm. This algorithm is called the *Extended Euclidean Algorithm*. $\qquad\square$

Session 3.14. *Here is a Mathematica implementation of the extended Euclidean algorithm which uses Mathematica's pattern matching capabilities for back substitution:*[19]

---

[19] In *Mathematica*, variables that are given as arguments of functions cannot be changed in the function module. Since we use x as a local variable, we utilize the argument name xx.

The warning which is generated by `General::"spell1"` is a typical message if *Mathematica* finds that a variable name is used which is similar to an already declared variable. Of course, you can ignore this message if you are sure that misspelling did not occur! In the future we will not print such messages.

```
In[1]:= Clear[extendedgcd]
        extendedgcd[xx_,y_] :=
            Module[{x,k,rule,r,q,X},
          x[0] = xx; x[1] = y; k = 0; rule = {};
              gcdmatrix = {};
          While[Not[x[k+1] == 0],
          k = k+1;
          r[k] = Mod[x[k-1],x[k]];
          q[k] = Quotient[x[k-1],x[k]];
          x[k+1] = r[k];
                AppendTo[rule,
                  X[k+1] →
                  Collect[X[k-1]-q[k]*X[k]/.rule,
                    {X[0],X[1]}]];
                AppendTo[gcdmatrix,
                 {k,"|",x[k-1]," = ",q[k]," * ",
                  x[k]," + ",r[k]}];
                ];
                {x[k],Coefficient[
                  X[k-2]-q[k-1]*X[k-1]/.rule,
                    {X[0],X[1]}]}
            ]/; IntegerQ[xx]&&IntegerQ[y]
```

General :: "*spell1*": *Possible spelling error: new symbol name*
"rule" is similar to existing symbol "Rule"

The local variable rule *collects (using* AppendTo*) the replacement rules that are necessary to express* x_k *as linear combination of* $x_0 = x$ *und* $x_1 = y$—*this can be forced by applying* rule— *and simplifies the result using* Collect*. Finally these replacement rules are substituted in* $x_{k-2} - q_{k-1} x_{k-1} = r_{k-1}$. *Observe that once the rule is applied, we have* $r_{k-1} = \gcd(x, y)$.

The function extendedgcd *results in a list whose first element is the greatest common divisor* $\gcd(x, y)$ *of* x *and* y *and whose second element is the list* $\{s, t\}$ *of the two Bézout coefficients of representation (3.6). Note that the built-in Mathematica function* ExtendedGCD[x, y] *has the same functionality.*

As a side effect, the call of the function extendedgcd *creates the variable* gcdmatrix[20] *which records all execution steps of the Euclidean algorithm and saves it in a matrix. For example,*

```
In[2]:= ext = extendedgcd[x = 1526757668, y = 7835626735736]
Out[2]= {4, {845922341123, -164826435}}
```

computes the sought representation

$$4 = sx + ty = 845922341123 \cdot 1526757668 - 164826435 \cdot 7835626735736.$$

The complete execution of the Euclidean algorithm is documented for this example by the following matrix:

```
In[3]:= gcdmatrix
```

---

[20] which is a global variable since it is not listed in the list of local variables. Otherwise the variable would not be accessible after the call of the function extendedgcd.

$$Out[3] = \begin{pmatrix}
1 & | & 1526757668 & = & 0 & * & 7835626735736 & + & 1526757668 \\
2 & | & 7835626735736 & = & 5132 & * & 1526757668 & + & 306383560 \\
3 & | & 1526757668 & = & 4 & * & 306383560 & + & 301223428 \\
4 & | & 306383560 & = & 1 & * & 301223428 & + & 5160132 \\
5 & | & 301223428 & = & 58 & * & 5160132 & + & 1935772 \\
6 & | & 5160132 & = & 2 & * & 1935772 & + & 1288588 \\
7 & | & 1935772 & = & 1 & * & 1288588 & + & 647184 \\
8 & | & 1288588 & = & 1 & * & 647184 & + & 641404 \\
9 & | & 647184 & = & 1 & * & 641404 & + & 5780 \\
10 & | & 641404 & = & 110 & * & 5780 & + & 5604 \\
11 & | & 5780 & = & 1 & * & 5604 & + & 176 \\
12 & | & 5604 & = & 31 & * & 176 & + & 148 \\
13 & | & 176 & = & 1 & * & 148 & + & 28 \\
14 & | & 148 & = & 5 & * & 28 & + & 8 \\
15 & | & 28 & = & 3 & * & 8 & + & 4 \\
16 & | & 8 & = & 2 & * & 4 & + & 0
\end{pmatrix}$$

We note that the computation of gcd(x, y) *needed* eucl$(x, y) = 16$ *steps.*

The defining equation $F_{n+1} = F_n + F_{n-1}$ of the Fibonacci numbers shows that F_n is an increasing sequence of positive integers for which furthermore the inequalities

$$1 < \frac{F_{n+1}}{F_n} = 1 + \frac{F_{n-1}}{F_n} < 2 \quad (n \in \mathbb{N})$$

are valid. Therefore we get quotient$(F_{n+1}, F_n) = 1$ *and* rem$(F_{n+1}, F_n) = F_{n-1}$. *Therefore, if we start the Euclidean algorithm with two successive Fibonacci numbers, then this process successively creates all smaller Fibonacci numbers as remainders.*

For example, we compute[21]

```
In[4]:= extendedgcd[Fibonacci[21],Fibonacci[20]]
Out[4]= {1,{-2584,4181}}
```

and get the computation matrix

```
In[5]:= gcdmatrix
```

$$Out[5] = \begin{pmatrix}
1 & | & 10946 & = & 1 & * & 6765 & + & 4181 \\
2 & | & 6765 & = & 1 & * & 4181 & + & 2584 \\
3 & | & 4181 & = & 1 & * & 2584 & + & 1597 \\
4 & | & 2584 & = & 1 & * & 1597 & + & 987 \\
5 & | & 1597 & = & 1 & * & 987 & + & 610 \\
6 & | & 987 & = & 1 & * & 610 & + & 377 \\
7 & | & 610 & = & 1 & * & 377 & + & 233 \\
8 & | & 377 & = & 1 & * & 233 & + & 144 \\
9 & | & 233 & = & 1 & * & 144 & + & 89 \\
10 & | & 144 & = & 1 & * & 89 & + & 55 \\
11 & | & 89 & = & 1 & * & 55 & + & 34 \\
12 & | & 55 & = & 1 & * & 34 & + & 21 \\
13 & | & 34 & = & 1 & * & 21 & + & 13 \\
14 & | & 21 & = & 1 & * & 13 & + & 8 \\
15 & | & 13 & = & 1 & * & 8 & + & 5 \\
16 & | & 8 & = & 1 & * & 5 & + & 3 \\
17 & | & 5 & = & 1 & * & 3 & + & 2 \\
18 & | & 3 & = & 1 & * & 2 & + & 1 \\
19 & | & 2 & = & 2 & * & 1 & + & 0
\end{pmatrix}$$

containing exclusively Fibonacci numbers. Note that eucl$(F_{n+1}, F_n) = n - 1$.

---

[21] Please check that the absolute values of the Bézout coefficients $|s|$ and $|t|$ are also Fibonacci numbers, see Exercise 3.10!

For the Fibonacci numbers the Euclidean length $\text{eucl}(x, y)$ is particularly large since in this case all quotients are as small as possible, namely 1. By mathematical induction one can show that $\text{eucl}(x, y)$ is maximal if the smaller input number is the Fibonacci number F_n and if $\gcd(x, y) = 1$, see Exercise 3.10. Since $F_n = O\left(\left(\frac{1+\sqrt{5}}{2}\right)^n\right)$, summation leads to the complexity measure for the Euclidean algorithm, see [GG1999], Chapter 3. The result is that if x and y are integers of length n and m, respectively, then the extended Euclidean algorithm has a complexity of $O(n \cdot m)$.

3.5 Unique Factorization

One of the consequences of the Euclidean algorithm is unique factorization in prime factors. First, we give a definition.

Definition 3.15. Let R be a commutative ring with 1. Then an element $a \in R$, which is not a unit, is called *reducible* or *composite* if there are $b, c \in R$, which are no units, such that $a = b \cdot c$. Otherwise, a is called *irreducible*. The irreducible elements in $R = \mathbb{Z}$ are called *prime numbers*.[22] We denote the set of positive prime numbers by $\mathbb{P}$. Obviously $\mathbb{P} = \{2, 3, 5, 7, 11, \ldots\}$.
△

Now we will confirm that every integer $x \in \mathbb{Z} \setminus \{0\}$ has a unique prime factor decomposition (up to units). Obviously it is sufficient to consider positive integers $x \in \mathbb{N}_{\geq 2}$.

First, we have to clarify what uniqueness means. Let $p_k \in \mathbb{Z}$ $(k = 1, \ldots, n)$. We do not exclude the possibility that some of the primes p_k agree with each other. By $\text{sort}(p_1, \ldots, p_n)$ we denote the *sorted list* (with respect to $<$) of the elements p_k $(k = 1, \ldots, n)$. Now assume that $x = p_1 \cdot p_2 \cdots p_n = q_1 \cdot q_2 \cdots q_m$ are two prime factorizations of x with $p_j \in \mathbb{P}$ $(j = 1, \ldots, n)$ and $q_j \in \mathbb{P}$ $(j = 1, \ldots, m)$. Then we say that the factorizations *agree* if the sorted lists $\text{sort}(p_1, p_2, \ldots, p_n)$ and $\text{sort}(q_1, q_2, \ldots, q_m)$ are identical. Hence two factorizations agree if they are the same *up to the order of the factors*. In particular, in this case $m = n$. We say that a factorization is unique if *all possible factorizations (without units) agree*.

Session 3.16. *Mathematica distinguishes between sets and lists. If we take a random list*

```
In[1]:= list = Table[Random[Integer, {1, 10}], {20}]
Out[1]= {6, 9, 10, 10, 9, 8, 5, 2, 1, 10, 7, 2, 1, 4, 1, 1, 7, 4, 1, 10}
```

then this list can be sorted by Sort:

```
In[2]:= Sort[list]
Out[2]= {1, 1, 1, 1, 1, 2, 2, 4, 4, 5, 6, 7, 7, 8, 9, 9, 10, 10, 10, 10}
```

The function Union *considers the list as a set, sorts it and deletes multiple entries.*

```
In[3]:= Union[list]
Out[3]= {1, 2, 4, 5, 6, 7, 8, 9, 10}
```

The following theorem shows the importance of unique factorizations in this context.

---

[22] In a more general setting prime elements are defined by the property of the following Lemma 3.18. In $\mathbb{Z}$ these properties coincide.

Theorem 3.17. (Fundamental Theorem of Number Theory) Every $x \in \mathbb{N}_{\geq 2}$ possesses a unique factorization.

Proof. For the proof we need the following lemma.

Lemma 3.18. Let $p \in \mathbb{P}$ be a divisor of $a \cdot b \, (a, b \in \mathbb{Z})$. Then either $p \mid a$ or $p \mid b$ is valid.

Proof. Define $g := \gcd(p, a)$. If $g \neq 1$, then $g = p$ since p is irreducible. Therefore in this case $p \mid a$.

If $g = 1$, then the extended Euclidean algorithm guarantees that

$$1 = a \cdot s + p \, t \qquad (s, t \in \mathbb{Z}),$$

and therefore we get

$$b = a \cdot b \cdot s + b \cdot p \cdot t. \tag{3.7}$$

Of course p is a divisor of p, and by hypothesis p is also a divisor of $a \cdot b$. From (3.7) it therefore follows that p must be a divisor of b. $\qquad\square$

Now we can prove Theorem 3.17. In the first step we will prove by mathematical induction that a prime factor decomposition exists. The induction basis for $x = 2$ is fulfilled since $x = 2$ is a prime number and therefore has the factorization $x = 2$. Now let $x > 2$. We assume (induction hypothesis) that a prime factorization exists for all $z < x$. Now either x is prime and therefore represents itself, or x is composite, i.e., $x = a \cdot b \, (a, b \in \mathbb{N}_{\geq 2})$. In this case both $a < x$ and $b < x$. By the induction hypothesis both a and b have a factorization, which leads to the sought factorization for $x = a \cdot b$. This completes the induction step.

Uniqueness is also proved using induction. Again the induction basis is clear. Now let the statement be true for every positive integer smaller than x, and let $x = p_1 \cdot p_2 \cdots p_n = q_1 \cdot q_2 \cdots q_m$ be two prime factor decompositions of x with $p_j \in \mathbb{P} \, (j = 1, \ldots, n)$ and $q_j \in \mathbb{P} \, (j = 1, \ldots, m)$.

Since therefore $p_1 \mid x = q_1 \cdot q_2 \cdots q_m$, Lemma 3.18 implies (again by induction) that $p_1 \mid q_j$ for some $j = 1, \ldots, m$. Since q_j is prime, we get furthermore $p_1 = q_j$. Therefore $x/p_1 \in \mathbb{N}$ with

$$x > \frac{x}{p_1} = p_2 \cdots p_n = q_1 \cdots q_{j-1} \cdot q_{j+1} \cdots q_m. \tag{3.8}$$

The induction hypothesis states that the two factorizations of x/p_1 in (3.8) agree, i.e., the prime lists $\text{sort}(p_2, \ldots, p_n)$ and $\text{sort}(q_1, \ldots, q_{j-1}, q_{j+1}, \ldots q_m)$ are identical. Since $p_1 = q_j$, we finally get $\text{sort}(p_1, p_2, \ldots, p_n) = \text{sort}(q_1, q_2, \ldots q_m)$ which completes the induction step. $\qquad\square$

Session 3.19. *In principle, greatest common divisors and least common multiples can be read off from the prime factor decompositions, see Exercise 3.18. This is the typical algorithm used at high school when "small numbers" are treated. However, whereas the computation of the gcd using the Euclidean algorithm is very fast (even for "large numbers") this is evidently not the case for prime factor decompositions.*

As an example, we choose two random integers x and y with 60 digits:

```
In[1]:= x = Random[Integer, {10^59, 10^60}]
        y = Random[Integer, {10^59, 10^60}]
Out[1]= 796512072083729182603369951295509873656833697293891874377709
Out[1]= 425170642064150171819227671312438691042876401245123358447956
```

The computation of the greatest common divisor is very fast:

```
In[2]:= GCD[x,y]//Timing
Out[2]= {0. Second, 1}
```

But the factorizations of x and y need much more time:[23]

```
In[3]:= FactorInteger[x]//Timing
```

$$Out[3]= \left\{0.421203 \; \text{Second}, \begin{pmatrix} 3 & 2 \\ 47 & 1 \\ 1285993 & 1 \\ 145007324521 & 1 \\ 1009772295002222178346356767511702346281 & 1 \end{pmatrix}\right\}$$

```
In[4]:= FactorInteger[y]//Timing
```

$$Out[4]= \left\{4.05603 \; \text{Second}, \begin{pmatrix} 2 & 2 \\ 11 & 1 \\ 169944719561579 & 1 \\ 568594844414700665156711205741422080428779 & 1 \end{pmatrix}\right\}$$

Note that the computation times vary heavily, depending on the size of the prime factors. In the most complicated case the prime factors have almost equal size. This is the case that we will test now. We use the function NextPrime *to find some prime numbers, and compute a 60-digit integer z with two random factors of 30 digits:*

```
In[5]:= z = NextPrime[Random[Integer, {10^29, 10^30}]]*
          NextPrime[Random[Integer, {10^29, 10^30}]]
Out[5]= 898257355512576902799120421325099465682586589545383256902223
```

The factorization of z needs about one minute:

```
In[6]:= FactorInteger[z]//Timing
```

$$Out[6]= \left\{54.2259 \; \text{Second}, \begin{pmatrix} 927061782305777483982199843739 & 1 \\ 968929334222409061431942596957 & 1 \end{pmatrix}\right\}$$

What can we say about the complexity of integer factorization? In the simplest algorithm for the complete factorization of a positive integer x of length n, we divide x iteratively by the primes 2, 3, 5, ..., until all divisors are found. This method is called factorization by trial division. It is implemented by

```
In[7]:= Clear[factorinteger]
        factorinteger[x_] :=
          Module[{z = x, divisor = 2, list = {}, z1},
             While[z > 1,
                While[IntegerQ[z1 = z/divisor],
                   AppendTo[list, divisor]; z = z1];
                divisor = NextPrime[divisor]];
             list
          ]/; IntegerQ[x] && x > 1
```

Let us choose a 20-digit number

```
In[8]:= x = 65176314651398250790
Out[8]= 65176314651398250790
```

and factorize it using the built-in Mathematica function:

---

[23] Depending on the random numbers, factorization might also be fast. Then you can choose different random numbers.

```
In[9]:= FactorInteger[x]//Timing
```

$$Out[9]= \left\{0.01 \text{ Second}, \begin{pmatrix} 2 & 1 \\ 5 & 1 \\ 7 & 1 \\ 23 & 1 \\ 29 & 1 \\ 439 & 1 \\ 3179811720469 & 1 \end{pmatrix}\right\}$$

However, our implementation

```
In[10]:= factorinteger[x]//Timing
```

Out[10]= *$Aborted*

must be aborted using **Evaluation, Abort Evaluation** *since the computation takes too long.*[24] *The complexity of this simple algorithm for an integer x of length n is obviously $O(x \cdot n^2) = O(n^2 \cdot 2^n)$, since the loop must be executed (potentially) for every odd number smaller than x, and each step needs several divisions. This is exponential in the length n of x, and the exponential complexity means that a duplication of the computing capacity results only in an additive improvement of the computing power.*[25]

We can improve our algorithm by checking in each step whether the remaining factor is prime. In this case, we do not have to search for further factors and the algorithm stops. In our concrete case this means that the algorithm terminates already after the division by 439. The fact that it really is more efficient to check irreducibility than finding a factor is considered in more detail in Section 4.6.

Therefore, we adapt our implementation accordingly using the built-in function PrimeQ:[26]

```
In[11]:= Clear[factorinteger]
         factorinteger[x_] :=
           Module[{z = x, divisor = 2, list = {},
               z1},
              While[z > 1&&Not[PrimeQ[z]],
                While[IntegerQ[z1 = z/divisor],
                  AppendTo[list, divisor]; z = z1];
                divisor = NextPrime[divisor]];
              If[PrimeQ[z], AppendTo[list, z]];
              list
           ]/; IntegerQ[x]&&x > 1
```

Now our implementation is successful:

```
In[12]:= factorinteger[x]//Timing
```

Out[12]= {0.01 Second, {2, 5, 7, 23, 29, 439, 3179811720469}}

However, the complexity still remains exponential in the length of x: It is now $O\left(\sqrt{x} \cdot (\log x)^2\right) = O(n^2 \cdot 2^{n/2})$ since in the worst case x might contain two prime factors of size $\sqrt{x}$.

Mathematica *cannot factorize integers with eighty or more decimal digits and large prime factors in a reasonable time. However, using faster software and better methods, larger integers can be factorized.*

---

[24] Of course our implementation works for simpler examples.

[25] If the loop runs only through the primes below x (like in our implementation), the complexity is still exponential, but this is much more difficult to prove. In fact, this result is the content of the *prime number theorem* which states that the number $\pi(x)$ of primes $\leq x$ is $O(\frac{x}{\ln x})$.

[26] The answer of a call like PrimeQ is sometimes called an *oracle*.

The current world record is the factorization of the 200-digit integer RSA-200 [Wei2005b]
by the team of Jens Franke from the University of Bonn, which was published on May 10,
2005.[27] For this purpose many computers were used in parallel, together with the best known
factorization algorithm (the General Number Field Sieve). Mathematica cannot factorize these
large integers, but we can easily check the published factorization:

```
In[13]:= RSA200 =
            27997833911221327870829467638722601621070446786955
            42853756000992932612840010760934567105295536085 6
            06182235191095136578863710595448200657677509858 0
            55761357909873495014417886317894629518723786922 1
            823983
Out[13]= 27997833911221327870829467638722601621070446786955428537569929
         3261284001076093456710529553608560618223519109513657886371 05
         95448200657677509858055761357909873495014417886317894629518 7
         237869221823983

In[14]:= RSA200-
            35324619344027701212726049781984643686711974001 97
            62502364930346877612125367942320005854795652 80
            88349*
            79258699544783330333470858414800596877379758573 6
            42199607343303414557678728181521353814093474 01
            85467
Out[14]= 0
```

Modern cryptography uses the high complexity of integer factorization as an essential ingredi-
ent. In Section 5.5 we will discuss such issues in more detail.

3.6 Rational Arithmetic

The arithmetic operations for rational numbers are declared by

$$\frac{a}{b} + \frac{c}{d} = \frac{a \cdot d + b \cdot c}{b \cdot d}$$

and

$$\frac{a}{b} \cdot \frac{c}{d} = \frac{a \cdot c}{b \cdot d},$$

where we can assume that $p = \frac{a}{b}$ and $q = \frac{c}{d}$ are given in lowest terms. How can we compute the
reduced forms of sum and product?

For the addition, the numerator and the denominator can be reduced by dividing by $\gcd(ad + bc, bd)$, whereas for the multiplication, the computation of $\gcd(a, d)$ and $\gcd(b, c)$ (or alternatively
$\gcd(a \cdot c, b \cdot d)$) leads to the desired result.

Let us assume that all four input numbers have length n. Therefore, to obtain the reduced form
of the sum, we need one addition and three multiplications of two numbers of length n in $\mathbb{Z}$

---

[27] On November 8th, 2005, the same group published the factorization of the 193-digit integer RSA-640
[Wei2005c], for which the company RSA Security had offered a prize money of 10 000 US-$. You can read
more about RSA numbers in [Wei2005a].

as well as two divisions and one gcd-computation of length $2n$. This has an overall complexity $O(n^2)$.[28]

For the multiplication, on the other hand, we need two gcd-computations of length n as well as two multiplications. Again, this has the complexity $O(n^2)$.

Session 3.20. *We check the efficiency of Mathematica's rational arithmetic. For this purpose, we declare four 1 000 000-digit integers a, b, c and d and their quotients $p = \frac{a}{b}$ and $q = \frac{c}{d}$:*

```
In[1]:= a = Random[Integer, {10^999999, 10^1000000}];
        b = Random[Integer, {10^999999, 10^1000000}];
        p = a/b;

        c = Random[Integer, {10^999999, 10^1000000}];
        d = Random[Integer, {10^999999, 10^1000000}];
        q = c/d;
```

Mathematica represents all rational numbers in lowest terms and does all necessary gcd-computations automatically. For the addition of p and q, the computation

```
In[2]:= Timing[p + q;]
Out[2]= {0.842405 Second, Null}
```

needs less time than for the computation of $\gcd(ad + bc, bd)$:

```
In[3]:= Timing[GCD[a*d+b*c, b*d];]
Out[3]= {1.65361 Second, Null}
```

Hence, Mathematica uses a better algorithm for computing the sum. For the product, the computation of $p \cdot q$

```
In[4]:= Timing[p*q;]
Out[4]= {1.38841 Second, Null}
```

needs about the same amount of time as the computation of the two necessary greatest common divisors:

```
In[5]:= Timing[GCD[a, d];]
Out[5]= {0.592804 Second, Null}
```

```
In[6]:= Timing[GCD[b, c];]
Out[6]= {0.608404 Second, Null}
```

The additional integer multiplications need much less time.

3.7 Additional Remarks

In ancient times the Babylonians used the basis $B = 60$. Within this system, they were able to compute the decimal expansion of $\sqrt{2}$ (in connection with the Pythagorean theorem) correctly up to six decimal digits, compare [FR1998] and [Wiki2016].

---

[28] If Karatsuba is used, the complexity is $O(n^{\log_2 3})$.

Karatsuba's algorithm was published in [Kar1962].

Iterative application of division with remainder also leads to the decimal expansions of real numbers, see e.g. [SK2000], Chapter 4.

An extensive consideration regarding the complexity of the Euclidean algorithm is given in [GG1999], Chapter 3, see also [Mig1992], Theorem 1.6. In [GG1999], Chapter 11, one can find more efficient algorithms for the computation of greatest common divisors. The connection with Fibonacci numbers was published in [Lam1844].

3.8 Exercises

Exercise 3.1. *Show that the multiplication of two integers of lengths n and m using the high school algorithm has the complexity $O(n \cdot m)$.*

Exercise 3.2. *Find a variant of Karatsuba's algorithm, which divides the integers in three or four equally long parts and which has a better complexity than the original Karatsuba algorithm. Implement this algorithm.*

Exercise 3.3. *Complete the implementation of Karatsuba's algorithm from Session 3.6 by implementing the addition* Add.

Exercise 3.4. *Show Theorem 3.8 for $x, y \in \mathbb{Z}$, $y \neq 0$, for the case $q \in \mathbb{Z}$ and $0 \leqq r < |y|$.*

Exercise 3.5. *Show that the set*

$$R := \{a + b\sqrt{-5} \mid a, b \in \mathbb{Z}\},$$

together with the natural addition and multiplication, forms a commutative ring with unity 1. Note that the identity

$$6 = 2 \cdot 3 = \left(1 + \sqrt{-5}\right)\left(1 - \sqrt{-5}\right)$$

shows that in this ring a greatest common divisor does not always exist.

Exercise 3.6. *Show that for $x, y \in \mathbb{N}$*

$$\gcd(x, y) \cdot \operatorname{lcm}(x, y) = x \cdot y. \tag{3.9}$$

Exercise 3.7. *Let R be a commutative ring with unity. Show that two elements $a, b \in R$ are associated, $a \sim b$, if and only if there is a unit $u \in R$ with $b = u \cdot a$. Show that $\sim$ forms an equivalence relation.*

Show further that the set of all units of R forms a group with respect to multiplication, called the group of units.

Exercise 3.8. *Perform the mathematical induction step in Theorem 3.12 explicitly.*

Exercise 3.9. *Perform the details of the proof of Theorem 3.13.*

Exercise 3.10. *Let* eucl(x, y) *be the number of iterations for the computation of* gcd(x, y), $x > y > 0$, *in Euclidean's algorithm.*

(a) *Compute* gcd(F_{n+1}, F_n) *and* eucl(F_{n+1}, F_n).
(b) *Let* $y < F_{n+1}$. *Then show that for every* $x > y$, *the relation* eucl$(x, y) \leq$ eucl(F_{n+1}, F_n) *holds.*
(c) *Compute* s *and* t *of the extended Euclidean algorithm for* F_{n+1} *and* F_n. *This gives a new identity for Fibonacci numbers.*

Exercise 3.11. (Euclidean Algorithm)

(a) *Implement the Euclidean algorithm* gcd[x,y] *using (3.5) recursively and test your implementation with the example* gcd$(2^{40}+3, 3^{30}+8)$. *Note that an efficient implementation needs no measurable time for this computation. Compare your result with the built-in functionality* GCD.
(b) *Define the function* extendedgcd *from Session 3.14 in Mathematica and apply it to* $x = 2^{40}+3$ *and* $y = 3^{30}+8$ *as well. Show the computation matrix.*

Exercise 3.12. (Continued Fractions) *A (simple) continued fraction is a fraction of the form*

$$a_1 + \cfrac{1}{a_2 + \cfrac{1}{a_3 + \cfrac{}{\ddots \cfrac{1}{a_{n-1} + \cfrac{1}{a_n}}}}} \tag{3.10}$$

Mathematica's command ContinuedFraction *generates the continued fraction of a rational number, whereas* FromContinuedFraction *converts a continued fraction decomposition into its corresponding rational number.*[29] *More precisely, the rational equivalent of the continued fraction (3.10) is computed by* FromContinuedFraction[$\{a_1, a_2, ..., a_n\}$].

(a) *Describe algorithms for both* ContinuedFraction *and* FromContinuedFraction.
(b) *Explain why* FromContinuedFraction[Range[n]], $n \in \mathbb{N}$, *creates only Fibonacci numbers in both the numerator and the denominator.*

Exercise 3.13. *Show: If* $a \in \mathbb{Z}$ *and* $b \in \mathbb{Z}$ *are divisors of* $c \in \mathbb{Z}$ *with* gcd$(a, b) = 1$, *then* $a \cdot b \,|\, c$.

Exercise 3.14. *Use suitable gcd-computations to find a proper divisor of the Fermat number* $F_5 = 2^{2^5} + 1$.

Exercise 3.15. (Horner Scheme) *Let* R *be a commutative ring with unity. Show that the direct evaluation of the polynomial*

$$p(x) = a_0 + a_1 x + a_2 x^2 + \cdots + a_n x^n$$

at some $x \in R$ *needs* $O(n^2)$ *ring operations. It is more efficient to evaluate* $p(x)$ *via the Horner formula*

$$p(x) = a_0 + x(a_1 + x(a_2 + \cdots x(a_{n-1} + x a_n))). \tag{3.11}$$

Prove the correctness of the Horner scheme and calculate its evaluation complexity.

---

[29] It is also possible to consider infinite continued fractions. This is supported by the commands ContinuedFraction and FromContinuedFraction,

Exercise 3.16. *Implement a Mathematica function* `Horner[p,x]`, *which converts a polynomial $p(x)$ into Horner form (3.11). Compare your function with the built-in function* `HornerForm`.

Exercise 3.17. *Let an integer a_B be given in a basis B which shall be converted towards a basis C. Present an algorithm for this task using division with remainder and implement your algorithm.*

Exercise 3.18. *Let $x = p_1^{e_1} \dots p_r^{e_r}$ and $y = p_1^{f_1} \dots p_r^{f_r}$ be the prime factor decompositions of the integers x and y where $p_1, \dots, p_r$ denote all prime factors of both x and y. The exponents e_i, f_i $(i = 1, \dots, r)$ satisfy $e_i, f_i \in \mathbb{N}_{\geq 0}$.*

Show that $\gcd(x, y) = p_1^{\min(e_1, f_1)} \cdots p_r^{\min(e_r, f_r)}$ *and* $\operatorname{lcm}(x, y) = p_1^{\max(e_1, f_1)} \cdots p_r^{\max(e_r, f_r)}$.

Exercise 3.19. (Sieve of Eratosthenes) *The sieve of Eratosthenes generates all primes up to a given integer $n \in \mathbb{N}_{\geq 2}$ by cancelling all multiples of the numbers $2, \dots, \lfloor \sqrt{n} \rfloor$ from the complete list $\{2, 3, \dots, n\}$. Prove that this algorithm generates all prime numbers between 2 and n.*

Implement a Mathematica procedure `Eratosthenes` *which uses this algorithm to compute an output of the form*

$In[1]:=$ **Eratosthenes[400]**

$$
Out[1]= \begin{pmatrix}
. & . & . & . & . & . & . & . & . & . & . & . & . & . & . & . & . & . & . & . \\
. & . & 23 & . & . & . & . & 29 & . & 31 & . & . & . & . & . & 37 & . & . & . \\
41 & . & 43 & . & . & 47 & . & . & . & . & 53 & . & . & . & . & . & 59 & . \\
61 & . & . & . & 67 & . & . & 71 & . & 73 & . & . & . & . & . & 79 & . \\
. & . & 83 & . & . & . & 89 & . & . & . & . & . & 97 & . & . & . \\
101 & . & 103 & . & . & 107 & . & 109 & . & . & . & 113 & . & . & . & . & . \\
. & . & . & . & 127 & . & . & 131 & . & . & . & 137 & . & 139 & . \\
. & . & . & . & . & 149 & . & 151 & . & . & . & 157 & . & . \\
. & . & 163 & . & . & 167 & . & . & . & 173 & . & . & . & 179 & . \\
181 & . & . & . & . & . & . & 191 & . & 193 & . & . & 197 & . & 199 & . \\
. & . & . & . & . & . & 211 & . & . & . & . & . & . \\
. & . & 223 & . & . & 227 & . & 229 & . & . & . & 233 & . & . & . & . & 239 & . \\
241 & . & . & . & . & . & . & 251 & . & . & . & 257 & . & . \\
. & . & 263 & . & . & . & 269 & . & 271 & . & . & . & 277 & . & . \\
281 & . & 283 & . & . & . & . & . & . & 293 & . & . & . & . \\
. & . & . & . & 307 & . & . & 311 & . & 313 & . & . & 317 & . & . \\
. & . & . & . & . & . & 331 & . & . & . & 337 & . & . \\
. & . & . & . & 347 & . & 349 & . & . & 353 & . & . & . & 359 & . \\
. & . & . & . & 367 & . & . & . & 373 & . & . & . & 379 & . \\
. & . & 383 & . & . & . & 389 & . & . & . & . & . & 397 & . & .
\end{pmatrix}
$$

Use the function `Partition` *to get an output of this type.*

Exercise 3.20. *Test the function* `factorinteger` *of Session 3.19, which was aborted there, for smaller input x.*

Exercise 3.21. *Rational numbers have periodic decimal representations, see e.g. [SK2000]. These can be computed using* `RealDigits`, *and* `FromDigits` *converts periodic representations into their corresponding rational numbers. Check this function with suitable examples and describe its mechanism.*

Chapter 4
Modular Arithmetic

4.1 Residue Class Rings

The integers with their operations addition and multiplication $(\mathbb{Z}, +, \cdot)$ form a commutative ring with unity 1. Now, we would like to translate this algebraic structure to finite subsets of $\mathbb{Z}$. This is done by identification of certain elements in $\mathbb{Z}$ that lie in common arithmetic progressions.

Definition 4.1. Let $p \in \mathbb{N}_{\geq 2}$. Two integers $a, b \in \mathbb{Z}$ are called *congruent modulo p*, denoted by $a \equiv b \pmod{p}$, if $p \mid b - a$ is valid. Therefore $a \equiv b \pmod{p}$ if and only if a and b have the same remainder after division by p:

$$\mathrm{rem}(a, p) = \mathrm{rem}(b, p). \tag{4.1}$$

Congruence is an *equivalence relation* (see Lemma 4.2) which partitions $\mathbb{Z}$ into subsets, called *equivalence classes*. The equivalence classes of similar numbers are the sets $[a]_p := \{a + k \cdot p \mid k \in \mathbb{Z}\}$. Hence every equivalence class possesses exactly one representative in the segment $0, 1, \ldots, p - 1$, namely $\mathrm{rem}(a, p)$. For this reason the equivalence classes are also called *remainder classes*. There are exactly p different remainder classes with respect to the *modulus* p which are described in a unique manner by the remainders (4.1). Therefore the set of remainder classes $\mathbb{Z}_p := \{[0]_p, [1]_p, \ldots, [p-1]_p\}$ can be identified with the set $\{0, 1, 2, \ldots, p-1\} \subset \mathbb{Z}$ of corresponding remainders. $\triangle$

We prove the following properties of the modular congruence.

Lemma 4.2. (Properties of Modular Arithmetic) Let $a, b, c, a', b', n \in \mathbb{Z}$ be given. Then the following properties hold:

(a) **(Reflexivity)** $a \equiv a \pmod{p}$;
(b) **(Symmetry)** $a \equiv b \pmod{p} \Rightarrow b \equiv a \pmod{p}$;
(c) **(Transitivity)** $a \equiv b \pmod{p}$ and $b \equiv c \pmod{p} \Rightarrow a \equiv c \pmod{p}$;
(d) **(Compatibility with Scalar Multiplikation)** $a \equiv b \pmod{p}$ and $n \in \mathbb{Z}$
$\Rightarrow n \cdot a \equiv n \cdot b \pmod{p}$;
(e) **(Compatibility with Addition)** $a \equiv a' \pmod{p}$ and $b \equiv b' \pmod{p}$
$\Rightarrow a + b \equiv a' + b' \pmod{p}$;
(f) **(Compatibility with Multiplication)** $a \equiv a' \pmod{p}$ and $b \equiv b' \pmod{p}$
$\Rightarrow a \cdot b \equiv a' \cdot b' \pmod{p}$.

Proof. (a) $p \mid a - a = 0$.
(b) $p \mid b - a \Rightarrow b - a = k \cdot p$ for $k \in \mathbb{Z}$. Hence $a - b = -k \cdot p \Rightarrow p \mid a - b$.

© Springer Nature Switzerland AG 2021
W. Koepf, *Computer Algebra*, Springer Undergraduate Texts in Mathematics and Technology,
https://doi.org/10.1007/978-3-030-78017-3_4

(c) By assumption we have $b - a = k \cdot p$ and $c - b = k' \cdot p$ for $k, k' \in \mathbb{Z}$. This implies $c - a = (c - b) + (b - a) = (k + k') \cdot p$.

By the way: (a)–(c) show that $\equiv$ defines an equivalence relation.

(d) We have $b - a = k \cdot p$. Therefore $n \cdot b - n \cdot a = n \cdot (b - a) = n \cdot k \cdot p$.

(e) The assumptions yield $a' - a = k \cdot p$ and $b' - b = k' \cdot p$ for $k, k' \in \mathbb{Z}$. This implies $(a' + b') - (a + b) = (a' - a) + (b' - b) = (k + k') \cdot p$.

(f) Using the same notations as in (e), we now get $a' = a + k \cdot p$ and $b' = b + k' \cdot p$ and therefore $a' \cdot b' = (a + k \cdot p) \cdot (b + k' \cdot p) = a \cdot b + p \cdot (ak' + bk + kk'p)$. Hence the difference $a' \cdot b' - a \cdot b$ is a multiple of p, which completes the proof.   □

Session 4.3. *If one chooses $a' = \mathrm{mod}(a, p)$ and $b' = \mathrm{mod}(b, p)$ in the compatibility conditions (e) and (f) in Theorem 4.2, we get the equations*

$$\mathrm{mod}(a + b, p) = \mathrm{mod}(\mathrm{mod}(a, p) + \mathrm{mod}(b, p), p) \tag{4.2}$$

and

$$\mathrm{mod}(a \cdot b, p) = \mathrm{mod}(\mathrm{mod}(a, p) \cdot \mathrm{mod}(b, p), p). \tag{4.3}$$

We check these relations with Mathematica. For this purpose, we choose random numbers a, b and p:

```
In[1]:= a = Random[Integer, {10^19, 10^20}]
        b = Random[Integer, {10^19, 10^20}]
        p = Random[Integer, {10^19, 10^20}]
```
Out[1]= 70785922828421004845
Out[1]= 58056464214342111922
Out[1]= 42478915027279858887

Now we compute the division remainder of the sum $a + b$,

```
In[2]:= Mod[a+b,p]
```
Out[2]= 1405641960923540106

and the sum of the division remainders,

```
In[3]:= Mod[a,p] + Mod[b,p]
```
Out[3]= 43884556988203398993

Typically the sum is too large and must be reduced modulo p:

```
In[4]:= Mod[Mod[a,p] + Mod[b,p],p]
```
Out[4]= 1405641960923540106

A similar computation for the product gives:

```
In[5]:= Mod[a*b,p]
```
Out[5]= 35127298486240420406

```
In[6]:= Mod[a,p] * Mod[b,p]
```
Out[6]= 44095380636083111303768094894 5263482530

The product is much too large, but:

```
In[7]:= Mod[Mod[a,p] * Mod[b,p],p]
```
Out[7]= 35127298486240420406

We will use these properties in many situations throughout the book.

Let $p \in \mathbb{N}_{\geq 2}$ be given. The map

$$\varphi: \begin{array}{ccc} \mathbb{Z} & \to & \mathbb{Z}_p \\ a & \mapsto & \varphi(a) := [a]_p \end{array}$$

defines in an obvious way an addition $\oplus$ and a multiplication $\otimes$ in $\mathbb{Z}_p$:[1]

$$\begin{aligned} [a]_p \oplus [b]_p &:= \varphi(a+b) = [a+b]_p, \\ [a]_p \otimes [b]_p &:= \varphi(a \cdot b) = [a \cdot b]_p. \end{aligned} \tag{4.4}$$

This makes $(\mathbb{Z}_p, \oplus, \otimes)$ again an algebraic structure. We will see that $\mathbb{Z}_p$ inherits the ring structure from $\mathbb{Z}$.

Theorem 4.4. $(\mathbb{Z}_p, \oplus, \otimes)$ forms a commutative ring with unity 1.

Proof. First we show that the operations $\oplus$ and $\otimes$ are *well defined*, i.e., the result of $\oplus$ and $\otimes$ is independent of the choice of the representing elements in the remainder classes.

For this purpose, let two different representatives a and a' of the remainder class $[a]_p$ and two different representatives b and b' of the remainder class $[b]_p$ be given. Then the compatibility properties (e)–(f) of Lemma 4.2 imply the relations

$$[a+b]_p = [a'+b']_p \qquad \text{and} \qquad [a \cdot b]_p = [a' \cdot b']_p.$$

This shows that the results are indeed independent of the choice of the representatives.

Now it is easy to see that the ring operations from $\mathbb{Z}$ are also valid in $\mathbb{Z}_p$. Clearly, the neutral element with respect to addition ($0 \in \mathbb{Z}$) makes the zero class $[0]_p \in \mathbb{Z}_p$ the neutral element in $\mathbb{Z}_p$. Similarly the neutral element with respect to multiplication ($1 \in \mathbb{Z}$) generates the neutral element $[1]_p \in \mathbb{Z}_p$. For the inverse with respect to addition we have

$$-[a]_p := [-a]_p = [p-a]_p$$

and Lemma 4.2 (d) for the case $n = -1$ shows that this definition is independent of the representative. $\square$

For convenience, we will denote the remainder classes shortly by $a = [a]_p$ in the sequel, hence $\mathbb{Z}_p = \{0, 1, \ldots, p-1\} \subset \mathbb{Z}$. Furthermore, we will use the usual notations $+$ and $\cdot$ for addition and multiplication instead of $\oplus$ and $\otimes$ in $\mathbb{Z}_p$.[2]

Let us consider the addition and multiplication tables in $\mathbb{Z}_p$ in more detail. For $p = 6$ and $p = 7$ they look as follows:

$(\mathbb{Z}_6, +)$	0	1	2	3	4	5
0	0	1	2	3	4	5
1	1	2	3	4	5	0
2	2	3	4	5	0	1
3	3	4	5	0	1	2
4	4	5	0	1	2	3
5	5	0	1	2	3	4

$(\mathbb{Z}_6, \cdot)$	0	1	2	3	4	5
0	0	0	0	0	0	0
1	0	1	2	3	4	5
2	0	2	4	0	2	4
3	0	3	0	3	0	3
4	0	4	2	0	4	2
5	0	5	4	3	2	1

---

[1] Therefore it creates a ring homomorphism.

[2] Therefore an equation like $-3 = 4$ is not at all false if it refers to $\mathbb{Z}_7$.

$(\mathbb{Z}_7, +)$	0	1	2	3	4	5	6
0	0	1	2	3	4	5	6
1	1	2	3	4	5	6	0
2	2	3	4	5	6	0	1
3	3	4	5	6	0	1	2
4	4	5	6	0	1	2	3
5	5	6	0	1	2	3	4
6	6	0	1	2	3	4	5

$(\mathbb{Z}_7, \cdot)$	0	1	2	3	4	5	6
0	0	0	0	0	0	0	0
1	0	1	2	3	4	5	6
2	0	2	4	6	1	3	5
3	0	3	6	2	5	1	4
4	0	4	1	5	2	6	3
5	0	5	3	1	6	4	2
6	0	6	5	4	3	2	1

Next, we will discuss under which conditions an element $a \in \mathbb{Z}_p$ is a unit, i.e., has a multiplicative inverse. If we find a multiplicative inverse for all $a \in \mathbb{Z}_p$, then $\mathbb{Z}_p$ is not only a ring, but a field. The above operation tables show immediately that $\mathbb{Z}_6$ is not a field since the line for $a = 2$ of the multiplication table does not contain the result 1 which is necessary for an inverse to exist. Therefore 2 does not have a multiplicative inverse in $\mathbb{Z}_6$. The same argument is valid for each divisor of p, if p is not prime. Furthermore the equation $2 \cdot 3 = 0$ in $\mathbb{Z}_6$ shows that this ring contains zero divisors (which do not exist in fields).

On the other hand the multiplication table for $\mathbb{Z}_7$ suggests that $\mathbb{Z}_7$ is a field. In general, $\mathbb{Z}_p$ is a field if p is prime.

Theorem 4.5. An element $a \in \mathbb{Z}_p$ is a unit if and only if $\gcd(a, p) = 1$. In particular, $(\mathbb{Z}_p, +, \cdot)$ is a field if and only if $p \in \mathbb{P}$.

Proof. Let $p \in \mathbb{N}_{\geq 2}$ and $a \in \mathbb{Z}_p$. If p is a unit, then there is a $b \in \mathbb{Z}$ with $a \cdot b \equiv 1 \pmod{p}$. This means that there is a $k \in \mathbb{Z}$ with $a \cdot b = 1 + k \cdot p$. From this equation we can deduce that the greatest common divisor of a and p must be 1.

Now assume that $\gcd(a, p) = 1$. Then the extended Euclidean algorithm guarantees the existence of $s, t \in \mathbb{Z}$ with $s \cdot a + t \cdot p = 1$. If we read this equation modulo p, we get $s \cdot a \equiv 1 \pmod{p}$, and therefore s is an inverse of a in $\mathbb{Z}_p$.

Now, if $p \in \mathbb{P}$, then the relation $\gcd(a, p) = 1$ is valid for all $a \in \mathbb{Z}_p$ since $a < p$ and p has no proper divisors. Therefore all elements of $\mathbb{Z}_p \setminus \{0\}$ are units, which implies that $\mathbb{Z}_p$ is a field.

However, if p is composite, then there is an $a \in \mathbb{Z}_p$ with $\gcd(a, p) \neq 1$, which is therefore not a unit. $\square$

Definition 4.6. The inverse $b = a^{-1} \pmod{p}$ of a modulo p is called the *modular inverse*. It exists—as shown—if p is prime. However, if p is composite, then $a^{-1} \pmod{p}$ only exists if a and p are coprime.

The *characteristic* of a commutative ring R with unity 1 is the smallest number c such that the c-fold addition of unity gives zero:

$$\underbrace{1 + 1 + \cdots + 1}_{c \text{ summands}} = 0 .$$

We write $c = \mathrm{char}(R)$. Since $p \cdot 1 \equiv 0 \pmod{p}$, the characteristic of $\mathbb{Z}_p$ is p. If such an integer $c \in \mathbb{N}$ does not exist, we say that R has the characteristic 0. $\mathbb{Z}$ is a ring of characteristic 0.

For $p \in \mathbb{P}$ we denote by $\mathbb{Z}_p^\star := \mathbb{Z}_p \setminus \{0\}$ the multiplicative group in $\mathbb{Z}_p$.[3] If $p \in \mathbb{P}$ is prime, then the field $\mathbb{Z}_p$ is also denoted by $\mathbb{F}_p$ or by $GF(p)$ (*Galois field*), see also Section 7.4. $\triangle$

Session 4.7. *In Mathematica we can use* Mod *to execute the basic operations in* $\mathbb{Z}_p$. *For example, the functions*[4]

```
In[1]:= AddZ[k_] :=
        Table[Mod[i+j,k],{i,0,k-1},{j,0,k-1}]//MatrixForm
        MultZ0[k_] :=
        Table[Mod[i*j,k],{i,0,k-1},{j,0,k-1}]//MatrixForm
        MultZ[k_] :=
        Table[Mod[i*j,k],{i,k-1},{j,k-1}]//MatrixForm
```

computes the addition and multiplication tables of $\mathbb{Z}_p$ *and* $\mathbb{Z}_p^\star$, *respectively, and returns them in matrix representations. The tables shown on pp. 67–68 can be computed very easily using these functions. In practice, Mathematica does not contain a data type for numbers modulo p, which is why we used* Mod *and the rules (4.4) to resort to the arithmetic in* $\mathbb{Z}$.

Modular inverses can be computed in different ways. A direct implementation is given by the function PowerMod *as*

$$a^{-1} \ (mod \ p) = \texttt{PowerMod[a,-1,p]} \ .$$

We get, for example,

```
In[2]:= PowerMod[a = 12345678, -1, p = 1234567891]
Out[2]= 908967567
```

We can also compute a^{-1} *(mod p) using the extended Euclidean algorithm as shown in the proof of Theorem 4.5:*

```
In[3]:= {g, {s, t}} = ExtendedGCD[a, p]
Out[3]= {1, {-325600324, 3256003}}
```

To be on the safe side, we can check the extended Euclidean algorithm

```
In[4]:= s*a+t*p
Out[4]= 1
```

and compute the modular inverse of a:

```
In[5]:= inv = Mod[s, p]
Out[5]= 908967567
```

Finally, one can also use Solve *for this purpose:*

```
In[6]:= Solve[a*x == 1, x, Modulus → p]
Out[6]= {x → 908967567}}
```

Note that the option Modulus$\to p$ *asks Mathematica to compute in* $\mathbb{Z}_p$.

---

[3] If p is not prime, then $\mathbb{Z}_p^\star$ denotes the set of units in $\mathbb{Z}_p$, which is also a group. However, we will not consider this case any further.

[4] The use of MatrixForm ensures that the pair AddZ[6],MultZ0[6] shows matrices.

4.2 Modulare Square Roots

Whereas linear equations

$$a \cdot x + b = c$$

in $\mathbb{Z}_p$—corresponding to the equation

$$a \cdot x + b \equiv c \pmod{p}$$

—by the group structure have the unique solution (for $\gcd(a, p) = 1$)

$$x = a^{-1} \pmod{p} \cdot (c - b),$$

it is already difficult to solve the simple quadratic equation

$$x^2 = a.$$

Any solution of the equation

$$x^2 \equiv a \pmod{p}$$

is called a *modular square root* and denoted by $x = \sqrt{a} \pmod{p}$.

Session 4.8. *By*

In[1]:= **Table[Mod[x², 11], {x, 0, 10}]**
Out[1]= {0, 1, 4, 9, 5, 3, 3, 5, 9, 4, 1}

we compute the modular squares modulo 11. *From the result, we can read off that for example* $1 \equiv \sqrt{1}$ *(mod* 11), $2 \equiv \sqrt{4}$ *(mod* 11), $3 \equiv \sqrt{9}$ *(mod* 11), $4 \equiv \sqrt{5}$ *(mod* 11), …

In particular, we realize that not all $a \in \mathbb{Z}_{11}$ *represent a modular square, but only the numbers* $0, 1, 3, 4, 5, 9$. *As a result, not every* $x \in \mathbb{Z}_p$ *possesses a modular square root. However, if a modular square root exists, then there are at least two of them, since[5]*

$$(p - x)^2 \equiv p^2 - 2xp + x^2 \equiv x^2 \pmod{p}. \tag{4.5}$$

This shows that with x *also* $p - x$ *is a modular square root of* a. *Therefore the graph of the modular square function is symmetric with respect to the vertical line* $x = \frac{p}{2}$. *Integers that possess a modular square root are called* quadratic residues.

The following list contains the quadratic residues modulo 123:

In[2]:= **Union[Table[Mod[x², 123], {x, 0, 122}]]**
Out[2]= {0, 1, 4, 9, 10, 16, 18, 21, 25, 31, 33, 36, 37, 39, 40, 42, 43, 45, 46, 49, 51, 57, 61,
            64, 66, 72, 73, 78, 81, 82, 84, 87, 90, 91, 100, 102, 103, 105, 114, 115, 118, 121}

Next, we draw the graph of the modular square function, first modulo 667,

In[3]:= **ListPlot[Table[Mod[x², 667], {x, 0, 666}]]**

---

[5] Therefore $\sqrt{a}$ (mod p) is multivalued.

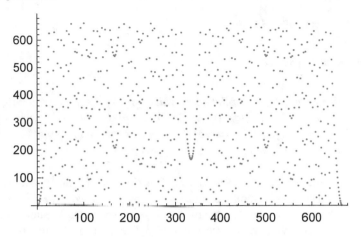

and then modulo 10007:

In[4]:= **ListPlot[Table[Mod[x^2, 10007], {x, 0, 10006}]]**

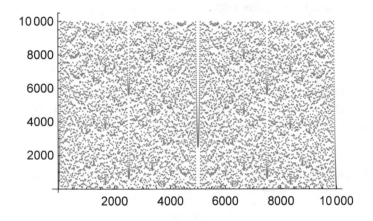

The symmetry with respect to p/2 resulting from (4.5) is clearly visible. Moreover, the graphs make it plausible that it seems to be very hard *to compute the modular square root—in contrast to the very regular situation in* $\mathbb{Z}$.

Next, we compute the modular square roots $\sqrt{a}$ *(mod 667) for a* = 500, ..., 510 *using Mathematica's* Solve *command:*

In[5]:= **Table[Solve[x^2 == a, x, Modulus → 667], {a, 500, 510}]**

Out[5]= {{}, {}, {}, {}, {}, {}, {x → 184, x → 483}, {}, {}, {x → 62, x → 315, x → 352, x → 605}, {}}

We see that only for a = 506 *and a* = 509 *we have quadratic residues. On the other hand, there are even four modular square roots* $\sqrt{509}$ *(mod 667).*

Finally, we build our own implementation for the computation of modular square roots. Given the list of all modular squares, the computation of the modular square roots—as inverses— is easy:

```
In[6]:= ModularSqrt[a_, p_] := Module[{list, x},
          list = Table[Mod[x^2, p], {x, 0, p - 1}];
          Flatten[Position[list, a] - 1]
        ]
```

$In[7]:=$ **ModularSqrt[506,667]**
$Out[7]=$ $\{184, 483\}$

$In[8]:=$ **ModularSqrt[509,667]**
$Out[8]=$ $\{62, 315, 352, 605\}$

For this computation we need p multiplications, which—using the high school algorithm—have the complexity $O((\log p)^2)$. This leads to an allover complexity $O(p \cdot (\log p)^2)$, which is again exponential in the length $\log p$ of p,[6] whereas the modular square of a number x of length n has a complexity of at most $O(n^2)$, hence $O((\log x)^2) \leqq O((\log p)^2)$.

With the modular square root we got to know a function which can be evaluated only very slowly, whereas its inverse—the modular square—can be computed very efficiently. It is practically impossible to compute modular square roots if p is large enough. Such functions play an important role in modern cryptography, see Section 5.5. Currently, no really efficient algorithms for the computation of modular square roots are known.

4.3 Chinese Remainder Theorem

If the values of a polynomial of degree $n-1$ is known at n points, then the polynomial can be uniquely reconstructed using *interpolation*. This will be treated in more detail in Section 6.5.

A similar situation is the following. Assume that an integer $x \in \mathbb{Z}$ is known modulo certain primes $p_1, \ldots, p_n \in \mathbb{P}$. How can we reconstruct x? We will see that this is uniquely possible if $0 \leq x < p_1 \cdots p_n$. We first consider the case $n = 2$. The general case will then be treated recursively.

Theorem 4.9. (Chinese Remainder Theorem for Two Equations) Let p and q be two integers with $\gcd(p, q) = 1$. Then the system of equations

$$x \equiv a \,(\mathrm{mod}\ p)$$
$$x \equiv b \,(\mathrm{mod}\ q)$$

has an integer solution $x \in \mathbb{Z}$, which is unique up to an additive multiple of $p \cdot q$.

Proof. We construct the solution. Since $\gcd(p, q) = 1$, the extended Euclidean algorithm generates integers $s, t \in \mathbb{Z}$ with

$$sp + tq = 1 .$$

Therefore the integer

$$x := bsp + atq$$

satisfies the equations

$$x \equiv atq \equiv a \cdot (1 - sp) \equiv a \,(\mathrm{mod}\ p)$$

and

$$x \equiv bsp \equiv b \cdot (1 - tq) \equiv b \,(\mathrm{mod}\ q)$$

and is thus a solution of the problem.
As for the uniqueness, let x and $\hat{x}$ be two solutions of the system. Then it follows that $x - \hat{x} \equiv$

---

[6] This result does not get much better if we use accelerated multiplication.

$0 \pmod{p}$ and $x - \hat{x} \equiv 0 \pmod{q}$. Because of $\gcd(p, q) = 1$ this implies that $x - \hat{x} \equiv 0 \pmod{p \cdot q}$, see Exercise 4.5. This proves the assertion. $\qquad\square$

It is simple to use this algorithm recursively to construct solutions of the system of equations

$$
\begin{aligned}
x &\equiv a_1 \pmod{p_1} \\
x &\equiv a_2 \pmod{p_2} \\
&\vdots \\
x &\equiv a_n \pmod{p_n}
\end{aligned}
\tag{4.6}
$$

with moduli $p_k\,(k = 1, \ldots, n)$ that are pairwise relatively prime. If l is the solution of the first pair of equations, then obviously

$$x \equiv l \pmod{p_1 \cdot p_2}$$

and we can replace the first two equations of (4.6) by this one and invoke the algorithm recursively. Of course the algorithm stops if only one equation remains. Therefore we have the following theorem.

Theorem 4.10. (Chinese Remainder Theorem) Let $p_1, \ldots, p_n$ be integers with $\gcd(p_j, p_k) = 1$ for $j \neq k$. Then the system of equations (4.6) has an integer solution $x \in \mathbb{Z}$, which is unique up to an additive multiple of $p_1 \cdots p_n$.

Proof. It remains to prove that $\gcd(p_1 \cdot p_2, p_k) = 1$ for $k = 3, \ldots, n$. Since neither p_1 nor p_2 have a common divisor with p_k, the product $p_1 \cdot p_2$ has no common divisor with p_k either. This follows from the Fundamental Theorem of Number Theory (Theorem 3.17). $\qquad\square$

Session 4.11. *The Mathematica function* `ChineseRemainder [{a_1, ..., a_n}, {p_1, ..., p_n}]` *computes the solution of the remainder problem (4.6) modulo* $\prod\limits_{k=1}^{n} p_k$.

We solve the system of equations

$$
\begin{aligned}
x &\equiv 17 \pmod{101} \\
x &\equiv 4 \pmod{97}
\end{aligned}
$$

by

```
In[1]:= ChineseRemainder[{17,4},{101,97}]
Out[1]= 2138
```

and the equations

$$
\begin{aligned}
x &\equiv 0 \pmod{2}, & \text{i.e., } x \text{ is even,} \\
x &\equiv 0 \pmod{3}, & \text{i.e., } x \text{ is a multiple of 3,} \\
x &\equiv 0 \pmod{5}, & \text{i.e., } x \text{ is a multiple of 5,} \\
x &\equiv 1 \pmod{7}
\end{aligned}
$$

with the call

```
In[2]:= ChineseRemainder[{0,0,0,1},{2,3,5,7}]
Out[2]= 120
```

Now we implement the algorithm given in the proof of Theorem 4.9:

```
In[3]:= CR[{a_,b_},{p_,q_}] := Module[{g,s,t},
         {g,{s,t}} = ExtendedGCD[p,q];
         Mod[b*s*p+a*t*q,p*q]
        ]
```

With this algorithm we can also solve the first problem directly:

```
In[4]:= CR[{17,4},{101,97}]
Out[4]= 2138
```

The second problem can be solved in three steps:

```
In[5]:= CR[{0,0},{2,3}]
Out[5]= 0
```

```
In[6]:= CR[{0,0},{2*3,5}]
Out[6]= 0
```

```
In[7]:= CR[{0,1},{2*3*5,7}]
Out[7]= 120
```

An implementation of the solution of the Chinese Remainder Theorem will be considered in Exercise 4.4.

Next, let us solve a more complicated problem:

```
In[8]:= a = Random[Integer,{10^99,10^100}];
        b = Random[Integer,{10^99,10^100}];
        p = NextPrime[a];
        q = NextPrime[b];
In[9]:= CR[{a,b},{p,q}]//Timing
Out[9]= {0. Second,
           1621441319261540414458241023833643796917557882591073764798471
           8799430728041010006428016021045162562063611597751035799521 7
           8031141686560407512934870999391008571681179665423581768819 5
           8691103508673600852 71}
```

Note that the application of the extended Euclidean algorithm solves our problem very efficiently for large integers as well.

4.4 Fermat's Little Theorem

A very important theorem from elementary number theory, which will have interesting applications in cryptography, is *Fermat's Little Theorem.*

Theorem 4.12. (Fermat's Little Theorem) For every prime $p \in \mathbb{P}$ and every $x \in \mathbb{N}_{\geq 0}$ the identity

$$x^p \equiv x \pmod{p}$$

is valid.

Proof. We use the *binomial formula*

$$(x+1)^p = x^p + \binom{p}{1} x^{p-1} + \binom{p}{2} x^{p-2} + \cdots + \binom{p}{p-1} x^1 + 1,$$

which is valid for $x \in \mathbb{Z}$, $p \in \mathbb{N}$, where

$$\binom{p}{k} = \frac{p \cdot (p-1) \cdot (p-2) \cdots (p-k+1)}{k \cdot (k-1) \cdot (k-2) \cdots 1} \in \mathbb{N}$$

are the integer-valued *binomial coefficients*.

Since $\binom{p}{k} \in \mathbb{N}$, for every prime $p \in \mathbb{P}$ and $0 < k < p$, the binomial coefficient $\binom{p}{k}$ is a multiple of p. We perform mathematical induction with respect to x.

Obviously Fermat's Little Theorem is valid for $x = 0$ and every prime $p \in \mathbb{P}$. As induction hypothesis we assume that the theorem is valid for some $x \in \mathbb{N}_{\geq 0}$, i.e., $x^p \equiv x \,(\text{mod } p)$. Then it follows that

$$(x+1)^p \equiv x^p + \binom{p}{1} x^{p-1} + \cdots + \binom{p}{p-1} x^1 + 1 \equiv x^p + 1 \equiv x + 1 \,(\text{mod } p), \qquad (4.7)$$

where we have used the induction hypothesis in the last step. We conclude that the assertion is also valid for $x + 1$. This completes the induction proof. $\qquad\qquad \square$

Similarly as (4.7), the equation

$$(a + b)^p \equiv a^p + b^p \,(\text{mod } p)$$

follows. This calculation rule will appear very frequently throughout the book.

Session 4.13. *Mathematica has an efficient implementation of modular powers: The command* PowerMod[a,n,p] *computes a^n (mod p). If you want to compute, e.g.,*

$$y := 123456789^{123456789} \mod 987654321$$

conventionally, then first the huge number $123456789^{123456789}$ (which has almost 10^9 decimal places) must be computed in $\mathbb{Z}$ and then this number must be reduced modulo 987654321.[7]

```
In[1]:= Mod[123456789^{123456789}, 987654321]
Out[1]= 598987215
```

Of course, this approach is completely inefficient. It is much better to reduce the computation modulo 987654321 after every multiplication. Therefore we use the following approach to compute the modular power by recursion:[8]

$$\text{mod}(a^n, p) = \begin{cases} \text{mod}(\text{mod}(a^{\frac{n}{2}}, p)^2, p) & \text{if } n \text{ even}. \\ \text{mod}(\text{mod}(a^{n-1}, p) \cdot a, p) & \text{if } n \text{ odd}. \end{cases} \qquad (4.8)$$

Obviously, this is a divide-and-conquer approach again, which is also used by PowerMod.

---

[7] This computation is very time-consuming. In older releases of *Mathematica* or if you don't have enough memory, the call Mod[123456789^123456789,987654321] results in an Overflow[].

[8] Here we use the relations (4.2)–(4.3) again.

The computation

```
In[2]:= PowerMod[123456789,123456789,987654321]
Out[2]= 598987215
```

is very fast. The given recursive algorithm for the computation of the modular power can then be easily implemented using (4.8):

```
In[3]:= $RecursionLimit = ∞;
        powermod[a_,0,p_] := 1
        powermod[a_,n_,p_] := Mod[powermod[a, n/2 ,p]^2 ,p] /; EvenQ[n]
        powermod[a_,n_,p_] := Mod[powermod[a,n-1,p] * a,p]/;
             OddQ[n]
```

Then the command

```
In[4]:= powermod[123456789,123456789,987654321]
Out[4]= 598987215
```

gives the above result in a relatively short time. The computation is very fast for integers with 200 digits as well:[9]

```
In[5]:= x = Random[Integer, {10^199, 10^200}];
        n = Random[Integer, {10^199, 10^200}];
        p = NextPrime[x];
In[6]:= powermod[x, n, p]//Timing
Out[6]= {0 Second,
            3532174365474810481600821826177575794087422026076928441602188893
            7248578405053721416266572107431219524836777571230590335188569
            2101374426797027902585458630935436890050415814496745498606102
            4548104902967}
```

Now, we can check Fermat's Little Theorem: The computation

```
In[7]:= Mod[powermod[x, p, p] - x, p]
Out[7]= 0
```

confirms that $x^p \equiv x \pmod{p}$, whereas the computation

```
In[8]:= Mod[powermod[x, p - 1, p], p]
Out[8]= 1
```

provides the equation $x^{p-1} \equiv 1 \pmod{p}$ reduced by the factor x.

As a consequence of Fermat's Little Theorem, we get for coprime x and p (to ensure that $x \neq 0 \pmod{p}$!) the following second formulation of Fermat's Little Theorem, when dividing by x:

$$x^{p-1} \equiv 1 \pmod{p}.$$

In particular, we have, for $a \in \mathbb{Z}_p^\star$, the relation $a^{p-1} = 1$. Since $\mathbb{Z}_p^\star$ contains only finitely many elements, it is not surprising that successive multiplication by a eventually yields the multiplicative unit 1. Fermat's Little Theorem provides an exponent for which this is always successful.

As a consequence, we get the following result.

----

[9] If a compiling programming language has an efficient integer arithmetic, then the computation times might be even better than in *Mathematica*.

Corollary 4.14. Let $p \in \mathbb{P}$ be prime. Then for every $a \in \mathbb{Z}_p^\star$ the sequence $(a^k \pmod{p})_{k \in \mathbb{N}_{\geq 0}}$ is periodic, and there is a smallest period which is a divisor of $p - 1$.

Proof. Fermat's Little Theorem implies $a^{p-1} \equiv 1 \pmod{p}$, so that the sequence $(a^0, a^1, a^2, \ldots) = (1, a, a^2, \ldots) \pmod{p}$ obviously has the period $p - 1$, since after $p - 1$ steps the value 1 appears again. Therefore every period must be either a divisor or a multiple of $p - 1$. If q and r are two coprime periods, then the extended Euclidean algorithm yields $s, t \in \mathbb{Z}$, with $1 = sq + tr$, so that

$$a \equiv a^1 \equiv a^{sq+tr} \equiv (a^q)^s \cdot (a^r)^t \equiv 1 \pmod{p},$$

hence $a \equiv 1 \pmod{p}$. In this latter case the period equals 1. $\qquad \square$

Definition 4.15. (Order, Generator and Primitive Element) Let $p \in \mathbb{P}$ and $a \in \mathbb{Z}_p^\star$. Then the smallest period according to Corollary 4.14 is called the (multiplicative) *order* of a in $\mathbb{Z}_p^\star$, denoted by $\mathrm{ord}(a)$. The order is the smallest $m \in \mathbb{N}$ such that $a^m = 1$. The order of every element $a \in \mathbb{Z}_p^\star$ is a divisor of $p - 1$.

In many cases, one wants the order to be as large as possible. Therefore an element $a \in \mathbb{Z}_p^\star$ with $\mathrm{ord}(a) = p - 1$ is called a *generator* or a *primitive element* of $\mathbb{Z}_p^\star$. $\triangle$

The following theorem justifies the above definition.

Theorem 4.16. Let a be a generator of $\mathbb{Z}_p^\star$. Then the powers of a generate the whole multiplicative group $\mathbb{Z}_p^\star$:

$$\{a^0, a^1, \ldots, a^{p-2}\} = \mathbb{Z}_p^\star.$$

Proof. The periodicity implies that there are exactly $p - 1$ potentially different powers $a^0, \ldots, a^{p-2}$. We will show that these powers all have different values. However, since $\mathbb{Z}_p^\star$ has only $p - 1$ elements, every element of $\mathbb{Z}_p^\star$ must be contained in the list $\{a^0, a^1, \ldots, a^{p-2}\}$.

In terms of the above mentioned claim, assume that there is an element which is generated twice: $a^q \equiv a^r \pmod{p}$ with $0 \leq q < r < p - 1$. This implies that $a^{r-q} \equiv 1 \pmod{p}$. Therefore the number $k := r - q$ contradicts the assumption. $\qquad \square$

Example 4.17. For $a = 2$ we have in $\mathbb{Z}_5$

$$a^1 = 2, \quad a^2 = 4, \quad a^3 = 3 \quad \text{and} \quad a^4 = 1.$$

Hence $\mathrm{ord}(2) = 4 = p - 1$, and 2 is a generator of $\mathbb{Z}_5^\star$.

Similarly, for $a = 2 \in \mathbb{Z}_{11}$, we get

$$a^1 = 2, \quad a^2 = 4, \quad a^3 = 8, \quad a^4 = 5, \quad a^5 = 10,$$

$$a^6 = 9, \quad a^7 = 7, \quad a^8 = 3, \quad a^9 = 6, \quad a^{10} = 1.$$

Again, 2 generates $\mathbb{Z}_{11}^\star$. On the other hand, for $a = 3$ we have

$$a^1 = 3, \quad a^2 = 9, \quad a^3 = 5, \quad a^4 = 4 \quad \text{and} \quad a^5 = 1.$$

Therefore $\mathrm{ord}(3) = 5 \mid 10 = p - 1$, and 3 does note generate $\mathbb{Z}_{11}^\star$. $\triangle$

Session 4.18. *Mathematica can find generating elements. The function*

```
In[1]:= MinGenerator[p_] := Module[{a = 2},
            While[Not[MultiplicativeOrder[a, p] == p - 1], a = a + 1];
            a
        ]
```

uses MultiplicativeOrder *for the computation of the multiplicative order of an element* $a \in \mathbb{Z}_p^\star$ *and generates the smallest generator of* $\mathbb{Z}_p^\star$.

For p = 10007, *we get for example*

```
In[2]:= MinGenerator[p = 10007]
Out[2]= 5
```

i.e., the smallest generator a = 5, *and the computation*

```
In[3]:= Length[Union[Table[PowerMod[5, n, p], {n, p - 1}]]]
Out[3]= 10006
```

confirms that the set of powers of a coincides with $\mathbb{Z}_p^\star$.

The following theorem—whose proof will be postponed until Section 7.4—shows the existence of a generator.

Theorem 4.19. Let $p \in \mathbb{P}$ be a prime. Then $\mathbb{Z}_p^\star$ possesses a generator. $\qquad\square$

4.5 Modular Logarithms

In this section we consider the problem of how to find a solution exponent x for the modular equation

$$a^x \equiv b \pmod{p}.$$

Such an exponent is called *modular logarithm* with respect to base a of b, modulo p, denoted by $x = \log_a b \pmod{p}$. This inverse function of the modular exponential function is often called *discrete logarithm*. In order to avoid confusion with the discrete logarithm in $\mathbb{Z}$, we prefer to speak of the modular logarithm in this book.

It turns out (as in the case of the modular square root) that the computation of the modular logarithm—in the actual state of knowledge—is a *very* difficult mathematical problem.

Session 4.20. *Let us first consider the discrete exponential function* 2^x *over* $\mathbb{Z}$. *This function has the following well-known regular (monotonously increasing) graph:*

```
In[1]:= ListPlot[Table[2^x, {x, 0, 20}]]
```

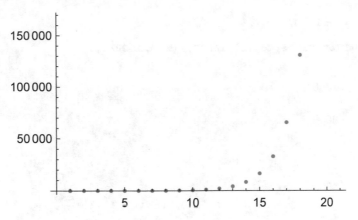

Therefore it is very easy to compute its inverse function:

```
In[2]:= Clear[DiscreteLogarithm]
        DiscreteLogarithm[y_, a_] := Module[{z, count},
          z = y; count = 0;
          While[IntegerQ[z] && z > 1,
            z = z/a; count = count + 1];
          If[Not[IntegerQ[z]],
            Return["discrete logarithm does not exist"]];
          count
        ]
```

Let us compute two examples:

```
In[3]:= DiscreteLogarithm[5^1000, 5]
Out[3]= 1000
```

```
In[4]:= DiscreteLogarithm[5^1000 + 1, 5]
Out[4]= discrete logarithm does not exist
```

The modular logarithm function (for example, with respect to the basis 2*) generates a completely different situation which can again be easily detected from its graph. For modulo* $p = 1009$, *we find:*

```
In[5]:= ListPlot[Table[Mod[2^x, 1009], {x, 0, 1008}]]
```

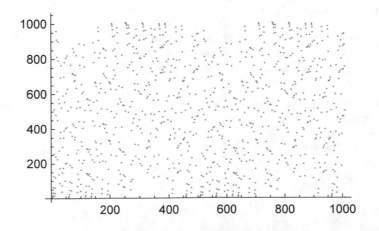

And for modulo p = 10007, *we find:*

In[6]:= **ListPlot[Table[Mod[2^x,10007],{x,0,10006}]]**

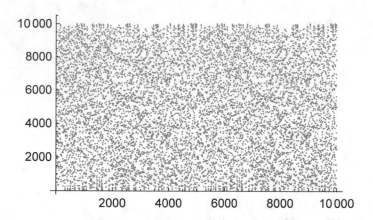

By Corollary 4.14, the graph is periodic with period P which is a divisor of p − 1. *In the case of p* = 1009, *we see that P* = $\frac{p-1}{2}$ = 504. *Therefore, for the computation of a modular logarithm we must generally compute O(p) values. As an example, we define a 20-digit prime p:*

In[7]:= **p = NextPrime[10^{20}]**
Out[7]= 100000000000000000039

We can compute the modular power b = mod(2^{12345678}, *p*) *very fast:*

In[8]:= **(b = PowerMod[2,12345678,p])//Timing**
Out[8]= {0. Second, 11341822793028139686}

Unlike the linear and quadratic cases, our current inverse problem cannot be solved using Solve:

In[9]:= **Solve[{2^x == b, x, Modulus → p}]**

...Solve: *This system cannot be solved with the methods available to Solve.*
Out[9]= *Solve*[2^x == 11341822793028139686, *x*, *Modulus* → 100000000000000000039]

The built-in function MultiplicativeOrder, *however, can compute modular logarithms:*[10]

In[10]:= **MultiplicativeOrder[2,p,b]//Timing**
Out[10]= {4.04043 Second, 12345678}

In our case the computation time is acceptable because the result has only 8 digits. However, if the result has 15 digits, the computation times get intolerably high, and it is absolutely impossible for Mathematica *to compute the modular logarithm of a 100-digit input.*

Note that the complexity of the modular logarithm computation is again exponential in the input length, and currently no really efficient algorithms are known.

---

[10] Since modular logarithms are not uniquely determined, MultiplicativeOrder computes the *smallest* solution.

4.6 Pseudoprimes

In Section 3.5 we considered the factorization of integers and realized that no efficient algorithm is known for this purpose. On the other hand, it is very easy to find out if a number is composite.

Session 4.21. *We would like to check whether* $p = 1234567891011121314151617189$ *is composite. To this end, let us declare*

```
In[1]:= p = 1234567891011121314151617189
Out[1]= 1234567891011121314151617189
```

We determine a random integer $a \in \mathbb{Z}_p^\star$,

```
In[2]:= a = Random[Integer, {1, p - 1}]
Out[2]= 203525390813282073701360409600
```

and compute a^{p-1} *(mod p)*:

```
In[3]:= PowerMod[a, p - 1, p]
Out[3]= 256295320614600045168282793000
```

Since the result does not equal 1, we have proved that p cannot be prime (because otherwise Fermat's Little Theorem would be violated). Hence p is composite.

In most cases, the choice $a = 2$ *is sufficient:*

```
In[4]:= PowerMod[2, p - 1, p]
Out[4]= 44139834945883877346112940080
```

Please note that we now know for sure that p is composite, but we have no idea about its factors since not a single trial division was performed.

The computation of all prime factors as in

```
In[5]:= FactorInteger[p]
```

$$
Out[5]= \begin{pmatrix}
13 & 1 \\
43 & 1 \\
79 & 1 \\
281 & 1 \\
1193 & 1 \\
833929457045867563 & 1
\end{pmatrix}
$$

uses completely different and typically much slower algorithms.

With the aid of Fermat's Little Theorem we have discovered that p is composite. The computation was fast since modular powers can be computed very efficiently. For obvious reasons this method is called the *Fermat test*. Next, we give a more detailed definition of this test.

Definition 4.22. (Fermat Test) The integer $a \in \mathbb{N}_{\geq 2}$ is called a *Fermat witness* for the compositeness of $p \in \mathbb{N}_{\geq 2}$ if $a^p \not\equiv a \pmod{p}$ is valid. An integer $p \in \mathbb{N}_{\geq 2} \setminus \mathbb{P}$ is called a *Fermat pseudoprime* with the base a if $a^p \equiv a \pmod{p}$ is valid. A composite integer is called a *Carmichael number* if the relation $a^p \equiv a \pmod{p}$ is valid for all $a \in \mathbb{Z}_p$. $\triangle$

If the Fermat test finds a witness for the compositeness of p, then p is composite. However, there is the question whether such a witness always exists. If no witness exists and p is not prime, it is called a Carmichael number. But do Carmichael numbers exist? The answer is yes, and we will search for them now.

Session 4.23. (Carmichael Numbers) *The following Mathematica function searches for the smallest Carmichael number $\geq n$ (if it exists). For this purpose we check for $p = n$, $n + 1, \ldots$ whether the Fermat criterion is satisfied for all $a < p$.*

```
In[1]:= Clear[NextCarmichael]
        NextCarmichael[n_] := Module[{done = False, p = n, list, a},
           While[Not[done],
             If[Not[PrimeQ[p]],
               list = Table[PowerMod[a, p, p] - a, {a, p - 1}];
               list = Union[list];
               If[list == {0}, done = True, p = p + 1], p = p + 1]];
           p]
```

Now we can search for the smallest Carmichael number:

```
In[2]:= (c = NextCarmichael[4])//Timing
Out[2]= {0.140401 Second, 561}
```

Hence we have found out that 561 *is the smallest Carmichael number. It has the decomposition*

```
In[3]:= FactorInteger[c]
```
$$Out[3] = \begin{pmatrix} 3 & 1 \\ 11 & 1 \\ 17 & 1 \end{pmatrix}$$

The next three Carmichael numbers are given as follows:

```
In[4]:= (c = NextCarmichael[c + 1])//Timing
Out[4]= {0.405603 Second, 1105}
```

```
In[5]:= FactorInteger[c]
```
$$Out[5] = \begin{pmatrix} 5 & 1 \\ 13 & 1 \\ 17 & 1 \end{pmatrix}$$

```
In[6]:= (c = NextCarmichael[c + 1])//Timing
Out[6]= {0.858006 Second, 1729}
```

```
In[7]:= FactorInteger[c]
```
$$Out[7] = \begin{pmatrix} 7 & 1 \\ 13 & 1 \\ 19 & 1 \end{pmatrix}$$

```
In[8]:= (c = NextCarmichael[c + 1])//Timing
Out[8]= {1.57561 Second, 2465}
```

```
In[9]:= FactorInteger[c]
```
$$Out[9] = \begin{pmatrix} 5 & 1 \\ 17 & 1 \\ 29 & 1 \end{pmatrix}$$

A continuation of these computations leads to more Carmichael numbers, but the computation times keep increasing. It is known that there are infinitely many Carmichael numbers [AGP1994].

The Fermat test is very simple and fast, and in most cases a single witness a is sufficient to verify that p is composite. In the negative case, we repeat the computation with a different random a. However, if p is a Carmichael number, then this method does not terminate and the Fermat test fails.

Next, we will show that Carmichael numbers have a very simple structure which can be easily mastered.

Theorem 4.24. (Structure of Carmichael Numbers) An integer $p \in \mathbb{N}_{\geq 4}$ is a Carmichael number if and only if[11]

(a) $p = p_1 \cdots p_n$ with primes $p_k \in \mathbb{P}$ that are pairwise different;
(b) $p_k - 1 \mid p - 1$ for all $k = 1, \ldots, n$.

Proof. Let us first assume that p is a Carmichael number. Hence $a^p \equiv a \pmod{p}$ for all $a = 0, \ldots, p-1$.
(a) We must show that p is *square free*, i.e., it does not contain a square. Assume on the contrary that q is a prime factor of p with $q^2 \mid p$. Then it follows from the assumption that $p \mid a^p - a = a \cdot (a^{p-1} - 1)$, and the choice $a = q$ leads to $q^2 \mid p \mid q \cdot (q^{p-1} - 1)$. Since $q \in \mathbb{P}$, we get $q \mid q^{p-1} - 1$, which is a contradiction. Hence p is square free, and all prime factors are pairwise different from each other.
(b) Since $p_k \in \mathbb{P}$ is prime, Theorem 4.19 guarantees the existence of a generating element $a_k \in \mathbb{Z}_{p_k}^\star$. This element satisfies $a_k^{p_k - 1} \equiv 1 \pmod{p_k}$, but for all $0 < r < p_k - 1$ the relation $a_k^r \not\equiv 1 \pmod{p_k}$ is valid. Since p is a multiple of p_k, the condition $a_k^p \equiv a_k \pmod{p}$ implies $a_k^p \equiv a_k \pmod{p_k}$ and therefore (by dividing by a_k, compare Exercise 4.3) $a_k^{p-1} \equiv 1 \pmod{p_k}$ since p_k is prime. Dividing $p - 1$ by $p_k - 1$ yields for the exponent $p - 1 = q \cdot (p_k - 1) + r$ with $0 \leq r < p_k - 1$. Therefore we are led to

$$1 \equiv a_k^{p-1} \equiv a_k^{q \cdot (p_k - 1) + r} \equiv (a_k^{p_k - 1})^q \cdot a_k^r \equiv a_k^r \pmod{p_k},$$

which contradicts the choice of a_k if $r \neq 0$. This implies $r = 0$ and therefore $p_k - 1 \mid p - 1$.

Now we prove the converse, i.e., that (a) and (b) imply that p is a Carmichael number. For this, suppose that $p = p_1 \cdots p_n$ with pairwise different primes $p_k \in \mathbb{P}$ and $p_k - 1 \mid p - 1$ for all $k = 1, \ldots, n$. Then Fermat's Little Theorem provides for all $a \in \mathbb{Z}_p^\star$ and all $k = 1, \ldots, n$,

$$a^{p-1} \equiv a^{(p_k - 1) \cdot q_k} \equiv (a^{p_k - 1})^{q_k} \equiv 1 \pmod{p_k}.$$

These equations imply $a^{p-1} \equiv 1 \pmod{p}$,[12] since the factors p_k of p do not have a common divisor (see Exercise 4.5). Therefore we finally get $a^p \equiv a \pmod{p}$. $\qquad\square$

Session 4.25. *With the aid of Theorem 4.24 we can compute the first Carmichael numbers much faster:*

---

[11] One can prove that moreover all p_k ($k = 1, \ldots, n$) are odd and $n \geq 3$.

[12] This follows also from the Chinese Remainder Theorem (Theorem 4.10) whose application is not necessary since all right-hand sides coincide.

```
In[1]:= Clear[FastNextCarmichael]
        FastNextCarmichael[n_] :=
          Module[{done = False, p = n},
            While[Not[done],
              If[PrimeQ[p], p = p + 1,
                factors = Transpose[FactorInteger[p]];
                If[Not[Union[factors[[2]]] == {1}] ||
                    Not[IntegerQ[
                      (p - 1)/Apply[LCM, factors[[1]] - 1]]],
                  p = p + 1, done = True]]
            ];
            p
          ]
```

```
In[2]:= (c = FastNextCarmichael[4])//Timing
Out[2]= {0. Second, 561}
```

```
In[3]:= (c = FastNextCarmichael[c + 1])//Timing
Out[3]= {0. Second, 1105}
```

```
In[4]:= (c = FastNextCarmichael[c + 1])//Timing
Out[4]= {0. Second, 1729}
```

```
In[5]:= (c = FastNextCarmichael[c + 1])//Timing
Out[5]= {0. Second, 2465}
```

In Exercises 4.15–4.16 you will find much larger Carmichael numbers.

In practice the Fermat test is replaced by a refined method which avoids the Carmichael numbers: the *Rabin-Miller test*. This test is also used by *Mathematica's* function `PrimeQ`.[13] Here, the Fermat test is enhanced in the following form.

Definition 4.26. (Rabin-Miller Test) An odd integer $p \in \mathbb{N}$ with a representation $p - 1 = 2^t \cdot u$, with u being odd, is called a *strong pseudoprime* with the base a if $\gcd(a, p) = 1$ and if $a^u \equiv 1 \pmod{p}$ or if there is an $s \in \{0, 1, ..., t-1\}$ such that $a^{2^s \cdot u} \equiv -1 \pmod{p}$.[14] If there is no strong pseudoprime with the base a, then we call a a *witness for the compositeness of p*. △

Next, we show that the above definition is meaningful.

Theorem 4.27. If p possesses a witness a for its compositeness, then p is composite.

Proof. We show the equivalent statement that every $p \in \mathbb{P}$ is a strong pseudoprime. Fermat's Little Theorem $a^p \equiv a \pmod{p}$ then implies $a^{p-1} \equiv 1 \pmod{p}$, since $\gcd(a, p) = 1$.

For this end, let u, s, t be given as in Definition 4.26. Now, if $a^u \equiv 1 \pmod{p}$, then p is a strong pseudoprime and we are done. Therefore we can assume that $a^u \not\equiv 1 \pmod{p}$. Because $a^{p-1} \equiv a^{2^t \cdot u} \equiv 1 \pmod{p}$, there is an $s \in \{0, 1, ..., t-1\}$ with[15]

---

[13] *Mathematica's* help pages state: `PrimeQ` first tests for divisibility using small primes, then uses the Miller-Rabin strong pseudoprime test with the base 2 and the base 3, and then uses a Lucas test.

[14] In $\mathbb{Z}_p$ the equation $x \equiv -1 \pmod{p}$ is equivalent to $x \equiv p - 1 \pmod{p}$. The largest nonnegative representative is therefore $p - 1 \in \mathbb{Z}_p$.

[15] Of course we have $s = \max\{m \in \mathbb{N}_{\geq 0} \mid a^{2^m \cdot u} \not\equiv 1 \pmod{p}\}$.

$$x := a^{2^s \cdot u} \not\equiv 1 \pmod{p}, \qquad \text{but} \qquad x^2 \equiv a^{2^{s+1} \cdot u} \equiv 1 \pmod{p}.$$

Hence, we get $x^2 \equiv 1 \pmod{p}$ and $x \not\equiv 1 \pmod{p}$ and therefore $x \equiv -1 \pmod{p}$, compare Exercise 4.6. ☐

Example 4.28. The proof tells us that a strong pseudoprime satisfies Fermat's criterion $a^p \equiv a \pmod{p}$, but has certain additional properties. There are much stronger facts as the following examples show (see [Rie1994]), that were confirmed by "hard computations":

- 2 is a witness for compositeness of all non-primes $< 2\,047$;

- $\{2, 3\}$ are witnesses (i.e., 2 or 3 is a witness) for compositeness of all non-primes $< 1\,373\,653$;

- $\{2, 3, 5\}$ are witnesses for compositeness of all non-primes $< 25\,326\,001$;

- $\{2, 3, 5, 7\}$ are witnesses for compositeness of all non-primes $< 3\,215\,031\,751$;

- $\{2, 3, 5, 7, 11\}$ are witnesses for compositeness of all non-primes $< 2\,152\,302\,898\,747$.

Session 4.29. *We implement the Rabin-Miller test:*

```
In[1]:= Clear[RabinMillerPrime]

        RabinMillerPrime :: "pseudoprime" =
          "strong pseudoprime base `1`";

        RabinMillerPrime :: "composite" =
          "`1` is witness for compositeness";

        RabinMillerPrime[p_, a_] := Module[{s, u, done, power},
          s = IntegerExponent[p - 1, 2];
          u = (p - 1)/2^s;
          power = PowerMod[a, u, p];
          If[power == 1,
            Message[RabinMillerPrime :: "pseudoprime", a]; Return[True]];
          done = False;
          While[Not[done] && u < p - 1,
            If[power == p - 1, done = True,
              u = 2 * u; power = PowerMod[power, 2, p]]];
          If[done, Message[RabinMillerPrime :: "pseudoprime", a],
            Message[RabinMillerPrime :: "composite", a]
          ];
          done
        ]

        RabinMillerPrime[p_] := Module[{a},
          a = Random[Integer, {2, p - 1}];
          RabinMillerPrime[p, a]
        ]
```

The result of `RabinMillerPrime[p,a]` *is* `False` *if a is a witness for the compositeness of p, and therefore p is composite. On the other hand,* `True` *means that p is a strong pseudoprime (with the base a). As a side effect, a message is created which specifies the witness or the pseudoprimeness, respectively.*

We test RabinMillerPrime:

In[2]:= **RabinMillerPrime[1234567,2]**

RabinMillerPrime :: *"composite"*: 2 *is witness for compositeness*
Out[2]= False

The function Prime[n] *computes the nth prime, so the following integer is clearly composite:*

In[3]:= **RabinMillerPrime[Prime[10^9] * Prime[2^{30}],2]**

RabinMillerPrime :: *"composite"*: 2 *is witness for compositeness*
Out[3]= False

Now we execute the Rabin-Miller test with some Carmichael numbers:

In[4]:= **RabinMillerPrime[2465,2]**

RabinMillerPrime :: *"composite"*: 2 *is witness for compositeness*
Out[4]= False

In[5]:= **RabinMillerPrime[41041,2]**

RabinMillerPrime :: *"composite"*: 2 *is witness for compositeness*
Out[5]= False

In[6]:= **RabinMillerPrime[825265,2]**

RabinMillerPrime :: *"composite"*: 2 *is witness for compositeness*
Out[6]= False

We see that 2 is always a witness for the compositeness.

Using NextPrime, *we generate a large prime p,*

In[7]:= **p = NextPrime[10^{100}]**
Out[7]= 100000000000000000000000000000000000000000000000000000000000000000
         00000000000000000267

and show that no witness $a \in \{2, ..., 10\}$ for the compositeness of p exists:

In[8]:= **Union[Table[RabinMillerPrime[p,a],{a,2,10}]]**

RabinMillerPrime :: *"pseudoprime"*: *strong pseudoprime base* 2

RabinMillerPrime :: *"pseudoprime"*: *strong pseudoprime base* 3

RabinMillerPrime :: *"pseudoprime"*: *strong pseudoprime base* 4

General :: *"stop"*: *Further output of* RabinMillerPrime :: *"pseudoprime" will be suppresse*
during this calculation
Out[8]= {True}

Finally we build the product p of two large primes:

In[9]:= **p = NextPrime[10^{100}] * NextPrime[10^{101}]**
Out[9]= 100000000000000000000000000000000000000000000000000000000000000000
         00000000000000002673000000000000000000000000000000000000000000000
         0000000000000801

If the function RabinMillerPrime[p] *is called with one argument only, then it is executed
with some random $a \in \{2, ..., p-1\}$. For our p we get*

In[10]:= **RabinMillerPrime[p]**

```
RabinMillerPrime :: "composite":
73665564733037105577 ≪160≫ 69185668341765554143 is witness for compositeness
Out[10]= False
```

hence we have found a witness for the compositeness of p immediately.

One can show that the probability that a composite integer satisfies the Rabin-Miller test for random a is at most 1/4. Therefore one can apply the Rabin-Miller test several times independently and can thus make the error probability arbitrarily small. For example, if one executes the Rabin-Miller test 100 times, the error probability is less than

```
In[11]:= 0.25^100
```
$$Out[11]= 6.22302 \times 10^{-61}$$

Such an algorithm that gives a correct answer only with a very (arbitrary) high probability is called a *probabilistic algorithm.*

4.7 Additional Remarks

In abstract algebra the method which we used to construct $\mathbb{Z}_p$ from $\mathbb{Z}$ is considered in a more general setting. $\mathbb{Z}_p$ is called the *quotient ring* $\mathbb{Z}_p = \mathbb{Z}/p\mathbb{Z}$. Another quotient ring will be considered in Section 7.1.

Fermat's Little Theorem is a consequence of Lagrange's theorem in group theory: The multiplicative group $\mathbb{Z}_p^\star$ has the order $p - 1$, and the order of every element a is a divisor of $p - 1$ (Corollary 4.14). Theorem 4.19 shows that $\mathbb{Z}_p^\star$ forms a cyclic group.

Algorithms whose complexity can be estimated by a polynomial in the input length are called *polynomial algorithms.* Up to this day, no polynomial algorithms are known for the factorization of integers as well as for the computation of modular square roots and of modular logarithms. Although there are good reasons to assume that such algorithms do not exist, proofs for these facts are not known. Such questions are considered in complexity theory ([BCS1997], [BDG1988]).

There are not only probabilistic algorithms for deciding whether or not an integer is prime, but also *deterministic algorithms* which have a better efficiency than the trivial algorithm by trial division. Such algorithms as well as further probabilistic algorithms are discussed in the books [Buc1999], [For1996], [GG1999] and [Rie1994].

If you want to be on the safe side when searching for primes in *Mathematica* you can use the much slower function `ProvablePrimeQ[n]` from the package `PrimalityProving` which can be loaded using `Needs["PrimalityProving`"]`.[16]

In 2002 three Indian mathematicians solved an old problem: In [AKS2004] they gave a deterministic polynomial time algorithm which decides whether a given integer $p \in \mathbb{Z}$ is prime. In practice, however, this algorithm is not (yet) as fast as other known non-polynomial deterministic algorithms, and it is not accessible in *Mathematica*.[17] But their main statement is: This problem can be solved in polynomial time.

---

[16] According to the *Mathematica* help pages, the certificate of primality used in this package for large n is based on the theory of elliptic curves.

[17] Asymptotic efficiency might not always be advantageous if one mainly deals with "very short integers".

4.8 Exercises

Exercise 4.1. *Prove in detail how the ring properties in Theorem 4.4 are transferred from $\mathbb{Z}$ to $\mathbb{Z}_p$.*

Exercise 4.2. *Show that a ring with a zero divisor cannot be a field.*

Exercise 4.3. *Prove that in $\mathbb{Z}_p$ the cancelation law*

$$a \cdot x \equiv a \cdot y \,(\mathrm{mod}\, p) \quad \Rightarrow \quad x \equiv y \,(\mathrm{mod}\, p)$$

is valid if and only if $\gcd(a, p) = 1$.

Exercise 4.4. *Implement the solution of the Chinese remainder problem given in Theorem 4.10.*

Exercise 4.5. *Show that for $\gcd(p, q) = 1$, we have*

$$x \equiv a \,(\mathrm{mod}\, p) \quad \text{and} \quad x \equiv a \,(\mathrm{mod}\, q) \quad \Leftrightarrow \quad x \equiv a \,(\mathrm{mod}\, p \cdot q).$$

How can this be generalized?

Exercise 4.6. *Show that a polynomial $p(x) = a_n x^n + a_{n-1} x^{n-1} + \cdots + a_0$ with coefficients $a_k \in \mathbb{K}$ of a field $\mathbb{K}$ has at most n zeros in $\mathbb{K}$.*

In particular, let $p \in \mathbb{P}$, and show that the equation $x^2 = 1$ has exactly two zeros $x \equiv \pm 1 \,(\mathrm{mod}\, p)$ in $\mathbb{Z}_p$.

Exercise 4.7. *Implement the function* powermod *iteratively.*

Exercise 4.8. *Implement—analogously to the function* powermod*—the fast matrix power* matrixpower$[A,n] = A^n$ *of matrices A and $n \in \mathbb{Z}$. Compare your result with the built-in function* MatrixPower*. Hint: Use . for matrix multiplication.*

Exercise 4.9. *Let G be a commutative group with unity $e \in G$, and let $a \in G$. Further, let $S := \{m \geq 1 \mid a^m = e\}$ and let the order of a be $\mathrm{ord}(a) = \min S$. Show that if $a^m = e$, then $\mathrm{ord}(a) \mid m$.*

Exercise 4.10. *Let $p \in \mathbb{P}$ and $a \in \mathbb{Z}_p^\star$. Prove that $\mathrm{ord}(a) = m$ implies $\mathrm{ord}(a^r) = \frac{m}{\gcd(r,m)}$.*

Exercise 4.11. *Let $\mathrm{GF}(9) = \{a + ib \mid a, b \in \mathbb{Z}_3, i^2 = -1\}$, a finite set with 9 elements. Show that $\mathrm{GF}(9)$ is a field. Find a generating element, i.e., an element $a \in \mathrm{GF}(9)$ with $\mathrm{ord}(a) = 8$.*

Exercise 4.12. *Set $p =$ NextPrime$[10^k]$ for $k = 2, \ldots, 5$ and compute in each case, how many generators exist in $\mathbb{Z}_p^\star$. (The number of generators of $\mathbb{Z}_p^\star$ is examined in detail in [Chi2000].)*

Exercise 4.13. *Show that the Fermat pseudoprimes form a subgroup of $\mathbb{Z}_p^\star$.*

Exercise 4.14. *Implement a function* ModularLog *which computes the modular logarithm, i.e., a solution (or the solutions) x of the equation $a^x \equiv y \,(\mathrm{mod}\, p)$. Solve the problem given in Session 4.20 with your implementation, and compare the efficiency of your code with the function* MultiplicativeOrder*.*

Exercise 4.15. *Compute the smallest Carmichael number with 4 or 5 different factors, respectively.*

Exercise 4.16. (Large Carmichael Numbers)

(a) *Let $k \in \mathbb{N}$. Consider*

$$N = (6k+1)(12k+1)(18k+1). \tag{4.9}$$

If each of these factors is prime, show that N is a Carmichael number.

(b) *Use (4.9) to find a Carmichael number with at least 120 decimal places.*

(c) *Compare the computation time of the algorithm in (b) with the algorithm given in Session 4.25. Explain the difference!*

Exercise 4.17. *Verify the first two statements given in Example 4.28 with Mathematica.*

Exercise 4.18. *Examine the Fermat numbers $F_n = 2^{2^n} + 1$ for $n = 1, \ldots, 10$ with the Rabin-Miller test for divisibility. Mathematica can find the prime factor decomposition up to $n = 8$ in reasonable time. Find these!*

Chapter 5
Coding Theory and Cryptography

5.1 Basic Concepts of Coding Theory

Coding Theory deals with the secure transfer of messages. We assume that the sender (source) of the message sends the message via a *communication channel* to the recipient. The type of coding depends on the practical situation, hence the communication channel might also be disturbed. Examples for such transmissions are given by

- the Morse code;
- playing a music CD, where the "digitized music" is converted into an analog signal and is transferred to an amplifier system;
- storing a file on a computer, where the content of the file is transferred from the computer memory to a hard disk;
- file compression, where the content of a file is converted using a compression method, e.g., the MP3 method for the encoding of music or by the WinZip program for the compression of data;
- sending an email;
- the transmission from the keyboard of a computer to the display;
- at the beginning of the electronic age, programs were sent to a central computer via punch-cards.

In these situations the following things matter:

- The message should be transferred efficiently.
- If the channel is disturbed, then transmission errors should be detected or even corrected, if possible.

Definition 5.1. (Alphabets and Codings) By *coding* of a message we mean the conversion from one alphabet into another. A method with which this coding is executed is called a *code*. By an *alphabet* $\mathcal{A} = \{z_1, z_2, ..., z_m\}$ we mean a finite set of *symbols* (or *characters* / *letters*) $z_1, z_2, ..., z_m$. A *word* or *string* of *length* n is an n-tuple $W \in \mathcal{A}^n$. Note that words are mostly denoted without parentheses or commas: $W = w_1 w_2 ... w_n$. The set of all words over the alphabet $\mathcal{A}$ is $\mathcal{W} = \bigcup_{n=0}^{\infty} \mathcal{A}^n$ where $\mathcal{A}^0$ contains the *empty word*. $\triangle$

© Springer Nature Switzerland AG 2021
W. Koepf, *Computer Algebra*, Springer Undergraduate Texts in Mathematics and Technology,
https://doi.org/10.1007/978-3-030-78017-3_5

Example 5.2. Positive integers with respect to the basis B are words over the alphabet $\mathbb{Z}_B = \{0, 1, ..., B - 1\}$. If a decimal number is converted from one basis B to another basis B', then this is a coding.

The binary alphabet contains the two symbols $\{0, 1\}$. The encoding of one of these symbols is called a *bit*. The *bit length* of a word is the number of bits in its binary representation. For a word of length n over an alphabet $\mathcal{A}$ with m symbols, one needs $n \cdot \log_2 m$ bits,[1] its bit length is therefore $n \cdot \log_2 m$, see Exercise 5.1. For instance, every symbol from the alphabet $\mathcal{A} = \{A, B, C, D, E, F, G, H\}$, consisting of 8 symbols, can be represented by $3 = \log_2 8$ bits, for example by

$$A \mapsto 000$$
$$B \mapsto 001$$
$$C \mapsto 010$$
$$D \mapsto 011$$
$$E \mapsto 100$$
$$F \mapsto 101$$
$$G \mapsto 110$$
$$H \mapsto 111$$

A word with 15 digits in $\mathcal{A}$ needs $15 \cdot 3 = 15 \cdot \log_2 8 = 45$ bits.

For written texts we usually take the alphabet $\{A, B, ..., Y, Z\}$, complemented by the lower case letters and some other symbols like blank (space), period etc.

For data storage in computers the *ASCII character set*[2] is used. This alphabet consists of 128 symbols, in particular all the "regular letters". Internally these 128 symbols are represented as binary data, whereby 7 bits are required for the representation of the $128 = 2^7$ different characters.

Since the standard ASCII character set does not contain international symbols like ö or é, Windows PCs use an *extended ASCII character set* consisting of 256 symbols. For the representation of $256 = 2^8$ different symbols one needs 8 bits. This storage unit is called a *byte*.

A slightly different example is the *Morse alphabet*, an alphabet for radio amateurs where every character is represented by a certain short-long pattern:[3]

$$
\begin{array}{lllll}
1 \mapsto \bullet\ {-}\,{-}\,{-}\,{-} & 2 \mapsto \bullet\,\bullet\ {-}\,{-}\,{-} & 3 \mapsto \bullet\,\bullet\,\bullet\ {-}\,{-} & 4 \mapsto \bullet\,\bullet\,\bullet\,\bullet\ {-} & 5 \mapsto \bullet\,\bullet\,\bullet\,\bullet\,\bullet \\
6 \mapsto {-}\,\bullet\,\bullet\,\bullet\,\bullet & 7 \mapsto {-}\,{-}\,\bullet\,\bullet\,\bullet & 8 \mapsto {-}\,{-}\,{-}\,\bullet\,\bullet & 9 \mapsto {-}\,{-}\,{-}\,{-}\,\bullet & 0 \mapsto {-}\,{-}\,{-}\,{-}\,{-} \\
A \mapsto \bullet\ {-} & B \mapsto {-}\,\bullet\,\bullet\,\bullet & C \mapsto {-}\,\bullet\,{-}\,\bullet & D \mapsto {-}\,\bullet\,\bullet & E \mapsto \bullet \\
F \mapsto \bullet\,\bullet\,{-}\,\bullet & G \mapsto {-}\,{-}\,\bullet & H \mapsto \bullet\,\bullet\,\bullet\,\bullet & I \mapsto \bullet\,\bullet & J \mapsto \bullet\ {-}\,{-}\,{-} \\
K \mapsto {-}\,\bullet\ {-} & L \mapsto \bullet\ {-}\,\bullet\,\bullet & M \mapsto {-}\,{-} & N \mapsto {-}\,\bullet & O \mapsto {-}\,{-}\,{-} \\
P \mapsto \bullet\ {-}\,{-}\,\bullet & Q \mapsto {-}\,{-}\,\bullet\ {-} & R \mapsto \bullet\ {-}\,\bullet & S \mapsto \bullet\,\bullet\,\bullet & T \mapsto {-} \\
U \mapsto \bullet\,\bullet\ {-} & V \mapsto \bullet\,\bullet\,\bullet\,{-} & W \mapsto \bullet\ {-}\,{-} & X \mapsto {-}\,\bullet\,\bullet\,{-} & Y \mapsto {-}\,\bullet\ {-}\,{-} \\
Z \mapsto {-}\,{-}\,\bullet\,\bullet & \ddot{A} \mapsto \bullet\ {-}\,\bullet\ {-} & \ddot{O} \mapsto {-}\,{-}\,{-}\,\bullet & \ddot{U} \mapsto \bullet\,\bullet\ {-}\,{-} & CH \mapsto {-}\,{-}\,{-}\,{-}
\end{array}
$$

Morse telegraphy is still in use by radio amateurs. The knowledge of the Morse alphabet can save lives: Even if radiocommunication is disturbed, a ship can still call SOS,[4] for example using spotlights. △

---

[1] maybe suitably rounded

[2] American Standard Code for Information Interchange

[3] The dot denotes "short", whereas the dash means "long".

[4] Save Our Souls

Session 5.3.

Mathematica contains a much larger character set. The first 128 symbols of the Mathematica character set correspond to the regular ASCII set whose first 32 symbols are functional characters. We get a table of the first 128 symbols by the call [5]

$In[1]:=$ **Partition[Table[FromCharacterCode[k],{k,0,127}],16]**

Our regular alphabet has ASCII numbers between 65 and 128:

$In[2]:=$ **Partition[Table[FromCharacterCode[k],{k,65,128}],16]**

$$Out[2]= \begin{pmatrix} A & B & C & D & E & F & G & H & I & J & K & L & M & N & O & P \\ Q & R & S & T & U & V & W & X & Y & Z & [& \backslash &] & \hat{} & \_ & \` \\ a & b & c & d & e & f & g & h & i & j & k & l & m & n & o & p \\ q & r & s & t & u & v & w & x & y & z & \{ & | & \} & \tilde{} & & \end{pmatrix}$$

Other character sets have larger numbers in Mathematica. For example, we get the Greek letters by

$In[3]:=$ **Partition[Table[FromCharacterCode[k],{k,913,976}],16]**

$$Out[3]= \begin{pmatrix} A & B & \Gamma & \Delta & E & Z & H & \Theta & I & K & \Lambda & M & N & \Xi & O & \Pi \\ P & & \Sigma & T & \Upsilon & \Phi & X & \Psi & \Omega & & & & & & & \\ \alpha & \beta & \gamma & \delta & \varepsilon & \zeta & \eta & \theta & \iota & \kappa & \lambda & \mu & \nu & \xi & o & \pi \\ \rho & \varsigma & \sigma & \tau & \upsilon & \phi & \chi & \psi & \omega & & & & & & & \end{pmatrix}$$

The coding between the ASCII characters and their corresponding ASCII numbers is executed by the Mathematica functions FromCharacterCode *and* ToCharacterCode. [6] *We get for example*

$In[4]:=$ **ToCharacterCode["α"]**

$Out[4]=$ {945}

The Greek letter α therefore has the number 945.

5.2 Prefix Codes

In this section we discuss an important property that codes can have.

Definition 5.4. (Code word) Let $\mathcal{A}$ and $\mathcal{B}$ denote two alphabets and $\mathcal{W} = \bigcup\limits_{n=0}^{\infty} \mathcal{A}^n$ and $\mathcal{W}' = \bigcup\limits_{n=0}^{\infty} \mathcal{B}^n$ the corresponding word sets. Then every injective mapping $c : \mathcal{A} \to \mathcal{W}'$, which therefore maps the different characters of $\mathcal{A}$ to different words in $\mathcal{W}'$, can be extended to a mapping $C : \mathcal{W} \to \mathcal{W}'$ by concatenating successively the images of the single characters to a new word:

$$C(x_1 x_2 \ldots x_n) := c(x_1)c(x_2)\ldots c(x_n).$$

C is a code, and the images under c are called *code words*. △

If a message is converted (encoded), then it should be possible to reconvert (decode) the encoding.

---

[5] We suppressed the output.

[6] The input of α is made by typing <ESC> a <ESC> or via one of the palettes.

Example 5.5. Let $\mathcal{A} = \{A, B, C\}$ and $\mathcal{B} = \{0, 1\}$ denote two alphabets and $\mathcal{W} = \overset{\infty}{\underset{n=0}{\cup}} \mathcal{A}^n$ and $\mathcal{W}' = \overset{\infty}{\underset{n=0}{\cup}} \mathcal{B}^n$ the corresponding word sets. Using Definition 5.4, we declare a code $C : \mathcal{W} \to \mathcal{W}'$ by $A \mapsto 1, B \mapsto 0, C \mapsto 01$. The word

$$W = ABCAACCBAC$$

is therefore encoded by

$$W' = 10011101010101 .$$

According to the mapping description this coding is clearly unique. However, now we want to generate W from W'. For this purpose, we start with the beginning word and interpret the characters successively. We realize that the first 1 of W' can be generated only by an A. Therefore the first character of W is an A. The second character of W' is 0, which could arise from B or from C. But since the third character of W' is 0, C is impossible and we know that the second character of W must be B. The next character combination in W' is 01. This yields a problem since this combination could be generated by C as well as by BA. Therefore a unique decoding is not possible.

A better code $C' : \mathcal{W} \to \mathcal{W}'$ is given by $A \mapsto 0, B \mapsto 10, C \mapsto 11$. In this case, W is encoded as

$$W' = 0101100111110011 .$$

The reader can easily check that now it is possible to generate W from W'. Which property of the code is responsible for that behavior? △

Definition 5.6. (Prefix Code) A word $x_1 x_2 \ldots x_k \in \mathcal{A}^k$ is called a *prefix* of a word $W \in \mathcal{A}^m$ if is is an initial part of W, i.e. $W = x_1 x_2 \ldots x_k x_{k+1} \ldots x_m$.

A *prefix code* is a code $C : \mathcal{W} \to \mathcal{W}'$ of the type presented in Definition 5.4 such that no code word is the prefix of another code word. △

Theorem 5.7. Prefix codes enable unique decoding.

Proof. Exercise 5.2   □

Example 5.8. Examples of prefix codes are *block codes* for which all images $c(x_k)$ have the same length.

One example is the Caesar code:[7] We set $\mathcal{A} = \{A, B, \ldots, Z\}$ and the code just shifts by 3 letters, i.e., $A \mapsto D, B \mapsto E, \ldots, W \mapsto Z, X \mapsto A, Y \mapsto B$ and $Z \mapsto C$.

An example of a block code with *block length* 2 is $c(A) = 00, c(B) = 01, c(C) = 11$.

However, more interesting are prefix codes which are no block codes. Such a code was considered in Example 5.5.

Note that the Morse code does not constitute a prefix code. To enable the decoding of Morse texts, any two Morse patterns are separated by a short pause. If these pause characters are included in the Morse alphabet, then the resulting code is a prefix code. △

---

[7] Caesar used this encoding as *cipher*, see Section 5.5. However, the code is not really suitable for this purpose. Why?

In practice one would like to generate codes in such a way that image words are as small as possible, i.e., they contain as little characters as possible. Any block code of length m encodes a word of length n by a word of length $m \cdot n$. Therefore block codes expand rather than compress a given message.

However, if we have some knowledge about the *letter frequency* of a text or of the words in $\mathcal{W}$, then we can use this information to generate a prefix code which assigns short code words to letters of high frequency and larger code words to letters of low frequency. This can lead to a *compression*, i.e., a reduction of the bit length.

For this purpose, let the alphabet $\mathcal{A} = \{A_1, A_2, \ldots, A_m\}$ be given together with a *frequency distribution*

$$\begin{pmatrix} A_1 & p_1 \\ A_2 & p_2 \\ \vdots & \vdots \\ A_m & p_m \end{pmatrix},$$

where $p_k \geq 0$ ($k = 1, \ldots, m$) and $\sum\limits_{k=1}^{m} p_k = 1$. The relative frequencies $(p_k)_{k=1,\ldots,m}$ therefore generate a *finite probability measure* on the alphabet $\mathcal{A}$. Every natural language like English, German or Russian has its own characteristic letter frequency, see e.g. [Lew2000]. Even every author has his or her own characteristic vocabulary resulting in a specific letter frequency by which he or she might be identified.[8]

The approximative letter distribution of the English language is given in the following table.

Table 5.1 Letter distribution in English language (see [Lew2000], p. 36)

E 12.702 %	T 9.056 %	A 8.167 %	O 7.507 %	I 6.966 %	N 6.749 %	S 6.327 %	H 6.094 %
R 5.987 %	D 4.253 %	L 4.025 %	C 2.782 %	U 2.758 %	M 2.406 %	W 2.360 %	F 2.228 %
G 2.015 %	Y 1.974 %	P 1.929 %	B 1.492 %	V 0.978 %	K 0.772 %	J 0.153 %	X 0.150 %
Q 0.095 %	Z 0.074 %						

Note that in English language, the space character is slightly more frequent than the top letter E.

A code which is based on letter frequencies is the *Huffman code* which uses the image alphabet $\{0, 1\}$. The Huffman code is generated in several steps. In the first steps, the two source symbols with the smallest frequencies are merged[9] and one considers the merged items with their total frequency as a new symbol. This process is iterated until only two symbols remain. For the assignment of the image bits, the iteration is run backwards and tells us how the bits must be selected. In all cases where merging occurred, the more frequent symbol list appends 1 to their code words, whereas the rarer symbols append 0.

---

[8] The Russian author Mikhail Sholokhov (en.wikipedia.org/wiki/Mikhail_Sholokhov) won the Nobel price in 1965 for the novel "And Quiet Flows the Don". In Russia there were rumours that he had plagiarized this work, and in an anonymous study the writer Fyodor Kryukov was given credit as possible author. Therefore the question arises whether Sholokhov may have received the Nobel prize unjustified? Based on statistical methods it was proved that with a very high probability Kryukov cannot be the author of "And Quiet Flows the Don", whereas nothing speaks against the authorship of Sholokhov [Erm1982].

[9] If there are several possibilities to do so, one chooses any one.

Example 5.9. (**Huffman Code**) Let the frequency distribution

$$\begin{pmatrix} A & 0.1 \\ B & 0.12 \\ C & 0.18 \\ D & 0.2 \\ E & 0.4 \end{pmatrix}$$

be given. Successive execution of the above method yields the iteration

$$\begin{pmatrix} A & 0.1 \\ B & 0.12 \\ C & 0.18 \\ D & 0.2 \\ E & 0.4 \end{pmatrix} \rightarrow \begin{pmatrix} C & 0.18 \\ D & 0.2 \\ AB & 0.22 \\ E & 0.4 \end{pmatrix} \rightarrow \begin{pmatrix} AB & 0.22 \\ CD & 0.38 \\ E & 0.4 \end{pmatrix} \rightarrow \begin{pmatrix} E & 0.4 \\ ABCD & 0.6 \end{pmatrix}$$

According to the last matrix we have $E \mapsto 0$, and the code words of the letters A, B, C and D start with 1. The previous step shows that for the merged letters $ABCD$ the combination AB has smaller frequency than CD, therefore we append 0 to the code words of A and B and 1 to the code words for C and D. Finally, the previous iteration step shows that $C \mapsto 110$ and $D \mapsto 111$, whereas the last step completes our computation with the assignments $A \mapsto 100$ and $B \mapsto 101$.

Therefore the complete Huffman code is given by the code words

$$E \mapsto 0$$
$$A \mapsto 100$$
$$B \mapsto 101$$
$$C \mapsto 110$$
$$D \mapsto 111$$

which can be represented by the following *code tree*:

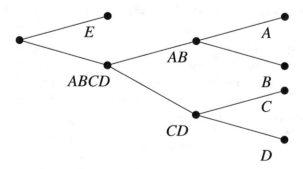

When you start from the root on the left, then going up means that you append a 0, and going down means that you append a 1. △

Session 5.10. *The Huffman code can be computed using the Mathematica function* Huffman *which calls the auxiliary function* HuffmanList, *which again generates the Huffman code words iteratively.*

```
In[1]:= Clear[Huffman, HuffmanList]

        Huffman[Alphabet_] := HuffmanList[Alphabet][[2]]

In[2]:= HuffmanList[Alphabet_] :=
          Module[{k, alphabet, createcode, first, second},
            If[
              !Abs[Sum[Alphabet[[k, 2]], {k, Length[Alphabet]}] -
                  1.] < 0.01, Return[{, "wrong input"}]];
            alphabet = Alphabet;
            createcode = Table[{alphabet[[k, 1]], {}},
              {k, Length[alphabet]}];
            Do[alphabet = Sort[alphabet, #1[[2]] ≤ #2[[2]]&];
              first = First[alphabet];
              alphabet = Rest[alphabet];
              second = First[alphabet];
              alphabet = Rest[alphabet];
              Do[If[!FreeQ[first[[1]], createcode[[k, 1]]],
                 PrependTo[createcode[[k, 2]], 0]],
                {k, Length[createcode]}];
              Do[If[!FreeQ[second[[1]], createcode[[k, 1]]],
                 PrependTo[createcode[[k, 2]], 1]],
                {k, Length[createcode]}];
              PrependTo[alphabet,
                {{first[[1]], second[[1]]}, first[[2]] + second[[2]]}],
              {j, Length[alphabet] - 1}];
            {alphabet, createcode}]
```

After entering our frequency distribution

```
In[3]:= alphabet = {{"A", 0.1}, {"B", 0.12}, {"C", 0.18},
          {"D", 0.2}, {"E", 0.4}}
```

$$Out[3] = \begin{pmatrix} A & 0.1 \\ B & 0.12 \\ C & 0.18 \\ D & 0.2 \\ E & 0.4 \end{pmatrix}$$

the Huffman code can be generated:

```
In[4]:= Huffman[alphabet]
```

$$Out[4] = \begin{pmatrix} A & \{1, 0, 0\} \\ B & \{1, 0, 1\} \\ C & \{1, 1, 0\} \\ D & \{1, 1, 1\} \\ E & \{0\} \end{pmatrix}$$

Note that the function HuffmanList *collects the decisive intermediate results for this computation.*

The Huffman coding always generates a prefix code. One can show that the Huffman code gives an optimal compression in some sense (see e.g. [Sch2003], Theorem 6.6, p. 44).

5.3 Check Digit Systems

In order to recognize transcription errors, data are often equipped with redundant information. The simplest way to add redundant information is a *check digit*. In UNIX systems, to every ASCII number z an extra bit is appended. The additional bit is 0 if the number of 1's in the binary representation of z is even (even *parity*), and 1 otherwise. This additional number is called check digit. Therefore every 7-bit ASCII number is stored with its check digit as a byte whose parity is always even by construction. If after transmission the parity is odd, then an error occurred, in which case the data can be transferred again. Note, however, that it is not possible to correct a transmission error directly.

A similar method applies to the UPC (Universal Product Code) which is used when scanning the product code in a supermarket. The UPC is a 13-digit decimal number $a_1 a_2 \ldots a_{13}$ whose last digit is a check digit which is computed according to the formula

$$1 \cdot a_1 + 3 \cdot a_2 + 1 \cdot a_3 + 3 \cdot a_4 + \cdots + 1 \cdot a_{11} + 3 \cdot a_{12} + 1 \cdot a_{13} \equiv 0 \pmod{10}. \tag{5.1}$$

Solving this equation for a_{13}, we get

$$a_{13} = \mathrm{mod}(-(1 \cdot a_1 + 3 \cdot a_2 + 1 \cdot a_3 + 3 \cdot a_4 + \cdots + 1 \cdot a_{11} + 3 \cdot a_{12}), 10).$$

If an error occurs during the scanning process, i.e., if (5.1) is not satisfied, then the check register will beep, and the scanning process must be repeated. Therefore the error is detected, but cannot be corrected. If you want to learn more about how the *bar codes* are interpreted as numbers, see e.g. `demonstrations.wolfram.com/UPCBarCode`.

In a similar manner the ISBN-10 (International Standard Book Number) contains a check digit. The ISBN-10 is a 9-digit decimal number extended by a check digit which is computed via the formula

$$\sum_{k=1}^{10} k\, a_k \equiv 0 \pmod{11}.$$

Solving for a_{10} yields the relation

$$a_{10} = \mathrm{mod}\left(\sum_{k=1}^{9} k\, a_k, 11 \right),$$

since $10 \equiv -1 \pmod{11}$. Because the computation is modulo 11, the check digit might be 10 in which case it is written as the Roman numeral X.

Starting in 2007 the ISBN-10 was replaced by the ISBN-13 which is a UPC starting with the digits 978 followed by the 9 ISBN-10 digits. Finally, the last digit of the ISBN-13 is the UPC check digit.

Check digits are widely used for passport numbers, personal identification numbers, account numbers such as IBAN (International Bank Account Number), banknotes, and many more.

5.4 Error Correcting Codes

In this section, we will present a simple example of an *error correcting code* which is a variant of the *Reed-Solomon code*.

Session 5.11. *We would like to equip a word with n letters with redundant information in such a way that we are able to correct up to one transmission error. As an example, we start with the word* `"WORD"`.

In the first step we replace the letters by their numbers in the alphabet, hence $W \mapsto 23$, $O \mapsto 15$, $R \mapsto 18$ *and* $D \mapsto 4$. *These replacements can be realized with* Mathematica:

In[1]:= `Digitize[word_] := ToCharacterCode⌊word] - 64`

In[2]:= `list = Digitize["WORD"]`
Out[2]= {23, 15, 18, 4}

The back transformation is done by

In[3]:= `Verbalize[list_] := FromCharacterCode[list + 64]`

In[4]:= `Verbalize[list]`
Out[4]= WORD

The considered replacement provides an n-tuple of numbers a_k *whose index is set to* $k = 2, \ldots, 5$:

In[5]:= `a[2] = 23; a[3] = 15; a[4] = 18; a[5] = 4;`

Next, we complement the numbers a_k $(k = 2, \ldots, n + 1)$ *by the two values* a_0 *and* a_1 *in such a way that the two equations*

$$\sum_{k=0}^{n+1} a_k \equiv 0 \ (mod \ 31) \tag{5.2}$$

and

$$\sum_{k=0}^{n+1} k \cdot a_k \equiv 0 \ (mod \ 31) \tag{5.3}$$

are satisfied. The idea is that we need two informations for correcting an error: the amount of the error and the place where it occurred. To find these informations, we will need two equations.

The equation (5.3) has a unique solution $a_1 \in \mathbb{Z}_{31}$, *and (5.2) finally yields a unique* $a_0 \in \mathbb{Z}_{31}$, *since* $\mathbb{Z}_{31}$ *is an additive group. Note that the choice* $p = 31$ *is arbitrary and guided by the minimum number of letters of our alphabet.*[10]

The following computation solves the given system of equations:

In[6]:= $\text{Solve}\left[\{\sum_{k=0}^{5} a[k] == 0, \sum_{k=0}^{5} k\,a[k] == 0\}, \{a[0], a[1]\}, \text{Modulus} \to 31\right]$
Out[6]= $\{\{a(0) \to 30, a(1) \to 3\}\}$

The complete coding is given by the function `ReedSolomon` *for an arbitrary* $p \in \mathbb{P}$ *with* $p > n$:

---

[10] For the decoding step we need—as we will see—a field. Hence every prime number p could be used. In practice p is a power of 2. This, however, requires the computation in a Galois field, see Section 7.4.

```
In[7]:= Clear[ReedSolomon]
        ReedSolomon[word_, p_] :=
          Module[{list, n, a0, a1, k, sol},
            list = Digitize[word];
            n = Length[list];
            sol = Solve[{a0 + a1 + Apply[Plus, list] == 0,
                  a1 + Sum[(k + 1) * list[[k]], {k, n}] == 0,
                  }, {a0, a1, Modulus → p}];
            list = Prepend[Prepend[list, a1], a0] /.
                sol[[1]];
            Verbalize[list]
          ]
```

For our example, we get

```
In[8]:= ReedSolomon["WORD", 31]
Out[8]= ^CWORD
```

Now, let us assume that an error occurred, more precisely, exactly one letter was transmitted incorrectly. Let us assume for our example that the word ^CLORD *was received:*

```
In[9]:= word = "^CLORD"
Out[9]= ^CLORD
```

When receiving this message, we certainly want to find out whether the word ^CLORD *indeed represents the "Lord". For this purpose, we convert the letter sequence toward their ASCII numbers:*

```
In[10]:= list = Digitize[word]
Out[10]= {30, 3, 12, 15, 18, 4}
```

We check the sums occurring in (5.2) and (5.3),

$$e := \sum_{k=0}^{5} a_k \,(mod\,31) \qquad and \qquad s := \sum_{k=0}^{5} k\,a_k \,(mod\,31) \qquad\qquad (5.4)$$

in the following way:

```
In[11]:= e = Mod[Apply[Plus, list], 31]
Out[11]= 20
```

```
In[12]:= s = Mod[ ∑_{k=1}^{6} (k - 1) list[[k]], 31]
Out[12]= 9
```

Since $e \neq 0$, *we can conclude, that the transmission was not error-free. Let us assume in the sequel that exactly one error occurred. (If more errors occurred, then we may not be able to correct them.)*

First, we would like to find out at which position x the error occurred. We know that the letter a_x *was replaced by* $a_x + e \,(mod\,31)$, *since this error produces in the sum* $\sum_{k=0}^{5} k\,a_k \,(mod\,31)$ *the difference* $x \cdot e \,(mod\,31)$. *Hence we get*

$$s \equiv x \cdot e \,(mod\,31)$$

or equivalently

$$x \equiv s \cdot e^{-1} \ (mod \ 31).$$

In our case, the computation

```
In[13]:= x = Mod[s * PowerMod[e, -1, 31], 31]
Out[13]= 2
```

shows that the error occurred at position a_2 representing the first letter of the original word. Now we can reconstruct this letter by

```
In[14]:= list[[x+1]] = Mod[list[[x+1]] - e, 31]
Out[14]= 23
```

and compute the corrected word

```
In[15]:= Verbalize[list]
Out[15]= ^CWORD
```

The following function combines all these steps:

```
In[16]:= Clear[InverseReedSolomon]
         InverseReedSolomon[word_, p_] :=
           Module[{list, n, e, s, x, k},
             list = Digitize[word];
             n = Length[list];
             e = Mod[Apply[Plus, list], p];
             s = Mod[Sum[(k-1) * list[[k]], {k, 1, n}], p];
             x = Mod[s * PowerMod[e, -1, p], p];
             list[[x+1]] = Mod[list[[x+1]] - e, p];
             list = Rest[Rest[list]];
             Verbalize[list]
           ]
```

With this implementation, error correction can be called in one step:

```
In[17]:= InverseReedSolomon["^CLORD", 31]
Out[17]= WORD
```

Finally, we consider another example:

```
In[18]:= code = ReedSolomon["MATHEMATICS", 31]
Out[18]= O\MATHEMATICS
```

We incorporate an error:

```
In[19]:= code = StringReplace[code, "S" → "A"]
Out[19]= O\MATHEMATICA
```

Next, we repair the error:

```
In[20]:= InverseReedSolomon[code, 31]
Out[20]= MATHEMATICS
```

However, if we incorporate a second error by

```
In[21]:= code = StringReplace[code, "C" → "K"]
Out[21]= O\MATHEMATIKA
```

then the error correction is not possible as expected:

```
In[22] := InverseReedSolomon[code,31]
```

Part :: **"partw"**: *Part* 20 *of* {15, 28, 13, 1, 20, 8, 5, 13, 1, 20, 9, 11, 1} *does not exist*

```
Out[22]= MATHEMATIKA
```

If the Reed-Solomon code is applied to 4-letter blocks, then these blocks are equipped with a redundancy of two additional letters. We say that such a code has the *information rate 4/6*. If a 6-tuple is transmitted error-free or with a single error, then we can correct this error and subsequently consider the text as error-free.

Assume that a given text has 1 million letters. If 2 out of 1 000 letters are false, then the text contains approximately 2 000 errors. Let us code the text with our Reed-Solomon code. Then the coded text has 1.5 million letters, and our letters are false with a probability $\frac{2}{1000} = 0.2\%$. Therefore a 6-letter block has *more than one error*[11] with probability 0.00006:

```
In[23] := w = ∑(k=2 to 6) Binomial[6,k] 0.002^k 0.998^(6-k)
```
$$In[23] := w = \sum_{k=2}^{6} \text{Binomial}[6,k]\, 0.002^k\, 0.998^{6-k}$$

```
Out[23]= 0.0000596807
```

Hence, we must expect in our text only approximately

```
In[24] := w*1.5*10^6
```

```
Out[24]= 89.5211
```

errors. This is already much better.

However, this rate can still be improved. For this purpose, we may develop a Reed-Solomon code whose blocks $a_4 a_5 \ldots a_{11}$ of lengths 8 are equipped with 4 redundant letters $a_0 a_1 a_2 a_3$ satisfying the relations

$$
\begin{aligned}
a_0 + a_1 + a_2 + \cdots + a_{11} &\equiv 0 \,(\text{mod }31) \\
a_1 + 2a_2 + \cdots + 11 a_{11} &\equiv 0 \,(\text{mod }31) \\
a_1 + 4a_2 + \cdots + 11^2 a_{11} &\equiv 0 \,(\text{mod }31) \\
a_1 + 8a_2 + \cdots + 11^3 a_{11} &\equiv 0 \,(\text{mod }31).
\end{aligned}
$$

This code has an information rate of 2/3 again. However, we will indicate how this code is suitable to correct two errors. Assume that after transmission we have

$$
\begin{aligned}
a_0 + a_1 + a_2 + \cdots + a_{11} &\equiv e \,(\text{mod }31) \\
a_1 + 2a_2 + \cdots + 11 a_{11} &\equiv s_1 \,(\text{mod }31) \\
a_1 + 4a_2 + \cdots + 11^2 a_{11} &\equiv s_2 \,(\text{mod }31) \\
a_1 + 8a_2 + \cdots + 11^3 a_{11} &\equiv s_3 \,(\text{mod }31).
\end{aligned}
$$

Assume further that not more than 2 errors occurred, with differences e_1 and e_2 at the positions i and j. This leads to the equations

$$
\begin{aligned}
e_1 + e_2 &\equiv e \,(\text{mod }31) \\
i e_1 + j e_2 &\equiv s_1 \,(\text{mod }31) \\
i^2 e_1 + j^2 e_2 &\equiv s_2 \,(\text{mod }31) \\
i^3 e_1 + j^3 e_2 &\equiv s_3 \,(\text{mod }31).
\end{aligned}
$$

---

[11] That is, 2, 3, 4, 5 or 6 errors. Note that we compute this probability with the binomial distribution.

Now we can eliminate e_1 and e_2 which leads to quadratic equations for i and j. After solving these equations for i and j one can determine e_1 and e_2.

The probability that a word of length 12 contains more than two errors is given by

$$In[25]:= \ w = \sum_{k=3}^{12} \text{Binomial}[12,k]0.002^k \ 0.998^{12-k}$$
$$Out[25]= \ 1.73639 \times 10^{-6}$$

therefore we can expect only

$$In[26]:= \ w*1.5*10^6$$
$$Out[26]= \ 2.60459$$

errors in our document. This is remarkably small if we keep in mind that our original document contained 2,000 errors before correction!

A music CD is rarely error-free. Actually it can contain hundreds of thousands of errors. In order to make sure that you can hear your music in reasonable quality, an error-correcting code with information rate 3/4 is used. Since errors on a CD do not occur arbitrarily, but typically in clusters (for example, if the CD has a scratch), a special Cross-Interleaved Reed-Solomon-Code (CIRC) in the field $GF(2^8)$ is applied.[12]

5.5 Asymmetric Ciphers

Whereas in the framework of coding theory we tried to detect or correct transmission errors, we now assume that error-free transmission is secured. The aim of this section is to introduce modern methods to encrypt messages, called *ciphers*.

Let us assume again that the sender sends his message to the recipient via a certain transmission channel. Since an unauthorized person might monitor the transmission, the message is *encrypted* (*ciphered*), and must be *decrypted* (*deciphered*) by the recipient.

$$\text{message/plaintext} \xrightarrow{\text{encryption}} \text{ciphertext/cryptogram}.$$

Cryptography is the science of ciphers, whereas *cryptanalytics* is the science of the successful breaking of ciphers. *Cryptology* combines the two disciplines.

Important applications of modern ciphers are

- the transmission of phone calls via satellite;
- pay-TV;
- internet banking;
- e-commerce;
- password techniques, for example for computer systems with multi-user operation;
- tap-proof remote login via data lines (secure shell);
- secure e-mail (see e.g. [PGP]).

---

[12] The Galois fields $GF(q)$ will be introduced in Section 7.4 on page 167.

Example 5.12. (**Caesar Cipher**) One method to encrypt a message is the Caesar cipher, compare Example 5.8. This cipher shifts every character by 3 positions in the alphabet. In this case the *encryption method* is Caesar and the *encryption key* of the method is the number 3. Encryption is very simple, and if you know the key, then decryption is also easy: One just uses the same method, but instead the key -3 for shifting back.

However, a cryptanalyst can decrypt every such message also *without prior knowledge of the key*. After all, if the underlying alphabet has m letters, then the Caesar method possesses just m different keys. And since a human alphabet has a small number of letters m, the cryptanalyst can try *all* keys and select the decryption which is a readable text. $\triangle$

Session 5.13. *The Mathematica function*

```
In[1]:= CaesarV[" ",n_] := " "

        CaesarV[x_String,n_] :=
          FromCharacterCode[
             Mod[ToCharacterCode[x]+n-65,26]+65]/;
           Length[ToCharacterCode[x]] === 1

        CaesarV[x_String,n_] :=
          Module[{length,first,rest},
             length = StringLength[x];
             first = StringTake[x,1];
             rest = StringDrop[x,1];
             StringJoin[CaesarV[first,n],
               CaesarV[rest,n]]]/;
           Length[ToCharacterCode[x]] > 1

        CaesarE[x_String,n_] := CaesarV[x,-n]
```

implements the Caesar code over the alphabet $\{A, B, \ldots, Z\}$ recursively, where by the first line the blank is encrypted as blank. Now we would like to break the method. For this purpose, let us assume that we received the cryptogram

```
In[2]:= test = "IEXTLX WXVBIAXK"
Out[2]= IEXTLX WXVBIAXK
```

Then we can generate the list of all possible plaintexts[13]

```
In[3]:= Do[Print[CaesarE[test,n]],{n,1,26}]
HDWSKW VWUAHZWJ
GCVRJV UVTZGYVI
FBUQIU TUSYFXUH
EATPHT STRXEWTG
DZSOGS RSQWDVSF
CYRNFR QRPVCURE
BXQMEQ PQOUBTQD
AWPLDP OPNTASPC
ZVOKCO NOMSZROB
YUNJBN MNLRYQNA
```

---

[13] Notice that the print command `Print` generates an output on the screen, but has the result `Null` just like `Do`.

```
XTMIAM LMKQXPMZ
WSLHZL KLJPWOLY
VRKGYK JKIOVNKX
UQJFXJ IJHNUMJW
TPIEWI HIGMTLIV
SOHDVH GHFLSKHU
RNGCUG FGEKRJGT
QMFBTF EFDJQIFS
PLEASE DECIPHER
OKDZRD CDBHOGDQ
NJCYQC BCAGNFCP
MIBXPB ABZFMEBO
LHAWOA ZAYELDAN
KGZVNZ YZXDKCZM
JFYUMY XYWCJBYL
IEXTLX WXVBIAXK
```

A short check of the resulting list reveals the correct key:

$In[4]:=$ **CaesarE[test,19]**
$Out[4]=$ PLEASE DECIPHER

Obviously this method does not possess enough keys.

We consider now the general case of a cryptosystem. Let N denote the message to be encrypted, E the encryption method and s the key used. Hence encryption is the mapping $C = E_s(N)$, and C denotes the cryptogram. The recipient needs the decryption method D and a key t. Using these, he can decrypt the cryptogram by $N = D_t(C)$. Altogether, the equation

$$D_t(E_s(N)) = N$$

must hold for *every message N*. Hence mathematically the function D_t is the inverse function of E_s, i.e., $D_t = E_s^{-1}$. Of course the inverse function is in principle uniquely determined by E_s. For many ciphers this holds not only in principle, but it is moreover well possible to compute D_t from E_s. In the Caesar cipher E_n (shift by n letters), for example, clearly $D_n = E_{-n}$. Therefore, in this case, decryption uses the same method. Such cryptosystems are called *symmetric*. In the Caesar code, decryption just uses a different key, and this key t can easily be determined from the encryption key s.

As noted before, the Caesar cipher is not a secure symmetric encryption cipher. Nevertheless there are very good approved symmetric ciphers which, however, are much more complicated. An approved, completely published, symmetric cipher with $D_t = E_s$, which is therefore completely symmetric (including the key pair), is the DES cipher (Data Encryption Standard), see e.g. [BS2011]. The original DES cipher from 1977 with its key length of 56 bits is not considered secure any more, therefore it is often used as "Triple-DES". However, we do not want to go into details here.

Instead, in the sequel we want to discuss *asymmetric cryptosystems*. In general, from every good encryption scheme we expect that encryption E_s and decryption D_t can be evaluated fast. Hence the cipher should be *efficient*. Furthermore it must be difficult to break (for *safety* reasons). An asymmetric cipher must have additional properties:

1.   The encryption method E and decryption method D are known to all potential participants.
2.   All participants have their own encryption-decryption key pair (s, t). The recipient's encryption key s is made public (for example in the internet or in a public key book) and can be used by every potential sender. However, the recipient's decryption key t is kept secret. Accordingly, the key s is called *public key* and t is called *private key*. This manifests the asymmetry of the method.
3.   The knowledge of the encryption method E_s does not allow to compute the decryption function D_t with justifiable effort. In particular, one cannot compute t with justifiable effort from s.[14]

Because of (2) such asymmetric ciphers are also called *public key ciphers*. Since the encryption key is public, the security condition (3) is unavoidable.

One of the main problems of conventional ciphers is the *handover of keys*. After all, if an intruder acquired the secret key on this occasion, every encryption would be completely worthless!

In a public-key cryptosystem this is completely different: Since the encryption keys are public anyway, a handover of keys is not necessary! Now, if Anna wants to send a message to Barbara, she encrypts her message with Barbara's public key s_B and sends the cryptogram to Barbara. Barbara uses her secret private key t_B to decrypt the cryptogram. As soon as the message is encrypted it can only be decrypted using Barbara's private key. Even Anna herself would not be able to decrypt her cryptogram.

If a public-key cryptosystem has the further property that $D = E$, then it can also be used for *user authentification*: Anna wants to send Barbara a message so that Barbara can verify that the sender was Anna. For this purpose, she generates a *signature* $(U = E_{t_A}(n))$ using her private key t_A which she sends to Barbara. Then Barbara can verify with Anna's public key s_A that the message came from her (by checking if $D_{s_A}(U) = n$). If this *digital signature procedure* is executed smartly by linking the signature with the message in a suitable way, it is even possible for the recipient to know that the message arrived in an unaltered state (*message integrity*).

Before we introduce the RSA cryptosystem [RSA1978], which is the most popular asymmetric cipher, we consider the *Diffie-Hellman key exchange* [DH1976] which enables two users to agree upon a common key without having to send it via a potentially insecure channel. Using this method, one can exchange a common key without running the risk that the common key gets stolen. The common key in the Diffie-Hellman system is a positive integer.

Example 5.14. (**Diffie-Hellman Key Exchange**) Anna and Barbara agree upon a positive integer $g \in \mathbb{N}$ and a prime number $p \in \mathbb{P}$, both of which are not assumed to be kept secret and can be made public in principal. Then Anna chooses a random integer $a < p$ and computes $\alpha := \mathrm{mod}(g^a, p)$. Similarly Barbara chooses a random integer $b < p$ and computes $\beta := \mathrm{mod}(g^b, p)$. Next, they exchange α and β in full awareness that an intruder could receive these informations. This fact should not influence the security of the system. Then Anna computes $s = \mathrm{mod}(\beta^a, p)$, and Barbara computes $t = \mathrm{mod}(\alpha^b, p)$. Since

$$s \equiv \beta^a \equiv (g^b)^a \equiv g^{ab} \equiv (g^a)^b \equiv \alpha^b \equiv t \pmod{p}$$

it happens that $s = t$, hence both keys agree. This shows that the method is *feasible*.

---

[14] If we can certify that it takes 100 years to compute t from s, then we would be satisfied.

Next, we would like to consider why an intruder has no chance to compute the common key t. An attack could consist in wiretapping the two numbers α and β. Furthermore we can expect that an intruder knows g and p. On the other hand Anna and Barbara kept a and b secure.

However, in order to compute, for example, t from $t \equiv \alpha^b \pmod{p}$ the intruder should know b and must therefore solve the equation

$$\beta \equiv g^b \pmod{p}$$

for b. This, however, is not feasible: As we saw, the computation of the modular logarithm is not possible in reasonable time,[15] given that g and p are chosen large enough. But this will be taken care of by Anna and Barbara! $\triangle$

The Diffie-Hellman key exchange is successful because the modular exponential function is a *one-way function*: The computation of $y = g^x \pmod{p}$ is efficiently possible, whereas the (multivalued) inverse function $x = \log_g y \pmod{p}$ is not computable in reasonable time. One-way functions play an important role in asymmetric ciphers.

Example 5.15. (**RSA Cryptosystem**) The leading asymmetric cryptosystem is the RSA system which also uses the modular exponential function in a clever way. We assume without loss of generality that the message is given as a positive integer. Then this is the *cryptographic protocol* of the RSA system:

1. Barbara computes
   a. two random primes p and q with at least 100 decimal digits;
   b. $n = p \cdot q$;[16]
   c. $\varphi = (p-1) \cdot (q-1)$;
   d. a random positive integer $e < \varphi$ with $\gcd(e, \varphi) = 1$. The pair $s_B = (e, n)$ is Barbara's public key and gets published;
   e. $d = e^{-1} \pmod{\varphi}$. The number $t_B = d$ constitutes Barbara's private key.

2. Barbara deletes p, q and φ (for security reasons).
3. Anna encrypts her message $N \in \mathbb{Z}_n$ using the computation

$$C = V_{s_B}(N) = \mathrm{mod}(N^e, n) \in \mathbb{Z}_n \,.$$

If $N \geqq n$, then the message in partitioned in blocks, and the text is encrypted blockwise.
4. Barbara decrypts (if necessary blockwise) by

$$E_{t_B}(C) = \mathrm{mod}(C^d, n) \,.$$

---

[15] If we could certify that the current state of the art needs 100 years for this computation, we would feel safe.

[16] This is sufficient for our purposes. Note, however, that the world record team can factorize 200-digit integers! These days, such a factorization needs heavy computing power and a very long time, see [Wei2005b], but once this becomes a feasible task, the method gets insecure and one should enlarge the size of p and q. A 200-digit decimal number has a bit length of $200 \log_2 10 \approx 665$. Nowadays the RSA method is mostly applied with $2^{10} = 1024$ bits.

First of all we want to show that the system works correctly. For this purpose, we have to show that (in $\mathbb{Z}_n$)

$$E_{t_B}(V_{s_B}(N)) = N$$

is valid. However, this is a consequence of the Little Fermat theorem: Since $d \cdot e \equiv 1 \pmod{\varphi}$, we have $d \cdot e = 1 + k\varphi$ for some $k \in \mathbb{N}_{\geq 0}$, and hence we get

$$E_{t_B}(V_{s_B}(N)) \equiv (N^e)^d \equiv N^{de} \equiv N^{1+k\varphi} \equiv N^{1+k \cdot (p-1) \cdot (q-1)} \pmod{n}.$$

Next, we compute modulo p and show by induction that

$$N^{1+K(p-1)} \equiv N \pmod{p} \qquad (K \in \mathbb{N}_{\geq 0}).$$

For $K = 0$ this is the case, and Fermat's theorem proves the induction step:[17]

$$N^{1+(K+1)(p-1)} \equiv N^p \cdot N^{K(p-1)} \equiv N \cdot N^{K(p-1)} \equiv N^{1+K(p-1)} \equiv N \pmod{p}.$$

Similarly we get modulo q:

$$N^{1+K(q-1)} \equiv N \pmod{q} \qquad (K \in \mathbb{N}_{\geq 0}).$$

Consequently we get

$$N^{de} \equiv N^{1+k(p-1)(q-1)} \equiv N \pmod{pq}$$

and therefore (in $\mathbb{Z}_n$)

$$E_{t_B}(V_{s_B}(N)) = N.$$

Hence we have proved that the RSA system decrypts an encrypted message correctly. But why should the RSA system be secure? In step (2) Barbara deleted the numbers p, q and φ. This makes sense since the knowledge of any of these numbers enables the computation of the private key d, see Exercise 5.9.

Since n is made public, the RSA system can be broken if n can be factored into $n = p \cdot q$. While it is quickly possible to check whether a positive integer (with high probability) is prime or not, see Section 4.6, it is not possible with current methods to factor a 300-digit decimal number, see Section 3.5. This explains the security of the RSA system: Here the product $n = p \cdot q$ acts as a one-way function.

What about the efficiency of the RSA system? If the modular exponential function is computed with the divide-and-conquer algorithm presented in Section 4.4 on page 75 or a similar approach, then the occurring modular powers can be computed in an efficient way. $\triangle$

Session 5.16. *We define the following auxiliary functions:*

---

[17] Using $N^{p-1} = 1$ we can also give a direct proof. But then the case $p \mid N$ must be considered separately.

```
In[1]:= ListToNumber[list_] := First[list] /; Length[list] == 1
        ListToNumber[list_] :=
          1000 * ListToNumber[Reverse[Rest[Reverse[list]]]] +
            Last[list]

        TextToNumber[string_] :=
          ListToNumber[ToCharacterCode[string]]

        NumberToList[number_] := {number} /; number < 1000
        NumberToList[number_] :=
          Append[NumberToList[(number - Mod[number, 1000]) / 1000],
            Mod[number, 1000]]

        NumberToText[number_] :=
          FromCharacterCode[NumberToList[number]]
In[2]:= Encipher[message_] :=
          PowerMod[TextToNumber[message], e, n]

        Decipher[number_] :=
          NumberToText[PowerMod[number, d, n]]
```

Using NextPrime, *we can initialize the RSA system as follows:*

```
In[3]:= InitializeRSA := Module[{},
          p = NextPrime[Random[Integer, {10^100, 10^101}]];
          q = NextPrime[Random[Integer, {10^100, 10^101}]];
          n = p * q;
          φ = (p - 1) * (q - 1);
          e = φ;
          While[Not[GCD[φ, e] === 1],
            e = NextPrime[
                Random[Integer, {10^50, 10^51}]]];
          d = PowerMod[e, -1, φ];
          ]
```

We call InitializeRSA

```
In[4]:= InitializeRSA
```

and get the following public key:

```
In[5]:= {e, n}
Out[5]= {40998536436223191577023612301326813733991005886029,
         17703595610997329310827973313390724504652355647464577868359302378840109843916609080910569408207600462317202347157855440179204731066636509323991051983269464484494147997662912857370089136440855363136589 43}
```

The readers are now invited to compute the corresponding private key d! With the current state of knowledge this should be only possible for specialized research teams.

Now, let us define a message:

```
In[6]:= message = "This is my message"
Out[6]= This is my message
```

We encrypt the message using the above public key:

In[7]:= **result = Encipher[message]**

Out[7]= 24160272612134484489820088587582508709681010955869795833220655256
  95732981330523354379632413521457746614738914847705292200155149
  43891477261578028469551530686919008306393183097792351183945237
  3003654137

The message can be easily reconstructed with the help of our private key:

In[8]:= **Decipher[result]**

Out[8]= This is my message

It is clear that this works only if no transmission error occurred. The smallest error disables the reconstruction:[18]

In[9]:= **Decipher[result + 1]**

One last example:

In[10]:= **result = Encipher["What about this message?"]**

Out[10]= 9425449873527351303001047817793880907458301440951061115999004653
  38949046408437323900198199466310977742638786293206508160284998
  2306373556390587075515381380663224939947766047167313859718525
  2390908968

In[11]:= **Decipher[result]**

Out[11]= What about this message?

Note that our RSA program only works if $N < 10^{200}$. For longer messages, we must partition the message in blocks of this size.

To conclude this chapter, let us mention that until now it is unproven whether or not one-way functions exist. This is the case if and only if $P \neq NP$ (see e.g. [BCS1997], [BDG1988]), a celebrated conjecture from complexity theory which is one of the 7 millennium problems for whose solution the Clay Mathematics Institute has offered a prize money of 1 million US-$, see www.claymath.org/millennium-problems.

5.6 Additional Remarks

For coding theory, see e.g. the book [Roth2006]. The presentation of the Reed-Solomon code in Section 5.4 was adopted from [Lint2000].

For cryptography, see e.g. [Lew2000]. Further asymmetric systems are treated for example in [Sal1990]. Crytosystems based on elliptic curves seem to be the most promising candidates.

---

[18] Since the result is typically not a representable message, we omitted the output.

5.7 Exercises

Exercise 5.1. (Bit Length) *Show: For a word of length n over an alphabet $\mathcal{A}$ with m letters, one needs $n \cdot \log_2 m$ bits. What is the correct rounding?*

Exercise 5.2. (Prefix Codes) *Prove constructively that prefix codes can be decoded in a unique way.*

Exercise 5.3. (Prefix Codes) *Implement the coding and decoding of the prefix code C :*
$$\bigcup_{n \in \mathbb{N}} \{A, B, C\}^n \to \bigcup_{n \in \mathbb{N}} \{0, 1\} \text{ given by}$$

$$A \mapsto 0, \quad B \mapsto 10, \quad C \mapsto 11.$$

Code and decode `"ABCAACCBAB"` *as well as a randomly generated word of length 1000. Check your results.*

Exercise 5.4. *Show that the introduction of the pause sign | in the Morse alphabet $\{\bullet, -\}$[19] makes the Morse code a prefix code. Decode the following text:*

$-|\bullet \bullet \bullet \bullet|\bullet \bullet|\bullet \bullet \bullet \bullet|--|\bullet|\bullet \bullet \bullet|\bullet \bullet \bullet|\bullet \, -|--\bullet|\bullet|\bullet \bullet|\bullet \bullet \bullet|\bullet \bullet|--|\bullet \, --\bullet|---|\bullet \, -\bullet|$

$-|\bullet \, -|-\bullet|-|\bullet \, --\bullet|\bullet \, -\bullet \bullet|\bullet|\bullet \, -|\bullet \bullet \bullet|\bullet|-\bullet \bullet|\bullet|-\bullet \, -\bullet|---|-\bullet \bullet|\bullet|\bullet \bullet|-|$

Exercise 5.5. (ISBN-10 Check Digit)

(a) *Implement a procedure* `ISBNCheckDigit[n]` *which computes the check digit, hence the last digit, from the first 9 digits of n.*

(b) *Compute the check digits of my books*
   - (i) *Mathematik mit DERIVE: ISBN-10 starts with 3-528-06549;*
   - (ii) *Höhere Analysis mit DERIVE: ISBN-10 starts with 3-528-06594;*
   - (iii) *DERIVE für den Mathematikunterricht: ISBN-10 starts with 3-528-06752;*
   - (iv) *Hypergeometric Summation: ISBN-10 starts with 3-528-06950;*
   - (v) *Die reellen Zahlen als Fundament und Baustein der Analysis: ISBN-10 starts with 3-486-24455.*

   Do the same check for three arbitrary other books.

(c) *Implement a procedure* `CheckISBNDigit[n]`, *which checks whether a complete ISBN-10 number n is correct. Note that a complete ISBN-10 number is either a 10-digit integer or a 9-digit integer followed by the letter X.*

(d) *Check your implementation using the above examples and 10 other random 10-digit integers.*

Exercise 5.6. (EAN Check Digit) *Implement a procedure* `EANCheckDigit[n]` *which computes the check digit from the first 12 digits of n. Compute the check digit of 402570000102.*

Exercise 5.7. (Huffman Code)

(a) *Let the following frequency distribution be given:*

---

[19] The Morse alphabet is therefore the set $\{\bullet, -, |\}$.

$$\begin{pmatrix} A & 0.4 \\ B & 0.12 \\ C & 0.2 \\ D & 0.18 \\ E & 0.1 \end{pmatrix}$$

Generate the corresponding Huffman code.

(b) Build the set of pairs of the distribution given in (a) and apply the Huffman algorithm. Explain the result.

(c) Compute the Huffman code for the frequency distribution

$$\begin{pmatrix} A & 0.08 \\ B & 0.13 \\ C & 0.08 \\ D & 0.06 \\ E & 0.1 \\ F & 0.06 \\ G & 0.06 \\ H & 0.05 \\ I & 0.12 \\ K & 0.03 \\ L & 0.01 \\ M & 0.02 \\ N & 0.06 \\ O & 0.03 \\ P & 0.03 \\ R & 0.01 \\ S & 0.02 \\ T & 0.04 \\ U & 0.01 \end{pmatrix}.$$

(d) Use this code to encode and decode the text HUFFMAN IS ALWAYS RIGHT.[20]

Exercise 5.8. (Reed-Solomon Code) *Check the functions* ReedSolomon *and* Inverse-ReedSolomon *with 10 examples. Generate your examples by introducing a random error at a random place.*

Exercise 5.9. (RSA Cryptosystem) *Show that the knowledge of one of the numbers p, q or φ enables an easy computation of the private key d.*

Exercise 5.10. (RSA Cryptosystem According To [Wie1999]–[Wie2000]**)**

(a) *Unfortunately the RSA cryptosystem has fixed points, i.e., there are texts whose cryptogram agrees with the original. However, the probability for this event is very small if n is large enough. Show that the unsuitable choice $e = 1 + \mathrm{lcm}(p-1, q-1)$ yields a fixed point for every encoding.*

(b) *Show that the RSA method remains correct if we define the modulus φ by $\varphi = \mathrm{lcm}(p-1, q-1)$. Reason why this is a better choice than the one chosen in the text.*

(c) *Show that the defect given in (a) also occurs if we define $\varphi = \mathrm{lcm}(p-1, q-1)$ as in (b). Why is this not as fatal in this case?*

---

[20] Ignore the blank spaces.

Chapter 6
Polynomial Arithmetic

6.1 Polynomial Rings

In the sequel (if not stated otherwise), we assume that R is a zero-divisor-free ring with unity 1. Such a ring is called an *integral domain*.

Example 6.1. (a) Obviously every field is an integral domain, see Exercise 4.2. In particular, $\mathbb{Q}$, $\mathbb{R}$, $\mathbb{C}$ and $\mathbb{Z}_p$, for $p \in \mathbb{P}$, are integral domains.
(b) The ring of integers $\mathbb{Z}$ is an integral domain.
(c) The set

$$\mathbb{Z}[i] := \{a + ib \mid a, b \in \mathbb{Z}\} \subset \mathbb{C}$$

of *Gaussian integers* also constitutes an integral domain. This ring has no zero divisors since from $(a, b, c, d \in \mathbb{Z})$

$$0 = (a + ib)(c + id) = (ac - bd) + i(ad + bc),$$

the two equations

$$ac - bd = 0 \qquad \text{and} \qquad ad + bc = 0 \tag{6.1}$$

follow. If one multiplies the first equation with c, and the second with d, then addition leads to

$$a(c^2 + d^2) = 0,$$

so that $a = 0$ or $c = d = 0$, hence $c + id = 0$. In the first case ($a = 0$, $c + id \neq 0$), it follows by substitution in (6.1) that also $b = 0$, hence $a + ib = 0$. Therefore there are no zero divisors.

Note that the proved result also follows from the fact that $\mathbb{Z}[i] \subset \mathbb{C}$ is a subring of the field $\mathbb{C}$ and therefore cannot possess zero divisors. △

Session 6.2. *Adapting the proof given in Section 3.5, one can show that every element in $\mathbb{Z}[i]$ possesses (up to units) a unique factorization. The units in $\mathbb{Z}[i]$ are given as $\{1, -1, i, -i\}$. However, a prime $p \in \mathbb{Z} \subset \mathbb{Z}[i]$ is not necessarily prime as element of $\mathbb{Z}[i]$:*

```
In[1]:= FactorInteger[5, GaussianIntegers → True]
```

$$Out[1] = \begin{pmatrix} -i & 1 \\ 1 + 2i & 1 \\ 2 + i & 1 \end{pmatrix}$$

If we multiply by suitable units, this factorization can be written as $5 = (1 + 2i)(1 - 2i)$.

In particular, we obtain

© Springer Nature Switzerland AG 2021
W. Koepf, *Computer Algebra*, Springer Undergraduate Texts in Mathematics and Technology,
https://doi.org/10.1007/978-3-030-78017-3_6

In[2]:= **PrimeQ[5,GaussianIntegers → True]**
Out[2]= False

The divisors of 5 in $\mathbb{Z}[i]$, which are uniquely determined up to units, are obtained by

In[3]:= **Divisors[5,GaussianIntegers → True]**
Out[3]= {1, 1 + 2*i*, 2 + *i*, 5}

We now consider *polynomials* over an integral domain R. The set of polynomials of the form

$$a(x) = a_n x^n + a_{n-1} x^{n-1} + \cdots + a_1 x + a_0 = \sum_{k=0}^{n} a_k x^k \tag{6.2}$$

with *coefficients* $a_k \in R$ $(k = 0, \ldots, n)$ is denoted by $R[x]$.[1]

It is easy to see that $R[x]$ with the usual addition and multiplication[2] is a ring. The neutral element with respect to addition is the *zero polynomial* $a(x) = 0$, for which $n = 0$ and $a_0 = 0$, and the neutral element with respect to multiplication is the *unity* $a(x) = 1$ with $n = 0$ and $a_0 = 1$. The ring $R[x]$ is called a *polynomial ring* over R, with respect to the variable x. If R is an integral domain, then so is $R[x]$, as we will see soon.

Let us first introduce some notation. If $a_n \neq 0$ in (6.2), we say that $a(x)$ has degree n and we write $\deg(a(x), x) = n$. The degree of the zero polynomial is set to $\deg(0, x) = -\infty$. If $a_n \neq 0$, then (6.2) is called the *standard representation* of the polynomial $a(x)$ and a_n is called the *leading coefficient* of $a(x)$, denoted by $\mathrm{lcoeff}(a(x), x)$. The product $a_n x^n$ is called *leading term* of $a(x)$ and denoted by $\mathrm{lterm}(a(x), x)$. By definition, two polynomials are equal if the coefficients of the corresponding powers of x in the polynomials are equal.

If $\mathrm{lcoeff}(a(x), x) = 1$, then $a(x)$ is called *monic*. If the degree of a polynomial $\deg(a(x), x) \leq 0$, then we speak about a *constant polynomial*. Both the zero polynomial and the unity are constant polynomials. Note that the constant polynomials form an isomorphic copy of R and therefore a subring of $R[x]$, i.e., $R \subset R[x]$.

A polynomial in standard representation is called *expanded*. The coefficients for the expanded product polynomial are given by the *Cauchy product* formula

$$\sum_{k=0}^{n} a_k x^k \cdot \sum_{k=0}^{m} b_k x^k = \sum_{k=0}^{m+n} c_k x^k, \tag{6.3}$$

with

$$c_k = \sum_{j=0}^{k} a_j b_{k-j}. \tag{6.4}$$

Here, for $n < m$ and $j = n+1, \ldots, m$, the coefficients are $a_j = 0$, and correspondingly $b_j = 0$ for $n > m$ and $j = m+1, \ldots, n$.

From the definitions of addition and multiplication, we get

$$\deg(a(x) + b(x), x) \leq \max(\deg(a(x), x), \deg(b(x), x)), \tag{6.5}$$

where equality holds (at least) for $\deg(a(x), x) \neq \deg(b(x), x)$, and

---

[1] The *square brackets* are essential in this notation!

[2] The resulting product polynomial can be brought in the form (6.2) by expanding (using the distributive law) and sorting (commutativity).

$$\deg(a(x) \cdot b(x), x) = \deg(a(x), x) + \deg(b(x), x) \tag{6.6}$$

with the convention $-\infty + n = -\infty$. To obtain the latter identity, we have used the following theorem.

Theorem 6.3. (Polynomial Ring) The polynomial ring $R[x]$ over an integral domain R is an integral domain.

Proof. We need to prove that $R[x]$ does not have zero divisors. For this purpose, let $a(x)$ have degree n and $b(x)$ degree m, and assume that the product $a(x) \cdot b(x) = 0$ is the zero polynomial. Since $a(x) \cdot b(x)$ is the zero polynomial, its leading term $a_n b_m x^{m+n}$ must also be zero, hence $a_n b_m = 0$. Since R is an integral domain, it follows that $a_n = 0$ or $b_m = 0$. In the first case, $a(x)$ must be the zero polynomial, since otherwise its leading coefficient could not be zero. In the latter case, $b(x)$ is the zero polynomial for the very same reason.   □

Please note that if R has zero divisors, then so has $R[x]$, and the important formula (6.6) is not valid. For example, the polynomials $a(x) = 2z$ and $b(x) = 3z^2$, considered as elements of $\mathbb{Z}_6[x]$, yield the product $a(x) \cdot b(x) = 2 \cdot 3 z^3 = 0$.

At this point, we would like to emphasize the fact that the elements of a polynomial ring are algebraic objects. Instead of writing $a(x)$ as in (6.2), it is therefore sufficient to define the polynomial by its coefficient list $a := (a_0, a_1, ..., a_n)$. Then addition and multiplication are certain operations with these lists. In particular, x is just a symbol for the representation of one particular polynomial representing the coefficient list $(0, 1)$ and should not be interpreted as an element of the integral domain R. Since the symbol x satisfies no algebraic relation with the elements in R, it is called a *transcendental element* over R, and $R[x]$ is a *transcendental ring extension* of R by *adjunction* of the symbol x.

Session 6.4. *We define a polynomial in Mathematica:*

$In[1]:= \mathbf{a} = \sum_{k=0}^{10} \mathbf{k\,x^k}$

$Out[1]= 10\,x^{10} + 9\,x^9 + 8\,x^8 + 7\,x^7 + 6\,x^6 + 5\,x^5 + 4\,x^4 + 3\,x^3 + 2\,x^2 + x$

The coefficient list is given by

$In[2]:=$ **CoefficientList[a,x]**

$Out[2]= \{0, 1, 2, 3, 4, 5, 6, 7, 8, 9, 10\}$

The computation

$In[3]:=$ **Coefficient[a,x,5]**

$Out[3]= 5$

yields the coefficient of x^5.

Although x is only considered as a symbol and not an element of R, we are entitled to substitute x by an element of R in the representation (6.2) of $a(x)$. For every $\alpha \in R$ and $a(x) \in R[x]$, the value $a(\alpha)$ is again an element of R. This element $a(\alpha) \in R$ is called the *value* of $a(x) \in R[x]$ at the point $\alpha \in R$. A *zero* of $a(x) \in R[x]$ is an element $\alpha \in R$ for which $a(\alpha) = 0$.

However, we must distinguish very clearly between the value $0 \in R$ and the zero polynomial $0 \in R[x]$. In this context, let us consider the following example.

Example 6.5. Let $R = \mathbb{Z}_2$. Obviously the polynomial $a(x) = x(x-1) = x^2 - x \in R[x]$ has a zero at both $x = 0$ and $x = 1$. Hence $a(\alpha) = 0$ for all $\alpha \in R$. Nevertheless $a(x)$ is not the zero polynomial.

Next, let $p \in \mathbb{P}$ and $R = \mathbb{Z}_p$. Fermat's Little Theorem shows that the polynomial $a(x) = x^p - x \in R[x]$ vanishes identically in R: Every $\alpha \in \mathbb{Z}_p$ is a zero of $a(x)$, and again $a(x)$ is not the zero polynomial.

In Theorem 6.26 we will see that this type of phenomenon is characteristic for finite rings R. $\triangle$

Our next step is to characterize the units in $R[x]$.

Lemma 6.6. (Units in Polynomial Rings) Let R be an integral domain. Then an element $u(x) \in R[x]$ is a unit in $R[x]$ if and only if it is constant and a unit in R.

Proof. Of course every unit $u \in R$, considered as a constant element of $R[x]$, is also a unit in the polynomial ring since, by assumption, there is an inverse v with $u \cdot v = 1$. This inverse is again a constant polynomial $v \in R[x]$.

Now let $u(x) \in R[x]$ be a unit of the polynomial ring. Then there is a $v(x) \in R[x]$ with $u(x) \cdot v(x) = 1$. When applying the degree formula (6.6) to this identity, we get

$$\deg(u(x), x) + \deg(v(x), x) = 0,$$

and since the degree of a polynomial is nonnegative,[3] it follows that both $\deg(u(x), x) = 0$ and $\deg(v(x), x) = 0$, hence $u(x)$ and $v(x)$ are constant polynomials. By our assumption, they are units in R. $\square$

Until now, we have considered polynomial rings in a single variable, but this concept can be easily extended to several variables. Depending on how extensively we use the distributive law, we get a *recursive* or a *distributive* representation.

By $R[x, y]$ we denote the ring of polynomials

$$a(x, y) = \sum_{j=0}^{n} \sum_{k=0}^{m} a_{j,k} x^j y^k \tag{6.7}$$

in two variables x and y with coefficients $a_{j,k} \in R$. We call (6.7) the distributive (completely expanded) representation of a polynomial in two variables. This representation results from an application of the distributive law to the recursive representation of a polynomial $a(x, y) \in R[x][y]$,

$$a(x, y) = \sum_{k=0}^{m} \left(\sum_{j=0}^{n} a_{j,k} x^j \right) y^k,$$

or of a polynomial $a(x, y) \in R[y][x]$

$$a(x, y) = \sum_{j=0}^{n} \left(\sum_{k=0}^{m} a_{j,k} y^k \right) x^j,$$

respectively. In this way we identify $R[x, y]$ with $R[x][y]$ as well as with $R[y][x]$.

By this method, we can iteratively generate polynomial rings with an arbitrary, but finite, number of variables.

---

[3] The degree formula does not allow that $u(x)$ or $v(x)$ is the zero polynomial.

Example 6.7. $\mathbb{Z}[x]$ is the ring of polynomials in x with integer coefficients; $\mathbb{Q}[x, y, z]$ is the ring of polynomials in the three variables x, y and z with rational coefficients; $\mathbb{Z}_p[z]$ is the polynomial ring in the variable z over the field $\mathbb{Z}_p$; finally $\mathbb{C}[x_1, x_2, \ldots, x_n]$ is the polynomial ring in the variables $x_1, x_2, \ldots, x_n$ over the field $\mathbb{C}$. △

Session 6.8. *We define the polynomial $a(x, y)$ in the two variables x and y as follows:*

```
In[1]:= a = (x+y+1)^5
```
$Out[1] = (x+y+1)^5$

Using `Expand`*, polynomials in several variables are completely expanded. Therefore we get*[4]

```
In[2]:= Expand[a]
```
$Out[2] = 10\,x^3\,y^2 + 10\,x^2\,y^3 + 30\,x^2\,y^2 + 5\,x^4\,y + 20\,x^3\,y + 30\,x^2\,y +$
$\qquad x^5 + 5\,x^4 + 10\,x^3 + 10\,x^2 + 5\,x\,y^4 + 20\,x\,y^3 + 30\,x\,y^2 +$
$\qquad 20\,x\,y + 5\,x + y^5 + 5\,y^4 + 10\,y^3 + 10\,y^2 + 5\,y + 1$

as the distributive form of $a(x, y) \in \mathbb{Z}[x, y]$. The recursive form can be obtained using `Collect`*. The call*

```
In[3]:= Collect[a,x]
```
$Out[3] = x^3\,(10\,y^2 + 20\,y + 10) + x^2\,(10\,y^3 + 30\,y^2 + 30\,y + 10) + x^4\,(5\,y + 5) +$
$\qquad x^5 + x\,(5\,y^4 + 20\,y^3 + 30\,y^2 + 20\,y + 5) + y^5 + 5\,y^4 + 10\,y^3 + 10\,y^2 + 5\,y + 1$

yields the recursive representation of $a(x, y)$ as element of $\mathbb{Q}[y][x]$.

It is also possible to consider polynomials over the fields $\mathbb{Z}_p$ ($p \in \mathbb{P}$). For example, with

```
In[4]:= Expand[a,Modulus → 5]
```
$Out[4] = x^5 + y^5 + 1$

we get the distributive standard representation of $a(x, y)$ over the field $\mathbb{Z}_5$.

6.2 Multiplication: The Karatsuba Algorithm

All addition and multiplication algorithms for integers which were considered in Chapter 3 can be directly translated into the polynomial case. Both the high school multiplication algorithm and Karatsuba's algorithm are simpler in the polynomial case, since carries are not necessary: Whereas in the transition from one power B^n of the basis B to the next power B^{n+1} carries are used, in the respective calculations for polynomials the results are never transferred from one power x^n to the next x^{n+1}. This simplifies the implementation, but also has a severe drawback: The coefficients of sum and product polynomials can get—in contrast to the bounded ciphers of integer representations—arbitrarily large. In the product case where the coefficients can be computed using the Cauchy product formula (6.4), this is particularly harmful. For example, although the polynomial $1 + x$ has very small coefficients, the largest coefficient of the multiple product $(1 + x)^{2n}$ is given by the binomial formula as $\binom{2n}{n}$. Using *Stirling's formula*[5]

$$n! \sim \left(\frac{n}{e}\right)^n \sqrt{2\pi n}$$

---

[4] The given order is produced using the output style $\boxed{\text{TraditionalForm}}$; in $\boxed{\text{InputForm}}$ and $\boxed{\text{FullForm}}$ the polynomials are sorted in ascending order. However, the order also depends on the *Mathematica* version used.

[5] The notation $L(n) \sim R(n)$ means that both sides are *asymptotically equivalent* for $n \to \infty$, i.e., $\lim\limits_{n \to \infty} \frac{L(n)}{R(n)} = 1$.

we see that $\binom{2n}{n}$ grows exponentially for large n, such as $O(4^n/\sqrt{n})$:

$$\binom{2n}{n} = \frac{(2n)!}{n!^2} \sim \frac{(2n)^{2n}}{e^{2n}} \cdot \frac{e^{2n}}{n^{2n}} \cdot \frac{\sqrt{4\pi n}}{2\pi n} \sim \frac{4^n}{\sqrt{\pi n}} \, .$$

Mathematica confirms this finding:

$$In[1]:= \text{Limit}\left[\frac{\text{Binomial}[2n, n]\sqrt{n}}{4^n}, n \to \infty\right]$$

$$Out[1]= \frac{1}{\sqrt{\pi}}$$

Whereas all coefficients of an integer are elements of the finite set $\{0, \ldots, B-1\}$, the coefficients of a polynomial over $R = \mathbb{Z}$ are arbitrary integers, and therefore not bounded. However, this is different in $\mathbb{Z}_p[x]$, where all coefficients are bounded by p.

The following theorem states the complexity results for addition and multiplication of polynomials.

Theorem 6.9. (Complexity of Basic Arithmetic Operations) Let $a(x), b(x)$ be two elements of the polynomial ring $R[x]$, and let $n = \deg(a(x), x)$ and $m = \deg(b(x), x)$ denote their respective degrees. Then addition of $a(x)$ and $b(x)$ needs $O(m+n)$ *ring operations*, i.e. arithmetic operations in R, and high school multiplication needs $O(m \cdot n)$ ring operations. In particular, if $m = n$, then addition and high school multiplication have a complexity of $O(n)$ and $O(n^2)$, respectively.

On the other hand, Karatsuba's algorithm for the multiplication of two polynomials of degree $\leq n$ needs only $O(n^{\log_2 3})$ ring operations.

Proof. The proofs of Chapter 3 can be directly adapted.   □

One should always be aware that—as mentioned above—for a ring R with infinitely many elements, not all ring operations are equally costly, but the computing time depends on the size of the ring elements.

Session 6.10. *We implement the polynomial Karatsuba algorithm. The function* List-Karatsuba *uses the coefficient lists of the input polynomials, whereas* Polynomial-Karatsuba *works directly with the polynomials and applies* ListKaratsuba.

```
In[1]:= Clear[ListKaratsuba, PolynomialKaratsuba]

        PolynomialKaratsuba[pol1_, pol2_, x_] :=
          Module[{list1, list2, list3, l},
            list1 = Reverse[CoefficientList[pol1, x]];
            list2 = Reverse[CoefficientList[pol2, x]];
            list3 = ListKaratsuba[list1, list2];
            l = Length[list3];
            Sum[list3[[l - k]] * x^k, {k, 0, l - 1}]
          ]
```

```
In[2]:= ListKaratsuba[list1_, list2_] :=
            {list1[[1]] * list2[[1]]}/;
              Min[Length[list1], Length[list2]] == 1
        ListKaratsuba[list1_, list2_] :=
          ListKaratsuba[list1, list2] =
            Module[{a, b, c, d, n, m, list1new, list2new,
                atimesc, btimesd, middleterm, tab},
              n = Length[list1];
              m = Length[list2];
              m = Max[n, m];
              n = 1; While[n < m, n = 2 * n];
              list1new = PadLeft[list1, n];
              list2new = PadLeft[list2, n];
              a = Take[list1new, n/2];
              b = Take[list1new, -n/2];
              c = Take[list2new, n/2];
              d = Take[list2new, -n/2];
              atimesc = ListKaratsuba[a, c];
              btimesd = ListKaratsuba[b, d];
              middleterm =
                atimesc + btimesd + ListKaratsuba[a - b, d - c];
              tab = Table[0, {n/2}];
              middleterm = Join[middleterm, tab];
              atimesc = Join[atimesc, tab, tab];
              atimesc = PadLeft[atimesc, 2 * n];
              middleterm = PadLeft[middleterm, 2 * n];
              btimesd = PadLeft[btimesd, 2 * n];
              atimesc + middleterm + btimesd
            ]
```

Now let us multiply $a(x) = 3x^2 + 2x - 1$ *and* $b(x) = 5x^3 + x + 2$ *with Karatsuba's algorithm:*

```
In[3]:= PolynomialKaratsuba[3x^2 + 2x - 1, 5x^3 + x + 2, x]
Out[3]= 15 x^5 + 10 x^4 - 2 x^3 + 8 x^2 + 3 x - 2
```

Of course Expand *yields the same result:*

```
In[4]:= Expand[(3x^2 + 2x - 1) (5x^3 + x + 2)]
Out[4]= 15 x^5 + 10 x^4 - 2 x^3 + 8 x^2 + 3 x - 2
```

Similarly as in the integer arithmetic case our high-level programs cannot reach the efficiency of Mathematica's *built-in capabilities. For comparison purposes, we multiply two random integer coefficient polynomials of degree 100,*

$$In[5]:= \mathtt{pol1} = \sum_{k=0}^{100} \mathtt{Random[Integer, \{1, 10\}] * x^k};$$

$$\mathtt{pol2} = \sum_{k=0}^{100} \mathtt{Random[Integer, \{1, 10\}] * x^k};$$

first using Mathematica's Expand,

```
In[6]:= Timing[prod1 = Expand[pol1 * pol2];]
Out[6]= {0. Second, Null}
```

and then using our implementation:

$In[7]:=$ **Timing[prod2 = PolynomialKaratsuba[pol1,pol2,x];]**
$Out[7]=$ {0.0936006 Second, Null}

Clearly both computations have the same results:

$In[8]:=$ **prod2 - prod1**
$Out[8]=$ 0

6.3 Fast Multiplication with FFT

Karatsuba's algorithm is asymptotically more efficient than the defining algorithm via the Cauchy product. This provokes the question whether its complexity can still be improved. The best possible complexity for the multiplication of two polynomials of degree n and m would be $O(m+n)$ ring operations, or $O(n)$ ring operations if $m = n$, since already the lookup of the $m+n+2$ coefficients of the two input polynomials has this cost. We will see that this complexity can almost be reached.

One option to decrease the complexity of Karatsuba's algorithm is to use more complex subdivisions: if we divide the input polynomial of degree n in r (instead of 2) equally long parts, then an algorithm can be justified whose complexitiy $K(n)$ satisfies the inequality $K(rn) \leq (2r+1)K(n)+C \cdot n$ for some constant C. If one chooses r large enough, then for every $\varepsilon > 0$ there is an algorithm of complexity $O(n^{1+\varepsilon})$, compare Exercise 3.2.[6]

In this section we will consider a completely different method, instead, with the help of which even the complexity $O(n\log_2 n)$ can be reached. Unfortunately this method cannot directly be applied to polynomials with integer coefficients, but it is suitable for polynomials with decimal coefficients. In the latter situation this yields a superb algorithm.

The method that we will consider is the *Fast Fourier Transform*, shortly called FFT. FFT is an efficient variant of the *discrete Fourier transform*. The idea of the Fourier transform is rather simple: instead of computing the coefficients of the product of two polynomials $a(x)$ and $b(x)$ directly, which is rather costly, we compute the products $c(x_k) = a(x_k)b(x_k)$ at certain points x_k, which is rather simple, and reconstruct the product polynomial $c(x)$ from its values $c(x_k)$. Since a polynomial $c(x)$ of degree g can be determined by $g+1$ values,[7] we need $g+1$ interpolation points x_k.

We are finally led to the following questions:

1. Which are suitable points x_k ($k = 0, ..., g$)?
2. How do we compute the values $a(x_k)$ and $b(x_k)$ with as little cost as possible?
3. How do we get the coefficients of the product $c(x)$ most efficiently from the values $c(x_k)$?

It turns out that the best choice of nodes (Question 1) is guaranteed by using primitive roots of unity since then all nodes can be represented as the powers of one unique ring element.

---

[6] Details can be found in [Mig1992, Chapter 1.6.]. For the computation of the complexity, the proof given in Section 3.2 can be adapted.

[7] This situation will be considered in the general case of polynomial interpolation, see Section 6.5.

Definition 6.11. (Primitive Roots of Unity) Let R be a commutative ring with unity 1, let further be $n \in \mathbb{N}_{\geq 2}$. The element $\zeta \in R$ is called *primitive nth root of unity* in R if $\zeta^n = 1$ (root of unity), and if additionally $\zeta^k \neq 1$ for $0 < k < n$ (primitive).

Example 6.12. (a) Let $R = \mathbb{C}$. Then $\zeta = e^{\frac{2\pi i}{n}}$ is a primitive root of unity, since

$$\zeta^n = e^{\frac{2\pi i n}{n}} = e^{2\pi i} = 1$$

and $\zeta^k \neq 1$ for $0 < k < n$. More precisely the points ζ^k for $0 < k \leq n$ form a regular n-gon on the unit circle of the Gauss plane.

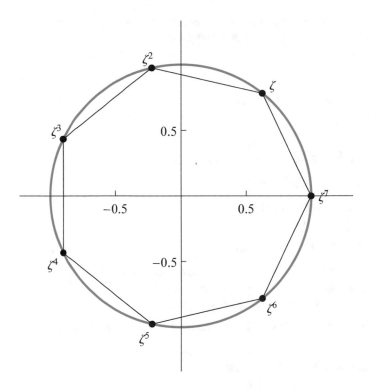

(b) The element $2 \in R = \mathbb{Z}_9$ is a primitive 6th root of unity in R since the powers 2^k ($k = 0, \ldots, 6$) are given by $(1, 2, 4, 8, 7, 5, 1)$. Furthermore, the table

```
In[1]:= Table[Mod[j^k,9],{j,2,8},{k,0,6}]
```

$$Out[1]= \begin{pmatrix} 1 & 2 & 4 & 8 & 7 & 5 & 1 \\ 1 & 3 & 0 & 0 & 0 & 0 & 0 \\ 1 & 4 & 7 & 1 & 4 & 7 & 1 \\ 1 & 5 & 7 & 8 & 4 & 2 & 1 \\ 1 & 6 & 0 & 0 & 0 & 0 & 0 \\ 1 & 7 & 4 & 1 & 7 & 4 & 1 \\ 1 & 8 & 1 & 8 & 1 & 8 & 1 \end{pmatrix}$$

shows that the element $5 = 2^{-1}$ (mod 9) is also a primitive 6th root of unity in R. On the other hand, 4 and $7 = 4^{-1}$ (mod 9) are primitive third roots of unity in R. Finally 8 is a square root of unity in R. The given inverses are confirmed by

```
In[2]:= PowerMod[{2,4,8},-1,9]
Out[2]= {5,7,8}
```

Definition 6.13. (Discrete Fourier Transform) Let $a(x) = a_{n-1}x^{n-1} + \cdots + a_1 x + a_0 \in R[x]$ be a polynomial of degree $n-1$ with coefficients in a ring R with a primitive nth root of unity ζ. Then the n-tuple of function values

$$\mathrm{DFT}(a(x), \zeta) = \mathrm{DFT}(a_0, \ldots, a_{n-1}, \zeta) := \left(a(1), a(\zeta), \ldots, a(\zeta^{n-1}) \right)$$

is called the *discrete Fourier transform* of $a(x)$. Similarly by

$$\mathrm{IDFT}(a(1), a(\zeta), \ldots, a(\zeta^{n-1}), \zeta) := (a_0, \ldots, a_{n-1})$$

we denote the *inverse discrete Fourier transform*.

Session 6.14. *We can execute the discrete Fourier transform of a polynomial with coefficients in $\mathbb{C}$ with Mathematica. We set $n = 8$ and therefore*

```
In[1]:= ζ = e^(2πi/8)
Out[1]= e^(iπ/4)
```

Next, we declare a polynomial $a(x)$ by

```
In[2]:= a = ∑_{k=0}^{7} (k+1) x^k
Out[2]= 8 x^7 + 7 x^6 + 6 x^5 + 5 x^4 + 4 x^3 + 3 x^2 + 2 x + 1
```

The polynomial has the coefficients

```
In[3]:= alist = CoefficientList[a, x]
Out[3]= {1, 2, 3, 4, 5, 6, 7, 8}
```

and the corresponding discrete Fourier transform can be computed using the command `Fourier`:[8]

```
In[4]:= φ = Chop[Fourier[alist, FourierParameters → {1,1}]]
Out[4]= {36., -4. - 9.65685 i, -4. - 4. i, -4. - 1.65685 i,
          -4., -4. + 1.65685 i, -4. + 4. i, -4. + 9.65685 i}
```

According to the definition of the discrete Fourier transform this yields the following table of values:

```
In[5]:= Table[a /. {x → ζ^k}, {k, 0, 7}] //N //Chop
Out[5]= {36., -4. - 9.65685 i, -4. - 4. i, -4. - 1.65685 i,
          -4., -4. + 1.65685 i, -4. + 4. i, -4. + 9.65685 i}
```

The command `InverseFourier` *computes the inverse discrete Fourier transform and therefore generates the polynomial coefficients again:*

```
In[6]:= InverseFourier[φ, FourierParameters → {1,1}]
Out[6]= {1., 2., 3., 4., 5., 6., 7., 8.}
```

Please note that the routines `Fourier` *and* `InverseFourier` *compute numeric decimal numbers and no integer values since the symbolic computation with complex roots of unity is relatively difficult. We will come back to this issue very soon.*

Now we use the routines `Fourier` *and* `InverseFourier` *as black boxes for the computation of the coefficients of the product of two polynomials $a(x)$ and $b(x)$. First we enter $b(x)$ as*

---

[8] *Mathematica* relies on several different definitions of the discrete Fourier transform. `FourierParameters` scales the `Fourier` function corresponding to our definition. With `Chop` very small decimal numbers are suppressed, for example, small imaginary parts ($< 10^{-10}$) that occur because of inaccuracies of decimal arithmetic.

$$In[7] := \mathbf{b} = \sum_{k=0}^{7} \mathbf{x}^k$$

$Out[7] = x^7 + x^6 + x^5 + x^4 + x^3 + x^2 + x + 1$

and compute the coefficient list of b(x):

$In[8] := \mathbf{blist = CoefficientList[b, x]}$

$Out[8] = \{1, 1, 1, 1, 1, 1, 1, 1\}$

Since our final result $c(x) = a(x) \cdot b(x)$ is a polynomial with almost twice as many nonzero coefficients,[9] we must double the number of coefficients:

$In[9] := \mathbf{alist = PadRight[alist, 16]}$

$Out[9] = \{1, 2, 3, 4, 5, 6, 7, 8, 0, 0, 0, 0, 0, 0, 0, 0\}$

$In[10] := \mathbf{blist = PadRight[blist, 16]}$

$Out[10] = \{1, 1, 1, 1, 1, 1, 1, 1, 0, 0, 0, 0, 0, 0, 0, 0\}$

Now we can compute the corresponding Fourier transforms:

$In[11] := \varphi = \mathbf{Chop[Fourier[alist, FourierParameters \rightarrow \{1, 1\}]]}$

$Out[11] = \{36., -8.13707 + 25.1367\ i, -4. - 9.65685\ i, 3.38009 + 7.48303\ i,$
$-4. - 4.\ i, 4.27677 + 3.34089\ i, -4. - 1.65685\ i,$
$4.48022 + 0.994562\ i, -4., 4.48022 - 0.994562\ i,$
$-4. + 1.65685\ i, 4.27677 - 3.34089\ i, -4. + 4.\ i,$
$3.38009 - 7.48303\ i, -4. + 9.65685\ i, -8.13707 - 25.1367\ i\}$

$In[12] := \psi = \mathbf{Chop[Fourier[blist, FourierParameters \rightarrow \{1, 1\}]]}$

$Out[12] = \{8., 1. + 5.02734\ i, 0, 1. + 1.49661\ i, 0,$
$1. + 0.668179\ i, 0, 1. + 0.198912\ i, 0, 1. - 0.198912\ i,$
$0, 1. - 0.668179\ i, 0, 1. - 1.49661\ i, 0, 1. - 5.02734\ i\}$

The product of the two Fourier transforms gives, for $k = 0, \ldots, 15$, the values $c(\zeta^k)$ as

$In[13] := \varphi * \psi$

$Out[13] = \{288., -134.508 - 15.7711\ i, 0, -7.81906 + 12.5417\ i, 0,$
$2.04446 + 6.19854\ i, 0, 4.28239 + 1.88573\ i, 0, 4.28239 - 1.88573\ i, 0,$
$2.04446 - 6.19854\ i, 0, -7.81906 - 12.5417\ i, 0, -134.508 + 15.7711\ i\}$

From these values we can finally detect the coefficients of c(x) using the inverse Fourier transform:

$In[14] := \mathbf{inv =}$
$\qquad \mathbf{Chop[InverseFourier[\varphi * \psi, FourierParameters \rightarrow \{1, 1\}]]}$

$Out[14] = \{1., 3., 6., 10., 15., 21., 28., 36., 35., 33., 30., 26., 21., 15., 8., 0\}$

A comparison with

$In[15] := \mathbf{CoefficientList[a * b, x]}$

$Out[15] = \{1, 3, 6, 10, 15, 21, 28, 36, 35, 33, 30, 26, 21, 15, 8\}$

shows that the result is correct.

Now we would like to move on to the realization of the Fast Fourier Transform. For this purpose we will need some properties of rings with roots of unity. The following theorem is valid.

---

[9] In general, the input polynomials have degree $n-1$ and $m-1$, and the product has therefore the degree $n+m-2$, so that we need $n + m - 1$ coefficients.

Theorem 6.15. Let R be a commutative ring with unity 1, let further $n \in \mathbb{N}_{\geq 2}$. If $\zeta \in R$ is a primitive nth root of unity and $x_k = \zeta^k$ $(k = 0, \ldots, n-1)$, then we get:

(a) The set $\{x_k \mid k = 0, \ldots, n-1\}$ consists of nth roots of unity and forms a group of n different elements with respect to multiplication.

(b) If ζ is a primitive nth root of unity in R, then $\zeta^{-1} \in R$, and ζ^{-1} is also a primitive nth root of unity.

(c) If n is even, then ζ^2 is a primitive $\frac{n}{2}$th root of unity in R.

(d) Let n be even and $\text{char}(R) \neq 2$. Then $x_{k+\frac{n}{2}} = -x_k$ $(k = 0, \ldots, \frac{n}{2} - 1)$.

Proof. (a) Every power of an nth root of unity is again an nth root of unity: $(\zeta^k)^n = (\zeta^n)^k = 1^k = 1$. Products of roots of unity are again roots of unity, since $\zeta^j \cdot \zeta^k = \zeta^{j+k}$. The neutral element is 1, associativity is clear, and the multiplicative inverse of ζ^j is the number ζ^{n-j}, since $\zeta^j \cdot \zeta^{n-j} = \zeta^n = 1$.

(b) As shown in (a), $\zeta^{-1} = \zeta^{n-1} \in R$. Assuming that ζ^{-1} is *not* primitive, then $\zeta^{-k} = 1$ for some $0 < k < n$ and therefore $\zeta^{n-k} = 1$, in contradiction to the primitivity of ζ.

(c) Obviously $(\zeta^2)^{\frac{n}{2}} = \zeta^n$. It is easy to prove the primitivity.

(d) Because of $\zeta^n = 1$, we get

$$x_{k+\frac{n}{2}} = \zeta^{k+\frac{n}{2}} = \zeta^k \cdot \zeta^{\frac{n}{2}} = \pm\zeta^k = \pm x_k,$$

since 1 possesses the two square roots ± 1. If the plus sign held, then $\zeta^{\frac{n}{2}} = 1$, and therefore the root of unity would not be primitive. In the case when $\text{char}(R) = 2$ we always have $1 = -1$, therefore we must postulate $\text{char}(R) \neq 2$. $\qquad\square$

The idea of the Fast Fourier Transform (FFT) is again a typical divide-and-conquer concept. Without loss of generality we assume that our input polynomial $a(x)$ has n coefficients and $n = 2^m$ is a power of 2. The polynomial $a(x) = a_0 + a_1 x + \cdots + a_{n-1} x^{n-1}$ corresponds to the coefficient sequence $(a_0, a_1, \ldots, a_{n-1})$. We will now partition this sequence into two equally long parts: the even coefficient sequence $(a_0, a_2, \ldots, a_{n-2})$ corresponds to the polynomial $a_e(x) = a_0 + a_2 x + \cdots + a_{n-2} x^{n/2-1}$ and the odd coefficient sequence $(a_1, a_3, \ldots, a_{n-1})$ to $a_o(x) = a_1 + a_3 x + \cdots + a_{n-1} x^{n/2-1}$. Obviously $a(x)$ can be written in the form

$$a(x) = a_e(x^2) + x a_o(x^2).$$

Next, we substitute powers of ζ in $a(x)$. Because of $\zeta^{k-n} = \zeta^k$, we obtain

$$a(\zeta^k) = \begin{cases} a_e(\zeta^{2k}) + \zeta^k a_o(\zeta^{2k}) & \text{if } k = 0, \ldots, \frac{n}{2} - 1, \\ a_e(\zeta^{2k-n}) + \zeta^k a_o(\zeta^{2k-n}) & \text{if } k = \frac{n}{2}, \ldots, n-1. \end{cases} \qquad (6.8)$$

The Fourier transforms $\text{DFT}(a_e(x), \zeta^2)$ and $\text{DFT}(a_o(x), \zeta^2)$ consist exactly of the values

$$\text{DFT}(a_e(x), \zeta^2) = (a_e(1), a_e(\zeta^2), \ldots, a_e(\zeta^{2(n/2-1)})),$$

and

$$\text{DFT}(a_o(x), \zeta^2) = (a_o(1), a_o(\zeta^2), \ldots, a_o(\zeta^{2(n/2-1)})).$$

The Fourier transform $\text{DFT}(a(x), \zeta)$ can therefore be decomposed into these two vectors thanks to (6.8). This leads to the following recursive scheme.

Theorem 6.16. (FFT) Let R be a ring with primitive nth root of unity ζ, and let n be a power of 2.

The following recursive algorithm computes the discrete Fourier transform $\mathrm{FFT}(a(x), \zeta) = \mathrm{DFT}(a(x), \zeta)$ of the polynomial $a(x) = a_0 + a_1 x + \cdots + a_{n-1} x^{n-1}$ (or the vector $(a_0, a_1, \ldots, a_{n-1})$):

(a)   Terminating condition: If $n = 1$, then $\mathrm{FFT}(a(x), \zeta) := (a_0)$.
(b)   Compute $a_e(x)$ and $a_o(x)$.
(c)   Set $(\varphi_0, \ldots, \varphi_{\frac{n}{2}-1}) := \mathrm{FFT}(a_e(x), \zeta^2)$ and $(\psi_0, \ldots, \psi_{\frac{n}{2}-1}) := \mathrm{FFT}(a_o(x), \zeta^2)$.
(d)   For $k = 0, \ldots, \frac{n}{2} - 1$, define $f_k := \varphi_k + \zeta^k \psi_k$ and $f_{\frac{n}{2}+k} := \varphi_k + \zeta^{\frac{n}{2}+k} \psi_k$.
(e)   Output: $(f_0, \ldots, f_{n-1})$.

Proof. The result follows from the explanations given above and Theorem 6.15. $\square$

Theorem 6.16 answers our second question. Before we implement the method, let us discuss our third question: How can we invert this procedure most efficiently? It turns out that the inverse function can be executed practically by the same code!

Theorem 6.17. (IFFT) Let R be an integral domain with primitive nth root of unity ζ, and let $n \in \mathbb{N}_{\geq 2}$. For the inverse Fourier transform the relation $n \cdot \mathrm{IDFT}(a(x), \zeta) = \mathrm{DFT}(a(x), \zeta^{-1})$ is valid.

Proof. Let IDFT be defined by $n \cdot \mathrm{IDFT}(a(x), \zeta) := \mathrm{DFT}(a(x), \zeta^{-1})$. Then we must prove that this definition computes the inverse Fourier transform. In other words, we must prove that for an arbitrary polynomial $a(x) = a_0 + a_1 x + \cdots + a_{n-1} x^{n-1}$ we indeed have $n \cdot \mathrm{IDFT}(\mathrm{DFT}(a(x), \zeta), \zeta) = n(a_0, \ldots, a_{n-1})$.

Let us first compute
$$\mathrm{DFT}(a(x), \zeta) = (a(1), a(\zeta), \ldots, a(\zeta^{n-1})) .$$

If this list is considered as the coefficient list of an image polynomial $A(t) \in R[t]$, then obviously

$$A(t) = \sum_{k=0}^{n-1} a(\zeta^k) t^k = \sum_{k=0}^{n-1} \sum_{j=0}^{n-1} a_j \zeta^{kj} t^k .$$

Applying $n \cdot \mathrm{IDFT}$ yields

$$
\begin{aligned}
n \cdot \mathrm{IDFT}(\mathrm{DFT}(a(x), \zeta), \zeta) &= \mathrm{DFT}(\mathrm{DFT}(a(x), \zeta), \zeta^{-1}) = \\
&= \left(A(1), A(\zeta^{-1}), \ldots, A(\zeta^{-n+1}) \right) \\
&= \left(\sum_{k=0}^{n-1} \sum_{j=0}^{n-1} a_j \zeta^{kj} \zeta^{-kl} \right)_{l=0,\ldots,n-1}
\end{aligned}
$$

If we change the order of summation, then we see that this vector equals $n(a_0, \ldots, a_{n-1})$ if we show that

$$\sum_{k=0}^{n-1} \zeta^{k(j-l)} = \begin{cases} n & \text{if } j = l , \\ 0 & \text{otherwise} . \end{cases}$$

However, this identity is clearly true for $j = l$, and for $j \neq l$ it follows from the division-free formula of the geometric sum

$$(1-\zeta) \sum_{k=0}^{n-1} \zeta^{k(j-l)} = (1-\zeta) \sum_{k=0}^{n-1} (\zeta^{j-l})^k = 1 - \zeta^{n(j-l)} = 0 ,$$

since $\zeta^n = 1$ and $\zeta \neq 1$. $\square$

Now we are able to implement the method given in Theorem 6.16.

Session 6.18. *We implement the procedures* `FFT` *for fast Fourier transform,*

```
In[1]:= Clear[FFT]
        FFT[list_] := list/; Length[list] === 1;
        FFT[list_] :=
          Module[{n = Length[list], ζ, fullist, even,
              odd, φ, ψ, tab1, tab2, i},
            ζ = e^(2πi/n);
            fullist = PadRight[list, n];
            fullist = Transpose[Partition[fullist, 2]];
            even = fullist[[1]];
            odd = fullist[[2]];
            φ = FFT[even];
            ψ = FFT[odd];
            tab1 = Table[φ[[i]] + ζ^(i-1) * ψ[[i]], {i, n/2}];
            tab2 = Table[φ[[i]] + ζ^(n/2+i-1) * ψ[[i]], {i, n/2}];
            Join[tab1, tab2]
          ]
```

as well as `IFFT` *for the inverse fast Fourier transform:*

```
In[2]:= Clear[IFFT, ifft]

        IFFT[list_] := ifft[list] / Length[list];
```

```
In[3]:= ifft[list_] := list/; Length[list] === 1;
        ifft[list_] :=
          Module[{n = Length[list], ζ, fullist, even,
              odd, φ, ψ, tab1, tab2, i},
            ζ = e^(-2πi/n);
            fullist = PadRight[list, n];
            fullist = Transpose[Partition[fullist, 2]];
            even = fullist[[1]];
            odd = fullist[[2]];
            φ = ifft[even];
            ψ = ifft[odd];
            tab1 = Table[φ[[i]] + ζ^(i-1) * ψ[[i]], {i, n/2}];
            tab2 = Table[φ[[i]] + ζ^(n/2+i-1) * ψ[[i]], {i, n/2}];
            Join[tab1, tab2]
          ]
```

For testing purposes, we define the polynomial $a(x)$ by the following coefficient list:

```
In[4]:= alist = Table[k, {k, 16}]
Out[4]= {1, 2, 3, 4, 5, 6, 7, 8, 9, 10, 11, 12, 13, 14, 15, 16}
```

Then we get its discrete Fourier transform by

```
In[5]:= fft = FFT[alist] //N
```

Out[5]= {136., −8. − 40.2187 i, −8. − 19.3137 i, −8. − 11.9728 i,

    −8. − 8. i, −8. − 5.34543 i, −8. − 3.31371 i, −8. − 1.5913 i,

    −8., −8. + 1.5913 i, −8. + 3.31371 i, −8. + 5.34543 i,

    −8. + 8. i, −8. + 11.9728 i, −8. + 19.3137 i, −8. + 40.2187 i}

This can be compared with the one computed by Mathematica:

In[6]:= **Chop[Fourier[alist, FourierParameters → {1, 1}]]**

Out[6]= {136., −8. − 40.2187 i, −8. − 19.3137 i, −8. − 11.9728 i,

    −8. − 8. i, −8. − 5.34543 i, −8. − 3.31371 i, −8. − 1.5913 i,

    −8., −8. + 1.5913 i, −8. + 3.31371 i, −8. + 5.34543 i,

    −8. + 8. i, −8. + 11.9728 i, −8. + 19.3137 i, −8. + 40.2187 i}

An application of the inverse Fourier transform yields the original coefficients again:

In[7]:= **IFFT[fft]//Chop**

Out[7]= {1., 2., 3., 4., 5., 6., 7., 8., 9., 10., 11., 12., 13., 14., 15., 16.}

The above computations were done numerically. Now, we would like to see what happens if we do symbolic computations with complex roots of unity. For this, let us start with our integer-valued coefficient list. Then we get the Fourier transform

In[8]:= **fft = FFT[alist]**

$$Out[8]= \left\{ 136, (-8-8\,i)-(8+8\,i)\,e^{\frac{i\pi}{4}}+e^{\frac{i\pi}{8}}\left((-8-8\,i)-(8+8\,i)\,e^{\frac{i\pi}{4}}\right), \right.$$

$$(-8-8\,i)-(8+8\,i)\,e^{\frac{i\pi}{4}},$$

$$(-8+8\,i)-(8-8\,i)\,e^{\frac{3i\pi}{4}}+e^{\frac{3i\pi}{8}}\left((-8+8\,i)-(8-8\,i)\,e^{\frac{3i\pi}{4}}\right), -8-8\,i,$$

$$(-8-8\,i)-(8+8\,i)\,e^{-\frac{3i\pi}{4}}+e^{\frac{5i\pi}{8}}\left((-8-8\,i)-(8+8\,i)\,e^{-\frac{3i\pi}{4}}\right),$$

$$(-8+8\,i)-(8-8\,i)\,e^{\frac{3i\pi}{4}},$$

$$(-8+8\,i)-(8-8\,i)\,e^{-\frac{i\pi}{4}}+e^{\frac{7i\pi}{8}}\left((-8+8\,i)-(8-8\,i)\,e^{-\frac{i\pi}{4}}\right),$$

$$-8, (-8-8\,i)-(8+8\,i)\,e^{\frac{i\pi}{4}}+e^{-\frac{7i\pi}{8}}\left((-8-8\,i)-(8+8\,i)\,e^{\frac{i\pi}{4}}\right),$$

$$(-8-8\,i)-(8+8\,i)\,e^{-\frac{3i\pi}{4}},$$

$$(-8+8\,i)-(8-8\,i)\,e^{\frac{3i\pi}{4}}+e^{-\frac{5i\pi}{8}}\left((-8+8\,i)-(8-8\,i)\,e^{\frac{3i\pi}{4}}\right), -8+8\,i,$$

$$(-8-8\,i)-(8+8\,i)\,e^{-\frac{3i\pi}{4}}+e^{-\frac{3i\pi}{8}}\left((-8-8\,i)-(8+8\,i)\,e^{-\frac{3i\pi}{4}}\right),$$

$$(-8+8\,i)-(8-8\,i)\,e^{-\frac{i\pi}{4}},$$

$$\left. (-8+8\,i)-(8-8\,i)\,e^{-\frac{i\pi}{4}}+e^{-\frac{i\pi}{8}}\left((-8+8\,i)-(8-8\,i)\,e^{-\frac{i\pi}{4}}\right)\right\}$$

which generates a list of complex numbers in terms of 8th roots of unity. The inverse Fourier transform is now a very complicated expression,[10] *which can be simplified with* Simplify, *though:*

In[9]:= **IFFT[fft] //Simplify//Timing**

Out[9]= {0.0625 Second, {1, 2, 3, 4, 5, 6, 7, 8, 9, 10, 11, 12, 13, 14, 15, 16}}

However, the necessary simplification slows down the "fast Fourier transform" considerably![11]

If we combine the previous results, we are led to the following algorithm for polynomial multiplication. Since the IFFT needs a division by n (by Theorem 6.17), we assume (for convenience) that the coefficient set is a field of characteristic 0.

---

[10] The suppressed output contains several pages!

[11] See Section 6.10 how this problem can be resolved.

Theorem 6.19. (Product by FFT) Let $a(x), b(x) \in \mathbb{K}[x]$ be two polynomials of degree $n - 1, m - 1 \leqq N/2$, respectively, where $\mathbb{K}$ is a field of characteristic 0, $N \in \mathbb{N}$ a power of 2 and $\zeta \in \mathbb{K}$ a primitive Nth root of unity. Then the following algorithm computes the coefficients $c_0, \dots, c_{N-1}$[12] of the product $c(x) = a(x) \cdot b(x)$:

(a)   Compute $\varphi := \mathrm{DFT}(a_0, \dots, a_{N-1}, \zeta)$ and $\psi := \mathrm{DFT}(b_0, \dots, b_{N-1}, \zeta)$.

(b)   Calculate $(c_0, \dots, c_{N-1}) := \mathrm{IDFT}(\varphi \cdot \psi, \zeta)$.

(c)   Output: $(c_0, \dots, c_{N-1})$ or $c(x) = \sum\limits_{k=0}^{N-1} c_k x^k$, respectively.

Note that in practice the number $N \in \mathbb{N}$ can be taken as the smallest power of 2 with $N \geqq n + m - 1$.   □

Finally we want to prove the above-mentioned complexity behavior.

Theorem 6.20. (Complexity of FFT) Let $\deg(a(x), x) = n$. Then the FFT algorithm has a complexity $K(n) = O(n \log_2 n)$. Similarly the multiplication of two polynomials of degree n, using Theorem 6.19, has this complexity.

Proof. If we denote the complexity for a vector of length n by $K(n)$, then the recursive application of the algorithm yields

$$K(n) = 2 \cdot K(n/2) + C \cdot n, \qquad K(1) = 1,$$

for a constant $C > 0$. Without loss of generality let n be a power of 2, hence $n = 2^m$. We will prove the relation

$$K(n) \leqq D \cdot n \cdot \log_2 n$$

for a constant $D > 0$ and all $n \in \mathbb{N}_{\geq 0}$ by induction with respect to m. The inequality is obviously correct for $m = 0$. Assuming that the relation holds for some m, it follows that

$$K(2n) = K(2^{m+1}) = 2 \cdot K(n) + 2C \cdot n \leqq 2D \cdot n \cdot \log_2 n + 2C \cdot n \leqq D \cdot (2n) \cdot \log_2(2n),$$

hence it is true for $m + 1$. FFT-based multiplication has the same complexity $O(n \log_2 n)$, since the two steps FFT and IFFT both have this complexity.   □

6.4 Division with Remainder

After having dealt with multiplication, we now move on to division, namely we will consider the division of polynomials of a single variable. When dividing two polynomials by each other, we must, in particular, divide their highest coefficients. For the sake of simplicity we therefore assume that R is a field $\mathbb{K}$. Similarly as in the ring R, there is also a *division with remainder* in the polynomial ring $\mathbb{K}[x]$ which executes the division of two polynomials $a(x) \in \mathbb{K}[x]$ and $b(x) \in \mathbb{K}[x]$ in the form

$$a(x) = q(x) b(x) + r(x),$$

---

[12] Only the first $n + m - 1$ coefficients are (potentially) different from zero.

where $q(x)$ denotes the *polynomial quotient* and $r(x)$ the *polynomial remainder*. The essential property which makes this representation unique is given by $\deg(r(x), x) < \deg(b(x), x)$, as the following theorem reveals.

Theorem 6.21. (Polynomial Division) Let $\mathbb{K}$ be a field and let $a(x), b(x) \in \mathbb{K}[x]$. Then there is exactly one pair $q(x), r(x) \in \mathbb{K}[x]$ such that the relation

$$a(x) = q(x) b(x) + r(x) \qquad \text{with} \qquad \deg(r(x), x) < \deg(b(x), x)$$

is valid.

Proof. The induction proof from Theorem 3.8 can be directly adapted. Induction is now with respect to the degree n of $a(x)$. $\square$

Similar to the situation in $\mathbb{Z}$, we denote the quotient by $q(x) = \text{quotient}(a(x), b(x), x)$ and the remainder by $r(x) = \text{rem}(a(x), b(x), x)$.

Example 6.22. For the execution of polynomial division, the high school algorithm can be used. For example, if $a(x) = x^3 - 2x^2 - 5x + 6$ and $b(x) = x - 1$, polynomial division leads to

$$
\begin{array}{l}
(x^3 \ -2x^2 \ -5x \ +6) : (x-1) = x^2 - x - 6 \\
\underline{- \ (x^3 \ -x^2)} \\
\qquad\quad -x^2 \ -5x \ +6 \\
\qquad\underline{- \ (-x^2 \ +x)} \\
\qquad\qquad\quad -6x \ +6 \\
\qquad\qquad\underline{- \ (-6x \ +6)} \\
\qquad\qquad\qquad\quad 0
\end{array}
$$

and hence $q(x) = x^2 - x - 6$ and $r(x) = 0$. Therefore $a(x)$ is divisible by $b(x)$, i.e., $b(x) \mid a(x)$, with

$$\frac{x^3 - 2x^2 - 5x + 6}{x - 1} = x^2 - x - 6.$$

Analogously, the computation

$$\frac{x^3 - 2x^2 - 5x + 7}{x - 1} = x^2 - x - 6 + \frac{1}{x - 1}$$

leads to $q(x) = x^2 - x - 6$ and $r(x) = 1$ for $a(x) = x^3 - 2x^2 - 5x + 7$ and $b(x) = x - 1$. $\triangle$

Session 6.23. *In Mathematica, polynomial division with remainder can be carried out by the corresponding functions* `PolynomialQuotient[a,b,x]` *and* `Polynomial-Remainder[a,b,x]`.

With their help, we can reproduce the above results:[13]

```
In[1]:= q = PolynomialQuotient[a = x³ - 2x² - 5x + 7, b = x - 1, x]
Out[1]= x² - x - 6

In[2]:= r = PolynomialRemainder[a, b, x]
Out[2]= 1
```

---

[13] Note that a and b are defined as side-effect in the arguments of `PolynomialQuotient`.

We check the results by computing

```
In[3]:= res = q*b+r
```
$Out[3]= (x-1)\left(x^2-x-6\right)+1$

which—after expansion—again leads to a(x):

```
In[4]:= res//Expand
```
$Out[4]= x^3-2\,x^2-5\,x+7$

By default, Mathematica works in the field $\mathbb{Q}$. The considered polynomials are therefore elements of $\mathbb{Q}[x]$.

Next, let us consider a bivariate example with $a(x,y), b(x,y) \in \mathbb{Q}[y][x]$.[14]

```
In[5]:= a = ∑ₖ₌₁⁵ xᵏy⁶⁻ᵏ
```
$$In[5]:= a = \sum_{k=1}^{5} x^k y^{6-k}$$
$Out[5]= x^4\,y^2+x^3\,y^3+x^2\,y^4+x^5\,y+x\,y^5$

$$In[6]:= b = \sum_{k=1}^{3} 2\,k\,x^k y^{3-k}$$
$Out[6]= 4\,x^2\,y+6\,x^3+2\,x\,y^2$

```
In[7]:= q = PolynomialQuotient [a,b,x]
```
$$Out[7]= \frac{x^2\,y}{6}+\frac{x\,y^2}{18}+\frac{2\,y^3}{27}$$

```
In[8]:= r = PolynomialRemainder [a,b,x]
```
$$Out[8]= \frac{16\,x^2\,y^4}{27}+\frac{23\,x\,y^5}{27}$$

For testing purposes, we compute

```
In[9]:= res = q*b+r
```
$$Out[9]= \frac{16\,x^2\,y^4}{27}+(4\,x^2\,y+6\,x^3+2\,x\,y^2)\left(\frac{x^2\,y}{6}+\frac{x\,y^2}{18}+\frac{2\,y^3}{27}\right)+\frac{23\,x\,y^5}{27}$$

which again yields the starting polynomial a(x, y):

```
In[10]:= res//Expand
```
$Out[10]= x^4\,y^2+x^3\,y^3+x^2\,y^4+x^5\,y+x\,y^5$

Note that the complexity of polynomial division with remainder is $\deg(b(x), x) \cdot (\deg(a(x), x) - \deg(b(x), x))$ field operations.

The division with remainder has some interesting consequences which we will now consider.

Lemma 6.24. Let $\mathbb{K}$ be a field and $a(x) \in \mathbb{K}[x]$ a non-constant polynomial. If α is a zero of $a(x)$, then $x - \alpha$ is a divisor of $a(x)$.

Proof. The division algorithm generates a representation $a(x) = q(x) \cdot (x - \alpha) + r(x)$ with $\deg(r(x), x) < 1$. Therefore $r(x)$ is constant, say $r(x) = r \in \mathbb{K}$, and

$$a(x) = q(x) \cdot (x - \alpha) + r\,.$$

If we substitute $x = \alpha$ in this equation, then $a(\alpha) = 0$ yields $r = 0$. Hence $x - \alpha \mid a(x)$. □

---

[14] The basic ring is here $\mathbb{Q}[y]$ which is no field. If divisions are necessary, then *Mathematica* will use the ground field $\mathbb{Q}(y)$, see Section 6.9.

Lemma 6.24 shows that linear factors of $a(x)$ are related to zeros of $a(x)$. Linear factors have degree 1 and are therefore automatically irreducible since every divisor must have degree 1 or 0. Every constant divisor (degree 0) is a unit and every divisor of degree 1 is itself linear and therefore associated with $a(x)$.

Corollary 6.25. Let $\mathbb{K}$ be a field and $a(x) \in \mathbb{K}[x]$ a polynomial of degree $n > 0$. Then $a(x)$ has at most n zeros in $\mathbb{K}$.

Proof. Since $n > 0$, $a(x)$ is not the zero polynomial. Therefore for every zero α_k we can apply Lemma 6.24 and divide by $x - \alpha_k$. This reduces the degree by 1, therefore such a step cannot be applied more than n times. $\square$

Finally we get the following theorem which was already announced in Section 6.1.

Theorem 6.26. Let $\mathbb{K}$ be a field with infinitely many elements. Then $a(x)$ is the zero polynomial if and only if $a(\alpha) = 0$ for every $\alpha \in \mathbb{K}$.

Proof. If $a(x)$ is the zero polynomial, then obviously $a(\alpha) = 0$ for every $\alpha \in \mathbb{K}$.

Now, let $a(\alpha) = 0$ for all $\alpha \in \mathbb{K}$. Since $\mathbb{K}$ has infinitely many elements, $a(x)$ has infinitely many zeros. Then, by Corollary 6.25, $a(x)$ must be the zero polynomial. $\square$

Corollary 6.25 also implies an identity theorem for polynomials. By definition, two polynomials agree if and only if they have the same coefficients. The identity theorem clarifies under which conditions two polynomials agree if a suitable number of values agree.

Theorem 6.27. (Identity Theorem for Polynomials) Let $\mathbb{K}$ be a field, $a(x), b(x) \in \mathbb{K}[x]$ with $\deg(a(x), x) = n$ and $\deg(b(x), x) = m$, and let $\mathbb{K}$ have at least $\max\{m, n\} + 1$ elements.[15] Then $a(x) = b(x)$ if and only if the values of $a(x)$ and $b(x)$ agree at $n + 1$ different points.

Proof. Obviously $a(x)$ and $b(x)$ do not agree if $m \neq n$. Hence we can now assume $m = n$.

If $a(x) = b(x)$, then the values agree for all $x \in \mathbb{K}$. Since $\mathbb{K}$ has enough elements, the conclusion follows.

To prove the converse assertion, assume now that $a(x)$ and $b(x)$ agree at $n + 1$ different points. Then we consider the polynomial $c(x) := a(x) - b(x) \in \mathbb{K}[x]$. This polynomial equals zero at $n + 1$ different points. Since its degree is at most n, Corollary 6.25 shows that $c(x)$ must be the zero polynomial. $\square$

For proving that two polynomials agree, it is therefore sufficient to check a finite number of values.

Example 6.28. The polynomials

```
In[1]:= p = ∏ (x - k)
             k=1
             5
Out[1]= (x - 5) (x - 4) (x - 3) (x - 2) (x - 1)
```

and

---

[15] The latter condition is necessary for the trivial direction of the theorem to be true!

```
In[2]:= q = Expand[p]
Out[2]= x^5 - 15 x^4 + 85 x^3 - 225 x^2 + 274 x - 120
```

agree since $p(0) = q(0) = -120$ and $p(k) = q(k) = 0$ for $k = 1, ..., 5$:

```
In[3]:= p/.x → 0
Out[3]= -120
```

```
In[4]:= Table[q, {x, 5}]
Out[4]= {0, 0, 0, 0, 0}
```

The remaining relations are obvious. △

6.5 Polynomial Interpolation

Let $\mathbb{K}$ be a field. If $n + 1$ values of the polynomial $a(x) \in \mathbb{K}[x]$ of degree $\leq n$ are known, then, by Theorem 6.27, it can be reconstructed from these values. This process is called *polynomial interpolation*.

Let $y_k = a(x_k)$ $(k = 0, 1, ..., n)$ be given and let us assume that the $x_k \in \mathbb{K}$ $(k = 0, 1, ..., n)$ are pairwise different. It is possible to explicitly write down the interpolation polynomial, using the *Lagrange polynomials*.[16]

For this, we first observe that the Lagrange polynomials

$$L_k(x) := \frac{(x - x_0)(x - x_1) \cdots (x - x_{k-1})}{(x_k - x_0)(x_k - x_1) \cdots (x_k - x_{k-1})} \frac{(x - x_{k+1})(x - x_{k+2}) \cdots (x - x_n)}{(x_k - x_{k+1})(x_k - x_{k+2}) \cdots (x_k - x_n)}$$

have the degree n and the values

$$L_k(x_j) = \begin{cases} 1 & \text{if } j = k \\ 0 & \text{if } j \neq k \end{cases}$$

at the *support points* x_j $(j = 0, ..., n)$. Hence the linear combination

$$L(x) := y_0 L_0(x) + y_1 L_1(x) + \cdots + y_n L_n(x) = \sum_{k=0}^{n} y_k L_k(x) \tag{6.9}$$

is a solution polynomial for the given problem, since direct comparison shows that

$$L(x_j) = y_0 L_0(x_j) + y_1 L_1(x_j) + \cdots + y_n L_n(x_j) = y_j$$

for all $j = 0, ..., n$. This solution—which is unique by Theorem 6.27—is called the *Lagrange interpolating polynomial*.

We collect these results in the following theorem.

Theorem 6.29. (Polynomial Interpolation) Assume that the field $\mathbb{K}$ has infinitely many elements, the polynomial $a(x) \in \mathbb{K}[x]$ has degree $\leq n$ and the values y_k at the support points

---

[16] It is also possible to compute the interpolating polynomial in *Newton form*. This representation has several advantages, in particular, that it is numerically more stable. But on the other hand, it is more complicated to obtain, see Exercise 6.2.

$x_k \in \mathbb{K}$ ($k = 0, \ldots, n$). Then $a(x)$ is uniquely determined and $a(x) = L(x)$, where $L(x)$ is given by (6.9). □

Session 6.30. *The polynomial $a(x) \in \mathbb{Q}[x]$,*

$In[1]:=$ **a $= \dfrac{4x^2}{3} - \dfrac{x^4}{3}$**

$Out[1]= \dfrac{4x^2}{3} - \dfrac{x^4}{3}$

has degree four and can therefore be reconstructed by five values. It has the following graphical representation:

$In[2]:=$ **Plot[a, {x, -2, 2}]**

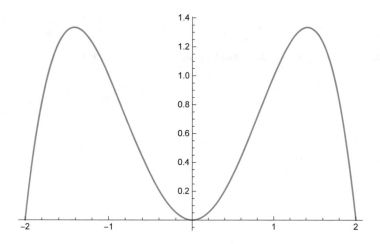

We compute the values of $a(x)$ at the points $x = -2, -1, 0, 1, 2$ and get the data points

$In[3]:=$ **list = Table[{x, a}, {x, -2, 2}]**

$Out[3]= \begin{pmatrix} -2 & 0 \\ -1 & 1 \\ 0 & 0 \\ 1 & 1 \\ 2 & 0 \end{pmatrix}$

The Lagrange polynomial, given by (6.9), is implemented by

```
In[4]:= Lagrange[list_, x_] := Sum[list[[k, 2]]*
            Product[x - list[[j, 1]], {j, 1, k - 1}]*
            Product[x - list[[j, 1]], {j, k + 1, Length[list]}]/
              (Product[list[[k, 1]] - list[[j, 1]],
                  {j, 1, k - 1}]*
                Product[list[[k, 1]] - list[[j, 1]],
              {j, k + 1, Length[list]}]), {k, 1, Length[list]}]
```

For our example, we get the following polynomial:

$In[5]:=$ **a = Lagrange[list, x]**

$Out[5]= \dfrac{1}{6}(2 - x)\,x\,(x + 1)(x + 2) - \dfrac{1}{6}(x - 2)(x - 1)\,x\,(x + 2)$

Expansion leads to our starting polynomial again:

In[6]:= **Expand[a]**

Out[6]= $\dfrac{4\,x^2}{3} - \dfrac{x^4}{3}$

Note that polynomial interpolation is already built in in Mathematica as `Interpolating-` `Polynomial`. *Applying this function to the same example, we get*

In[7]:= **a = InterpolatingPolynomial[list,x]**

Out[7]= $(x+2)\left((x+1)\left(\left(\dfrac{1-x}{3}+\dfrac{2}{3}\right)x-1\right)+1\right)$

Mathematica uses the Newton interpolation polynomial, see Exercise 6.2. Of course, in expanded form we get the same result as before:

In[8]:= **Expand[a]**

Out[8]= $\dfrac{4\,x^2}{3} - \dfrac{x^4}{3}$

Finally, the data from Example 6.28 yield

In[9]:= **list = Join[{{0,-120}},Table[{k,0},{k,1,5}]]**

Out[9]= $\begin{pmatrix} 0 & -120 \\ 1 & 0 \\ 2 & 0 \\ 3 & 0 \\ 4 & 0 \\ 5 & 0 \end{pmatrix}$

In[10]:= **Lagrange[list,x]**

Out[10]= $(x-5)\,(x-4)\,(x-3)\,(x-2)\,(x-1)$

as expected.

6.6 The Extended Euclidean Algorithm

Similarly to the case of integers, we can compute greatest common divisors in a polynomial ring. The greatest common divisor is only unique up to units, and by Lemma 6.6 the units of a polynomial ring agree with the units of the underlying coefficient ring. Hence the greatest common divisor $\gcd(a(x), b(x))$ of two polynomials $a(x), b(x) \in R[x]$ is unique up to a constant factor. If $R = \mathbb{K}$ is a field, then this factor can be chosen such that the gcd is a monic polynomial.

Session 6.31. *In Mathematica the greatest common divisor of a finite number of polynomials can be computed with the command* `PolynomialGCD`. *For example, the computation*

In[1]:= **PolynomialGCD[6x^3 - 17x^2 + 14x - 3,**
            4x^4 - 14x^3 + 12x^2 + 2x - 3, -8x^3 + 14x^2 - x - 3]

Out[1]= $2x-3$

provides the greatest common divisor $2x - 3$. *In this computation Mathematica's assumption is that the operands are polynomials from* $\mathbb{Z}[x]$. *Therefore the greatest common divisor is also given as an element of* $\mathbb{Z}[x]$. *The greatest common divisor in* $\mathbb{Z}[x]$ *is uniquely determined if the leading coefficient is chosen positive and as small as possible, since here the units are* ± 1.

However, if the input polynomials are considered as elements of $\mathbb{Q}[x]$, *then the (monic) greatest common divisor is* $x - \dfrac{3}{2}$.

If at least one of the input polynomials contains a rational coefficient, then Mathematica works with the ground field $\mathbb{Q}$. Therefore the result of the computation

$In[2]:=$ **PolynomialGCD** $\left[6\,x^3 - 16\,x^2 + \dfrac{23\,x}{2} - \dfrac{3}{2},\right.$

$$4\,x^4 - 14\,x^3 + 12\,x^2 + x - \frac{3}{2}, -8\,x^3 + 14\,x^2 - 2\,x - \frac{3}{2}\Big]$$

$Out[2]=\ x - \dfrac{3}{2}$

is the monic greatest common divisor as calculated earlier.

If we write $\gcd(a(x), b(x)) = 1$, then this means that $a(x)$ and $b(x)$ do not possess a nontrivial common divisor, where trivial common divisors are units.[17]

The *Euclidean algorithm for polynomials* is based on the property of polynomial division which was formulated in Theorem 6.21. The same construction that we used in the Euclidean algorithm and its extension, see Section 3.4 on page 52, can also be applied to polynomials. This time the algorithm terminates since by $\deg(r(x), x) < \deg(b(x), x)$ the degree is reduced in every step, which finally yields the zero polynomial. The extended Euclidean algorithm of Section 3.4 can be directly converted to the polynomial case as well, as the following result shows.

Theorem 6.32. (Extended Euclidean Algorithm) If the extended Euclidean algorithm is applied to two polynomials $a(x), b(x) \in \mathbb{K}[x]$, then we get the greatest common divisor $g(x) = \gcd(a(x), b(x))$ as well as two polynomials $s(x), t(x) \in \mathbb{K}[x]$ such that the polynomial equation

$$g(x) = s(x)\,a(x) + t(x)\,b(x)$$

with the *Bézout coefficients* $s(x), t(x) \in \mathbb{K}[x]$ is valid.

Proof. Adaption of the proofs of Theorems 3.12 and 3.13. $\qquad\qquad\square$

Session 6.33. *We give an example for the application of the polynomial extended Euclidean algorithm. For this purpose we slightly modify the implementation of* extendedgcd *from page 54:*

```
In[1]:= extendedgcd[m_,n_,var_] :=
          Module[{x,k,rule,r,q,X},x[0] = m; x[1] = n;
            k = 0; rule = {}; gcdmatrix = {};
            While[Not[x[k+1] === 0],k = k+1;
              r[k] = PolynomialRemainder[x[k-1],x[k],var];
              q[k] = PolynomialQuotient[x[k-1],x[k],var];
              x[k+1] = r[k];
              AppendTo[rule,
                Collect[X[k+1] → X[k-1] - q[k]*X[k]/.rule,
                  {X[0],X[1]}]];
              AppendTo[gcdmatrix,
                {k,x[k-1],q[k],x[k],r[k]}];];
            {x[k],Coefficient[
              X[k-2] - q[k-1]*X[k-1]//.rule,
              {X[0],X[1]}]}]
```

---

[17] Hence the correct *Mathematica* formula for this property is not PolynomialGCD[a,b]==1, but FreeQ[PolynomialGCD[a,b],x].

Please observe that the question posed in the `While`*-loop had to be slightly adapted from* `Equal` *(==) to* `SameQ` *(===) which refers to syntactical equality, since otherwise the equality is not identified correctly, compare Session 9.1. Furthermore, we would like to emphasize the fact that the old and the new version of* `extendedgcd` *can be used in parallel since the first one has two arguments and the second one has three arguments. This clearly produces different patterns!*

For instance, we get

$In[2]:=$ `{g,{s,t}} =`
      `extendedgcd[a = 6x`^{3}` + 19x`^{2}` + 19x + 6,`
        `b = 6x`^{5}` + 13x`^{4}` + 12x`^{3}` + 13x`^{2}` + 6x, x]`
$Out[2]= \left\{ -12\,x^2 - 26\,x - 12, \left\{ -x^2 + x - 2, 1 \right\} \right\}$

and we can check the Bézout identity:

$In[3]:=$ `s∗a+t∗b-g//Expand`
$Out[3]=$ 0

As before, `gcdmatrix` *generates a complete record of the process:*

$In[4]:=$ `gcdmatrix`

$$Out[4]= \begin{pmatrix} 1 & 6x^3+19x^2+19x+6 & 0 & 6x^5+13x^4+12x^3+13x^2+6x & 6x^3+19x^2+19x+6 \\ 2 & 6x^5+13x^4+12x^3+13x^2+6x & x^2-x+2 & 6x^3+19x^2+19x+6 & -12x^2-26x-12 \\ 3 & 6x^3+19x^2+19x+6 & -\frac{x}{2}-\frac{1}{2} & -12x^2-26x-12 & 0 \end{pmatrix}$$

The application of the polynomial extended Euclidean algorithm demonstrates the intermediate expression swell very well, which is typical for many symbolic algorithms: even if the output is very small (for example, if the gcd is 1), the intermediate expressions that had to be computed can grow in an unregulated way, and this is the normal case rather than the exception. We consider an example:

$In[5]:=$ `res = extendedgcd[`

$$a = \sum_{k=0}^{5} \text{Random[Integer, {1,10}]x}^k,$$

$$b = \sum_{k=1}^{6} \text{Random[Integer, {1,10}]x}^k, x]$$

$$Out[5]= \left\{ \frac{2507959199728808515431}{28334208625805848249}, \right.$$

$$\left\{ \frac{18053006966792\,x^5}{952814976047} + \frac{959058548245451198300\,x^4}{28334208625805848249} + \right.$$

$$\frac{767683960567299635034\,x^3}{28334208625805848249} + \frac{919570529119225802\,x^2}{1231922114165471663} -$$

$$\frac{6019387033427006372\,x}{28334208625805848249} + \frac{278662133303200946159}{28334208625805848249},$$

$$-\frac{6769877612547\,x^4}{952814976047} - \frac{661624458256135114986\,x^3}{28334208625805848249} -$$

$$\frac{333457091683615032\,x^2}{10181174497235303} - \frac{896389712204198139585\,x}{28334208625805848249} -$$

$$\left. \left. \frac{224487650002357888811}{28334208625805848249} \right\} \right\}$$

The greatest common divisor of these random polynomials is "1" as expected, but the progression of the algorithm generates as gcd a very complicated (clearly associated) rational number, and the Bézout coefficients of the extended Euclidean algorithm are rational numbers with large numerators and denominators. The gcdmatrix *(which we suppressed) shows the details.*

Mathematica's built-in procedure PolynomialExtendedGCD *uses a more efficient method which avoids the computation of these rational numbers:*

$In[6]:=$ **PolynomialExtendedGCD[a, b]**

$Out[6]= \left\{1, \left\{\dfrac{1}{34927000}\,(6775140\,x^5 + 6713800\,x^4 + 3336680\,x^3 + 11220760\,x^2 + 1842820\,x + 3492700),\right.\right.$

$\left.\left.\dfrac{1}{34927000}\,(-2258380\,x^4 - 10518660\,x^3 - 20108020\,x^2 - 2116600\,x - 26677600)\right\}\right\}$

Note that this command generates $\gcd(a(x), b(x)) = 1$ *as desired, and the Bézout coefficients also look friendlier.*

We would like to mention that it is possible—for example using modular arithmetic—to reduce the complexity for the computation of the polynomial gcd substantially, see also [GG1999], Chapter 7. As a principal phenomenon, however, intermediate expression swell cannot be turned off. Note that our complexity computations only count ring or field operations and do not record the size of the arising ring or field elements.

6.7 Unique Factorization

Now we are interested in irreducible elements and the factorization in polynomial rings. As usual, let $\mathbb{K}$ be a field.

Theorem 6.34. (Unique Factorization) Every $a(x) \in \mathbb{K}[x]$ possesses a representation as product of irreducible elements $p_k(x)$ $(k = 1, \ldots, n)$:

$$a(x) = \prod_{k=1}^{n} p_k(x). \tag{6.10}$$

This representation is unique in the following sense: every further representation

$$a(x) = \prod_{k=1}^{m} q_k(x)$$

satisfies $m = n$, and there is a bijective map $p_j \mapsto q_k$ between the polynomials $p_1, \ldots, p_n$ and $q_1, \ldots, q_n$ such that p_j and q_k are associated elements, hence they differ only by a multiplicative unit.

Proof. This result follows by adaptation of the proofs of Theorem 3.17 and Lemma 3.18. The induction is executed with respect to the degree of $a(x)$. □

The uniqueness statement in Theorem 6.34 is simpler if all prime factors are taken monic so that p_j and q_k are not only associated, but equal: $p_j = q_k$. This is possible if $\mathbb{K}$ is a field, in which case we can write

$$a(x) = a \prod_{j=1}^{m} p_j(x)^{e_j}, \quad \text{with } \gcd(p_j(x), p_k(x)) = 1 \ (j \neq k) \text{ and } e_j \in \mathbb{N}. \tag{6.11}$$

Since we will also consider the case $\mathbb{Z}[x] \subset \mathbb{Q}[x]$, we will not assume the latter representation in general.

If a factorization (6.10) of a polynomial $p(x) \in \mathbb{K}[x]$ is known, then its zeros $x_k \in \mathbb{K}$ can be read off immediately. By Lemma 6.24 they correspond in a unique way to the linear factors:

$$p_k(x) = a \cdot x + b \quad \Leftrightarrow \quad x_k = -\frac{b}{a} \text{ is a zero}.$$

Now let $x_j \in \mathbb{K}$ be a zero of $p(x) \in \mathbb{K}[x]$. Then $p(x)$ has the linear factor $x - x_j$ (or an associated one) in (6.10). Of course this factor can appear multiple times in (6.10), and the number of its appearances is called the *order of the zero*. If the order of a zero is 1, then the zero is called simple, otherwise multiple.

How can we find the factorization of a polynomial? In the remaining part of this section, we will consider some examples.

If $\mathbb{K}$ is a finite field, for example $\mathbb{K} = \mathbb{Z}_p$ for $p \in \mathbb{P}$, and if $a(x) \in \mathbb{K}[x]$, then—since $\mathbb{K}$ possesses only finitely many elements—there are only finitely many possible divisor polynomials $b(x) \in \mathbb{K}[x]$ of $a(x)$. We can generate the corresponding candidate polynomials and decide by division which of them are divisors of $a(x)$. Obviously with this method we can completely factorize $a(x)$, although in a quite inefficient way.[18]

Session 6.35. (Factorization in $\mathbb{Z}_p[x]$) *We implement the above algorithm.*

The function Factors *computes the monic factors of $a(x)$ up to degree s according to the given algorithm:*

```
In[1]:= Clear[Factors]

         Factors[a_,x_,p_,s_] :=
           Module[{tab,k,blist,remainderlist,pos},
             tab = Table[k,{k,0,p-1}];
             blist = tab;
             Do[
               blist =
                 Flatten[Outer[Plus,blist,tab*x^k]],
               {k,s}];
             blist = Complement[blist,{0}];
             remainderlist =
               Map[PolynomialRemainder[a,#,x,
                   Modulus->p]&,blist];
             pos = Position[remainderlist,0];
             tab = Table[blist[[pos[[k,1]]]],
                 {k,1,Length[pos]}];
             Select[tab,
               Coefficient[#,x,Exponent[#,x]] == 1&]
           ]
```

---

[18] If s is the degree of the divisor polynomial, then we must divide by p^s test polynomials. Since the size of the divisor can be $s = \frac{n}{2}$, the complexity of the considered algorithm is $O(p^{n/2})$. For small p this is possible; for large p, however, this algorithm cannot be executed.

In the first Do *loop the list of potential divisor polynomials is generated using* Outer. *Afterwards a(x) is divided by these test polynomials to detect the divisors of a(x). The monic ones form the output of the procedure.*

Let us consider an example:

```
In[2]:= a = x^5 + x^3 + x
```
$Out[2] = x^5 + x^3 + x$

```
In[3]:= p = 11
```
$Out[3] = 11$

Now we generate all monic divisors of degree ≤ 2 of $a(x) = x^5 + x^3 + x \in \mathbb{Z}_{11}[x]$ using Factors:

```
In[4]:= Factors[a, x, 11, 2]
```
$Out[4] = \{1, x, x^2 + x + 1, x^2 + 10x + 1\}$

Indeed, we get

```
In[5]:= PolynomialMod[x(x^2 + x + 1)(x^2 + 10x + 1), p]
```
$Out[5] = x^5 + x^3 + x$

We can also divide successively:

```
In[6]:= q = PolynomialQuotient[a, x^2 + x + 1, x, Modulus → p]
```
$Out[6] = x^3 + 10x^2 + x$

```
In[7]:= PolynomialQuotient[q, x, x, Modulus → p]
```
$Out[7] = x^2 + 10x + 1$

Of course our implementation can also detect whether a polynomial is irreducible modulo p. In such a case there are only trivial (hence constant) factors up to degree $s = \lfloor n/2 \rfloor$:

```
In[8]:= Factors[x^4 + 2, x, 5, 2]
```
$Out[8] = \{1\}$

Mathematica's built-in function Factor *is also able to factorize over $\mathbb{Z}_p$:*

```
In[9]:= Factor[a, Modulus → p]
```
$Out[9] = x(x^2 + x + 1)(x^2 + 10x + 1)$

This implementation is much faster: the computation

```
In[10]:= Factor[a, Modulus → 10007]
```
$Out[10] = x(x^2 + x + 1)(x^2 + 10006x + 1)$

is clearly beyond the capabilities of our implementation. We will consider more efficient factorization algorithms in Chapter 8.

Next, we consider the special case $\mathbb{K} = \mathbb{Q}$, i.e. polynomials with rational coefficients. If $a(x) \in \mathbb{Q}[x]$ is given, then by multiplication with an integer $z \in \mathbb{N}$ we can generate a polynomial $b(x) = z \cdot a(x) \in \mathbb{Z}[x]$.[19] Hence we can assume without loss of generality that $a(x) \in \mathbb{Z}[x]$. Therefore in the sequel we search for the divisors $b(x)$ of $a(x) \in \mathbb{Z}[x]$.[20]

---

[19] The best number z is the least common multiple of the denominators of the coefficients of $a(x)$.

[20] In Section 8.1 we will show that for factorization in $\mathbb{Q}[x]$ it is generally sufficient to solve this subproblem.

For the factorization of a polynomial $a(x) \in \mathbb{Z}[x]$ over $\mathbb{Z}$, Kronecker [Kro1882]–[Kro1883] gave the following algorithm:

If $a(x)$ has degree n and possesses a nontrivial factor, then at least one of the divisors must have degree $\leq s := \lfloor n/2 \rfloor$. We now build the function values $a(x_0), a(x_1), \ldots, a(x_s)$ at $s+1$ arbitrarily chosen integer points $x_j \in \mathbb{Z}$. Since $a(x)$ is divisible by $b(x)$, we get for the chosen points x_j ($j = 0, \ldots, s$) the divisibility conditions $b(x_j) \mid a(x_j)$. Since the values $a(x_j) \in \mathbb{Z}$ are given, there is only a finite number of potential divisors $b(x_j) \in \mathbb{Z}$. For every possible choice $b(x_j) \in \mathbb{Z}$ ($j = 0, \ldots, s$), by Theorem 6.29 there is exactly one possible candidate $b(x)$ of degree $\leq s$ to be found by interpolation. For every candidate polynomial $b(x)$ we can check by division with remainder (trial division) whether $b(x) \mid a(x)$. In this way we can find all divisors $b(x) \in \mathbb{Z}[x]$ of degree $\leq s$ of $a(x)$. Divisors of degree $> s$ have the form $a(x)/b(x)$ and can be found by polynomial division.

Of course this algorithm is rather inefficient in practice. At a later stage of this book, we will discuss more efficient algorithms. However, the efficiency of the above *Kronecker algorithm* can be enhanced by the following measures:

- In a first step the greatest common divisor of the coefficients of $a(x)$ should be factored out. This number is called the *content* of the polynomial and is denoted by content($a(x), x$). This step avoids unneccesary divisors of the values of $a(x_j)$.

- By a smart choice of the values x_j ($j = 0, \ldots, s$) the number of divisors of $a(x_j)$ and therefore the number of case distinctions can be kept small.

- Candidate polynomials $b(x) \in \mathbb{Q}[x] \setminus \mathbb{Z}[x]$ can be skipped.

Session 6.36. (Kronecker Algorithm) *We implement the Kronecker algorithm in Mathematica. The following implementation for the factorization of $a(x) \in \mathbb{Z}[x]$ is valid under the assumption that $a(0) \neq 0$. Otherwise a power of x can be factored out.*

We consider again $a(x) = x^5 + x^3 + x \in \mathbb{Z}[x]$, this time as an element of $\mathbb{Z}[x]$. In a first step we can factor out x. Hence it remains to search for the divisors of $a(x) = x^4 + x^2 + 1$. We set

```
In[1]:= a = x^4 + x^2 + 1
Out[1]= x^4 + x^2 + 1
```

and define the interpolating points x_j:

```
In[2]:= points = {-1, 0, 1}
Out[2]= {-1, 0, 1}
```

Next, we compute the values $a(x_j)$:

```
In[3]:= values = a /. x → points
Out[3]= {3, 1, 3}
```

The function `divisors` *finds the positive and negative divisors of an integer:*

```
In[4]:= divisors[z_] := Module[{d = Divisors[z]}, Join[d, -d]]
```

We compute the divisor lists of the values $a(x_j)$:

```
In[5]:= divisorlist = Table[divisors[values[[k]]], {k, Length[values]}]
Out[5]= {{1, 3, -1, -3}, {1, -1}, {1, 3, -1, -3}}
```

Now we can—again using Outer—*collect all possible point triples which are divisors of* $a(x_j)$. *Note that the command* Apply[Sequence, ...] *converts a list towards a sequence of arguments (without the list braces).*

```
In[6]:= table =
          Partition[
            Flatten[
              Outer[List, Apply[Sequence, divisorlist]]], 3]
```

$$
Out[6]= \begin{pmatrix}
1 & 1 & 1 \\
1 & 1 & 3 \\
1 & 1 & -1 \\
1 & 1 & -3 \\
1 & -1 & 1 \\
1 & -1 & 3 \\
1 & -1 & -1 \\
1 & -1 & -3 \\
3 & 1 & 1 \\
3 & 1 & 3 \\
3 & 1 & -1 \\
3 & 1 & -3 \\
3 & -1 & 1 \\
3 & -1 & 3 \\
3 & -1 & -1 \\
3 & -1 & -3 \\
-1 & 1 & 1 \\
-1 & 1 & 3 \\
-1 & 1 & -1 \\
-1 & 1 & -3 \\
-1 & -1 & 1 \\
-1 & -1 & 3 \\
-1 & -1 & -1 \\
-1 & -1 & -3 \\
-3 & 1 & 1 \\
-3 & 1 & 3 \\
-3 & 1 & -1 \\
-3 & 1 & -3 \\
-3 & -1 & 1 \\
-3 & -1 & 3 \\
-3 & -1 & -1 \\
-3 & -1 & -3
\end{pmatrix}
$$

In the next step, we generate the table of potential divisor polynomials of $a(x)$ *by polynomial interpolation:*

```
In[7]:= poltable =
          Table[
            {Expand[InterpolatingPolynomial[
                Transpose[{{-1, 0, 1}, table[[k]]}], x]]},
            {k, Length[table]}]
```

$$
Out[7]= \begin{pmatrix}
1 \\
x^2+x+1 \\
-x^2-x+1 \\
-2\,x^2-2\,x+1 \\
2\,x^2-1 \\
3\,x^2+x-1 \\
x^2-x-1 \\
-2\,x-1 \\
x^2-x+1 \\
2\,x^2+1 \\
1-2\,x \\
-x^2-3\,x+1 \\
3\,x^2-x-1 \\
4\,x^2-1 \\
2\,x^2-2\,x-1 \\
x^2-3\,x-1 \\
-x^2+x+1 \\
2\,x+1 \\
1-2\,x^2 \\
-3\,x^2-x+1 \\
x^2+x-1 \\
2\,x^2+2\,x-1 \\
-1 \\
-x^2-x-1 \\
-2\,x^2+2\,x+1 \\
-x^2+3\,x+1 \\
-3\,x^2+x+1 \\
1-4\,x^2 \\
2\,x-1 \\
x^2+3\,x-1 \\
-x^2+x-1 \\
-2\,x^2-1
\end{pmatrix}
$$

The division of $a(x)$ by the candidate polynomials leads to the following remainders:

```
In[8]:= remainders =
            Table[PolynomialRemainder[a,poltable[[k,1]],
            x],{k,Length[poltable]}]
```

$$
Out[8]= \Big\{0,0,4-4\,x,\frac{9}{4}-3\,x,\frac{7}{4},\frac{40}{27}-\frac{16\,x}{27},4\,x+4,\frac{21}{16},0,\frac{3}{4},\frac{21}{16},12-36\,x,
$$

$$
\frac{16\,x}{27}+\frac{40}{27},\frac{21}{16},3\,x+\frac{9}{4},36\,x+12,4\,x+4,\frac{21}{16},\frac{7}{4},\frac{40}{27}-\frac{16\,x}{27},4-4\,x,
$$

$$
\frac{9}{4}-3\,x,0,0,3\,x+\frac{9}{4},36\,x+12,\frac{16\,x}{27}+\frac{40}{27},\frac{21}{16},\frac{21}{16},12-36\,x,0,\frac{3}{4}\Big\}
$$

The remainders are zero at the positions

```
In[9]:= pos = Position[remainders,0]
```

$$
Out[9]= \begin{pmatrix}
1 \\
2 \\
9 \\
23 \\
24 \\
31
\end{pmatrix}
$$

Hence the list command

```
In[10]:= Table[poltable[[pos[[k]],1]],{k,Length[pos]}]
```

$$Out[10] = \begin{pmatrix} 1 \\ x^2 + x + 1 \\ x^2 - x + 1 \\ -1 \\ -x^2 - x - 1 \\ -x^2 + x - 1 \end{pmatrix}$$

creates a complete list of the divisors of $a(x)$ in $\mathbb{Z}[x]$. Our computations are confirmed by the Mathematica function `Factor`

`In[11]:=` **`Factor[a]`**
`Out[11]=` $(x^2 - x + 1)(x^2 + x + 1)$

much faster.

Please note that our implementation has proved the irreducibility of the factors $x^2 - x + 1$ and $x^2 + x + 1$ over $\mathbb{Z}$. This general capability is already quite interesting.

The complexity of the Kronecker algorithm is even worse than the one considered for $\mathbb{Z}_p[x]$, since the values $a(x_j)$ can have arbitrary many divisors. We do not give a detailed complexity analysis, though. There are more efficient algorithms for polynomial factorization, some of which we will consider in the next section and in Chapter 8.

6.8 Squarefree Factorization

We have seen that the complete factorization of polynomials is relatively expensive. But there is a type of factorization which can be realized with much less effort, namely the *squarefree factorization*. Luckily this type of factorization often suffices, for example for rational integration, see Chapter 12.

Definition 6.37. (Squarefree Factorization) Let $a(x) \in \mathbb{K}[x]$. Then $a(x)$ is called *squarefree* if there is no $b(x) \in \mathbb{K}[x]$ with $\deg(b(x), x) > 0$ and $b^2(x) \mid a(x)$.

The *squarefree factorization* of $a(x)$ is given by

$$a(x) = \prod_{k=1}^{m} a_k^k(x), \tag{6.12}$$

where every $a_k(x)$ is squarefree and $\gcd(a_j(x), a_k(x)) = 1$ for $j \neq k$. Note that some of the $a_k(x)$ may be constant.

The *squarefree part* of $a(x)$ is given by

$$\mathrm{sfp}(a(x), x) := \prod_{k=1}^{m} a_k(x),$$

i.e., the simple product of all irreducible factors of $a(x)$. $\triangle$

From the complete factorization of $a(x)$ (into irreducible factors $p_j(x)$ as in (6.11)),

$$a(x) = a \prod_{j=1}^{m} p_j(x)^{e_j}, \quad \text{with } \gcd(p_j(x), p_k(x)) = 1 \ (j \neq k) \text{ and } e_j \in \mathbb{N},$$

we can read off that a polynomial is squarefree if and only if it has no multiple factors, i.e., $e_j = 1$ ($j = 1, \ldots, m$). Therefore the factor $a_k(x)$ in (6.12) contains those factors of $a(x)$ that occur exactly k times in (6.11), hence $e_j = k$.

Example 6.38. Let
$$a(x) := (x^2 - 25)^5 (x^2 - 9)^3 (x^2 - 1).$$
Then we have $m = 5$ and $a_1(x) = x^2 - 1$, $a_2(x) = 1$, $a_3(x) = x^2 - 9$, $a_4(x) = 1$ and $a_5(x) = x^2 - 25$.

The squarefree part of $a(x)$ is $(x^2 - 25)(x^2 - 9)(x^2 - 1)$. △

For the examination of squarefreeness we use derivatives. If we introduce the following definitions, we can do so without using limits.[21]

Definition 6.39. Let $\mathbb{K}$ be a field. A mapping $' : \mathbb{K}[x] \to \mathbb{K}[x]$ of the polynomial ring $\mathbb{K}[x]$ into itself with the properties

(A_1) **(Linearity)** $(a(x) + b(x))' = a'(x) + b'(x)$ and $(c \cdot a(x))' = c \cdot a'(x)$ for $c \in \mathbb{K}$;
(A_2) **(Product Rule)** $(a(x) \cdot b(x))' = a'(x) \cdot b(x) + a(x) \cdot b'(x)$

with $x' \neq 0$ is called a *derivation* (or *derivative*). △

Theorem 6.40. Every derivation has the following well-known properties:

(A_3) **(Constant Rule)** $c' = 0$ for $c \in \mathbb{K}$;
(A_4) **(Power Rule)** $(a(x)^n)' = n a(x)^{n-1} \cdot a'(x)$;
(A_5) **(Product Rule)** $(a_1(x) \cdots a_n(x))' = \sum_{k=1}^{n} a_1(x) \cdots a_{k-1}(x) \cdot a_k'(x) \cdot a_{k+1}(x) \cdots a_n(x)$.

Proof. From now on, we will also use the operator notation $D(a)$ for $a'(x)$.
(A_3): From (A_2) it follows that

$$D(a) = D(a \cdot 1) = D(a) \cdot 1 + a \cdot D(1),$$

hence $D(1) = 0$. Using (A_1) we then get $c' = c \cdot D(1) = 0$ for all $c \in \mathbb{K}$.
(A_4) and (A_5) follow by a simple induction proof.   □

In the sequel we will consider the "regular" derivative which is uniquely determined by the additional property $x' := 1$. Using (A_4), this property implies $(x^n)' = nx^{n-1}$, and finally the linearity yields the usual derivative for every polynomial. In particular, we get the very important property that the degree of $a'(x)$ is smaller than the degree of the polynomial $a(x) \neq 0$.

Next, we consider a relation between the multiplicity of a factor of a polynomial and its derivative.

Lemma 6.41. Let $\mathbb{K}$ be a field and $n \in \mathbb{N}_{\geq 1}$. If $p(x)$ is an n-fold divisor of $a(x)$, then $p(x)$ is an $n-1$-fold divisor of $a'(x)$, if $\mathrm{char}(\mathbb{K})$ is not a divisor of n. In particular, if $\alpha \in \mathbb{K}$ is an n-fold zero of $a(x)$, then α is an $n-1$-fold zero of $a'(x)$.

Proof. Since $a(x)$ has the n-fold divisor $p(x)$, the representation $a(x) = p(x)^n \cdot b(x)$ is valid with $b(x) \in \mathbb{K}[x]$, $\gcd(p(x), b(x)) = 1$. Now we apply the derivative operator and get by the product rule and (A_4):

---

[21] We will also use derivatives in Chapter 12 on integration. There the notion of a differential field is studied exhaustively.

$$a'(x) = n\,p(x)^{n-1} \cdot p'(x) \cdot b(x) + p(x)^n \cdot b'(x) = p(x)^{n-1} \cdot \Big(n\,b(x) \cdot p'(x) + p(x) \cdot b'(x)\Big).$$

Hence the assumption follows since $\gcd(p(x), b(x)) = 1$ and $\deg(p'(x), x) < \deg(p(x), x)$.   □

Please note that $a'(x) = 0$ does not always imply that $a(x)$ is constant. This needs the extra assumption $\operatorname{char}\mathbb{K} = 0$. For example, for $a(x) = x^5 \in \mathbb{Z}_5[x]$ we get the derivative $a'(x) = 5 \cdot x^4 = 0 \cdot x^4 = 0$.

However, if $\operatorname{char}\mathbb{K} = 0$, then $a'(x) = 0$ always implies that $a(x)$ is constant, and we get the following characterization of squarefreeness.

Theorem 6.42. (Characterization of Squarefreeness) Let $a(x) \in \mathbb{K}[x]$ and let $\mathbb{K}$ be a field with $\operatorname{char}\mathbb{K} = 0$. Then the polynomial $a(x)$ is squarefree if and only if $\gcd(a(x), a'(x)) = 1$.

Proof. Let us assume first that $a(x)$ is not squarefree. Then there is a $b(x) \in \mathbb{K}[x]$ with $\deg(b(x), x) > 0$ and

$$a(x) = b(x)^2 \cdot c(x).$$

It follows from (A_4) that

$$a'(x) = 2\,b(x) \cdot b'(x) \cdot c(x) + b(x)^2 \cdot c'(x) = b(x) \cdot \Big(2\,b'(x) \cdot c(x) + b(x) \cdot c'(x)\Big).$$

Hence, $a'(x)$ and $a(x)$ possess the non-constant common divisor $b(x)$, and we get

$$\gcd(a(x), a'(x)) \neq 1.$$

Now let $a(x)$ be squarefree. We further assume that $\gcd(a(x), a'(x)) \neq 1$, and we want to construct a contradiction. By assumption there is a factorized representation

$$a(x) = p_1(x) \cdot p_2(x) \cdots p_m(x) \tag{6.13}$$

of $a(x)$ where $p_k(x)$ are pairwise irreducible and have positive degree. Then we can use (A_5) to compute the derivative

$$a'(x) = p_1'(x) \cdot p_2(x) \cdots p_m(x) + p_1(x) \cdot p_2'(x) \cdots p_m(x) + \cdots + p_1(x) \cdot p_2(x) \cdots p_m'(x). \tag{6.14}$$

Since the polynomials $p_k(x)$ in (6.13) are irreducible, we must have that $p_k(x) \mid \gcd(a(x), a'(x))$ for at least one $k = 1, \ldots, m$. Without loss of generality we assume that this is the case for $k = 1$. Then it follows that $p_1(x) \mid a'(x)$, and using (6.14) we get

$$p_1(x) \mid p_1'(x) \cdot p_2(x) \cdots p_m(x).$$

Since $\gcd(p_k(x), p_1(x)) = 1$ $(k > 1)$ and $p_1(x)$ is irreducible over $\mathbb{K}[x]$, $p_1(x) \mid p_1'(x)$ must be valid. Since $\deg(p_1(x), x) > 0$, the degree of $p_1'(x)$ is less than the degree of $p_1(x)$, and this can only happen if $p_1'(x) = 0$. However, in a field of characteristic 0 this implies that $p_1(x)$ is constant (see Exercise 6.18), which contradicts the fact that $\deg(p_1(x), x) > 0$.   □

After these preparations we can now compute the squarefree factorization. Let $a(x)$ have the squarefree factorization (6.12). Since

$$a'(x) = \sum_{k=1}^{m} a_1(x) \cdots k\, a_k(x)^{k-1}\, a_k'(x) \cdots a_m(x)^m$$

we obtain

$$g(x) := \gcd(a(x), a'(x)) = \prod_{k=2}^{m} a_k(x)^{k-1} .$$

It follows that the squarefree part of $a(x)$ is given by

$$h(x) := \frac{a(x)}{g(x)} = \prod_{k=1}^{m} a_k(x) .$$

Next, we can compute

$$b(x) := \gcd(g(x), h(x)) = \prod_{k=2}^{m} a_k(x) ,$$

and get

$$a_1(x) = \frac{h(x)}{b(x)} .$$

If we iterate this process by continuing with $g(x)$, we find $a_1(x), a_2(x), \ldots, a_m(x)$ successively. The following theorem sums up this process.

Theorem 6.43. (Squarefree Part and Squarefree Factorization) Let $a(x) \in \mathbb{K}[x]$ where $\mathbb{K}$ is a field of characteristic 0. Then the squarefree part of $a(x)$ is given by

$$\text{sfp}(a(x), x) = \frac{a(x)}{\gcd(a(x), a'(x))} .$$

With the above described iterative algorithm, we can compute the squarefree factorization (6.12) of $a(x)$.

Proof. The result follows from the above explanations.   □

Session 6.44. *The Mathematica function* `FactorSquareFree[a]` *provides a squarefree factorization of the polynomial $a(x)$. Let*

```
In[1]:= a = Product[((x+k)(x-k))^k, {k, 1, 5, 2}]
```
$Out[1] = (x-5)^5 (x-3)^3 (x-1)(x+1)(x+3)^3 (x+5)^5$

For the expanded polynomial

```
In[2]:= a = Expand[a]
```
$Out[2] = x^{18} - 153 x^{16} + 10020 x^{14} - 365972 x^{12} +$
$\qquad 8137854 x^{10} - 112806750 x^8 + 957212500 x^6 -$
$\qquad 4649062500 x^4 + 10916015625 x^2 - 7119140625$

we get, for example, the factorization

```
In[3]:= b = FactorSquareFree[a]
```
$Out[3] = (x^2 - 25)^5 (x^2 - 9)^3 (x^2 - 1)$

Of course we can read off the complete factorization from the input

$$a(x) = (x+5)^5 (x-5)^5 (x+3)^3 (x-3)^3 (x+1)(x-1) ,$$

from which we can determine the squarefree factorization by collecting identical powers.

The squarefree part of $a(x)$ is given by

$In[4]:=$ **c = b/. {3 → 1, 5 → 1}**
$Out[4]=$ $(x^2 - 25)(x^2 - 9)(x^2 - 1)$

or in expanded form

$In[5]:=$ **Expand[c]**
$Out[5]=$ $x^6 - 35 x^4 + 259 x^2 - 225$

The squarefree part can also be determined by the algorithm given in Theorem 6.43 without having to compute a complete factorization:

$In[6]:=$ **Together** $\left[\dfrac{\mathbf{a}}{\textbf{PolynomialGCD[a, D[a, x]]}}\right]$
$Out[6]=$ $x^6 - 35 x^4 + 259 x^2 - 225$

6.9 Rational Functions

When we start with the ring $\mathbb{Z}$ of integers, the field $\mathbb{Q}$ of rational numbers can be generated by building quotients modulo the equivalence relation

$$\frac{a}{b} \sim \frac{c}{d} \qquad \Leftrightarrow \qquad ad - bc = 0.$$

In the same manner the *field of rational functions* $\mathbb{K}(x)$ over the field $\mathbb{K}$ is generated from the polynomial ring $\mathbb{K}[x]$ by building quotients modulo the equivalence relation

$$\frac{a(x)}{b(x)} \sim \frac{c(x)}{d(x)} \qquad \Leftrightarrow \qquad a(x)d(x) - b(x)c(x) = 0.$$

Please note that in this case it is essential to distinguish between square brackets (denoting rings) and round brackets (denoting fields)! Of course, via $b(x) = 1$, $\mathbb{K}[x] \subset \mathbb{K}(x)$ is a subring of $\mathbb{K}(x)$. Analogously, the field of rational functions in several variables $\mathbb{K}(x_1, x_2, \ldots, x_n)$ is obtained.

Now, let the rational function

$$r(x) = \frac{a(x)}{b(x)} \in \mathbb{K}(x)$$

be given. Since the Euclidean algorithm determines the greatest common divisor $g(x) = \gcd(a(x), b(x))$, we can bring the rational function into lowest terms by dividing both the numerator and the denominator by $g(x)$. Therefore, without loss of generality we can assume that $\gcd(a(x), b(x)) = 1$, and we call

$$r(x) = \frac{a(x)}{b(x)} \in \mathbb{K}(x) \qquad (\gcd(a(x), b(x)) = 1) \tag{6.15}$$

the *standard representation* or *normal form* of a rational function.

Session 6.45. *Whereas Mathematica automatically reduces every rational number into lowest terms, e.g.*

$In[1]:=$ $\dfrac{\mathbf{123}}{\mathbf{456}}$
$Out[1]=$ $\dfrac{41}{152}$

for efficiency reasons rational functions are not automatically simplified:

$In[2]:=$ $\mathbf{r} = \dfrac{\mathbf{x^3 - 2x^2 - 5x + 6}}{\mathbf{x - 1}}$

$Out[2]=$ $\dfrac{x^3 - 2x^2 - 5x + 6}{x - 1}$

Rather, a simplification with the command `Together` *is necessary in order to bring rational expressions (with the aid of the Euclidean algorithm or more efficient methods) into rational normal form:*

$In[3]:=$ **Together[r]**

$Out[3]=$ $x^2 - x - 6$

This works similarly in the multivariate case:

$In[4]:=$ **Together** $\left[\dfrac{\mathbf{x^{10} - y^{10}}}{\mathbf{x^4 - y^4}}\right]$

$Out[4]=$ $\dfrac{x^6 y^2 + x^4 y^4 + x^2 y^6 + x^8 + y^8}{x^2 + y^2}$

In particular, `Together` *can decide whether a rational expression is the zero polynomial:*

$In[5]:=$ **Together** $\left[\dfrac{\mathbf{x}}{\mathbf{(x-y)(x-z)}} + \dfrac{\mathbf{y}}{\mathbf{(y-z)(y-x)}} + \dfrac{\mathbf{z}}{\mathbf{(z-x)(z-y)}}\right]$

$Out[5]=$ 0

This property will be discussed in more detail in Section 9.3.

6.10 Additional Remarks

The substitution of a ring element in a polynomial, which plays a role in the definition of a zero, is called the *substitution homomorphism*.

In the division with remainder we referred to a ground field. If R is an integral domain, then—by multiplying the first polynomial by a suitable ring element—one can execute a *pseudo division* in $R[x]$. Many consequences remain the same, in particular, Lemma 6.24 and Corollary 6.25.

Since the FFT-based multiplication algorithms cannot be directly applied for the multiplication in $\mathbb{Z}[x]$ (or in $\mathbb{Z}$), Schönhage and Strassen developed a method [SS1971] to eliminate this defect by adjunction of virtual roots of unity, see e.g., [GG1999, Section 8.3]. This *Schönhage-Strassen algorithm* is the fastest up-to-date multiplication algorithm.

In rings with a *Euclidean norm* ($|z|$ for $z \in \mathbb{Z}$ or $\deg(a(x), x)$ for $a(x) \in \mathbb{K}(x)$), the Euclidean algorithm can be used to compute greatest common divisors. These rings are always *unique factorization domains* (UFD),[22] hence their elements possess unique factorizations.

More efficient factorization algorithms will be considered in Chapter 8.

---

[22] Sometimes they are also called *factorial rings*.

6.11 Exercises

Exercise 6.1. *Prove the degree formulas (6.5) and (6.6). Give an example where (6.5) is not an equality.*

Exercise 6.2. *The Newton form of the interpolating polynomial is given by*

$$P_n(x) = c_0 + c_1(x - x_0) + c_2(x - x_0)(x - x_1) + \cdots + c_n(x - x_0)\cdots(x - x_n),$$

where $x_0, \ldots, x_n$ denote the $n + 1$ support points.

The interpolation conditions $P_n(x_k) = y_k$ ($k = 0, \ldots, n + 1$) lead to a linear system in triangular form for the constants $c_0, \ldots, c_n$.[23] Implement Newton interpolation by solving the corresponding linear system. The main advantage of Newton interpolation is—besides numerical stability— that an already computed interpolation polynomial with n support points can be directly used to compute the interpolation polynomial with $n + 1$ support points, once a support point is added. This is not possible with Lagrange interpolation.

Apply your implementation to some suitable examples and compare your results with those of Mathematica's `InterpolatingPolynomial`.

Exercise 6.3. *Design a variation of the extended Euclidean algorithm which avoids unnecessary rational coefficients, see Session 6.33. Compare your code with* `PolynomialExtendedGCD`.

Exercise 6.4. *Spell out the proof of Theorem 6.34. The following lemma is necessary: Let $p(x) \in \mathbb{K}[x]$ be irreducible and $p(x) \mid a(x) \cdot b(x)$. Then either $p(x) \mid a(x)$ or $p(x) \mid b(x)$.*

Exercise 6.5. (Factorization in $\mathbb{Z}_p[x]$) *Factorize*

(a) $x^6 - x^5 + x^2 + 1$ *modulo p for the first 20 primes p;*
(b) $5x^4 - 4x^3 - 48x^2 + 44x + 3$ *modulo p for the first 20 primes.*

Which consequences for a possible factorization over $\mathbb{Z}$ can you see from your computations?

Exercise 6.6. (Factorization in $\mathbb{Z}_p[x]$) *Factorize*

(a) $x^5 - x \in \mathbb{Z}_5[x]$;
(b) $x^8 + x^7 + x^6 + x^4 + 1 \in \mathbb{Z}_2[x]$;
(c) $x^8 + x^7 + x^6 + x^4 + 1 \in \mathbb{Z}_{11}[x]$;
(d) $x^8 + x^7 + x^6 + x^4 + 1 \in \mathbb{Z}_{10007}[x]$.

Exercise 6.7. (Eisenstein Irreducibility Criterion) *Assume that $a(x) = a_n x^n + a_{n-1} x^{n-1} + \cdots + a_1 x + a_0 \in \mathbb{Z}[x]$ and there exists a prime number $p \in \mathbb{P}$ which is not a divisor of a_n, but $p \mid a_k$ for all $k = 0, \ldots, n - 1$, and let p^2 be no divisor of a_0. Show that $a(x)$ is irreducible in $\mathbb{Z}[x]$. Hint: Assume reducibility to reach a contradiction.*

Exercise 6.8. *Show, using Eisenstein's irreducibility criterion (Exercise 6.7), that in $\mathbb{Z}[x]$ there are irreducible polynomials of every degree $n \in \mathbb{N}$.*

Exercise 6.9. *Implement Eisenstein's irreducibility criterion (Exercise 6.7) as Mathematica function* `Eisenstein`. *Check the irreducibility of $a(x) \in \mathbb{Z}[x]$ for*

---

[23] They can also be computed efficiently by using *divided differences*.

(a) $a(x) = x^5 - 2$;
(b) $a(x) = x^5 - 1$;
(c) $a(x) = x^5 - 4x + 2$;
(d) $a(x) = x^5 + 4x^4 + 2x^3 + 3x^2 - x + 5$.

Exercise 6.10. *Implement a Mathematica function* Content *for the computation of the content of a polynomial $a(x)$.*

Exercise 6.11. *Implement the Kronecker algorithm from Session 6.36 as Mathematica function* Kronecker *and test your function with 10 suitable polynomials. For efficiency purposes you can use the function* Content *from Exercise 6.10.*

Exercise 6.12. (Factorization in $\mathbb{Z}[x]$) *Factorize the following polynomials $a(x) \in \mathbb{Z}[x]$:*

(a) $a(x) = x^4 + x + 1$;
(b) $a(x) = 3x^4 + 5x - 1$;
(c) $a(x) = 6x^4 - 2x^3 + 11x^2 - 3x + 3$;
(d) $a(x) = 12x^6 - 8x^5 + 17x^4 + 2x^3 - 2x^2 - 2x + 5$;
(e) $a(x) = 12x^6 - 8x^5 + 17x^4 + 5x^3 - 4x^2 + 3x + 5$.

Exercise 6.13. *Show by induction that in a polynomial ring with several variables $\mathbb{K}[x_1, x_2, \ldots, x_n]$ over a field $\mathbb{K}$ there is a unique factorization.*

Exercise 6.14. *Show that the polynomial $x^2 - 2 \in \mathbb{Q}[x]$ is irreducible. Show that this is equivalent to the fact that $\sqrt{2}$ is irrational. In which polynomial ring is $x^2 - 2$ reducible?*

Exercise 6.15. *Let be $p \in \mathbb{P}$ and $m \in \mathbb{N}_{\geq 2}$. Show that $\sqrt[m]{p}$ is irrational. Furthermore, show that if $p_1, \ldots, p_n \in \mathbb{P}$ are pairwise different, then $\sqrt[m]{p_1 \cdots p_n}$ is irrational, too.*

Exercise 6.16. *Show: If $x \in \mathbb{R}$ and $x^n + a_{n-1}x_{n-1} + \cdots + a_1 x + a_0 = 0$ with integer coefficients $a_0, \ldots, a_{n-1} \in \mathbb{Z}$, then either $x \in \mathbb{Z}$ or x is irrational. Hint: Use a proof by contradiction.*

Exercise 6.17. (Squarefree Factorization)

(a) *Implement a Mathematica function* SquarefreePart[$a(x),x$] *which computes the squarefree part of the polynomial $a(x) \in \mathbb{Q}[x]$.*
(b) *Test your implementation for*

$$a(x) := x^8 - 17x^7 + 103x^6 - 241x^5 + 41x^4 + 533x^3 - 395x^2 - 275x + 250$$

and plot both $a(x)$ and its square free part over $\mathbb{R}$. Explain the differences and judge them!
(c) *Implement the function* SquarefreeFactorization[$a(x),x$] *which computes the squarefree factorization of $a(x) \in \mathbb{Q}[x]$.*
(d) *Test your function* SquarefreeFactorization *for the following polynomials and compare with Mathematica's built-in routine.*
    (i) $a(x) = x^8 - 2x^6 + 2x^2 - 1$;
    (ii) $a(x) = x^8 - 17x^7 + 103x^6 - 241x^5 + 41x^4 + 533x^3 - 395x^2 - 275x + 250$;
    (iii) $a(x) = \text{Expand}\left(\prod_{k=1}^{20}(x-k)^k\right)$.

Exercise 6.18. *Let $\mathbb{K}$ be a field of characteristic 0 and let $a(x) \in \mathbb{K}[x]$. Prove that $a'(x) = 0$ implies that $a(x)$ is constant.*

Chapter 7

Algebraic Numbers

7.1 Polynomial Quotient Rings

In the same way as we defined the computation modulo an integer $p \in \mathbb{Z}$ and the resulting remainder class rings in Chapter 4, the computation modulo a polynomial $p(x) \in \mathbb{K}[x]$ can be defined. Again, this results in remainder class rings, which are now called polynomial quotient rings.　•

Definition 7.1. Let $\mathbb{K}$ be a field and let the polynomial $p(x) \in \mathbb{K}[x]$ have positive degree. Two polynomials $a(x), b(x) \in \mathbb{K}[x]$ are called *congruent modulo $p(x)$*, denoted by

$$a(x) \equiv b(x) \pmod{p(x)},$$

if $p(x) \mid b(x) - a(x)$. Hence $a(x) \equiv b(x) \pmod{p(x)}$ if and only if $a(x)$ and $b(x)$ have the same remainder when dividing by $p(x)$:

$$r(x) := \operatorname{rem}(a(x), p(x), x) = \operatorname{rem}(b(x), p(x), x). \tag{7.1}$$

This congruence is again an equivalence relation which yields a classification in $\mathbb{K}[x]$. The classes of congruent polynomials are the sets $[a(x)]_{p(x)} := \{a(x) + k(x) \cdot p(x) \mid k(x) \in \mathbb{K}[x]\}$. Every equivalence class therefore possesses exactly one element $r(x) \in \mathbb{K}[x]$ with $\deg(r(x), x) < \deg(p(x), x)$, which is deduced by (7.1). The equivalence classes $[a(x)]_{p(x)}$ are again called remainder classes. The resulting set of remainder classes is denoted by $\mathbb{K}[x]/\langle p(x)\rangle$, in words $\mathbb{K}[x]$ modulo $p(x)$, whose elements are again identified via (7.1). △

As in $\mathbb{Z}_p$, we can now define addition and multiplication in $\mathbb{K}[x]/\langle p(x)\rangle$, which makes $\mathbb{K}[x]/\langle p(x)\rangle$ a commutative ring with unity 1 (compare Theorem 4.4). The statements and proofs are analogous to the ones in Section 4.1.

We are mostly interested under which conditions this construction of $\mathbb{K}[x]/\langle p(x)\rangle$ yields a field.

Theorem 7.2. An element $a(x) \in \mathbb{K}[x]/\langle p(x)\rangle$ is a unit if and only if $\gcd(a(x), p(x)) = 1$. In particular, $(\mathbb{K}[x]/\langle p(x)\rangle, +, \cdot)$ is a field if and only if $p(x) \in \mathbb{K}[x]$ is irreducible.

Proof. The proof corresponds to the proof of Theorem 4.5, but we would like to spell it out once again because it contains an algorithm for the computation of the reciprocal.

© Springer Nature Switzerland AG 2021
W. Koepf, *Computer Algebra*, Springer Undergraduate Texts in Mathematics and Technology,
https://doi.org/10.1007/978-3-030-78017-3_7

Let $p(x) \in \mathbb{K}[x]$ and $a(x) \in \mathbb{K}[x]/\langle p(x) \rangle$. If $a(x)$ is a unit then there is a $b(x) \in \mathbb{K}[x]/\langle p(x) \rangle$ with $a(x) \cdot b(x) = 1 + k(x) \cdot p(x)$. However, from this equation it follows that the greatest common divisor of $a(x)$ and $p(x)$ equals 1.

If, on the other hand, $\gcd(a(x), p(x)) = 1$, then the extended Euclidean algorithm yields $s(x), t(x) \in \mathbb{K}[x]$ with $s(x) \cdot a(x) + t(x) \cdot p(x) = 1$. If we read this equation modulo $p(x)$, we get $s(x) \cdot a(x) \equiv 1 \pmod{p(x)}$ and therefore $\mathrm{rem}(s(x), p(x), x)$ is a multiplicative inverse of $a(x)$ in $\mathbb{K}[x]/\langle p(x) \rangle$.

If $p(x) \in \mathbb{K}[x]$ is irreducible, then we get for all nonzero $a(x) \in \mathbb{K}[x]/\langle p(x) \rangle$ the relation $\gcd(a(x), p(x)) = 1$, since $p(x)$ has no nontrivial divisors. Therefore, all elements of $\mathbb{K}[x]/\langle p(x) \rangle \setminus \{0\}$ are units and $\mathbb{K}[x]/\langle p(x) \rangle$ is a field.

However, if $p(x)$ is composite, then $a(x) \in \mathbb{K}[x]/\langle p(x) \rangle \setminus \{0\}$ with $\gcd(a(x), p(x)) \neq 1$ and $a(x)$ is no unit. $\square$

Now let us look at some examples.

Example 7.3. We consider the congruence modulo $x^2 - 2$. It is easy to see that $x^2 - 2 \in \mathbb{Q}[x]$ is irreducible. This is equivalent to the fact that the zero $\sqrt{2}$ is irrational, see also Exercise 6.14. We get

$$
\begin{aligned}
x^2 &\equiv 2 \pmod{x^2 - 2} \\
x^3 &\equiv 2x \pmod{x^2 - 2} \\
&\vdots \\
x^n &\equiv 2x^{n-2} \pmod{x^2 - 2} .
\end{aligned}
$$

If we apply these replacement rules recursively, then every polynomial $a(x) \in \mathbb{Q}[x]/\langle x^2 - 2 \rangle$ will be reduced to a linear one.

For the elements $a(x), b(x) \in \mathbb{Q}[x]/\langle x^2 - 2 \rangle$ addition and multiplication are defined by

$$
\begin{aligned}
[a(x)]_{x^2-2} + [b(x)]_{x^2-2} &:= [a(x) + b(x)]_{x^2-2} \\
[a(x)]_{x^2-2} \cdot [b(x)]_{x^2-2} &:= [a(x) \cdot b(x)]_{x^2-2}
\end{aligned}
$$

and $\mathbb{Q}[x]/\langle x^2 - 2 \rangle$ is a commutative ring with $1 = [1]_{x^2-2}$. Every equivalence class in $\mathbb{Q}[x]/\langle x^2 - 2 \rangle$ has exactly one element of degree < 2 which can be computed by the division algorithm. If we identify the equivalence classes with these representatives again, then the elements of $\mathbb{Q}[x]/\langle x^2 - 2 \rangle$ consist of the linear polynomials and $\mathbb{Q}$ is therefore embedded in $\mathbb{Q}[x]/\langle x^2 - 2 \rangle$, i.e. $\mathbb{Q} \subset \mathbb{Q}[x]/\langle x^2 - 2 \rangle$. Addition and multiplication of $a + bx, c + dx \in \mathbb{Q}[x]/\langle x^2 - 2 \rangle$ is defined by

$$
\begin{aligned}
(a + bx) + (c + dx) &= (a + c) + (b + d)x \\
(a + bx) \cdot (c + dx) &= (ac + 2bd) + (bc + ad)x ,
\end{aligned}
$$

since

$$
(a + bx) \cdot (c + dx) \equiv ac + (bc + ad)x + bdx^2 \equiv (ac + 2bd) + (bc + ad)x \pmod{x^2 - 2} . \tag{7.2}
$$

Note that one can also interpret the elements of $\mathbb{Q}[x]/\langle x^2 - 2 \rangle$ as polynomials in $\mathbb{Q}[x]$ which are evaluated at the point $\alpha = [x]_{x^2-2}$ since

$$
\alpha^2 = [x]_{x^2-2}^2 = [x^2]_{x^2-2} = [x^2 - 2 + 2]_{x^2-2} = [x^2 - 2]_{x^2-2} + [2]_{x^2-2} = 2 .
$$

The symbol α is no element of the base ring $\mathbb{Q}$, but a newly[1] "invented" element with the property ($\alpha^2 = 2$).[2] Equation (7.2) therefore reads as

$$(a + b\alpha) \cdot (c + d\alpha) = ac + (bc + ad)\alpha + bd\,\alpha^2 = (ac + 2bd) + (bc + ad)\alpha, \qquad (7.3)$$

hence we compute, as in $\mathbb{Q}$, under the side condition $\alpha^2 = 2$. Therefore, the set[3]

$$\mathbb{Q}(\alpha) := \{a + b\alpha \mid a, b \in \mathbb{Q}\} \supset \mathbb{Q}$$

with the multiplication (7.3), which is based on the equation $\alpha^2 - 2 = 0$, is isomorphic to $\mathbb{Q}[x]/\langle x^2 - 2\rangle$ and is furthermore a vector space of dimension 2 over $\mathbb{Q}$. Since $\sqrt{2}$ is the common notation for a solution of the equation $\alpha^2 - 2 = 0$, we also write $\mathbb{Q}(\sqrt{2})$ for this set.

In Theorem 7.2 it was proved that $\mathbb{Q}[x]/\langle x^2 - 2\rangle$ is a field. Hence $\mathbb{Q}(\sqrt{2})$ is also a field. Let us check this with the help of another method. For this purpose we search for a reciprocal of $a + b\alpha$ ($a, b \neq 0$) by using the ansatz

$$(a + b\alpha) \cdot (c + d\alpha) = 1 \,.$$

Hence by (7.3) we find

$$bc + ad = 0 \quad \text{and} \quad ac + 2bd = 1 \,.$$

This leads to a linear system of equations for the computation of c and d with the result

$In[1]:=$ **sol = Solve[{b c + a d == 0, a c + 2 b d == 1}, {c, d}]**

$Out[1]=$ $\left\{\left\{c \rightarrow \dfrac{a}{a^2 - 2b^2}, d \rightarrow \dfrac{b}{2b^2 - a^2}\right\}\right\}$

which we can verify by the computation

$In[2]:=$ $\left(a + b\,\sqrt{2}\right) * \left(c + d\,\sqrt{2}\right)$ **/.sol[[1]]//Simplify**

$Out[2]=$ 1

This particular result can be also obtained by "smart expansion" (i.e. by rationalizing the denominator):

$$\frac{1}{a + b\sqrt{2}} = \frac{a - b\sqrt{2}}{(a + b\sqrt{2})(a - b\sqrt{2})} = \frac{a}{a^2 - 2b^2} - \frac{b}{a^2 - 2b^2}\,\sqrt{2}\,.$$

From the last representation we can see that the reciprocal is again an element of $\mathbb{Q}(\sqrt{2})$.

Hence we have found a reciprocal of $a + b\alpha$ if $a^2 - 2b^2 \neq 0$. However, if $a^2 - 2b^2 = 0$, then $\frac{a}{b}$ would be a rational solution of the equation $x^2 - 2 = 0$ which does not exist by the irreducibility of $x^2 - 2 \in \mathbb{Q}[x]$. △

Example 7.4. As a second, similar, example we consider the congruence modulo $p(x) = x^2 + 1 \in \mathbb{R}[x]$ over $\mathbb{K} = \mathbb{R}$. The polynomial $p(x)$ is irreducible since the equation $x^2 + 1 = 0$ has no real solution. Again every polynomial $a(x) \in \mathbb{R}[x]$ can be reduced to a linear polynomial by the replacement rules $x^n \mapsto -x^{n-2}$ ($n \geq 2$) or with the division algorithm (dividing by $x^2 + 1$), respectively. Hence the elements of $\mathbb{R}[x]/\langle x^2 + 1\rangle$ have the form $a + b[x]_{x^2+1}$. By introducing

---

[1] Our invention is $\alpha = [x]_{x^2-2}$.

[2] Of course we know that this number $\alpha = \sqrt{2}$ is an element of $\mathbb{R} \supset \mathbb{Q}$, for example. By the way, there is no priority which of the two zeros $\pm\sqrt{2}$ of the given polynomial $x^2 - 2$ is denoted by α. From an algebraic perspective, they cannot be distinguished.

[3] For the notation we use round brackets since we get a field. In this case, however, the notation in the literature is not unique, see e.g. [Chi2000], [GG1999].

the abbreviation $i = [x]_{x^2+1}$, we get the usual representation of complex numbers $z = a + b \cdot i$. Using this notation, the set $\mathbb{R}[x]/\langle x^2 + 1 \rangle$ therefore contains the same elements as $\mathbb{C}$, and the computational rules agree, since

$$i^2 = [x]^2_{x^2+1} = [x^2]_{x^2+1} = [-1]_{x^2+1} = -1$$

and

$$
\begin{aligned}
(a + bi) + (c + di) &= (a + c) + (b + d) \cdot i \\
(a + bi) \cdot (c + di) &= (ac - bd) + (ad + bc) \cdot i.
\end{aligned}
$$

Hence $\mathbb{R}[x]/\langle x^2 + 1 \rangle$ and $\mathbb{C} = \mathbb{R}(i)$ are isomorphic to each other. This isomorphism is denoted by $\mathbb{R}[x]/\langle x^2 + 1 \rangle \cong \mathbb{C}$. As in Example 7.3 we find that $\mathbb{C}$ is a field since the polynomial $x^2 + 1 \in \mathbb{R}[x]$ is irreducible. $\triangle$

Session 7.5. *Computations modulo a polynomial can be done using* `PolynomialMod`. *To this end, we reduce a generic polynomial of degree 10 modulo $x^2 - 2$:*

$In[1]:=$ **PolynomialMod** $\left[\sum_{k=0}^{10} a_k x^k, x^2 - 2 \right]$

$Out[1]=$ $a_1 x + 2 a_3 x + 4 a_5 x + 8 a_7 x + 16 a_9 x + a_0 + 2 a_2 + 4 a_4 + 8 a_6 + 16 a_8 + 32 a_{10}$

The resulting polynomial of first degree can also be obtained by a recursive application of the rule $x^2 \mapsto 2$:

$In[2]:=$ $\sum_{k=0}^{10} a_k x^k$ `//.{x^n_ /; n > 1 -> 2 x^{n-2}}`

$Out[2]=$ $a_1 x + 2 a_3 x + 4 a_5 x + 8 a_7 x + 16 a_9 x + a_0 + 2 a_2 + 4 a_4 + 8 a_6 + 16 a_8 + 32 a_{10}$

The division algorithm generates the same result:

$In[3]:=$ **PolynomialRemainder** $\left[\sum_{k=0}^{10} a_k x^k, x^2 - 2, x \right]$

$Out[3]=$ $(a_1 + 2 a_3 + 4 a_5 + 8 a_7 + 16 a_9) x + a_0 + 2 a_2 + 4 a_4 + 8 a_6 + 16 a_8 + 32 a_{10}$

And finally—knowing the solution of the equation $x^2 - 2 = 0$—we can substitute $x = \sqrt{2}$ in the extension field[4] $\mathbb{R} \supset \mathbb{Q}$. Mathematica's built-in simplifier executes the desired reduction:

$In[4]:=$ $\sum_{k=0}^{10} a_k \sqrt{2}^k$

$Out[4]=$ $a_0 + \sqrt{2} a_1 + 2 a_2 + 2 \sqrt{2} a_3 + 4 a_4 + 4 \sqrt{2} a_5 + 8 a_6 + 8 \sqrt{2} a_7 + 16 a_8 + 16 \sqrt{2} a_9 + 32 a_{10}$

In contrast to `PolynomialRemainder`, *the command* `PolynomialMod` *can also be used over $\mathbb{K} = \mathbb{Z}_p$:[5]*

$In[5]:=$ **PolynomialMod** $[x^9, x^2 - 2, \text{Modulus} \to 5]$

$Out[5]=$ x

We continue with the same computations modulo $x^2 + 1$. Contrary to Example 7.4, however, we use the base field $\mathbb{Q}$ and therefore compute in $\mathbb{Q}/\langle x^2 + 1 \rangle$.

---

[4] If $\mathbb{K}$ and $\mathbb{E}$ are fields with $\mathbb{K} \subset \mathbb{E}$, and if $\mathbb{K}$ is a subfield of $\mathbb{E}$, then $\mathbb{E}$ is called an extension field of $\mathbb{K}$.

[5] *Mathematica's* online help system additionally states: "Unlike `PolynomialRemainder`, `PolynomialMod` never performs divisions in generating its results."

$In[6]:=$ **PolynomialMod $\left[\sum_{k=0}^{10} a_k x^k, x^2+1\right]$**

$Out[6]= a_1\,x - a_3\,x + a_5\,x - a_7\,x + a_9\,x + a_0 - a_2 + a_4 - a_6 + a_8 - a_{10}$

$In[7]:= \sum_{k=0}^{10} a_k x^k\,//\,. \{x^n\_ /; n>1 \rightarrow -x^{n-2}\}$

$Out[7]= a_1\,x - a_3\,x + a_5\,x - a_7\,x + a_9\,x + a_0 - a_2 + a_4 - a_6 + a_8 - a_{10}$

$In[8]:=$ **PolynomialRemainder $\left[\sum_{k=0}^{10} a_k x^k, x^2+1, x\right]$**

$Out[8]= \left(a_1 - a_3 + a_5 - a_7 + a_9\right) x + a_0 - a_2 + a_4 - a_6 + a_8 - a_{10}$

Using the imaginary unit i, we get the same result:

$In[9]:= \sum_{k=0}^{10} a_k i^k$

$Out[9]= a_0 + i\,a_1 - a_2 - i\,a_3 + a_4 + i\,a_5 - a_6 - i\,a_7 + a_8 + i\,a_9 - a_{10}$

7.2 Chinese Remainder Theorem

The following theorem can be established in complete analogy to the Chinese Remainder Theorem in Chapter 4.

Theorem 7.6. (Chinese Remainder Theorem for Polynomials) Let $p_k(x) \in \mathbb{K}[x]$ be pairwise coprime polynomials for $k = 1, \dots, n$. Then the system of equations

$$f(x) \equiv a_1(x) \,(\text{mod } p_1(x))$$
$$f(x) \equiv a_2(x) \,(\text{mod } p_2(x))$$
$$\vdots$$
$$f(x) \equiv a_n(x) \,(\text{mod } p_n(x))$$

has exactly one solution $f(x) \in \mathbb{K}[x]$ of degree $\deg(f(x), x) < \deg(p_1(x) \cdots p_n(x), x)$. All other solutions differ from $f(x)$ by an additive multiple of $p_1(x) \cdots p_n(x)$.

Proof. The proof of Theorems 4.9–4.10 can be easily modified. By an application of the extended Euclidean algorithm for polynomials, it leads again to an algorithm for the computation of $f(x)$. □

Session 7.7. *The implementation of Session 4.11 can be adapted for polynomials as follows:*

```
In[1]:= Clear[PolynomialCR]
        PolynomialCR[{a_,b_},{p_,q_},x_] := Module[{g,s,t},
          {g,{s,t}} = PolynomialExtendedGCD[p,q,x];
          PolynomialMod[b*s*p+a*t*q,p*q]
        ]
```

As a first example we consider the remainder problem

$$f(x) \equiv 0 \ (mod \ x)$$
$$f(x) \equiv 1 \ (mod \ x-1).$$

The first condition is equivalent to $x \mid f(x)$, and therefore to $f(0) = 0$. Similarly the second condition means $x - 1 \mid f(x) - 1$ or $f(1) = 1$, respectively. The solution

```
In[2]:= PolynomialCR[{0,1}, {x,x-1},x]
Out[2]= x
```

can therefore be easily checked by mental arithmetic. In the result

```
In[3]:= cr = PolynomialCR[{0,1}, {x² - 2, x² + 1},x]
```
$$Out[3]= \frac{2}{3} - \frac{x^2}{3}$$

the first condition, namely $x^2 - 2 \mid f(x)$, is also visible, and we can test the second condition:

```
In[4]:= PolynomialMod[cr, x² + 1]
Out[4]= 1
```

We implement the general case of Theorem 7.6 recursively:

```
In[5]:= PolynomialCR[alist_List,plist_List] :=
            Module[{a,p},
                a = PolynomialCR[
                    {alist[[1]],alist[[2]]},
                    {plist[[1]],plist[[2]]}];
                p = plist[[1]] * plist[[2]];
                PolynomialCR[
                  Join[{a},Rest[Rest[alist]]],
                  Join[{p},Rest[Rest[plist]]]]
            ]/; Length[alist] == Length[plist]&&
            Length[alist] > 2
```

As an example, let us consider the following problem:

$$f(x) \equiv 0 \ (mod \ x^2 - 2) \qquad resp. \ x^2 - 2 \mid f(x)$$
$$f(x) \equiv 0 \ (mod \ x^3 + 1) \qquad resp. \ x^3 + 1 \mid f(x)$$
$$f(x) \equiv 0 \ (mod \ x + 2) \qquad resp. \ x + 2 \mid f(x)$$
$$f(x) \equiv 1 \ (mod \ x) \qquad resp. \ f(0) = 1$$

We get the solution

```
In[6]:= cr = PolynomialCR[{0,0,0,1}, {x² - 2, x³ + 1, x + 2, x}]
```
$$Out[6]= -\frac{x^6}{4} - \frac{x^5}{2} + \frac{x^4}{2} + \frac{3 x^3}{4} - \frac{x^2}{2} + \frac{x}{2} + 1$$

and check the result by

```
In[7]:= Map[PolynomialRemainder[cr,#,x]&, {x² - 2, x³ + 1, x + 2, x}]
Out[7]= {0,0,0,1}
```

The factorization[6]

```
In[8]:= Factor[cr]
```

---

[6] Of course the verification does not need a factorization.

$$Out[8] = -\frac{1}{4}(x+1)(x+2)(x^2-2)(x^2-x+1)$$

shows, again, the validity of the three equations.

7.3 Algebraic Numbers

In the Examples 7.3 and 7.4 we had defined fields containing "new numbers". Let us repeat these developments shortly.

The irreducibility of $x^2 - 2 \in \mathbb{Q}[x]$ implies that $\mathbb{Q}[x]/\langle x^2 - 2\rangle$ is a field. For this field we use the abbreviation $\mathbb{Q}(\sqrt{2})$, where we adapted the usual notation $\alpha = \sqrt{2}$ for a number with the property $\alpha^2 - 2 = 0$. Obviously the field $\mathbb{Q}(\sqrt{2})$ is a field extension of $\mathbb{Q}$ since it consists of numbers of the form $\{a + b \cdot \sqrt{2} \mid a, b \in \mathbb{Q}\}$. Therefore, the notation $\mathbb{Q}(\sqrt{2})$ expresses the fact that the new element $\sqrt{2}$ is *adjoined* to $\mathbb{Q}$. $\mathbb{Q}(\sqrt{2})$ constitutes a $\mathbb{Q}$-vector space of dimension 2.

Analogously the irreducibility of $x^2 + 1 \in \mathbb{R}[x]$ identifies $\mathbb{R}[x]/\langle x^2 + 1\rangle$ as field. We use the usual abbreviation $\mathbb{C} = \mathbb{R}(i)$, where we have adjoined to $\mathbb{R}$ the imaginary number i. Again $\mathbb{C}$ is a field extension of $\mathbb{R}$, since $\mathbb{R}(i)$ contains numbers of the form $\{a + b \cdot i \mid a, b \in \mathbb{R}\}$. Obviously $\mathbb{R}(i)$ forms an $\mathbb{R}$-vector space of dimension 2.

In the general case we find that if $p(x) \in \mathbb{K}[x]$ is irreducible of degree n, then $\mathbb{K}(\alpha) = \mathbb{K}[x]/\langle p(x)\rangle$ is a field containing the $\mathbb{K}$-vector space V which is generated by the powers of α. The equation $p(\alpha) = 0$ expresses the linear dependency of the elements $1, \alpha, ..., \alpha^n$ of V, so that only the powers $1, \alpha, ..., \alpha^{n-1}$ are linearly independent.

Definition 7.8. Let $\mathbb{K}$ be a field, and let $p(x) \in \mathbb{K}[x]$ be an irreducible polynomial of degree $\deg(p(x), x) = n$. Then $\mathbb{K}[x]/\langle p(x)\rangle$ is isomorphic to the field

$$\mathbb{K}(\alpha) := \{c_0 + c_1\alpha + \cdots + c_{n-1}\alpha^{n-1} \mid c_k \in \mathbb{K}, \ k = 0, ..., n-1\}, \tag{7.4}$$

where the multiplication is performed in the usual way, and powers α^k are reduced for $k \geq n$ using the equation $p(\alpha) = 0$. Field extensions which are generated by this type of *adjunction* of an element α to $\mathbb{K}$ are called *simple field extensions*, and the presentation in (7.4) is called the *standard representation* of the elements of $\mathbb{K}(\alpha)$.

An iterative adjunction of elements generates multiple field extensions which are denoted by $\mathbb{K}(\alpha, \beta, ...) = \mathbb{K}(\alpha)(\beta)....$ △

Session 7.9. *Let us consider the simple field extension* $\mathbb{Q}(\sqrt[3]{5})$. *In this example,* $p(x) = x^3 - 5 \in \mathbb{Q}[x]$. *All elements of* $\mathbb{Q}(\sqrt[3]{5})$ *have the form* $a + b\sqrt[3]{5} + c\sqrt[3]{25}$ $(a, b, c \in \mathbb{Q})$. *The multiplication of two elements of* $\mathbb{Q}(\sqrt[3]{5})$ *yields an element of* $\mathbb{Q}(\sqrt[3]{5})$, *again:*

$In[1] :=$ **x = a + b $\sqrt[3]{5}$ + c $\sqrt[3]{25}$;**

      y = d + e $\sqrt[3]{5}$ + f $\sqrt[3]{25}$;

$In[2] :=$ **x * y // Expand**

$Out[2] = ad + \sqrt[3]{5}\,ae + 5^{2/3}\,af + \sqrt[3]{5}\,bd + 5^{2/3}\,be + 5\,bf + 5^{2/3}\,cd + 5\,ce + 5\sqrt[3]{5}\,cf$

In this case it is more difficult to compute the standard representation of the reciprocal

$$\frac{1}{a + b\sqrt[3]{5} + c\sqrt[3]{25}}.$$

An obvious trick as in Example 7.3 does not exist, and Mathematica does not compute the standard representation:

```
In[3]:= 1/x//FullSimplify
```
$$Out[3]= \frac{1}{a + \sqrt[3]{5}\, b + 5^{2/3}\, c}$$

For this purpose we can use the algorithm given in Theorem 7.2. However, we would like to present another method with Mathematica. Let $x = a + b\sqrt[3]{5} + c\sqrt[3]{25}$ be given. Then we search for $y = d + e\sqrt[3]{5} + f\sqrt[3]{25}$ such that the product $x \cdot y$ equals 1. For this purpose, let us compute $x \cdot y - 1$ and substitute $\sqrt[3]{5}$ by α:

```
In[4]:= expr = ((x*y-1)//Expand)/.{∛5 → α, ∛25 → α²}
```
$$Out[4]= a\,d + a\,\alpha\,e + a\,\alpha^2\,f + \alpha\,b\,d + \alpha^2\,b\,e + 5\,b\,f + \alpha^2\,c\,d + 5\,c\,e + 5\,a\,c\,f - 1$$

By the linear independence[7] of $\{1, \alpha, \alpha^2\}$ over $\mathbb{Q}$ this expression is identical to zero if and only if all coefficients with respect to α are zero:

```
In[5]:= list = CoefficientList[expr,α];
        sol = Solve[list == 0, {d,e,f}]
```
$$Out[5]= \left\{\left\{d \to -\frac{5\,b\,c - a^2}{a^3 - 15\,b\,c\,a + 5\,b^3 + 25\,c^3}, \right.\right.$$
$$\left.\left. e \to -\frac{a\,b - 5\,c^2}{a^3 - 15\,b\,c\,a + 5\,b^3 + 25\,c^3}, f \to -\frac{a\,c - b^2}{a^3 - 15\,b\,c\,a + 5\,b^3 + 25\,c^3}\right\}\right\}$$

Hence we have solved our problem. We can check the result by

```
In[6]:= (x*y/.sol[[1]])//Simplify
Out[6]= 1
```

Note that the "correct command" for the simplification of algebraic numbers is RootReduce, *but unfortunately* RootReduce *is not able to simplify our expression.*

Definition 7.10. A field $\mathbb{E}$ is called a *finite field extension of degree n* of the field $\mathbb{K} \subset \mathbb{E}$, if every $x \in \mathbb{E}$ can be written as a linear combination of finitely many elements $\alpha_1, \ldots, \alpha_n \in \mathbb{E}$ with coefficients in $\mathbb{K}$,

$$x = \sum_{k=1}^{n} c_k \alpha_k \qquad (c_k \in \mathbb{K},\ k = 1, \ldots, n),$$

but at least one element cannot be represented as a linear combination of only $n - 1$ elements.

In this case $\mathbb{E}$ is an n-dimensional $\mathbb{K}$-vector space with the basis $\alpha_1, \ldots, \alpha_n$, and n is the degree of the field extension, i.e. $[\mathbb{E} : \mathbb{K}] = n$. $\triangle$

Example 7.11. Every simple field extension $\mathbb{K}(\alpha) = \mathbb{K}[x]/\langle p(x)\rangle$ is finite. If $n = \deg(p(x), x)$ is the degree of the corresponding irreducible polynomial $p(x) \in \mathbb{K}[x]$, then the elements $1, \alpha, \alpha^2, \ldots, \alpha^{n-1}$ form a basis of $\mathbb{K}(\alpha)$ over $\mathbb{K}$, and we have $[\mathbb{K}(\alpha) : \mathbb{K}] = n$.

Every multiple field extension $\mathbb{K}(\alpha, \beta)$, where α has degree n and β has degree m, forms an $n \cdot m$-dimensional vector space over $\mathbb{K}$ with the basis

$$\{\alpha^j \cdot \beta^k \mid j = 0, \ldots, n-1,\ k = 0, \ldots, m-1\}.$$

---

[7] Note that the linear independence follows from the irreducibility of the polynomial $x^3 - 5 \in \mathbb{Q}[x]$.

Obviously the resulting vector space does not depend on the order of adjunction, so that $\mathbb{K}(\alpha, \beta) = \mathbb{K}(\beta, \alpha)$. $\triangle$

A simple linear algebra argument shows: if two finite field extensions $\mathbb{K} \subset \mathbb{E} \subset \mathbb{F}$ satisfy $[\mathbb{E} : \mathbb{K}] = n$ and $[\mathbb{F} : \mathbb{E}] = m$, then the field extension $\mathbb{K} \subset \mathbb{F}$ is also finite and has degree $[\mathbb{F} : \mathbb{K}] = n \cdot m$, see Exercise 7.7.

Next we would like to study those extension fields that are connected to zeros of irreducible polynomials in more detail. We already know the following theorem.

Theorem 7.12. Let $\mathbb{K}$ be a field and $p(x) \in \mathbb{K}[x]$ irreducible. Then there is an extension field $\mathbb{E} \supset \mathbb{K}$, which contains a zero of $p(x)$.

Proof. Obviously we can take $\mathbb{E} = \mathbb{K}[x]/\langle p(x) \rangle$. The sought zero of $p(x)$ is the element $\alpha = [x]_{p(x)}$. $\qquad\square$

This theorem allows the construction of an extension field containing *all* zeros of a given polynomial $a(x) \in \mathbb{K}[x]$.

Theorem 7.13. (Splitting Field) Let $\mathbb{K}$ be a field and $a(x) \in \mathbb{K}[x]$ a polynomial of degree $n = \deg(a(x), x) \geqq 1$. Then there is an extension field $\mathbb{E} \supset \mathbb{K}$ such that $a(x)$ can be decomposed within $\mathbb{E}[x]$ into linear factors. In this case, $\mathbb{E}$ is called a *splitting field*[8] of the polynomial $a(x)$ over $\mathbb{K}$.

Proof. We perform mathematical induction with respect to n. For $n = 1$ the statement is clearly true. Now let $n > 1$ and suppose that the statement is true for polynomials of degree $< n$. Let $a(x)$ have the decomposition $a(x) = p_1(x) \cdots p_m(x)$ in the ground field $\mathbb{K}$ into (over $\mathbb{K}$) irreducible polynomials $p_k(x)$ $(k = 1, \ldots, m)$. If now $\deg(p_k(x), x) = 1$ for all $k = 1, \ldots, m$, then $\mathbb{K}$ itself is a splitting field of $a(x)$ and the conclusion is true. Hence let $\deg(p_1(x), x) > 1$ without loss of generality. Let further $\mathbb{K}_1 = \mathbb{K}[x]/\langle p_1(x) \rangle = \mathbb{K}(\alpha)$ be a field extension according to Theorem 7.12, which adjoins a zero α of $p_1(x)$. Then $\mathbb{K}_1$ is an extension field of $\mathbb{K}$ and α is a zero of $p_1(x)$ which is contained in $\mathbb{K}_1$. Using Lemma 6.24 we can write $p_1(x) = (x - \alpha) q_1(x)$ within $\mathbb{K}_1[x]$, with $q_1(x) \in \mathbb{K}_1[x]$ or $a(x) = (x - \alpha) q_1(x) p_2(x) \cdots p_m(x)$. Since the degree of $b(x) := q_1(x) p_2(x) \cdots p_m(x) \in \mathbb{K}_1[x]$ is $n - 1$, we know from the induction hypothesis that there is a field $\mathbb{E} \supset \mathbb{K}_1$ such that $b(x)$ can be decomposed into linear factors over $\mathbb{E}$. This is obviously the wanted splitting field. $\qquad\square$

Note that for every polynomial $a(x) \in \mathbb{Q}[x]$ the field $\mathbb{C}$ of complex numbers is a splitting field. This is a consequence of the *fundamental theorem of algebra* which states that every polynomial $a(x) \in \mathbb{C}[x]$ possesses a zero $\alpha \in \mathbb{C}$ (and therefore exactly n zeros by Lemma 6.24).

Whereas $\mathbb{C}$ cannot be generated from $\mathbb{Q}$ by an iteration of simple field extensions, the proof of Theorem 7.13 shows how to construct a splitting field by a finite field extension. Unfortunately the degree of the field extension can be rather large. In the worst case, the first step needs an extension of degree n, the next induction step needs another one of degree $n - 1$ and so on, so that the degree of the splitting field can be up to $n(n - 1) \cdots 1 = n!$, see Exercise 7.7. One can show that this situation occurs for every n indeed, see [Her1975], Theorem 5.8.1.

Session 7.14. *We consider an example. The polynomial* $a(x) = x^3 - 4 \in \mathbb{Q}[x]$

---

[8] Some authors assume that the splitting field is the smallest such extension.

```
In[1]:= a = x³ - 4
```
$Out[1] = x^3 - 4$

is irreducible. We recognize a zero $\alpha = \sqrt[3]{4} \in \mathbb{R} \setminus \mathbb{Q}$. *Hence polynomial division yields*

```
In[2]:= PolynomialQuotient [a, x - ∛4, x]
```
$Out[2] = x^2 + 2^{2/3} x + 2 \sqrt[3]{2}$

and the remainder is zero:

```
In[3]:= PolynomialRemainder [a, x - ∛4, x]
```
$Out[3] = 0$

The degree of the given irreducible polynomial could therefore be decreased by one in the extension field $\mathbb{Q}(\alpha)$.

However, the reduction looks simpler if we set—as in the proof of Theorem 7.13—one arbitrary zero of $a(x)$ *equal to* α *and continue our computation in* $\mathbb{Q}(\alpha)$:

```
In[4]:= q₁ = PolynomialQuotient [a, x - α, x]
```
$Out[4] = \alpha^2 + x^2 + \alpha x$

The division remainder does not have to (and even cannot) be checked, it is zero by definition: this is how we defined α.

Next we adjoin a second zero β *of* $a(x)$, *hence a zero of* $b(x) = x^2 + \alpha x + \alpha^2$. *Then after another polynomial division, we obtain*

```
In[5]:= q₂ = PolynomialQuotient [q₁, x - β, x]
```
$Out[5] = \alpha + \beta + x$

Hence, by two divisions we have decomposed the starting polynomial into linear factors:

$$x^3 - 4 = (x - \alpha)(x - \beta)(x + \alpha + \beta).$$

In particular, we have discovered the relation $\gamma = -(\alpha + \beta)$ *for the third zero* γ *of* $a(x)$. *Therefore another field extension is not necessary. This follows also from* Vieta's *theorem: if the (monic) polynomial has the factorization* $a(x) = (x - \alpha)(x - \beta)(x - \gamma)$, *then expansion yields*

```
In[6]:= Collect [ (x - α) (x - β) (x - γ), x]
```
$Out[6] = -\alpha\beta\gamma + x^2 (-\alpha - \beta - \gamma) + x^3 + x (\alpha\beta + \alpha\gamma + \beta\gamma)$

and therefore the coefficients of $a(x)$ *depend in a unique way on its zeros.*[9] *In our case the following relations hold:*

$$-\alpha - \beta - \gamma = 0, \qquad \alpha\beta + \gamma\beta + \alpha\gamma = 0, \qquad -\alpha\beta\gamma = -4.$$

Note that the first identity is the same as the one determined for the third root.

Mathematica has a built-in function Root *for the representation of the zeros of a polynomial, hence for algebraic numbers. For example, we can substitute* x *by a zero of* $a(x)$ *and get:*

```
In[7]:= a/.x → Root[#³ - 4&, 1]//Simplify
```

$Out[7] = 0$

The notation Root [a(#)&, 1] *denotes the first root of* $a(x)$.[10]

---

[9] The functions appearing in the case of a polynomial of degree n are called the *elementary symmetric polynomials.*

[10] Mathematica sorts the zeros by their location in the complex plane. Real roots are sorted in ascending order on the real line.

*What can we say about the degree of the field extension? In the first step a field extension
of degree 3 was necessary to adjoin $\sqrt[3]{4}$, and in the second step another field extension of
degree 2 was necessary. Therefore, the complete field extension was of degree $3 \cdot 2 = 6$, see also
Exercise 7.7.*

This is also shown by the computation

$In[8]:=$ **Solve[q_1 == 0, x]**

$Out[8]=$ $\left\{ \left\{ x \to -\dfrac{1}{2} i \left(\sqrt{3}\alpha - i\alpha \right) \right\}, \left\{ x \to \dfrac{1}{2} i \left(\sqrt{3}\alpha + i\alpha \right) \right\} \right\}$

which reveals that the splitting field of $x^3 - 4$ over $\mathbb{Q}$ can be represented by $\mathbb{Q}(\sqrt[3]{4}, \sqrt{3}i)$

Definition 7.15. (Algebraic Number) Let $\mathbb{K}$ be a field, $\mathbb{E} \supset \mathbb{K}$ an extension field and $\alpha \in \mathbb{E}$. If
α is a zero of a polynomial $p(x) \in \mathbb{K}[x]$, then α is called *algebraic over* $\mathbb{K}$. △

For example, for $\mathbb{E} = \mathbb{K}(\alpha) = \mathbb{K}/\langle p(x) \rangle$, the number $\alpha \in \mathbb{E}$ is a zero of the—in $\mathbb{K}$ irreducible—
polynomial $p(x)$, compare Exercise 7.6.

Like in the latter case, every algebraic number is characterized by its *minimal polynomial*.

Lemma 7.16. (Minimal Polynomial) Let $\alpha \in \mathbb{E}$ be algebraic over $\mathbb{K}$. Then there is exactly one
irreducible monic polynomial $p(x) \in \mathbb{K}[x]$ with zero α. This polynomial is called the *minimal
polynomial* of the algebraic number α over $\mathbb{K}$ and is denoted by $\mathrm{minpol}(\alpha, x)$ or, in more detail,
$\mathrm{minpol}_{\mathbb{K}}(\alpha, x)$. Its degree n is called the *degree of the algebraic number* α over $\mathbb{K}$ and we write
$\mathrm{grad}(\alpha) = n$ or, in more detail, $\mathrm{grad}_{\mathbb{K}}(\alpha) = n$.

The minimal polynomial $p(x)$ is the monic polynomial of lowest degree in $\mathbb{K}[x]$, which has α
as zero. Every polnomial $a(x) \in \mathbb{K}[x]$ with zero α is a multiple of $p(x)$.

Proof. Since α is a zero of a polynomial $a(x) \in \mathbb{K}[x]$, there is a unique polynomial of lowest degree with zero α. Otherwise, if there were two different monic polynomials $a(x) =
x^n + a_{n-1} x^{n-1} + \cdots + a_0 \in \mathbb{K}[x]$ and $b(x) = x^n + b_{n-1} x^{n-1} + \cdots + b_0 \in \mathbb{K}[x]$ of degree n with zero
α, then α would also be a zero of the polynomial $a(x) - b(x) \neq 0$ of degree $< n$, contradicting the
minimality. Hence there is exactly one monic polynomial of lowest degree with zero α.

Next we show that this minimal polynomial $p(x)$ of degree n is irreducible. Assume that $p(x) =
a(x)b(x)$ with $a(x), b(x) \in \mathbb{K}[x]$, both of degree > 0, then it follows that $0 = p(\alpha) = a(\alpha)b(\alpha)$.
Since $\mathbb{K}$ has no zero divisors, it follows that either $a(\alpha) = 0$ or $b(\alpha) = 0$. Since both $a(x)$ and $b(x)$
have a degree lower than n, this is a contradiction. Hence the minimal polynomial is irreducible.

Finally, we want to show that every polynomial $a(x) \in \mathbb{K}[x]$ with zero α is a multiple of
$p(x)$. Hence, let $a(x) \in \mathbb{K}[x]$ be given with zero α. By the above we know that $\deg(a(x), x) \geqq
\deg(p(x), x)$. We use the division algorithm to get

$$a(x) = p(x) \cdot q(x) + r(x)$$

with $\deg(r(x), x) < \deg(p(x), x)$. Substituting $x = \alpha$ gives $r(\alpha) = 0$. However, since besides the
zero polynomial no polynomial of degree $< \deg(p(x), x)$ has the zero α, we deduce that $r(x)$
must be the zero polynomial, which shows that $p(x) \mid a(x)$.   □

Example 7.17. (a) Every rational number $\alpha \in \mathbb{Q}$ is algebraic over $\mathbb{Q}$ of degree 1.
(b) The minimal polynomial of $\sqrt{2} \in \mathbb{Q}(\sqrt{2})$ is $\mathrm{minpol}_{\mathbb{Q}}(\sqrt{2}, x) = x^2 - 2 \in \mathbb{Q}[x]$. Hence $\sqrt{2}$ is
algebraic over $\mathbb{Q}$ of degree 2.

(c) The minimal polynomial of $i \in \mathbb{R}(i) = \mathbb{C}$ is $\mathrm{minpol}_{\mathbb{R}}(i, x) = x^2 + 1 \in \mathbb{R}[x]$. We also have $\mathrm{minpol}_{\mathbb{Q}}(i, x) = x^2 + 1$. Therefore i is algebraic over $\mathbb{Q}$ and $\mathbb{R}$ of degree 2.

(d) Let $\mathbb{K} = \mathbb{R}$, $\mathbb{E} = \mathbb{C}$ and $\alpha \in \mathbb{C} \setminus \mathbb{R}$ be arbitrary. Then $\alpha = a + b \cdot i$ with $b \neq 0$. Since $\alpha \notin \mathbb{R}$ there is no minimal polynomial of degree 1. We construct a monic polynomial $p(x) \in \mathbb{R}[x]$ of degree 2 with zero α. By the existence and uniqueness (see Lemma 7.16), this must be the minimal polynomial of α over $\mathbb{R}$. The sought polynomial is given by[11]

$$\mathrm{minpol}_{\mathbb{R}}(\alpha, x) = (x - \alpha)(x - \overline{\alpha}) = x^2 - 2\,\mathrm{Re}\,(\alpha)x + |\alpha|^2 = x^2 - 2ax + (a^2 + b^2).$$

(e) If we set $\mathbb{K} = \mathbb{C}$ in Example (d), then the fundamental theorem of algebra shows that

$$\mathrm{minpol}_{\mathbb{C}}(\alpha, x) = x - \alpha.$$

(f) A number α is algebraic over $\mathbb{K}$ if and only if the set of powers $\{1, \alpha, \alpha^2, \ldots\}$ spans a finite-dimensional vector space V over $\mathbb{K}$. Why? Well, if $p(x)$ is the minimal polynomial of α of degree n, then the equation $p(\alpha) = 0$ shows that $1, \alpha, \alpha^2, \ldots, \alpha^n$ are linearly dependent. By multiplying the equation $p(\alpha) = 0$ with powers of α, we additionally see that all higher powers α^k with $k \geq n$ are also linearly dependent of $1, \alpha, \alpha^2, \ldots, \alpha^{n-1}$. On the other hand, the minimality of $p(x)$ shows that the elements $\{1, \alpha, \alpha^2, \ldots, \alpha^{n-1}\}$ are linearly independent. Hence the dimension of V is n. However, if the dimension of V is n, then in particular $\{1, \alpha, \alpha^2, \ldots, \alpha^n\}$ are linearly dependent. This yields the minimal polynomial. △

For the characterization of an algebraic number, its minimal polynomial is most important. In our above Example 7.17 (b) with $\alpha = \sqrt{2}$, we have $\mathrm{minpol}(\alpha, x) = x^2 - 2$ and $\mathrm{grad}(\alpha) = 2$. These properties are clearly non-trivial since they contain the irrationality of $\sqrt{2}$: if $\sqrt{2}$ was rational, then its minimal polynomial would be of degree 1.

The latter scenario with a minimal polynomial of degree 1 occurs, for example, for $\beta = \sqrt{4}$. By definition, the algebraic number β is a solution of the quadratic equation $p(x) = x^2 - 4 = 0$. However, this polynomial can be factorized over $\mathbb{Q}$ as $p(x) = x^2 - 4 = (x - 2)(x + 2)$. Therefore it follows from Lemma 7.16 that the minimal polynomial must be one of those factors. In our case $\mathrm{minpol}(\beta, x) = x - 2$ and therefore $\beta = 2$. Please note that these considerations led to a simplification of our input.

How can we deal with more complicated nested roots? Let for example

$$\gamma = \sqrt{4 + 2\sqrt{3}},$$

where we interpret γ as algebraic over $\mathbb{Q}$. We try to *denest* the nested root by algebraic transformations. If we square the above equation, we get

$$\gamma^2 = 4 + 2\sqrt{3}$$

and therefore

$$\gamma^2 - 4 = 2\sqrt{3}.$$

Squaring again yields

$$(\gamma^2 - 4)^2 = 12,$$

hence γ satisfies the polynomial equation

---

[11] Let $\alpha = a + bi \in \mathbb{C}$, $a, b \in \mathbb{R}$. Then by $\overline{\alpha} = a - bi$ we denote its *conjugate*, by $\mathrm{Re}\,\alpha = a$ its *real part*, by $\mathrm{Im}\,\alpha = b$ its *imaginary part* and by $|\alpha|$ its *absolute value* for which the relation $|\alpha|^2 = \alpha \cdot \overline{\alpha} = a^2 + b^2$ is valid.

$$p(x) = x^4 - 8x^2 + 4 = 0.$$

However, is $p(x)$ really the minimal polynomial of γ? It turns out that this is not the case.

Session 7.18. *We use Mathematica for the determination of the minimal polynomial of γ. To answer the question if the polynomial $p(x) = x^4 - 8x^2 + 4 \in \mathbb{Q}[x]$ is irreducible, we use* `Factor` *and get*

```
In[1]:= fac = x⁴ - 8x² + 4//Factor
Out[1]= (x² - 2 x - 2)(x² + 2 x - 2)
```

Hence the polynomial $p(x)$ is reducible over $\mathbb{Q}$, and by Lemma 7.16 one of its resulting factors must be the minimal polynomial of γ. As a test, we substitute γ in the first factor and get[12]

```
In[2]:= fac[[1]]/.{x → √(4+2√3)}
Out[2]= 2 + 2 √3 - 2 √(4+2 √3)
```

This was not helpful. But for computations with algebraic numbers, Mathematica offers the simplification function `RootReduce`*:*

```
In[3]:= fac[[1]]/.{x → √(4+2√3)}//RootReduce
Out[3]= 0
```

Hence $\text{minpol}(\gamma, x) = x^2 - 2x - 2$. *In this case it is easy to compute the two solutions of this quadratic equation and to select the correct one. It turns out that $\gamma = 1 + \sqrt{3}$, therefore we have deduced the identity*

$$\sqrt{4 + 2\sqrt{3}} = 1 + \sqrt{3}.$$

This simplification is also done by Mathematica itself:

```
In[4]:= √(4+2√3)//RootReduce
Out[4]= 1 + √3
```

Summing up, the given nested root was denested by computing its minimal polynomial and a factorization.

If a nested root γ is given by sums, products and reciprocals of algebraic numbers, then the method which we just described can obviously be used to compute its minimal polynomial.

Session 7.19. *Let $\alpha = \sqrt{2 + 3\sqrt[3]{2}}$. We compute the minimal polynomial of α over $\mathbb{Q}$ by the given algorithm. We start by entering the defining equation:*

```
In[1]:= eq1 = α == √(2+3 ∛2)
Out[1]= α == √(2+3 ∛2)
```

Next we square:

```
In[2]:= eq2 = Map[#²&, eq1]
Out[2]= α² == 2 + 3 ∛2
```

Now we isolate the root,

---

[12] Note that we chose this factor randomly. With the second factor we would not have been successful.

In[3]:= **eq3 = Map[# - 2&, eq2]**
Out[3]= $\alpha^2 - 2 == 3\sqrt[3]{2}$

take the third power,

In[4]:= **eq4 = Map[#³&, eq3]**
Out[4]= $(\alpha^2 - 2)^3 == 54$

and bring everything to one side:

In[5]:= **eq5 = Map[# - 54&, eq4]**
Out[5]= $(\alpha^2 - 2)^3 - 54 == 0$

Finally, we receive the polynomial

In[6]:= **minpol = Expand[eq5[[1]] /. α → x]**
Out[6]= $x^6 - 6x^4 + 12x^2 - 62$

which has the zero α.[13]

The only candidates for the minimal polynomial of α are the irreducible monic factors of the resulting polynomial. A rational factorization yields

In[7]:= **Factor[minpol]**
Out[7]= $x^6 - 6x^4 + 12x^2 - 62$

Since minpol *is irreducible, it is the sought minimal polynomial of α over $\mathbb{Q}$. You can find further examples of this type in the Exercises 7.9–7.10.*

Definition 7.20. A field extension $\mathbb{E} \supset \mathbb{K}$ is called *algebraic* if every element of $\mathbb{E}$ is algebraic over $\mathbb{K}$. $\triangle$

Next we show that finite field extensions are algebraic.

Theorem 7.21. (Algebraic Field Extension)

(a) Every finite field extension $\mathbb{E} \supset \mathbb{K}$ is algebraic and can be generated by adjoining finitely many algebraic elements.
(b) Every field extension $\mathbb{E} \supset \mathbb{K}$ which is generated by adjoining finitely many algebraic elements, is finite (and therefore algebraic by (a)).

Proof. (a) Let $\mathbb{E} \supset \mathbb{K}$ be finite of degree n and $\alpha \in \mathbb{E}$. Since $\mathbb{E}$ forms an n-dimension vector space over $\mathbb{K}$, there are at most n linearly dependent powers $1, \alpha, \alpha^2, \ldots$. Therefore a relation of the form

$$\sum_{k=0}^{n} c_k \alpha^k = 0$$

must be valid, hence α is algebraic. Therefore the field extension is algebraic.

Assume that the vector space $\mathbb{E}$ is generated by the elements α_k ($k = 1, \ldots, n$). Then obviously it is sufficient to adjoin those elements.
(b) Adjoining an algebraic number α of degree n generates a finite field extension with basis $1, \alpha, \alpha^2, \ldots, \alpha^{n-1}$. Any finite iteration of finite field extensions, however, is again a finite field extension. $\qquad \square$

---

[13] The irreducibility of $p(x) := x^6 - 6x^4 + 12x^2 - 62$ can be easily checked by Eisenstein's irreducibilty criterion with $p = 2$, see Exercise 6.7. The polynomial $p(\sqrt{x}) \in \mathbb{Q}[x]$ is the minimal polynomial of $\alpha^2 = 2 + 3\sqrt[3]{2}$. Why?

Corollary 7.22. The sum, the difference, the product and the quotient of algebraic numbers $\neq 0$ are algebraic. $\triangle$

Note that Corollary 7.22 predicts for algebraic numbers α and β the existence of a minimal polynomial for $\alpha \pm \beta$, $\alpha \cdot \beta$ and α/β.

Since $[\mathbb{K}(\alpha, \beta) : \mathbb{K}] = [\mathbb{K}(\alpha, \beta) : \mathbb{K}(\alpha)] \cdot [\mathbb{K}(\alpha) : \mathbb{K}]$, see Exercise 7.7, we get that $\mathrm{grad}(\alpha + \beta) \leqq \mathrm{grad}(\alpha) \cdot \mathrm{grad}(\beta)$ and $\mathrm{grad}(\alpha \cdot \beta) \leqq \mathrm{grad}(\alpha) \cdot \mathrm{grad}(\beta)$. In the next theorem we want to show this using a different method, namely by presenting algorithms which compute the sought minimal polynomials.

Theorem 7.23. Let α and β be two algebraic numbers over $\mathbb{K}$ with minimal polynomials $p(x), q(x) \in \mathbb{K}[x]$. Then $\alpha \pm \beta$ and $\alpha \cdot \beta$ are also algebraic over $\mathbb{K}$ and there is an algorithm to compute the respective minimal polynomials. The degree of $\alpha \pm \beta$ and of $\alpha \cdot \beta$ is at most $\mathrm{grad}(\alpha) \cdot \mathrm{grad}(\beta)$.

If $p(x) = x^n + a_{n-1} x^{n-1} + \cdots + a_1 x + a_0$ is the minimal polynomial of α, then the minimal polynomial of $1/\alpha$ is given by

$$\frac{1}{a_0} \left(a_0 x^n + a_1 x^{n-1} + \cdots + a_{n-1} x + a_n \right) = \frac{1}{a_0} x^n p\left(\frac{1}{x}\right).$$

Proof. We give a constructive proof by computing the sought minimal polynomials, hence we present the postulated algorithm. We consider here the algorithm for the sum, the other algorithms are quite similar.

For the minimal polynomials $p(x)$ and $q(x)$ of α and β, respectively, we have

$$p(\alpha) = \alpha^n + a_{n-1} \alpha^{n-1} + \cdots + a_1 \alpha + a_0 = 0$$

and

$$q(\beta) = \beta^m + b_{m-1} \beta^{m-1} + \cdots + b_1 \beta + b_0 = 0.$$

These equations are equivalent to the replacement rules

$$\alpha^n = -(a_{n-1} \alpha^{n-1} + \cdots + a_1 \alpha + a_0) \tag{7.5}$$

and

$$\beta^m = -(b_{m-1} \beta^{m-1} + \cdots + b_1 \beta + b_0) \tag{7.6}$$

for certain powers of α and β. Note that these replacement rules can be iteratively applied for every power of α and β so that all powers of α can be replaced by powers with exponents $< n$ and all powers of β can be replaced by powers with exponents $< m$.

The number which we are interested in is $\gamma = \alpha + \beta$. We can expand every power $\gamma^k = (\alpha + \beta)^k$, $(k = 0, \ldots, m \cdot n)$ and replace high powers of α and β according to (7.5) and (7.6), respectively. Iteratively, all high powers of α and β can be written as linear combinations of powers of α and β with exponents $< n$ and $< m$, respectively. Eventually we get a linear combination with terms of the type $a_{ij} = \alpha^i \beta^j$ $(i = 0, \ldots, n-1, j = 0, \ldots, m-1)$. There are $m \cdot n$ terms a_{ij} of that type.

If we set

$$F(\gamma) = \sum_{k=0}^{m \cdot n} c_k \gamma^k = 0,$$

with undetermined coefficients c_k $(k = 0, \ldots, m \cdot n)$, then the powers of γ in $F(\gamma)$ can be represented by the $m \cdot n$ variables a_{ij}. If all coefficients are set to zero, then the resulting sum is obviously also zero.

This method yields a homogeneous linear system of $m \cdot n$ equations in $m \cdot n + 1$ variables c_k $(k = 0, \ldots, m \cdot n)$ which has a nontrivial solution. Therefore, we get a polynomial for γ of degree $\leq m \cdot n$, which is the minimal polynomial or a multiple of it. As before, factorization can be used to find the minimal polynomial.[14]

As for the minimal polynomial of the reciprocal, let α be a zero of $p(x)$. Then obviously, the reciprocal $1/\alpha$ is a zero of $p(1/x)$, and therefore of the monic polynomial $\frac{1}{a_0} x^n p\left(\frac{1}{x}\right)$. $\square$

If we want to compute the minimal polynomial *without* factorization, then we can adapt the above algorithm iteratively: We set, for some $J \geq 1$,

$$F(\gamma) = \sum_{k=0}^{J} c_k \gamma^k = 0,$$

with the indeterminates c_k $(k = 0, \ldots, J)$, starting with $J = 1$. Whenever the corresponding linear system has no solution, we increase J by 1. This iteration stops with $J = m \cdot n$ at the latest.

Session 7.24. *Let us compute minimal polynomials of sums, products and reciprocals of algebraic numbers algorithmically.*

Let $\alpha = \sqrt{2}$ and $\beta = \sqrt[3]{5}$. Then $p(x) = x^2 - 2$ and $q(x) = x^3 - 5$ are the corresponding minimal polynomials. We will compute the minimal polynomial of $\gamma = \alpha + \beta$.

For this purpose we declare the recursive substitution rules

```
In[1]:= rule = {αⁿ⁻ :→ 2αⁿ⁻²/;n ≥ 2,  βᵐ⁻ :→ 5βᵐ⁻³/;m ≥ 3};
```

and apply these rules to $F(\gamma)$:

```
In[2]:= F = ∑_{k=0}^{6} Expand[cₖ(α+β)ᵏ]//.rule
```
$Out[2]=$ $c_2 \beta^2 + 3 \alpha c_3 \beta^2 + 12 c_4 \beta^2 + 20 \alpha c_5 \beta^2 +$
              $5 c_5 \beta^2 + 30 \alpha c_6 \beta^2 + 60 c_6 \beta^2 + c_1 \beta + 2 \alpha c_2 \beta +$
              $6 c_3 \beta + 8 \alpha c_4 \beta + 5 c_4 \beta + 25 \alpha c_5 \beta + 20 c_5 \beta +$
              $24 \alpha c_6 \beta + 150 c_6 \beta + c_0 + \alpha c_1 + 2 c_2 + 2 \alpha c_3 + 5 c_3 +$
              $20 \alpha c_4 + 4 c_4 + 4 \alpha c_5 + 100 c_5 + 200 \alpha c_6 + 33 c_6$

Next, we equate the coefficients of $F(\gamma)$ and solve for the variables c_k:

```
In[3]:= sol =
           Solve[Flatten[CoefficientList[F, {α, β}]] ==
              0, Table[cₖ, {k, 0, 5}]]
```
Solve :: **"svars"**: *Equations may not give solutions for all "solve" variables.*
$Out[3]=$ $\{\{c_0 \to 17 c_6, c_1 \to -60 c_6, c_2 \to 12 c_6, c_3 \to -10 c_6, c_4 \to -6 c_6, c_5 \to 0\}\}$

Finally we get the associated monic polynomial

```
In[4]:= F = ∑_{k=0}^{6} cₖxᵏ/.sol[[1]]/.c₆ → 1
```

---

[14] This, however, requires a factorization algorithm in $\mathbb{K}[x]$. For $\mathbb{K} = \mathbb{Q}$ and $\mathbb{K} = \mathbb{Z}_p$ we have already seen such an algorithm.

$Out[4] = x^6 - 6x^4 - 10x^3 + 12x^2 - 60x + 17$

This is the minimal polynomial of $\gamma = \alpha + \beta$, *as the computation*

$In[5] :=$ **Factor[F]**
$Out[5] = x^6 - 6x^4 - 10x^3 + 12x^2 - 60x + 17$

verifies.

Mathematica contains an algorithm for the computation of minimal polynomials. For this purpose, we use the Root *construct (to declare algebraic numbers) and the simplification procedure* RootReduce.

We define α *and* β *by*

$In[6] :=$ $\alpha =$ **Root[#^{2} - 2&, 2];**
          $\beta =$ **Root[#^{3} - 5&, 1];**

as zeros of the minimal polynomials $p(x) = x^2 - 2$ *and* $q(x) = x^3 - 5$, *respectively. Since the algorithm uses the minimal polynomials only, the second argument of* Root *is not really essential.*

Now, RootReduce *is called,* [15]

$In[7] :=$ **result = $\alpha + \beta$//RootReduce**
$Out[7] =$ $\text{Root}[\#1^6 - 6\,\#1^4 - 10\,\#1^3 + 12\,\#1^2 - 60\,\#1 + 17\&, 2]$

and provides the minimal polynomial of $\alpha + \beta$:

$In[8] :=$ **minpol = result[[1]][x]**
$Out[8] = x^6 - 6x^4 - 10x^3 + 12x^2 - 60x + 17$

Similarly, we get for the product $\alpha \cdot \beta$

$In[9] :=$ $\alpha * \beta$**//RootReduce**
$Out[9] =$ $\text{Root}[\#1^6 - 200\&, 2]$

and for the reciprocal of $\alpha \cdot \beta$

$In[10] := \dfrac{1}{\alpha + \beta}$ **//RootReduce**
$Out[10] =$ $\text{Root}[17\,\#1^6 - 60\,\#1^5 + 12\,\#1^4 - 10\,\#1^3 - 6\,\#1^2 + 1\&, 1]$

thus verifying Theorem 7.23.

7.4 Finite Fields

With $\mathbb{Z}_p$ we have already encountered finite fields, if p is prime. This takes us to the question whether there are further finite fields.

The following procedure will always generate a finite field, starting with the finite field $\mathbb{Z}_p$ with $p \in \mathbb{P}$ elements. Let $a(x) = a_n x^n + \cdots + a_1 x + a_0 \in \mathbb{Z}_p[x]$ be the minimal polynomial of an algebraic number α of degree n over $\mathbb{Z}_p$. Then we can construct the algebraic extension field $\mathbb{Z}_p(\alpha)$ which is an n-dimensional vector field over $\mathbb{Z}_p$ and therefore contains p^n elements. We have shown that $\mathbb{Z}_p(\alpha)$ is a field if p is prime, hence a finite field with $q = p^n$ elements.

Next we will successively show that

---

[15] In newer versions of *Mathematica* you have to move the mouse over the output to see the rest in its full form.

- **(Structure)** finite fields always have $q = p^n$ elements,

- **(Existence)** such a field with p^n elements indeed exists for all $p \in \mathbb{P}$ and $n \in \mathbb{N}$,

- **(Uniqueness)** up to an isomorphism, for every $p \in \mathbb{P}$ and $n \in \mathbb{N}$ there is only one such field,

- **(Structure)** every finite field with p^n elements can be constructed as described by a simple field extension.

These results will authorize us to give a name to this uniquely determined field with p^n elements.

Definition 7.25. In all cases, the described method will generate a model of the field with $q = p^n$ elements which is called—after its discoverer Évariste Galois—the *Galois field* with q elements and is denoted by $\mathrm{GF}(q)$ or $\mathbb{F}_q$. The elements of $\mathrm{GF}(p^n)$ can be written in the form

$$\sum_{k=0}^{n-1} c_k \alpha^k \qquad (c_k \in \mathbb{Z}_p),$$

where α is a zero of an irreducible polynomial of degree n in $\mathbb{Z}_p[x]$, and therefore generate a vector space of dimension n over $\mathbb{Z}_p$. $\triangle$

In this section we will prove the above statements. But let us start with an example.

Example 7.26. We consider the case $p = 3$ and $n = 2$. The elements of $\mathrm{GF}(9)$ have the form $a + b\alpha$, $a, b \in \{0, 1, 2\}$. We get the following addition and multiplication tables for $\mathrm{GF}(9)$:

+	0	1	2	α	$\alpha + 1$	$\alpha + 2$	2α	$2\alpha + 1$	$2\alpha + 2$
0	0	1	2	α	$\alpha + 1$	$\alpha + 2$	2α	$2\alpha + 1$	$2\alpha + 2$
1	1	2	0	$\alpha + 1$	$\alpha + 2$	α	$2\alpha + 1$	$2\alpha + 2$	2α
2	2	0	1	$\alpha + 2$	α	$\alpha + 1$	$2\alpha + 2$	2α	$2\alpha + 1$
α	α	$\alpha + 1$	$\alpha + 2$	2α	$2\alpha + 1$	$2\alpha + 2$	0	1	2
$\alpha + 1$	$\alpha + 1$	$\alpha + 2$	α	$2\alpha + 1$	$2\alpha + 2$	2α	1	2	0
$\alpha + 2$	$\alpha + 2$	α	$\alpha + 1$	$2\alpha + 2$	2α	$2\alpha + 1$	2	0	1
2α	2α	$2\alpha + 1$	$2\alpha + 2$	0	1	2	α	$\alpha + 1$	$\alpha + 2$
$2\alpha + 1$	$2\alpha + 1$	$2\alpha + 2$	2α	1	2	0	$\alpha + 1$	$\alpha + 2$	α
$2\alpha + 2$	$2\alpha + 2$	2α	$2\alpha + 1$	2	0	1	$\alpha + 2$	α	$\alpha + 1$

and

$\cdot$	0	1	2	α	$\alpha + 1$	$\alpha + 2$	2α	$2\alpha + 1$	$2\alpha + 2$
0	0	0	0	0	0	0	0	0	0
1	0	1	2	α	$\alpha + 1$	$\alpha + 2$	2α	$2\alpha + 1$	$2\alpha + 2$
2	0	2	1	2α	$2\alpha + 2$	$2\alpha + 1$	α	$\alpha + 2$	$\alpha + 1$
α	0	α	2α	$2\alpha + 1$	1	$\alpha + 1$	$\alpha + 2$	$2\alpha + 2$	2
$\alpha + 1$	0	$\alpha + 1$	$2\alpha + 2$	1	$\alpha + 2$	2α	2	α	$2\alpha + 1$
$\alpha + 2$	0	$\alpha + 2$	$2\alpha + 1$	$\alpha + 1$	2α	2	$2\alpha + 2$	1	α
2α	0	2α	α	$\alpha + 2$	2	$2\alpha + 2$	$2\alpha + 1$	$\alpha + 1$	1
$2\alpha + 1$	0	$2\alpha + 1$	$\alpha + 2$	$2\alpha + 2$	α	1	$\alpha + 1$	2	2α
$2\alpha + 2$	0	$2\alpha + 2$	$\alpha + 1$	2	$2\alpha + 1$	α	1	2α	$\alpha + 2$

The content of both tables can be easily checked using the minimal polynomial $a(x) = x^2 + x + 2$ modulo 3. Note that there are several irreducible polynomials of degree 2. The chosen one—which is also used by *Mathematica*—leads to the replacement rule $\alpha^2 \mapsto 2\alpha + 1$. $\triangle$

Session 7.27. *The Mathematica package* `FiniteFields` *enables the computation in Galois fields. After loading the package by*

In[1]:= **Needs["FiniteFields`"]**

we can access the package. It supports several different representations for the elements of GF(q). *We will use one of these representations to get the above addition and multiplication tables. Using the command*

In[2]:= **SetFieldFormat[GF[9],FormatType → FunctionOfCode[GF9]]**

we declare that the elements of GF(9) *are numbered by* $0, 1, \ldots, 8$ *and can be selected by the function call* GF9(n). *Then the following command reproduces the above multiplication table:*

In[3]:= **Table[ElementToPolynomial[GF9[j] * GF9[k],**
            α], {j,0,8}, {k,0,8}]/.
            ElementToPolynomial[0,α] → 0

$$
Out[3]= \begin{pmatrix}
0 & 0 & 0 & 0 & 0 & 0 & 0 & 0 & 0 \\
0 & 1 & 2 & \alpha & \alpha+1 & \alpha+2 & 2\alpha & 2\alpha+1 & 2\alpha+2 \\
0 & 2 & 1 & 2\alpha & 2\alpha+2 & 2\alpha+1 & \alpha & \alpha+2 & \alpha+1 \\
0 & \alpha & 2\alpha & 2\alpha+1 & 1 & \alpha+1 & \alpha+2 & 2\alpha+2 & 2 \\
0 & \alpha+1 & 2\alpha+2 & 1 & \alpha+2 & 2\alpha & 2 & \alpha & 2\alpha+1 \\
0 & \alpha+2 & 2\alpha+1 & \alpha+1 & 2\alpha & 2 & 2\alpha+2 & 1 & \alpha \\
0 & 2\alpha & \alpha & \alpha+2 & 2 & 2\alpha+2 & 2\alpha+1 & \alpha+1 & 1 \\
0 & 2\alpha+1 & \alpha+2 & 2\alpha+2 & \alpha & 1 & \alpha+1 & 2 & 2\alpha \\
0 & 2\alpha+2 & \alpha+1 & 2 & 2\alpha+1 & \alpha & 1 & 2\alpha & \alpha+2
\end{pmatrix}
$$

Note that the call `ElementToPolynomial` *converts these elements to our standard representation. Of course an analogous command generates the addition table.*

The (irreducible) minimal polynomial which is used by the package can be discovered by

In[4]:= **FieldIrreducible[GF[9],x]**
Out[4]= $x^2 + x + 2$

Using the characteristic 3 *and the degree of the field extension* 2, *the minimal polynomial can be also computed by*

In[5]:= **IrreduciblePolynomial[x,3,2]**
Out[5]= $x^2 + x + 2$

Of course the individual multiplications of the multiplication table can also be computed by `PolynomialMod`. *For example, the computation* $(2\alpha + 1)(\alpha + 1)$ *(mod* $\alpha^2 + \alpha + 2$*) in* $\mathbb{Z}_3$ *confirms one of the entries of the above multiplication table:*

In[6]:= **PolynomialMod[(2α+1)(α+2),α² +α+2,Modulus → 3]**
Out[6]= 1

In a first step, we will now prove that every finite field has prime characteristic.

Lemma 7.28. (Characteristic of a Finite Field) Let $\mathbb{K}$ be a finite field. Then there is some prime $p \in \mathbb{P}$ with char($\mathbb{K}$) $= p$, and an isomorphic copy of $\mathbb{Z}_p$ is contained in $\mathbb{K}$.

Proof. Let

$$n \cdot 1 = \underbrace{1 + \cdots + 1}_{n \text{ times}} . \tag{7.7}$$

We consider the set $M := \{1, 2 \cdot 1, \ldots, n \cdot 1, \ldots\} \subset \mathbb{K}$. Since $\mathbb{K}$ is finite, $M \subset \mathbb{K}$ is also finite. Hence two of the elements from M coincide: $m \cdot 1 = n \cdot 1$. This implies that $(n - m) \cdot 1 = 0$. Therefore $\mathbb{K}$ has finite characteristic. However, the characteristic must be prime, since the identity[16] $(p \cdot q) \cdot 1 = (p \cdot 1) \cdot (q \cdot 1) = 0$ implies either $p = 0$ or $q = 0$.

If we identify $n \cdot 1 \in \mathbb{K}$ with $[n]_p \in \mathbb{Z}_p$, then we see that $\mathbb{Z}_p \subset \mathbb{K}$ is a subfield of $\mathbb{K}$. $\square$

Next we get the first main result about the structure of finite fields.

Theorem 7.29. (Structure of Finite Fields) Let $\mathbb{K}$ be a finite field. Then $\mathbb{K}$ is isomorphic to a finite field extension of $\mathbb{Z}_p$ and therefore has $q = p^n$ elements, where n is the degree of the field extension.

Proof. Let $\mathbb{K}$ be given and (corresponding to Lemma 7.28) let the characteristic $p \in \mathbb{P}$, and $\mathbb{Z}_p \subset \mathbb{K}$. Since $\mathbb{K}$ has only finitely many elements, only n of them $\alpha_1, \ldots, \alpha_n \in \mathbb{K}$ are linearly independent over $\mathbb{Z}_p$, and every element $a \in \mathbb{K}$ can be represented as

$$a = \sum_{k=1}^{n} c_k \alpha_k \tag{7.8}$$

with uniquely determined coefficients $c_k \in \mathbb{Z}_p$. Every coefficient c_k has p possible values, hence there are exactly p^n different elements of the form (7.8). Since all field elements can be generated in this way, $\mathbb{K}$ must be a finite field extension of $\mathbb{K}$ with p^n elements. $\square$

Corollary 7.30. If $q \in \mathbb{N}$ is not the power of a prime, then there is no finite field with q elements. $\square$

In the next step, we show the existence of finite fields.

Theorem 7.31. (Existence of Finite Fields) For every $p \in \mathbb{P}$ and every $n \in \mathbb{N}$ there is a finite field with exactly p^n elements.

Proof. For $n = 1$ we have the finite field $\mathbb{Z}_p$. Hence in the sequel we assume that $n \geq 2$.

We consider the polynomial $f(x) := x^{p^n} - x \in \mathbb{Z}_p[x]$. Then according to Theorem 7.13 there is a splitting field $\mathbb{E} \supset \mathbb{Z}_p$, in which the polynomial $f(x)$ factors into linear factors. Now let us consider the subset $\mathbb{K} \subset \mathbb{E}$ consisting of all zeros of $f(x)$:

$$\mathbb{K} = \{a \in \mathbb{E} \mid a^{p^n} = a\} .$$

We will show that $\mathbb{K}$ is a field with exactly $q = p^n$ elements.

$\mathbb{K}$ is a field since with every $a, b \in \mathbb{K}$:

(1) $a + b \in \mathbb{K}$:

---

[16] Please note that this equation uses three different multiplication signs: the multiplication in $\mathbb{Z}$, the multiplication in $\mathbb{K}$ and the scalar multiplication declared in (7.7).

$$(a+b)^{p^n} = \sum_{k=0}^{p^n} \binom{p^n}{k} a^k b^{p^n-k} = a^{p^n} + b^{p^n} \, ,$$

because all occurring binomial coefficients except the first and the last contain the characteristic p as a factor.

(2) $-a \in \mathbb{K}$: $-(a^{p^n}) = -a$.

(3) $a \cdot b \in \mathbb{K}$: $(a \cdot b)^{p^n} = a^{p^n} \cdot b^{p^n}$.

(4) $a^{-1} \in \mathbb{K}$: $(a^{-1})^{p^n} = (a^{p^n})^{-1} = a^{-1}$.

Since $0 \in \mathbb{K}$ and $1 \in \mathbb{K}$, we see that $\mathbb{K}$ is a subfield of $\mathbb{E}$.

It remains to prove that the constructed field $\mathbb{K}$ has exactly p^n elements. For this purpose, we prove that all zeros of $f(x)$ are simple. Since the degree of $f(x)$ is p^n, this implies the result.

Since the characteristic of $\mathbb{Z}_p$ equals p, we get $f'(x) = p^n x^{p^n-1} - 1 = -1 \neq 0$, and the proof of Lemma 6.41 shows that $f(x)$ only has simple zeros, as announced. $\qquad \square$

Next we prove the uniqueness of finite fields.

Theorem 7.32. (Uniqueness of Finite Fields) Let $p \in \mathbb{P}$ and $n \in \mathbb{N}$. A finite field with $q = p^n$ elements is uniquely determined up to an isomorphism.

Proof. Let $\mathbb{K}$ be an arbitrary finite field. We first show that for all $a \in \mathbb{K}^* = \mathbb{K} \setminus \{0\}$ the relation

$$a^{q-1} = 1 \tag{7.9}$$

is valid. Let $a \in \mathbb{K}^*$ be arbitrary. Since $\mathbb{K}^*$ is a multiplicative group, the mapping $x \mapsto x \cdot a$ is obviously bijective. Now we multiply all $q - 1$ elements of $\mathbb{K}^*$ in two ways and get

$$\prod_{x \in \mathbb{K}^*} x = \prod_{x \in \mathbb{K}^*} (a \cdot x) = a^{q-1} \prod_{x \in \mathbb{K}^*} x \, ,$$

since $\mathbb{K}^*$ is commutative. This shows (7.9). The resulting equation $a^q - a = 0$ is also valid for $a = 0$ and therefore for all $a \in \mathbb{K}$. Hence all field elements are zeros of the polynomial $x^q - x \in \mathbb{Z}_p[x]$ whose coefficients are considered as elements of $\mathbb{Z}_p$.

Let $\mathbb{K} = \{a_1, a_2, \ldots, a_q\}$. Then the product

$$\prod_{k=1}^{q} (x - a_k)$$

is a divisor of $x^q - x$, and since both polynomials have the same degree and are monic, we finally get

$$x^q - x = \prod_{k=1}^{q} (x - a_k) = \prod_{a \in \mathbb{K}} (x - a) \, . \tag{7.10}$$

All elements of $\mathbb{K}$ are therefore zeros of the same polynomial $x^q - x \in \mathbb{Z}_p[x]$. By adjoining all zeros which do not lie in $\mathbb{Z}_p$ to $\mathbb{Z}_p$, we get an algebraic extension field of $\mathbb{Z}_p$ which is uniquely determined up to an isomorphism, since the order of adjunction does not matter. $\qquad \square$

Theorem 7.32 therefore states that $\mathrm{GF}(p^n)$ is unique up to an isomorphism and that it can be generated by a *finite field extension* from $\mathbb{Z}_p$.

This is already quite a nice result. However, the icing on the cake of our considerations will be our proof that even a *simple field extension* of $\mathbb{Z}_p$ is sufficient to generate $GF(p^n)$.

For this purpose we need the next theorem which was already formulated (without proof) in Theorem 4.19 and was used in Section 4.6 for the structure of Carmichal numbers.

Theorem 7.33. (Existence of a Primitive Element) Let $p \in \mathbb{P}$, $n \in \mathbb{N}$ and let $\mathbb{K}$ be a field with $q = p^n$ elements. Then $\mathbb{K}$ has a generator (or primitive element), i.e. an element a of order $\mathrm{ord}(a) = q - 1$.

Proof. In the proof the following set of all possible orders of the elements of $\mathbb{K}^\star$ is considered:

$$S := \{\mathrm{ord}(a) \mid a \in \mathbb{K}^\star\} \subset \mathbb{N}.$$

We first prove the following lemma about the structure of S.

Lemma 7.34. Let $j, k \in S$. Then the following statements are valid:

(a) $\gcd(j, k) = 1 \quad \Rightarrow \quad j \cdot k \in S$;
(b) $l \mid k$ and $l > 0 \quad \Rightarrow \quad l \in S$;
(c) $\mathrm{lcm}(j, k) \in S$.
(d) Let $m := \max S$. Then every number $j \in S$ is a divisor of m.

Proof. (a) Assume that a has order j and b has order k with $\gcd(j, k) = 1$. Then we get

$$(a \cdot b)^{j \cdot k} = (a^j)^k \cdot (b^k)^j = 1. \tag{7.11}$$

We show that $a \cdot b$ has order $j \cdot k$. For this purpose, we study for which $l \in \mathbb{N}$ the relation $(a \cdot b)^l = 1$ is valid. This relation implies

$$1 = \left((a \cdot b)^l\right)^j = (a \cdot b)^{l \cdot j} = (a^j)^l \cdot b^{j \cdot l} = b^{j \cdot l}.$$

Since b has order k, it follows that $k \mid j \cdot l$. Because $\gcd(j, k) = 1$, it follows that $k \mid l$. Similarly we can show that $j \mid l$, hence we get $j \cdot k \mid l$, and therefore (7.11) implies that $a \cdot b$ has order $j \cdot k$.
(b) Let b have order k, let $l \mid k$ and $l > 0$. Then obviously $j := \frac{k}{l} \in \mathbb{N}$. We set $c := b^j$ and get

$$c^l = b^{j \cdot l} = b^k = 1,$$

and no smaller power of c is equal to 1; therefore the order of c is l.
(c) Let $j = p_1^{e_1} \cdots p_r^{e_r}$ and $k = p_1^{f_1} \cdots p_r^{f_r}$ be the prime factor decompositions of j and k where $p_1, \ldots, p_r$ denote all the prime factors of j and k. The exponents e_i, f_i ($i = 1, \ldots, r$) are the orders of the factors and satisfy $e_i, f_i \in \mathbb{N}_{\geq 0}$. Part (b) implies that with j and k every divisor $p_i^{\max(e_i, f_i)} \in S$. By part (a) this is also valid for the product

$$l := p_1^{\max(e_1, f_1)} \cdots p_r^{\max(e_r, f_r)} \in S.$$

However, it is easy to see that $l = \mathrm{lcm}(j, k)$, see Exercise 3.18.
(d) Let $j \in S$. Then by part (c) we have $\mathrm{lcm}(j, m) \in S$. Since m is the maximum of S, we find $\mathrm{lcm}(j, m) = m$ and therefore $j \mid m$. $\qquad\square$

Now we can finalize the proof of Theorem 7.33. Let again $m = \max S$. The conclusion of the theorem is the identity $m = q - 1$. To prove it, we will prove the two corresponding inequalities.

(i) $m \leq q-1$: We already know from (7.9) that for all $a \in \mathbb{K}^\star$ the relation $a^{q-1} = 1$ is valid. This immediately yields $m \leq q-1$.

(ii) $m \geq q-1$: We consider $a \in \mathbb{K}^\star$. Let j be the order of a. Then $a^j = 1$. Because of Lemma 7.34 (d) this implies $a^m = 1$ for all $a \in \mathbb{K}^\star$. Since $\mathbb{K}^\star$ contains exactly $q-1$ elements, the polynomial $x^m - 1$ must have exactly $q-1$ zeros in $\mathbb{K}^\star$. However, Corollary 6.25 states that a polynomial of degree m has at most m zeros, therefore we find $q-1 \leq m$. □

In *Mathematica*, generators can be computed with the help of the command `Multiplicative-Order`. This was already treated in Session 4.18 on page 78.

As a consequence of Theorem 7.33 we get the following result.

Theorem 7.35. (Structure of Finite Fields) Let $p \in \mathbb{P}$ and $n \in \mathbb{N}$. Every finite field $\mathbb{K}$ with $q = p^n$ elements is isomorphic to a simple field extension of $\mathbb{Z}_p$ of degree n.

Proof. From Theorem 7.33 we know that $\mathbb{K}$ contains a primitive element $\alpha \in \mathbb{K}$. Because of (7.10) this element is a zero of the polynomial $x^{q-1} - 1 \in \mathbb{Z}_p[x]$. By Lemma 7.16 α possesses an irreducible minimal polynomial $a(x) \in \mathbb{Z}_p[x]$, and the algebraic extension $\mathbb{Z}_p(\alpha) = \mathbb{Z}_p[x]/\langle a(x) \rangle$ is a field with $\alpha \in \mathbb{Z}_p(\alpha)$. Since the set of powers $\{1, \alpha, \alpha^2, \ldots, \alpha^{q-1}\}$ generates the whole multiplicative group $\mathbb{K}^\star$, we must have $\mathbb{K} = \mathbb{Z}_p(\alpha)$, and the isomorphism $\mathbb{K} \cong \mathbb{Z}_p(\alpha)$ is easily detected. □

Corollary 7.36. (Existence of Irreducible Polynomials) For every $p \in \mathbb{P}$ and $n \in \mathbb{N}_{\geq 2}$ there is an irreducible polynomial $a(x) \in \mathbb{Z}_p[x]$ of degree n.

Proof. By Theorem 7.35 the Galois field $GF(p^n) = \mathbb{Z}_p/\langle a(x) \rangle$ is generated by a simple field extension, therefore there must be an irreducible polynomial $a(x) \in \mathbb{Z}_p[x]$. □

Be aware how difficult it was to prove this existence theorem![17] In reality there are *extremely many* irreducible polynomials $a(x) \in \mathbb{Z}_p[x]$, see [Chi2000], Chapter III 13, and Exercise 7.16.

7.5 Resultants

In this section, let R again be an integral domain. We assume that two polynomials $a(x), b(x) \in R[x] \setminus \{0\}$ are given, and we wish to investigate the question if they a have common zero, no matter if this zero is a member of R: in fact, it may also lie in a suitable splitting field.[18] For example, the polynomials $a(x) = x^2 - 2 \in \mathbb{Z}[x]$ and $b(x) = x^3 - 2x \in \mathbb{Z}[x]$ have the common zero $\sqrt{2} \in \mathbb{R}$ which, however, is not a member of $\mathbb{Z}$. On the other hand, the common zero yields the factorizations $a(x) = x^2 - 2$ and $b(x) = x(x^2 - 2)$, hence $a(x)$ and $b(x)$ have the common divisor $x^2 - 2 \in \mathbb{Z}[x]$. By considering common zeros of $a(x)$ and $b(x)$—which will lead us to the definition of the resultant—the greatest common divisor can be viewed from a different perspective.

The polynomials may have the representations

---

[17] In contrast to the case $R = \mathbb{Z}$, see Exercise 6.8.

[18] Every integral domain lies in a suitable quotient field which has a splitting field.

$$a(x) = a_n x^n + a_{n-1} x^{n-1} + \cdots + a_1 x + a_0 \qquad (a_n \neq 0)$$

and

$$b(x) = b_m x^m + b_{m-1} x^{m-1} + \cdots + b_1 x + b_0 \qquad (b_m \neq 0).$$

A common zero α of $a(x)$ and $b(x)$ therefore satisfies the two equations

$$a_n \alpha^n + a_{n-1} \alpha^{n-1} + \cdots + a_1 \alpha + a_0 = 0$$

and

$$b_m \alpha^m + b_{m-1} \alpha^{m-1} + \cdots + b_1 \alpha + b_0 = 0.$$

If we now define the new variables $x_j := \alpha^j$ ($j \geq 0$), then we get

$$a_n x_n + a_{n-1} x_{n-1} + \cdots + a_1 x_1 + a_0 x_0 = 0$$

and

$$b_m x_m + b_{m-1} x_{m-1} + \cdots + b_1 x_1 + b_0 x_0 = 0.$$

These two equations have many solutions. In order to make the solution unique, we use the fact that besides the equations $a(\alpha) = 0$ and $b(\alpha) = 0$, the following additional equations are automatically valid as well:

$$
\begin{aligned}
a(\alpha) &= 0, & b(\alpha) &= 0, \\
\alpha\, a(\alpha) &= 0, & \alpha\, b(\alpha) &= 0, \\
\alpha^2 a(\alpha) &= 0, & \alpha^2 b(\alpha) &= 0, \\
&\;\vdots & &\;\vdots \\
\alpha^{m-1} a(\alpha) &= 0, & \alpha^{n-1} b(\alpha) &= 0.
\end{aligned}
$$

This gives $m + n$ linear equations in the $m + n$ variables $x_j = \alpha^j$ ($j = 0, \ldots, m + n - 1$):

$$
\begin{aligned}
a_n x_{m+n-1} + a_{n-1} x_{m+n-2} + \cdots + a_0 x_{m-1} &&&= 0 \\
a_n x_{m+n-2} + \cdots + a_1 x_{m-1} + a_0 x_{m-2} &&&= 0 \\
\vdots \qquad\qquad\qquad && \\
a_n x_n + \cdots + a_0 x_0 &&&= 0 \\
b_m x_{m+n-1} + b_{m-1} x_{m+n-2} + \cdots + b_0 x_{n-1} &&&= 0 \\
b_m x_{m+n-2} + \cdots + b_1 x_{n-1} + b_0 x_{n-2} &&&= 0 \\
\vdots \qquad\qquad\qquad && \\
b_m x_m + \cdots + b_0 x_0 &&&= 0
\end{aligned}
$$

The associated $(m + n) \times (m + n)$ matrix S of this homogeneous linear system is given by

$$
S(a(x), b(x), x) := \left.\left(
\begin{array}{ccccccccc}
a_n & a_{n-1} & \cdots & a_1 & a_0 & 0 & \cdots & 0 \\
0 & a_n & a_{n-1} & \cdots & a_1 & a_0 & \cdots & 0 \\
\cdots & \cdots & \cdots & \cdots & \cdots & \cdots & \cdots & \cdots \\
0 & \cdots & 0 & a_n & a_{n-1} & \cdots & a_1 & a_0 \\
b_m & b_{m-1} & \cdots & b_1 & b_0 & 0 & \cdots & 0 \\
0 & b_m & b_{m-1} & \cdots & b_1 & b_0 & \cdots & 0 \\
\cdots & \cdots & \cdots & \cdots & \cdots & \cdots & \cdots & \cdots \\
0 & \cdots & 0 & b_m & b_{m-1} & \cdots & b_1 & b_0
\end{array}
\right)\right\}
\begin{array}{l}
\\ \\ m \text{ rows} \\ \\ \\ \\ n \text{ rows}
\end{array}
$$

and is called the *Sylvester matrix*.[19] If $a(x)$ and $b(x)$ have at least one common zero, then this linear system has a non-trivial solution. This, however, is only possible if the corresponding determinant of S is zero, which motivates the following definition.

Definition 7.37. Let R be an integral domain and let $a(x), b(x) \in R[x] \setminus \{0\}$ with

$$a(x) = \sum_{k=0}^{n} a_k x^k \quad \text{and} \quad b(x) = \sum_{k=0}^{m} b_k x^k .$$

Then the determinant of the Sylvester matrix

$$\operatorname{res}(a(x), b(x), x) := \begin{vmatrix} a_n & a_{n-1} & \cdots & a_1 & a_0 & 0 & \cdots & 0 \\ 0 & a_n & a_{n-1} & \cdots & a_1 & a_0 & \cdots & 0 \\ \cdots & \cdots & \cdots & \cdots & \cdots & \cdots & \cdots & \cdots \\ 0 & \cdots & 0 & a_n & a_{n-1} & \cdots & a_1 & a_0 \\ b_m & b_{m-1} & \cdots & b_1 & b_0 & 0 & \cdots & 0 \\ 0 & b_m & b_{m-1} & \cdots & b_1 & b_0 & \cdots & 0 \\ \cdots & \cdots & \cdots & \cdots & \cdots & \cdots & \cdots & \cdots \\ 0 & \cdots & 0 & b_m & b_{m-1} & \cdots & b_1 & b_0 \end{vmatrix}$$

is called the *resultant* of $a(x)$ and $b(x)$. If both $a(x)$ and $b(x)$ are constant—in which case the Sylvester matrix is empty—we define $\operatorname{res}(a, b, x) := 1$. Obviously $\operatorname{res}(a(x), b(x), x) \in R$. △

As we already deduced, the resultant of $a(x)$ and $b(x)$ equals zero if and only if $a(x)$ and $b(x)$ have a common zero in a suitable splitting field. Further important properties are given in the following theorem where—for simplicity—we consider a field $R = \mathbb{K}$.

Theorem 7.38. (Properties of Resultants) Let $\mathbb{K}$ be a field and let $a(x), b(x) \in \mathbb{K}[x] \setminus \{0\}$ with

$$a(x) = \sum_{k=0}^{n} a_k x^k \quad \text{and} \quad b(x) = \sum_{k=0}^{m} b_k x^k \quad (a_n, b_m \neq 0).$$

Then the following statements are valid:

(a) $\operatorname{res}(a(x), c, x) = c^n$ for a constant polynomial $c \in \mathbb{K}$;
(b) $\operatorname{res}(b(x), a(x), x) = (-1)^{mn} \operatorname{res}(a(x), b(x), x)$;
(c) Let $\deg(a(x), x) = n \geq m = \deg(b(x), x) > 0$ and let $r(x) = \operatorname{rem}(a(x), b(x), x)$ with $\deg(r(x), x) = k$, then $\operatorname{res}(a(x), b(x), x) = (-1)^{mn} b_m^{n-k} \operatorname{res}(b(x), r(x), x)$;
(d) $\operatorname{res}(a(x), b(x), x) = 0$ if and only if $a(x)$ and $b(x)$ have a non-trivial common divisor $\gcd(a(x), b(x)) \in \mathbb{K}[x]$;
(e) Let α_i $(i = 1, \ldots, n)$ denote the zeros of $a(x)$ and let β_j $(j = 1, \ldots, m)$ denote the zeros of $b(x)$, counted by multiplicity, in a suitable splitting field of $a(x) \cdot b(x)$ over $\mathbb{K}$, then

$$\operatorname{res}(a(x), b(x), x) = a_n^m \prod_{i=1}^{n} b(\alpha_i) = (-1)^{mn} b_m^n \prod_{j=1}^{m} a(\beta_j) = a_n^m b_m^n \prod_{i=1}^{n} \prod_{j=1}^{m} (\alpha_i - \beta_j). \quad (7.12)$$

Proof. (a) The result follows directly from the definition. In this case, the Sylvester matrix is c times the unit matrix.
(b) It is well known that the determinant changes sign if two neighboring rows are switched.

---

[19] The name is misleading, since this matrix was first considered by Leonhard Euler and not by J.J. Sylvester.

Since any of the first m rows is shifted down by n rows, the result follows.

(c) Let $m > 0$. We write $a(x) = q(x) b(x) + r(x)$. Then $\deg(q(x), x) \leqq n - m$, hence $q(x) = \sum_{j=0}^{n-m} q_j x^j$.
We now consider $\mathrm{res}(b(x), a(x), x) = (-1)^{mn} \mathrm{res}(a(x), b(x), x)$ and convert the corresponding Sylvester matrix

$$
\begin{pmatrix}
b_m & b_{m-1} & \cdots & b_1 & b_0 & 0 & \cdots & 0 \\
0 & b_m & b_{m-1} & \cdots & b_1 & b_0 & \cdots & 0 \\
\cdots & \cdots & \cdots & \cdots & \cdots & \cdots & \cdots & \cdots \\
0 & \cdots & 0 & b_m & b_{m-1} & \cdots & b_1 & b_0 \\
a_n & a_{n-1} & \cdots & a_1 & a_0 & 0 & \cdots & 0 \\
0 & a_n & a_{n-1} & \cdots & a_1 & a_0 & \cdots & 0 \\
\cdots & \cdots & \cdots & \cdots & \cdots & \cdots & \cdots & \cdots \\
0 & \cdots & 0 & a_n & a_{n-1} & \cdots & a_1 & a_0
\end{pmatrix}
= \begin{pmatrix} B \\ A \end{pmatrix}
$$

in such a way that its determinant does not change. For this purpose we subtract from any of the last m rows—representing $a(x)$—a suitable linear combination with coefficients q_j of the first m rows—representing $b(x)$. The underlying idea is to subtract the product $q(x) \cdot b(x)$. This results in a matrix

$$
T = \begin{pmatrix} B \\ O \ R \end{pmatrix},
$$

which has the same determinant as the Sylvester matrix. Here the rows of R correspond to the polynomial $r(x) = a(x) - q(x) \cdot b(x)$ and O corresponds to a zero matrix with m rows and $n - k$ columns. The conclusion then follows by $n - k$-fold expansion via Laplace's formula along the respective first column.

(d) This follows from (a)–(c) and the Euclidean algorithm.

(e) Obviously both the second and the third expression equal the fourth expression. Let us denote this common value by S. The properties (a)–(c) define the resultant uniquely: obviously one can compute the resultant using these three rules recursively.[20] Hence it remains to prove that S satisfies the three conditions (a)–(c). Property (a) is obvious, and (b) is equivalent to the equality of the expressions in (7.12). To show (c), we use the third representation in (7.12) to compute

$$
\mathrm{res}(a(x), b(x), x) = (-1)^{mn} b_m^n \prod_{j=1}^{m} a(\beta_j).
$$

On the other hand, using the second representation in (7.12), we get

$$
(-1)^{mn} b_m^{n-k} \mathrm{res}(b(x), r(x), x) = (-1)^{mn} b_m^{n-k} b_m^k \prod_{j=1}^{m} r(\beta_j).
$$

From $a(x) = q(x) b(x) + r(x)$, however, the relation $a(\beta_j) = r(\beta_j)$ follows. Therefore the statement is valid. $\square$

Please note that (7.12) generally is a representation in a suitable splitting field of $\mathbb{K}$. For $\mathbb{K} = \mathbb{Q}$ such a splitting field is $\mathbb{C}$, hence α_i and β_k are complex numbers in general. Nevertheless, the resultant—by definition—is an element of $\mathbb{Q}$. If $\mathbb{K} = R$ is an integral domain, for example $R = \mathbb{Z}$, then the resultant is an element of R despite the representations (7.12).

---

[20] Rule (b) is used to ensure that the degree of $a(x)$ is not smaller than the degree of $b(x)$. Rule (c) is used to reduce the degree successively until the termination condition (a) applies.

Session 7.39. *Mathematica computes the resultant of two polynomials by the function* `Resultant`. *We set*

```
In[1]:= a = x² + 2x;
        b = x - c;
```

Obviously $a(x)$ and $b(x)$ have a common zero—namely $x = 0$ or $x = -2$—if and only if $c = 0$ or $c = -2$. The computation

```
In[2]:= Resultant[a, b, x]
Out[2]= -((-c - 2) c)
```

confirms this consideration since the resultant is zero if and only if c has one of those values.

We can compute the resultant according to Theorem 7.38 as follows:

```
In[3]:= Clear[resultant]

        resultant[a_, b_, x_] := b^Exponent[a,x] /; FreeQ[b, x]

        resultant[a_, b_, x_] :=
          (-1)^(Exponent[a,x]*Exponent[b,x]) *
            resultant[b, a, x] /;
          Exponent[a, x] < Exponent[b, x]

        resultant[a_, b_, x_] :=
          Module[{m, n, r},
            n = Exponent[a, x];
            m = Exponent[b, x];
            r = PolynomialRemainder[a, b, x];
            (-1)^(n*m) * Coefficient[b, x, m]^(n-Exponent[r,x]) *
              resultant[b, r, x]
          ]
```

This implementation again yields

```
In[4]:= resultant[a, b, x]
Out[4]= c² + 2 c
```

With the aid of resultants one can compute minimal polynomials of algebraic numbers. For this, let α and β be two algebraic numbers over $\mathbb{K}$ with the minimal polynomials $p(x)$ and $q(x)$, respectively. Then the pair $(\alpha + \beta, \beta)$ is a common zero of the polynomials $p(x - y)$ and $q(y) \in \mathbb{K}[x, y]$ and therefore also of

$$r(x) := \mathrm{res}(p(x - y), q(y), y) \in \mathbb{K}[x].$$

Hence the minimal polynomial $\mathrm{minpol}(\alpha + \beta, x)$ *of the sum is a divisor of $r(x)$. In Exercise 7.21 we will give further examples for the computation of minimal polynomials of algebraic numbers using resultants.*

For example, the computation

```
In[5]:= res = Resultant[(x - y)² - 2, y³ - 5, y]
Out[5]= x⁶ - 6 x⁴ - 10 x³ + 12 x² - 60 x + 17
```

again yields the minimal polynomial of $\sqrt{2} + \sqrt[3]{5}$ as in Session 7.24. The irreducibility is shown by

$In[6]:=$ **Factor[res]**

$Out[6]=$ $x^6 - 6\,x^4 - 10\,x^3 + 12\,x^2 - 60\,x + 17$

However, please note that our self-defined function yields

$In[7]:=$ **res = resultant[(x-y)^2 - 2, y^3 - 5, y]**

$Out[7]=$ $(3\,x^2+2)^2\left(\dfrac{4\,x^6}{(3\,x^2+2)^2} - \dfrac{4\,x^4}{3\,x^2+2} - \dfrac{16\,x^4}{(3\,x^2+2)^2} + \dfrac{20\,x^3}{(3\,x^2+2)^2} + \right.$

$\left. \dfrac{8\,x^2}{3\,x^2+2} + \dfrac{16\,x^2}{(3\,x^2+2)^2} + x^2 - \dfrac{10\,x}{3\,x^2+2} - \dfrac{40\,x}{(3\,x^2+2)^2} + \dfrac{25}{(3\,x^2+2)^2} - 2\right)$

i.e., the computations are executed without simplification in $\mathbb{Q}(x)[y]$. The reason is that in Theorem 7.38 we have assumed a coefficient field. After simplification we obtain the same result again:

$In[8]:=$ **Together[res]**

$Out[8]=$ $x^6 - 6\,x^4 - 10\,x^3 + 12\,x^2 - 60\,x + 17$

Finally we want to simulate the decisive step in the proof of Theorem 7.38 (c) by an example. For this purpose, we first declare

$In[9]:=$ **Clear[SylvesterMatrix]**

```
SylvesterMatrix[a_, b_, x_] :=
  Module[{n, m, alist, blist, amat, bmat},
    n = Exponent[a, x];
    m = Exponent[b, x];
    alist =
      PadRight[Reverse[CoefficientList[a, x]], n+m];
    blist =
      PadRight[Reverse[CoefficientList[b, x]], n+m];
    If[m === 0, amat = {},
      amat = NestList[RotateRight, alist, m-1]];
    If[n === 0, bmat = {},
      bmat = NestList[RotateRight, blist, n-1]];
    Join[amat, bmat]
  ]
```

for the computation of the Sylvester matrix.[21] *Let the two polynomials*

$In[10]:=$ **a =** $\displaystyle\sum_{k=0}^{7}$ **(k+1)^3 x^k**

$Out[10]=$ $512\,x^7 + 343\,x^6 + 216\,x^5 + 125\,x^4 + 64\,x^3 + 27\,x^2 + 8\,x + 1$

$In[11]:=$ **b =** $\displaystyle\sum_{k=0}^{5}$ **k^2 x^k**

$Out[11]=$ $25\,x^5 + 16\,x^4 + 9\,x^3 + 4\,x^2 + x$

be given. Then we get the corresponding Sylvester matrix

$In[12]:=$ **S = SylvesterMatrix[a, b, x]**

---

[21] Older *Mathematica* versions require an additional command Unprotect[SylvesterMatrix] to remove an internal protection.

$$Out[12] = \begin{pmatrix} 512 & 343 & 216 & 125 & 64 & 27 & 8 & 1 & 0 & 0 & 0 & 0 \\ 0 & 512 & 343 & 216 & 125 & 64 & 27 & 8 & 1 & 0 & 0 & 0 \\ 0 & 0 & 512 & 343 & 216 & 125 & 64 & 27 & 8 & 1 & 0 & 0 \\ 0 & 0 & 0 & 512 & 343 & 216 & 125 & 64 & 27 & 8 & 1 & 0 \\ 0 & 0 & 0 & 0 & 512 & 343 & 216 & 125 & 64 & 27 & 8 & 1 \\ 25 & 16 & 9 & 4 & 1 & 0 & 0 & 0 & 0 & 0 & 0 & 0 \\ 0 & 25 & 16 & 9 & 4 & 1 & 0 & 0 & 0 & 0 & 0 & 0 \\ 0 & 0 & 25 & 16 & 9 & 4 & 1 & 0 & 0 & 0 & 0 & 0 \\ 0 & 0 & 0 & 25 & 16 & 9 & 4 & 1 & 0 & 0 & 0 & 0 \\ 0 & 0 & 0 & 0 & 25 & 16 & 9 & 4 & 1 & 0 & 0 & 0 \\ 0 & 0 & 0 & 0 & 0 & 25 & 16 & 9 & 4 & 1 & 0 & 0 \\ 0 & 0 & 0 & 0 & 0 & 0 & 25 & 16 & 9 & 4 & 1 & 0 \end{pmatrix}$$

with the determinant

$In[13] :=$ **Det[S]**

$Out[13] = -2385878833$

Obviously this result corresponds to

$In[14] :=$ **Resultant[a, b, x]**

$Out[14] = -2385878833$

Let us use this example to perform the essential proof steps of Theorem 7.38 (c). When dividing
$a(x)$ *by* $b(x)$, *then the division remainder is given by*

$In[15] :=$ **r = PolynomialRemainder[a, b, x]**

$$Out[15] = \frac{368198\, x^4}{15625} + \frac{518652\, x^3}{15625} + \frac{357612\, x^2}{15625} + \frac{111328\, x}{15625} + 1$$

If we replace $a(x)$ *by* $r(x)$ *in the Sylvester matrix, we obtain the following matrix* T:

$In[16] :=$ **T = S/.** $\left\{ 512 \to 0, 343 \to 0, 216 \to 0, \right.$

$$125 \to \frac{368198}{15625}, \ 64 \to \frac{518652}{15625},$$

$$\left. 27 \to \frac{357612}{15625}, \ 8 \to \frac{111328}{15625} \right\}$$

$$Out[16] = \begin{pmatrix} 0 & 0 & 0 & \frac{368198}{15625} & \frac{518652}{15625} & \frac{357612}{15625} & \frac{111328}{15625} & 1 & 0 & 0 & 0 & 0 \\ 0 & 0 & 0 & 0 & \frac{368198}{15625} & \frac{518652}{15625} & \frac{357612}{15625} & \frac{111328}{15625} & 1 & 0 & 0 & 0 \\ 0 & 0 & 0 & 0 & 0 & \frac{368198}{15625} & \frac{518652}{15625} & \frac{357612}{15625} & \frac{111328}{15625} & 1 & 0 & 0 \\ 0 & 0 & 0 & 0 & 0 & 0 & \frac{368198}{15625} & \frac{518652}{15625} & \frac{357612}{15625} & \frac{111328}{15625} & 1 & 0 \\ 0 & 0 & 0 & 0 & 0 & 0 & 0 & \frac{368198}{15625} & \frac{518652}{15625} & \frac{357612}{15625} & \frac{111328}{15625} & 1 \\ 25 & 16 & 9 & 4 & 1 & 0 & 0 & 0 & 0 & 0 & 0 & 0 \\ 0 & 25 & 16 & 9 & 4 & 1 & 0 & 0 & 0 & 0 & 0 & 0 \\ 0 & 0 & 25 & 16 & 9 & 4 & 1 & 0 & 0 & 0 & 0 & 0 \\ 0 & 0 & 0 & 25 & 16 & 9 & 4 & 1 & 0 & 0 & 0 & 0 \\ 0 & 0 & 0 & 0 & 25 & 16 & 9 & 4 & 1 & 0 & 0 & 0 \\ 0 & 0 & 0 & 0 & 0 & 25 & 16 & 9 & 4 & 1 & 0 & 0 \\ 0 & 0 & 0 & 0 & 0 & 0 & 25 & 16 & 9 & 4 & 1 & 0 \end{pmatrix}$$

This matrix has the same determinant as S:

$In[17] :=$ **Det[T]**

$Out[17] = -2385878833$

From this representation it is easy to see how T can be converted into the matrix $S(b(x), r(x), x)$ by a triple expansion via Laplace's formula along the first column.

Resultants can be used to solve certain polynomial systems of equations. This will be considered in the next section.

7.6 Polynomial Systems of Equations

In this section we first consider a system G of n *linear* equations $G_1, G_2, ..., G_n$ in the m variables $x_1, ..., x_m$ with coefficients in a field $\mathbb{K}$. In such a case the *complete solution set*, also called *general solution*, can be determined using *Gaussian elimination*[22].

Gaussian elimination successively eliminates variables and thus transforms the system into an equivalent system $G' = \tilde{G}_1, \tilde{G}_2, ..., \tilde{G}_{n'}$ in triangular form. Equivalent means that the solution sets of G and G' coincide. If triangulation produces trivial equations[23], they are deleted. Therefore the number of equations n' of G' can be smaller than n. Starting with the last equation $\tilde{G}_{n'}$ which has the maximum number of eliminated variables, one can compute the general solution by *backward substitution*.

Example 7.40. We consider the system of linear equations

$$
\begin{array}{rcrcrcl}
x & + & y & + & z & = & 1 \\
x & - & y & - & 2z & = & 0
\end{array}
$$

over $\mathbb{Q}$. Gaussian elimination generates the equivalent system

$$
\begin{array}{rcrcrcl}
x & + & y & + & z & = & 1 \\
 & & 2y & + & 3z & = & 1
\end{array} .
$$

The last row yields, for arbitrary $z \in \mathbb{Q}$,

$$
y = \frac{1}{2} - \frac{3}{2}z .
$$

If this equation is substituted into the first equation, we get, after solving for x,

$$
x = \frac{1}{2}z + \frac{1}{2} .
$$

Therefore the complete solution is given by the vector

$$
\begin{pmatrix} x \\ y \\ z \end{pmatrix} = \begin{pmatrix} \frac{1}{2}z + \frac{1}{2} \\ \frac{1}{2} - \frac{3}{2}z \\ z \end{pmatrix} \qquad (z \in \mathbb{Q}) .
$$

If we add the third equation $x - y = 1$ to our system, we reach the triangular form

---

[22] There are also other algorithms for this purpose. However, variants of Gaussian elimination are particularly efficient.

[23] Trivial equations are of the form $0 = 0$.

$$x \ + \ y \ + \ z \ = \ 1$$
$$2y \ + \ 3z \ = \ 1 \ .$$
$$2z \ = \ 1$$

In this case we first get $z = \frac{1}{2}$ and after backward substitution we find the unique solution

$$\begin{pmatrix} x \\ y \\ z \end{pmatrix} = \begin{pmatrix} \frac{3}{4} \\ -\frac{1}{4} \\ \frac{1}{2} \end{pmatrix} . \quad \triangle$$

Session 7.41. *Mathematica can solve linear systems and work with matrices. For example, if we declare*

$In[1] :=$ **A = {{1, 1, 1}, {1, -1, -2}, {1, -1, 0}}**

$Out[1] = \begin{pmatrix} 1 & 1 & 1 \\ 1 & -1 & -2 \\ 1 & -1 & 0 \end{pmatrix}$

and

$In[2] :=$ **b = {1, 0, 1}**
$Out[2] = \{1, 0, 1\}$

then we can solve the linear system $A \cdot x = b$ as follows:

$In[3] :=$ **LinearSolve[A, b]**
$Out[3] = \{\frac{3}{4}, -\frac{1}{4}, \frac{1}{2}\}$

The interface which is given by the Solve *command, however, is more general and simpler. Here we can enter the system of equations and get the solution directly in the form of replacement rules:*

$In[4] :=$ **sol = Solve[{G1 = x + y + z == 1, G2 = x - y - 2z == 0,**
           G3 = x - y == 1}, {x, y, z}]
$Out[4] = \{\{x \to \frac{3}{4}, y \to -\frac{1}{4}, z \to \frac{1}{2}\}\}$

As an example, we can substitute those rules in the original equations:

$In[5] :=$ **{G1, G2, G3}/.sol[[1]]**
$Out[5] = \{\text{True, True, True}\}$

Analogously the command

$In[6] :=$ **sol = Solve[{x + y + z == 1, x - y - 2z == 0}, {x, y, z}]**

yields the complete solution

Solve :: **"svars"**: *Equations may not give solutions for all "solve" variables.*
$Out[6] = \{\{y \to 2 - 3x, \to 2x - 1\}\}$

whereas

$In[7] :=$ **lin = LinearSolve[{{1, 1, 1}, {1, -1, -2}}, {1, 0}]**
$Out[7] = \{\frac{1}{2}, \frac{1}{2}, 0\}$

gives only one solution. The general solution is then found by using NullSpace *which computes the kernel or nullspace of a linear map. By using*

In[8]:= **null = NullSpace[{{1, 1, 1}, {1, -1, -2}}]**
Out[8]= $(1 \quad -3 \quad 2)$

we get the general solution:

In[9]:= **lin + t * null[[1]]**
Out[9]= $\{t + \frac{1}{2}, \frac{1}{2} - 3t, 2t\}$

which—for $t = \frac{z}{2}$—obviously agrees with the one given by the Solve *command.*

Next, we would like to extend this algorithm for $m = n$ to systems of polynomial equations. For this purpose, let a system G of n *polynomial* equations $G_1, G_2, ..., G_n$ in n variables $x_1, ..., x_n$, with coefficients in a field $\mathbb{K}$, be given. We further assume that the system has only finitely many solutions.[24]

The equations can be brought into the form $p_1(x_1, ..., x_n) = 0, p_2(x_1, ..., x_n) = 0, ..., p_n(x_1, ..., x_n) = 0$ with polynomials $p_k \in \mathbb{K}[x_1, ..., x_n]$ $(k = 1, ..., n)$. We would like to find the set of common zeros of the polynomials $p_1, p_2, ..., p_n$ in a suitable algebraic extension field of $\mathbb{K}$.

The main idea is again to eliminate some of the variables. Since the resultant of two polynomials equals zero if and only if they have a common zero, we can add the polynomial

$$q_2(x_2, ..., x_n) := \text{res}(p_1(x_1, ..., x_n), p_2(x_1, ..., x_n), x_1) \in \mathbb{K}[x_2, ..., x_n]$$

to the system $p_1, p_2, ..., p_n$ without changing the zeros. Please note that $q_2(x_2, ..., x_n)$ does no longer contain the variable x_1 that was eliminated. One can show (see, e.g., [GCL1992], Theorem 9.5, p. 414) that it is allowed to replace p_2 by q_2 without changing the solutions of the system.

Similarly, for $k = 2, ..., n$, we can compute the resultants

$$q_k(x_2, ..., x_n) := \text{res}(p_1(x_1, ..., x_n), p_k(x_1, ..., x_n), x_1) \in \mathbb{K}[x_2, ..., x_n]$$

and subsequently replace p_k by q_k. The result is a new system of polynomial equations in which the variable x_1 is only contained in p_1, while it is eliminated in all other polynomials q_k $(k = 2, ..., n)$.

If we iterate this process, we successively obtain equations for which yet another variable is eliminated. Since we start with n equations in n variables, the last equation will contain only a single variable, namely x_n. The new system has the triangular form

$$\begin{aligned} p_1(x_1, x_2, ..., x_n) &= 0 \\ p_2(x_2, ..., x_n) &= 0 \\ &\vdots \\ p_n(x_n) &= 0, \end{aligned}$$

where for simplicity we have denoted the new polynomials again by p_k $(k = 1, ..., n)$. If one of the computed resultants is zero, then we can split our problem into several partial problems by a factorization or by a gcd-computation of the corresponding polynomials.

Once our system is in triangular form, it is easy to find its general solution. First, we compute the solutions of the equation $p_n(x_n) = 0$. This can be done, if applicable, by a factorization over $\mathbb{K}$. If this does not generate the factors, then we know at least that all solutions satisfy the

---

[24] This is called the zero-dimensional or *generic case.*

irreducible equation $p_n(x_n) = 0$, and hence are algebraic numbers whose minimal polynomial we know, and we can do our computations in a suitable algebraic extension field.

An iterative backward substitution successively generates—possibly by considering certain case distinctions—the possible values for $x_{n-1}, x_{n-2}, \ldots, x_1$. For this purpose it may be necessary to consider further algebraic extensions. Finally we must confirm the solutions by substitution.

This process yields the following theorem.

Theorem 7.42. (Solving Polynomial Systems of Equations) Let G be a system of polynomial equations $G_1, G_2, \ldots, G_n$ in n variables $x_1, \ldots, x_n$ with coefficients in a field $\mathbb{K}$ that has only finitely many solutions.

Then the algorithm as described above finds the solutions of the system G. The solution values $(x_1, \ldots, x_n)$ lie in an algebraic extension field of $\mathbb{K}$. $\qquad\qquad\square$

Session 7.43. *We apply the above algorithm to the system of equations*

$$
\begin{aligned}
x + y + z &= 6 \\
x^2 + y^2 + z^2 &= 14 \\
x^4 + y^4 + z^4 &= 98
\end{aligned}
$$

For this purpose we define the polynomials

```
In[1]:= p₁ = x + y + z - 6;
        p₂ = x² + y² + z² - 14;
        p₃ = x⁴ + y⁴ + z⁴ - 98;
```

whose common zeros are sought. To eliminate x, we compute

```
In[2]:= q₂ = Resultant[p₁, p₂, x]
Out[2]= 2 y² + 2 z y - 12 y + 2 z² - 12 z + 22
```

```
In[3]:= q₃ = Resultant[p₁, p₃, x]
Out[3]= 6 y² z² + 4 y³ z - 72 y² z + 2 y⁴ - 24 y³ + 216 y² + 4 y z³ -
        72 y z² + 432 y z - 864 y + 2 z⁴ - 24 z³ + 216 z² - 864 z + 1198
```

and check with the `Factor` *command whether the resulting polynomials are irreducible. This turns out to be the case. We can therefore eliminate the variable y in the next step:*

```
In[4]:= r₃ = Resultant[q₂, q₃, y]
Out[4]= 9216 z⁶ - 110592 z⁵ + 534528 z⁴ - 1327104 z³ + 1778688 z² - 1216512 z + 331776
```

What remains is a univariate polynomial in z for which we get the following factorization:

```
In[5]:= Factor[r₃]
Out[5]= 9216 (z - 3)² (z - 2)² (z - 1)²
```

Hence either $z = 1$, $z = 2$ or $z = 3$. To continue the computation, we consider all three cases and substitute them in q_2:

```
In[6]:= Factor[q₂ /. {z → 1}]
Out[6]= 2 (y - 3) (y - 2)
```

```
In[7]:= Factor[q₂ /. {z → 2}]
Out[7]= 2 (y - 3) (y - 1)
```

```
In[8]:= Factor[q₂ /. {z → 3}]
```

Out[8]= $2(y-2)(y-1)$

Finally, using p_1, we get

In[9]:= `Factor[p`$_1$`/.{y`$\to$`2,z`$\to$`1}]`
Out[9]= $x-3$

In[10]:= `Factor[p`$_1$`/.{y`$\to$`3,z`$\to$`1}]`
Out[10]= $x-2$

The remaining solutions can be computed similarly. However, by the symmetry of the starting polynomials they are already clear.

Mathematica's `Solve` *command can solve polynomial systems of equations in a single step:*

In[11]:= `Solve[{p`$_1$` == 0,p`$_2$` == 0,p`$_3$` == 0},{x,y,z}]`
Out[11]= $\{\{x\to 1, y\to 2, z\to 3\}, \{x\to 1, y\to 3, z\to 2\}, \{x\to 2, y\to 1, z\to 3\},$
$\{x\to 2, y\to 3, z\to 1\}, \{x\to 3, y\to 1, z\to 2\}, \{x\to 3, y\to 2, z\to 1\}\}$

Our system therefore has 6 solutions which, by the symmetry of the starting polynomials, are all equivalent.

Next we want to consider an example system which needs an algebraic extension. For this, let

In[12]:= p_1 `=x`^{3}`+y`^{2}` - 2;`
$\qquad\quad$ p_2 ` = (x-y)`^{2}` - 3;`

This time we eliminate the variable y by:

In[13]:= `f = Factor[Resultant[p`$_1$`,p`$_2$`,y]]`
Out[13]= $x^6 + 2x^5 + x^4 + 2x^3 - 10x^2 + 1$

and substitute the solution in p_1:

In[14]:= `r = Root[f/.{x`$\to$`#}&,1];`

The factorization

In[15]:= `Factor[p`$_1$`/.{x`$\to$`r}]`
Out[15]= $\text{Root}[\#1^6 + 2\#1^5 + \#1^4 + 2\#1^3 - 10\#1^2 + 1\&, 1]^3 + y^2 - 2$

does not work. But, attention: we factored over $\mathbb{Q}$ *although we should have factored over* $\mathbb{Q}(r)$*! The latter factorization provides the "relatively simple" factorizations of p_1 and p_2:*

In[16]:= `Factor[p`$_1$`/.{x`$\to$`r},Extension`$\to$`{r}]/.{r`$\to$`$\alpha$}`
Out[16]= $\frac{1}{4}\left(-\alpha^5 - 2\alpha^4 - \alpha^3 - \alpha^2 + 9\alpha + 2y\right)\left(\alpha^5 + 2\alpha^4 + \alpha^3 + \alpha^2 - 9\alpha + 2y\right)$

In[17]:= `Factor[p`$_2$`/.{x`$\to$`r},Extension`$\to$`{r}]/.{r`$\to$`$\alpha$}`
Out[17]= $\frac{1}{4}\left(-\alpha^5 - 2\alpha^4 - \alpha^3 - \alpha^2 + 9\alpha + 2y\right)\left(\alpha^5 + 2\alpha^4 + \alpha^3 + \alpha^2 - 13\alpha + 2y\right)$

Note that—for the sake of simplicity—we have used the symbol α for $r = $ Root[...].

The above computation explains the output of the `Solve` *command:*

In[18]:= `Solve[{p`$_1$` == 0,p`$_2$` == 0},{x,y}]`

$Out[18]= \left\{\left\{y \to \frac{1}{2} \text{Root}\left[\#1^6 + 2\#1^5 + \#1^4 + 2\#1^3 - 10\#1^2 + 1\&, 1\right]\right.\right.$
$\left(-9 + \text{Root}\left[\#1^6 + 2\#1^5 + \#1^4 + 2\#1^3 - 10\#1^2 + 1\&, 1\right] + \right.$
$\text{Root}\left[\#1^6 + 2\#1^5 + \#1^4 + 2\#1^3 - 10\#1^2 + 1\&, 1\right]^2 +$
$2\,\text{Root}\left[\#1^6 + 2\#1^5 + \#1^4 + 2\#1^3 - 10\#1^2 + 1\&, 1\right]^3 +$
$\left.\text{Root}\left[\#1^6 + 2\#1^5 + \#1^4 + 2\#1^3 - 10\#1^2 + 1\&, 1\right]^4\right),$
$\left.x \to \text{Root}\left[\#1^6 + 2\#1^5 + \#1^4 + 2\#1^3 - 10\#1^2 + 1\&, 1\right]\right\},$

$\left\{y \to \frac{1}{2} \text{Root}\left[\#1^6 + 2\#1^5 + \#1^4 + 2\#1^3 - 10\#1^2 + 1\&, 2\right]\right.$
$\left(-9 + \text{Root}\left[\#1^6 + 2\#1^5 + \#1^4 + 2\#1^3 - 10\#1^2 + 1\&, 2\right] + \right.$
$\text{Root}\left[\#1^6 + 2\#1^5 + \#1^4 + 2\#1^3 - 10\#1^2 + 1\&, 2\right]^2 +$
$2\,\text{Root}\left[\#1^6 + 2\#1^5 + \#1^4 + 2\#1^3 - 10\#1^2 + 1\&, 2\right]^3 +$
$\left.\text{Root}\left[\#1^6 + 2\#1^5 + \#1^4 + 2\#1^3 - 10\#1^2 + 1\&, 2\right]^4\right),$
$\left.x \to \text{Root}\left[\#1^6 + 2\#1^5 + \#1^4 + 2\#1^3 - 10\#1^2 + 1\&, 2\right]\right\},$

$\left\{y \to \frac{1}{2} \text{Root}\left[\#1^6 + 2\#1^5 + \#1^4 + 2\#1^3 - 10\#1^2 + 1\&, 3\right]\right.$
$\left(-9 + \text{Root}\left[\#1^6 + 2\#1^5 + \#1^4 + 2\#1^3 - 10\#1^2 + 1\&, 3\right] + \right.$
$\text{Root}\left[\#1^6 + 2\#1^5 + \#1^4 + 2\#1^3 - 10\#1^2 + 1\&, 3\right]^2 +$
$2\,\text{Root}\left[\#1^6 + 2\#1^5 + \#1^4 + 2\#1^3 - 10\#1^2 + 1\&, 3\right]^3 +$
$\left.\text{Root}\left[\#1^6 + 2\#1^5 + \#1^4 + 2\#1^3 - 10\#1^2 + 1\&, 3\right]^4\right),$
$\left.x \to \text{Root}\left[\#1^6 + 2\#1^5 + \#1^4 + 2\#1^3 - 10\#1^2 + 1\&, 3\right]\right\},$

$\left\{y \to \frac{1}{2} \text{Root}\left[\#1^6 + 2\#1^5 + \#1^4 + 2\#1^3 - 10\#1^2 + 1\&, 4\right]\right.$
$\left(-9 + \text{Root}\left[\#1^6 + 2\#1^5 + \#1^4 + 2\#1^3 - 10\#1^2 + 1\&, 4\right] + \right.$
$\text{Root}\left[\#1^6 + 2\#1^5 + \#1^4 + 2\#1^3 - 10\#1^2 + 1\&, 4\right]^2 +$
$2\,\text{Root}\left[\#1^6 + 2\#1^5 + \#1^4 + 2\#1^3 - 10\#1^2 + 1\&, 4\right]^3 +$
$\left.\text{Root}\left[\#1^6 + 2\#1^5 + \#1^4 + 2\#1^3 - 10\#1^2 + 1\&, 4\right]^4\right),$
$\left.x \to \text{Root}\left[\#1^6 + 2\#1^5 + \#1^4 + 2\#1^3 - 10\#1^2 + 1\&, 4\right]\right\},$

$\left\{y \to \frac{1}{2} \text{Root}\left[\#1^6 + 2\#1^5 + \#1^4 + 2\#1^3 - 10\#1^2 + 1\&, 5\right]\right.$
$\left(-9 + \text{Root}\left[\#1^6 + 2\#1^5 + \#1^4 + 2\#1^3 - 10\#1^2 + 1\&, 5\right] + \right.$
$\text{Root}\left[\#1^6 + 2\#1^5 + \#1^4 + 2\#1^3 - 10\#1^2 + 1\&, 5\right]^2 +$
$2\,\text{Root}\left[\#1^6 + 2\#1^5 + \#1^4 + 2\#1^3 - 10\#1^2 + 1\&, 5\right]^3 +$
$\left.\text{Root}\left[\#1^6 + 2\#1^5 + \#1^4 + 2\#1^3 - 10\#1^2 + 1\&, 5\right]^4\right),$
$\left.x \to \text{Root}\left[\#1^6 + 2\#1^5 + \#1^4 + 2\#1^3 - 10\#1^2 + 1\&, 5\right]\right\},$

$\left\{y \to \frac{1}{2} \text{Root}\left[\#1^6 + 2\#1^5 + \#1^4 + 2\#1^3 - 10\#1^2 + 1\&, 6\right]\right.$
$\left(-9 + \text{Root}\left[\#1^6 + 2\#1^5 + \#1^4 + 2\#1^3 - 10\#1^2 + 1\&, 6\right] + \right.$
$\text{Root}\left[\#1^6 + 2\#1^5 + \#1^4 + 2\#1^3 - 10\#1^2 + 1\&, 6\right]^2 +$
$2\,\text{Root}\left[\#1^6 + 2\#1^5 + \#1^4 + 2\#1^3 - 10\#1^2 + 1\&, 6\right]^3 +$
$\left.\text{Root}\left[\#1^6 + 2\#1^5 + \#1^4 + 2\#1^3 - 10\#1^2 + 1\&, 6\right]^4\right),$
$\left.\left.\left.x \to \text{Root}\left[\#1^6 + 2\#1^5 + \#1^4 + 2\#1^3 - 10\#1^2 + 1\&, 6\right]\right\}\right\}\right\}$

Obviously it specifies the required algebraic extensions in a more complicated form.

Before concluding this section, let us remark that

- the efficiency of this method might heavily depend on the *order* in which the variables are eliminated.

- in principle factorizations are *not necessary*; if a resultant computation yields zero, then we know that there must be a common factor which can be found using a gcd-computation. However, the last univariate polynomial should be factorized in order to avoid unnecessary algebraic extensions.

- the complexity of this method is typically rather high. In our example we have seen that—despite the *very simple* final result—there was quite an intermediate expression swell. This problem already occurred in the polynomial Euclidean algorithm (Section 6.6, see page 136).

- this method can be readily extended to m equations in n variables.

- there are other methods for the computation of the solution set of a polynomial system, the most prominent being the use of *Groebner bases* where the *ideal* generated by the polynomial system $p_1, \ldots, p_n$ is examined.

7.7 Additional Remarks

In the language of group theory, Theorem 7.33 about the existence of a primitive element in a finite field $\mathrm{GF}(q)$ tells that the multiplicative group $\mathrm{GF}(q)^\star$ of a finite field is *cyclic*. In contrast to the cyclicity of the additive group in $\mathbb{Z}_p$ this is a very remarkable result.

As shown, resultants are used for elimination purposes and for the computation of minimal polynomials of algebraic numbers. However, they are also of theoretical interest and they are used in the consideration of efficient algorithms for the modular computation of greatest common divisors of polynomials, see for example [GCL1992] and [GG1999].

Groebner bases play a very important role in computer algebra and current research. However, their detailed consideration was beyond the scope of this book. On the other hand, there are quite a few recommendable textbooks. A particularly nice introduction is given in [CLO1997].

7.8 Exercises

Exercise 7.1. *For every $a(x) \in \mathbb{K}[x]$ and every $\alpha \in \mathbb{K}$, show that $a(x) \equiv a(\alpha) \pmod{x - \alpha}$.*

Exercise 7.2. *Let $a(x), b(x) \in \mathbb{Z}_p[x]$. Show that $a(\alpha) \equiv b(\alpha) \pmod{p}$ for all $\alpha \in \mathbb{Z}_p$[25] if and only if $a(x) \equiv b(x) \pmod{x^p - x}$.*

Exercise 7.3. *Show that Theorem 6.29 about polynomial interpolation follows from the Chinese Remainder Theorem for polynomials (Theorem 7.6). Hint: Choose $p_k(x) = x - x_k$.*

Use this connection to implement an alternative approach for computing the interpolating polynomial and check your code.

Exercise 7.4. *Implement the Chinese remainder algorithm over $\mathbb{Z}_p[x]$.*

Exercise 7.5. *Implement a Mathematica function* `PolynomialModInverse` *for the computation of the inverse of $a(x)$ modulo $p(x)$.*

Exercise 7.6. *Let $\mathbb{K}$ be a field and α a zero of the irreducible monic polynomial $p(x) \in \mathbb{K}[x]$ (in a suitable extension field). Show that $p(x)$ is the minimal polynomial of α over $\mathbb{K}$.*

---

[25] In other words, $a(x)$ and $b(x)$ are identical as functions in $\mathbb{Z}_p$

Exercise 7.7. (Dimension Formula) *Show that if two finite field extensions* $\mathbb{K} \subset \mathbb{E} \subset \mathbb{F}$ *have the degrees* $[\mathbb{E} : \mathbb{K}] = n$ *and* $[\mathbb{F} : \mathbb{E}] = m$, *then the field extension* $\mathbb{K} \subset \mathbb{F}$ *is also finite and has degree* $[\mathbb{F} : \mathbb{K}] = n \cdot m$.

Exercise 7.8. *Factorize* $a(x) = x^4 + 1$ *in the field* $\mathbb{Q}(\alpha) = \mathbb{Q}[x]/\langle x^4 + 1 \rangle$ *by manual computation. Specify the field extension of the splitting field of* $x^4 + 1$ *over* $\mathbb{Q}$. *What is the degree of this extension?*

Exercise 7.9. (Minimal Polynomials of Algebraic Numbers) *Compute the minimal polynomials of the following algebraic numbers without* RootReduce. *If possible, simplify the given terms using this information.*

(a) $\alpha = \sqrt{11 + 6\sqrt{2}} + \sqrt{11 - 6\sqrt{2}}$;

(b) $\beta = \sqrt[3]{14\sqrt{\frac{93}{243}} + \frac{5}{9}}$;

(c) $\gamma = \sqrt[3]{\frac{14\sqrt{93}}{243} + \frac{5}{9}}$.

Exercise 7.10. (Minimal Polynomials of Algebraic Numbers) *Compute the minimal polynomials of the following algebraic numbers using* RootReduce.[26]

(a) $\sqrt[4]{3 + 2\sqrt{5} + 3\sqrt[3]{2}} + \sqrt[4]{3 - 2\sqrt{5} + 3\sqrt[3]{2}}$;

(b) $\sqrt[4]{3 + 2\sqrt{5} + 3\sqrt[3]{2}} \cdot \sqrt[4]{3 - 2\sqrt{5} + 3\sqrt[3]{2}}$;

(c) $\sqrt[4]{3 + 2\sqrt{5} + 3\sqrt[3]{2}} \cdot \dfrac{1}{\sqrt[4]{3 - 2\sqrt{5} + 3\sqrt[3]{2}}}$.

Exercise 7.11. (Construction of Regular 17**-gon)** *In order to draw the regular* n-*gon, which we assume to be inscribed in the unit circle, one must construct the side length* $l_n = 2\sin\frac{\pi}{n}$ *or* $s_n = \sin\frac{\pi}{n}$.

If we want to achieve this using ruler and compass, then the numbers s_n *und* $c_n = \sqrt{1 - s_n^2} = \cos\frac{\pi}{n}$ *must be algebraic over* $\mathbb{Q}$ *of a degree which is a power of 2, since equations of lines and of circles are linear or quadratic (see for example [Henn2003]). It is exactly in this case that they can be represented as nested square roots which can be constructed iteratively—for example using the theorem of Pythagoras—using ruler and compass.*

It can be proved that the regular 5-gon and the regular 17-gon have this property, whereas this is not true for the regular 7-gon, for example. This should be proved now. Furthermore we search for a representation of the side length of the regular 5-gon and the regular 17-gon.

For this purpose, find representations of

(a) $\alpha = \cos\frac{\pi}{5}$ *and*
(b) $\beta = \cos\frac{\pi}{17}$

as nested square roots. On the other hand you must prove that

(c) $\gamma = \cos\frac{\pi}{7}$

is an algebraic number whose degree is not a power of 2.

---

[26] Of course you can also try to get these results without RootReduce!

Hint: RootReduce *can find the minimal polynomials F. For the factorization of F for β (in a suitable extension field) you can use that $\sqrt{17}$ must be adjoined.*[27] *Finally use* Solve *for the remaining polynomials of degree 4. This will lead you to a nested square root representation for β.*

Exercise 7.12. (Standard Representation) *Compute the minimal polynomials and the standard representations of the following algebraic numbers:*

(a) $\alpha = \frac{1}{\sqrt{2}+\sqrt{3}}$

(b) $\beta = \frac{1}{\sqrt{2}+\sqrt{3}+\sqrt{5}+\sqrt{7}}$

(c) $\gamma = \frac{\sqrt[3]{4}-2\sqrt[3]{2}+1}{\sqrt[3]{4}+2\sqrt[3]{2}+1}$

(d) $\delta = \frac{1}{a+b\sqrt[3]{n}+c\sqrt[3]{n^2}}$ *for* $a,b,c \in \mathbb{Q}$

Exercise 7.13. (Galois Fields) *Load the package* FiniteFields.

(a) *Examine the different forms of representation in the Galois fields GF(16), GF(25) and GF(27) and describe these.*

(b) *Which minimal polynomials are used for GF(16), GF(25) and GF(27)? Prove their irreducibility.*

(c) *Check the equation*

$$x^q - x = \prod_{a \in GF(q)} (x-a)$$

in GF(16), GF(25) and GF(27). If the computing times are too slow, implement the necessary computations yourself.

Exercise 7.14. *Adjust the programs* ReedSolomon *and* InverseReedSolomon *of the Reed-Solomon code from Section 5.4 to the field GF(2^8) and test your implementation.*

Exercise 7.15. (Factorization in Finite Fields) *Factorize*

(a) $x^{12} - 2x^8 + 4x^4 - 8$ *over* $\mathbb{Q}$;

(b) $x^{12} - 2x^8 + 4x^4 - 8$ *over* $\mathbb{Q}(\sqrt[4]{2})$;

(c) $x^6 - 2x^4 + 4x^2 - 8$ *over* $\mathbb{Q}$;

(d) $x^6 - 2x^4 + 4x^2 - 8$ *over* $\mathbb{Q}(\sqrt[4]{2})$;

(e) $x^6 - 2x^4 + 4x^2 - 8$ *over* $\mathbb{Z}_5(\sqrt[4]{2}) = GF(q)$. *Compute q.*

Exercise 7.16. (Irreducible Polynomials in $Z_p[x]$) *Implement (with the use of* Factor*) a function* NumberIrreducible[p]*, which for $p \in \mathbb{P}$ computes the number of monic irreducible polynomials in $\mathbb{Z}_p[x]$. Apply your function to several values of $p \in \mathbb{P}$ to observe how fast this number is growing with increasing p.*

Exercise 7.17. (Sylvester Matrix and Resultant)

(a) *Implement the Mathematica function* SylvesterMatrix[a,b,x] *from Session 7.39 for the computation of the Sylvester matrix of the polynomials $a(x)$ and $b(x)$ with respect to the variable x.*

(b) *Compute the Sylvester matrix and the resultant for the following pairs of polynomials:*

---

[27] Since we deal with the construction of the regular 17-gon, this is highly plausible.

(i) $a(x) = x^8 + x^6 - 3x^4 - 3x^3 + 8x^2 + 2x - 5, \ b(x) = 3x^6 + 5x^4 - 4x^2 - 9x + 21;$

(ii) $a(x) = x^5 + x^4 - x^3 - x^2 - 2x - 2, \ b(x) = 3x^5 + 2x^4 - 5x^3 - 4x^2 - 2x.$

Which pairs of polynomials have common zeros? Check your result by a factorization.

(c)   In the theory of summation the dispersion $\mathrm{disp}(a(x), b(x), x)$ of two polynomials $a(x), b(x) \in \mathbb{Q}[x]$ plays an important role, see Section 11.4. The dispersion is defined as follows:[28]

$$\mathrm{disp}(a(x), b(x), x) := \max\{j \in \mathbb{N}_0 \mid \gcd(a(x), b(x+j)) \neq 1\}.$$

Find an algorithm with the help of which the dispersion can be computed, and use it to compute

$$\mathrm{disp}(3x^7 - 7608x^6 + cx^5 - 2533cx^4 - 7608cx^3 + c^2x^2 - 2536c^2x,$$

$$3x^4 - 11748x^3 + cx^2 + 10016730x^2 - 3226cx - 3209390100x + 993945c + 354912910875, x).$$

How can these results be read off from the rational factorizations of $a(x)$ and $b(x)$?

Exercise 7.18. Show that, for $a(x) = c(x) \cdot d(x)$, we have

$$\mathrm{res}(a(x), b(x), x) = \mathrm{res}(c(x), b(x), x) \cdot \mathrm{res}(d(x), b(x), x).$$

Exercise 7.19. (Discriminant) Let $\mathbb{K}$ be a field of characteristic 0. The discriminant of a polynomial $a(x) \in \mathbb{K}[x]$ is defined as the resultant of $a(x)$ and $a'(x)$:

$$\mathrm{disc}\,(a(x), x) := \mathrm{res}(a(x), a'(x), x).$$

Show:

(a) $\mathrm{disc}\,(a(x), x) = 0$ if and only if $a(x)$ has a multiple price factor.

(b)   If $\alpha_k, k = 1, \ldots, n$, are the zeros of $a(x)$, counted by their multiplicity, in a suitable extension field $\mathbb{K}$, then

$$\mathrm{disc}\,(a(x), x) = \prod_{k=1}^{n} \prod_{j<k} (\alpha_k - \alpha_j)^2.$$

(c)   Compute the discriminant in the special case of a generic quadratic polynomial $a(x) = ax^2 + bx + c$.

Exercise 7.20. (Logistic Iteration) The logistic map $f : \mathbb{R} \to \mathbb{R}, \ x \mapsto \alpha x(1-x)$ defines an iteration scheme

$$x_{n+1} = f(x_n)$$

which is well defined for $\alpha \in (0, 4)$, but shows more and more chaotic behavior for $\alpha \to 4$.

(a)   The following Mathematica function displays the iteration $x_{n+1} = f(x_n)$ graphically:

---

[28] If the maximum does not exist, we set the dispersion as -1.

```
In[1]:= FixedPointGraph[f_,{x_,x1_,x2_},x0_,
            n_,options___] :=
        Module[{list,table},
          list = FixedPointList[N[(f/.x→#)]&,
              x0,n];
          table = Graphics[{Thickness[0.001],
              Table[
                Line[{{list[[j]],list[[j]]},
                      {list[[j]],list[[j+1]]},
                      {list[[j+1]],list[[j+1]]}}],
                {j,1,Length[list]-1}]}];
          plot = Plot[{f,x},{x,x1,x2},
              DisplayFunction→Identity];
          Show[plot,table,
            DisplayFunction→$DisplayFunction,
            options]]

In[2]:= FixedPointGraph[f_,{x_,x1_,x2_},x0_] :=
        Module[{list,table},
          list = FixedPointList[N[(f/.x→#)]&,
              x0];
          table = Graphics[{Thickness[0.001],
              Table[
                Line[{{list[[j]],list[[j]]},
                      {list[[j]],list[[j+1]]},
                      {list[[j+1]],list[[j+1]]}}],
                {j,1,Length[list]-1}]}];
          plot = Plot[{f,x},{x,x1,x2},
              DisplayFunction→Identity];
          Show[plot,table,
            DisplayFunction→$DisplayFunction]]
```

Describe the functionality of this procedure and define this function in your *Mathematica* notebook.

(b) *Apply* FixedPointGraph *to the logistic iteration and generate iteration pictures for several different values of* α.

(c) *Implement an analogous function* FixedPointRestGraph[f, {$x, x1, x2$}, $x0, n$] *which prints only the "end" of the iteration, i.e., for example the last 100 iteration steps. Apply your function to the examples of (b). Give an interpretation of the resulting pictures.*

(d) *The logistic map possesses—starting from a certain value* α_2—*a 2-cycle which corresponds to a multiple zero of* $g(x) := f(f(x)) - x$. *Compute the number* α_2 *symbolically by using resultants, and visualize the situation using* FixedPointRestGraph.

(e) *Solve the question analogous to (d) for* 3- *and for* 4-*cycles and compute* α_3 *and* α_4.[29] *What is the degree with respect to* α *of the resulting polynomials? Represent these situations graphically.*

Exercise 7.21. (Minimal Polynomials via Resultants) *Minimal polynomials of algebraic numbers can be computed with the aid of resultants.*

---

[29] The computation of the corresponding resultant for 5-cycles is very time consuming.

(a) Prove that if α and β are two algebraic numbers with minimal polynomials $p(x)$ and $q(x)$, then the following table is valid:

algebraic number	zero polynomial
$\alpha + \beta$	$\mathrm{res}(p(x-y), q(y), y)$
$\alpha - \beta$	$\mathrm{res}(p(x+y), q(y), y)$
$\alpha \cdot \beta$	$\mathrm{res}(p(x/y), q(y), y)$
α/β	$\mathrm{res}(p(x \cdot y), q(y), y)$
$\sqrt[n]{\alpha}$	$\mathrm{res}(p(y), x^n - y, y)$

The minimal polynomial can then be computed by a factorization of the above zero polynomial.
(b) Use the above table for three personal examples.
(c) Repeat the computations of Exercise 7.10 with this new method.

Exercise 7.22. (Polynomial Systems of Equations) *Solve the following systems of equations* $p_k(x, y) = 0$, $(k = 1, 2, 3)$ *iteratively using resultants. Find the complete solutions and then check your results using* `Solve`.

(a) $p_1(x, y) = (x+y)^2 + 4xy$, $p_2(x, y) = (x-y)^2 - 1$;
(b) $p_1(x, y) = (x+y)^4 + 4xy$, $p_2(x, y) = (x-y)^2 - 1$;
(c) $p_1(x, y) = x^2 c + xy - yc - 1$, $p_2(x, y) = 2xy^2 + yc^2 - c^2 - 2$, $p_3(x, y) = x + y^2 - 2$.

Chapter 8
Factorization in Polynomial Rings

8.1 Preliminary Considerations

The aim of this chapter is the development of efficient algorithms for factorization in $\mathbb{Q}[x]$. For this purpose, efficient factorization algorithms for $\mathbb{Z}_p[x]$ are used which, of course, are interesting by themselves.

In Section 6.7 we had presented Kronecker's algorithm for the computation of the factors of a polynomial $a(x) \in \mathbb{Z}[x]$. In this section we will see that for factorization in $\mathbb{Q}[x]$ it is sufficient to compute factorizations in $\mathbb{Z}[x]$. For this purpose we repeat the following definition from Section 6.7.

Definition 8.1. Let $a(x) = \sum\limits_{k=0}^{n} a_k x^k \in \mathbb{Z}[x]$. Then

$$\text{content}(a(x), x) := \gcd(a_0, a_1, \ldots, a_n)$$

is called the *content* of the polynomial $a(x)$. A polynomial $a(x) \in \mathbb{Z}[x]$ with $\text{content}(a(x), x) = 1$ is called *primitive*. △

We first show the following lemma.

Lemma 8.2. For every $a(x) \in \mathbb{Q}[x]$ there is a primitive polynomial $b(x) \in \mathbb{Z}[x]$ which is associated with $a(x)$. This polynomial is given uniquely up to units in $\mathbb{Z}$, hence up to a factor ± 1.

Proof. Existence: Let $a(x) = \sum\limits_{k=0}^{n} a_k x^k \in \mathbb{Q}[x]$ and let $a_k = \frac{b_k}{c_k}$ ($k = 0, \ldots, n$) be the coefficients in lowest terms. If we multiply $a(x)$ by the least common multiple $\text{lcm}(c_0, \ldots, c_n)$, then we get a polynomial $c(x) \in \mathbb{Z}[x]$ associated with $a(x)$ whose coefficients are integers. Finally the polynomial $b(x) := c(x)/\text{content}(c(x), x)$ is primitive and also associated with $a(x)$.

Uniqueness: Let $b(x)$ and $c(x)$ be both primitive and associated with $a(x)$. Then $b(x)$ and $c(x)$ are also associated with each other. Hence $t \cdot c(x) = s \cdot b(x)$ with $s, t \in \mathbb{Z}$ where at least one of these numbers does not equal ± 1 if $b(x)$ and $c(x)$ do not differ just by a unit. Without loss of generality we can assume that $s \neq \pm 1$. The primitivity of $b(x)$ implies that the greatest common divisor of the coefficients of $c(x)$ contains the factor s. This, however, is a contradiction to the primitivity of $c(x)$. □

© Springer Nature Switzerland AG 2021
W. Koepf, *Computer Algebra*, Springer Undergraduate Texts in Mathematics and Technology,
https://doi.org/10.1007/978-3-030-78017-3_8

Definition 8.3. As a result of the lemma we define the *primitive part* of a polynomial $a(x) \in \mathbb{Q}[x]$ as the (essentially unique) primitive polynomial which is associated with $a(x)$, and we write primpart($a(x), x$) for it. $\triangle$

Session 8.4. *We implement the given algorithm. The function*

```
In[1]:= Clear[PrimitivePart]

        PrimitivePart[a_,x_] := Module[
            {list,denominator,lcm,gcd,k},
            CoefficientList[a,x];
            denominator = Map[Denominator,list];
            lcm = LCM[Apply[Sequence,denominator]];
            list = list*lcm;
            gcd = GCD[Apply[Sequence,list]];
            list = list/gcd;
            Sum[list[[k]] x^(k-1),{k,1,Length[list]}]
        ]
```

generates the primitive part of the polynomial $a(x) \in \mathbb{Q}[x]$. *For*

$$In[2]:= \mathbf{a} = \sum_{k=0}^{10} \frac{\mathbf{Random[Integer,\{0,9\}]}}{\mathbf{Random[Integer,\{1,9\}]}} \mathbf{x}^{\mathbf{k}}$$

$$Out[2]= \frac{x^{10}}{5} + \frac{6x^7}{7} + 2x^6 + \frac{x^5}{8} + \frac{4x^4}{9} + x^3 + 4x^2 + \frac{1}{4}$$

the command

$$In[3]:= \mathbf{PrimitivePart[a,x]}$$
$$Out[3]= 504x^{10} + 2160x^7 + 5040x^6 + 315x^5 + 1120x^4 + 2520x^3 + 10080x^2 + 630$$

computes the associated primitive polynomial.

We can also apply the algorithm to polynomials with integer coefficients:

$$In[4]:= \mathbf{PrimitivePart[144x^2 + 60x - 24, x]}$$
$$Out[4]= 12x^2 + 5x - 2$$

In this case, the algorithm divides by the content of the polynomial.

Since associated polynomials have the same factorizations up to a constant, it is reasonable to restrict ourselves to primitive polynomials in the factorization process. To show that this process will always give us the opportunity to work with polynomials with integer coefficients—and that it is therefore sufficient to find factorizations over $\mathbb{Z}$ instead of $\mathbb{Q}$—we will need another lemma.

Lemma 8.5. (Product of Primitive Polynomials) The product of two primitive polynomials is primitive, too.

Proof. First note that the product $c(x) = a(x)b(x)$ of two polynomials with integer coefficients $a(x) \in \mathbb{Z}[x]$ and $b(x) \in \mathbb{Z}[x]$ again yields a polynomial with integer coefficients. Now let

$$a(x) = \sum_{j=0}^{n} a_j x^j \quad \text{and} \quad b(x) = \sum_{k=0}^{m} b_k x^k$$

be two primitive polynomials, hence

$$\gcd(a_0, a_1, ..., a_n) = 1 \quad \text{and} \quad \gcd(b_0, b_1, ..., b_m) = 1. \tag{8.1}$$

The product then has the representation

$$c(x) = \sum_{l=0}^{m+n} c_l x^l \quad \text{with} \quad c_l = \sum_{j+k=l} a_j b_k.$$

We now assume that $c(x)$ is not primitive. Then there exists a $p \in \mathbb{P}$ which is a divisor of all coefficients of $c(x)$: $p \mid c_l$ for $l = 0, ..., m+n$. Now let $A := \{j \mid p$ is no divisor of $a_j\}$ and $B := \{k \mid p$ is no divisor of $b_k\}$. Relation (8.1) implies that $A \neq \emptyset$ and $B \neq \emptyset$, hence there are $j_0 := \min A$ and $k_0 := \min B$. By definition of j_0 and k_0 we therefore get

$$p \mid a_j \text{ for all } j < j_0 \quad \text{and} \quad p \mid b_k \text{ for all } k < k_0 \tag{8.2}$$

as well as

$$p \text{ is no divisor of } a_{j_0} \quad \text{and} \quad p \text{ is no divisor of } b_{k_0}. \tag{8.3}$$

Next we set $l_0 := j_0 + k_0$ and consider $c_{l_0} = \sum_{j+k=l_0} a_j b_k$. We can split this sum in the following form:

$$c_{l_0} = \sum_{j+k=l_0}^{j<j_0, k>k_0} a_j b_k + \sum_{j+k=l_0}^{j>j_0, k<k_0} a_j b_k + a_{j_0} b_{k_0}, \tag{8.4}$$

where the remaining sums are allowed to be empty. Since $p \mid c_{l_0}$, it follows from (8.2) that p is a divisor of the first and the second sum in (8.4). This implies that $p \mid a_{j_0} b_{k_0}$, contrary to (8.3). This proves our assertion. $\qquad\qquad\square$

Session 8.6. *We continue with Session 8.4 and check the lemma. First we declare a second polynomial*

$$In[5] := \mathbf{b} = \sum_{k=0}^{10} \frac{\texttt{Random[Integer, \{0,9\}]}}{\texttt{Random[Integer, \{1,9\}]}} x^k$$

$$Out[5] = 2x^{10} + \frac{x^9}{7} + \frac{5x^8}{3} + \frac{5x^7}{8} + \frac{4x^6}{3} + \frac{2x^5}{3} + \frac{7x^4}{6} + \frac{x^3}{9} + \frac{3x}{4} + \frac{3}{4}$$

and compute the product of the primitive parts of $a(x)$ and $b(x)$:

$In[6] := \mathbf{c} = \texttt{PrimitivePart[a, x]PrimitivePart[b, x]//Expand}$

$Out[6] = 317520 x^{20} + 5103000 x^{19} + 5199768 x^{18} + 4659417 x^{17} + 7423920 x^{16} +$
$\qquad 7309080 x^{15} + 10410372 x^{14} + 12306336 x^{13} + 18563328 x^{12} + 9197706 x^{11} +$
$\qquad 18375630 x^{10} + 13476288 x^9 + 14293888 x^8 + 8214150 x^7 + 9194640 x^6 +$
$\qquad 3804360 x^5 + 4476640 x^4 + 5821200 x^3 + 4021920 x^2 + 926100 x + 714420$

The computation

$In[7] := \texttt{GCD[Apply[Sequence, CoefficientList[c, x]]]}$
$Out[7] = 1$

confirms that the product is primitive again.

The essential theorem of this section is *Gauss's lemma*.

Theorem 8.7. (Gauss's Lemma) Let $a(x) \in \mathbb{Z}[x]$ and let $a(x) = b(x) \cdot c(x)$ be a factorization with $b(x), c(x) \in \mathbb{Q}[x]$. Then there is a factorization $a(x) = \tilde{b}(x) \cdot \tilde{c}(x)$ with $\tilde{b}(x), \tilde{c}(x) \in \mathbb{Z}[x]$ such that $b(x)$ and $\tilde{b}(x)$ are associated with each other and $c(x)$ and $\tilde{c}(x)$ are associated with each other.

Proof. By Lemma 8.2 there is an $r \in \mathbb{Q}$ such that $r \cdot a(x)$ is primitive. Since by $r \cdot a(x) = r \cdot b(x) \cdot c(x)$ the assertion is valid for $a(x)$ if and only if it is valid for $r \cdot a(x)$, we can assume without loss of generality that $a(x)$ is primitive.

Lemma 8.2 further implies that there are $s, t \in \mathbb{Q}$ such that $\tilde{b}(x) := s \cdot b(x)$ and $\tilde{c}(x) := t \cdot c(x)$ are primitive. Lemma 8.5 therefore implies that $st \cdot b(x) \cdot c(x)$ is primitive, too. From $a(x) = b(x) \cdot c(x)$ and the primitivity of $a(x)$ it follows (by the uniqueness in Lemma 8.2) that $st = \pm 1$ and therefore $a(x) = \pm \tilde{b}(x) \cdot \tilde{c}(x)$. $\qquad\square$

Gauss's lemma proves that for the factorization of a polynomial $a(x) \in \mathbb{Q}[x]$ it is sufficient to factorize the polynomial primpart$(a(x), x) \in \mathbb{Z}[x]$. However, this factorization lies in $\mathbb{Z}[x]$, and we have thus transformed the factorization problem—as mentioned previously—from $\mathbb{Q}[x]$ to $\mathbb{Z}[x]$. In particular we now know that, for example, by an application of Kronecker's algorithm, we can find the factorization of every polynomial $a(x) \in \mathbb{Q}[x]$.

Modern factorization algorithms for $a(x) \in \mathbb{Z}[x]$ proceed as follows: the given polynomial is reduced modulo a suitable prime $p \in \mathbb{P}$ to a polynomial $a_p(x) \in \mathbb{Z}_p[x]$ which is then factorized. Ultimately the sought factorization for $a(x) \in \mathbb{Z}[x]$ is reconstructed from the factorization found for $a_p(x) := \mathrm{mod}(a(x), p) \in \mathbb{Z}_p[x]$. This last step is called *lifting*.

In particular: if we want to prove the irreducibility of the polynomial $a(x) \in \mathbb{Z}[x]$, then it is sufficient to prove the irreducibility of $a_p(x) \in \mathbb{Z}_p[x]$ for *a single* $p \in \mathbb{P}$, since every proper factorization of $a(x) \in \mathbb{Z}[x]$ over $\mathbb{Z}$ automatically generates proper factorizations of $a_p(x) \in \mathbb{Z}_p[x]$ for all $p \in \mathbb{P}$. Therefore, this information can be lifted very easily.

For example, the factorization $3x^3 + 28x^2 - 35x + 10 = (x^2 + 10x - 5)(3x - 2)$ over $\mathbb{Z}$ automatically provides the decomposition $x^3 + x = (x^2 + 1)x$ modulo 2, but does not generate the full factorization $x^3 + x = (x + 1)^2 x$.

Session 8.8. *We load the function* `Factors` *from Session 6.35 with the help of which we can (principally) compute all factorizations of a polynomial $b(x) \in \mathbb{Z}_p[x]$.*

The computation

```
In[1]:= Factors[x⁴+x+1,x,2,2]
Out[1]= {1}
```

shows that the polynomial $x^4 + x + 1 \in \mathbb{Z}_2[x]$ is irreducible, hence it is also irreducible over $\mathbb{Z}$. This is confirmed by the command

```
In[2]:= Factor[x⁴+x+1]
Out[2]= x⁴+x+1
```

To realize the proposed program, we therefore first need an efficient algorithm for the factorization of polynomials over finite fields. In the next section we consider the factorization in $\mathbb{Z}_p[x]$.

8.2 Efficient Factorization in $\mathbb{Z}_p[x]$

As observed earlier, the factorization of a univariate polynomial can be reduced to one (or several) factorizations of the polynomial over finite fields. We do so for two reasons: first, in finite fields there is no *intermediate expression swell* since all coefficients in $GF(p^n)$ are bounded by p. This is a very remarkable advantage. Secondly, the special structure of finite fields allows us to find very efficient factorization methods.

In the sequel we consider an algorithm for the factorization in $\mathbb{Z}_p[x]$, which goes back to Berlekamp [Ber1967]. This algorithm uses only computations of greatest common divisors and linear algebra and is therefore very efficient for relatively small $p \in \mathbb{P}$.

Let $a(x) = a_n x^n + a_{n-1} x^{n-1} + \cdots + a_1 x + a_0 \in \mathbb{Z}_p[x]$, $a_n \neq 0$. We assume that there is a non-constant polynomial $b(x) \in \mathbb{Z}_p[x]$ of degree smaller than n, $1 \leq m := \deg(b(x), x) < n$, such that $a(x) \mid b(x)^p - b(x)$. Obviously the polynomial $b(x)^p - b(x) \in \mathbb{Z}_p[x]$ is not the zero polynomial since the coefficient of x^{mp} does not equal zero.

Because of the identity

$$y^p - y = \prod_{k=0}^{p-1} (y - k)$$

(see (7.10)) which is valid for $y \in \mathbb{Z}_p$ as a consequence of Fermat's Little Theorem, we therefore get

$$a(x) \mid \prod_{k=0}^{p-1} (b(x) - k). \tag{8.5}$$

From this relation the following lemma arises.

Lemma 8.9. Let $a(x) \in \mathbb{Z}_p[x]$ with $\deg(a(x), x) = n$ and let $b(x) \in \mathbb{Z}_p[x]$ with $1 \leq m := \deg(b(x), x) < n$ such that $a(x) \mid b(x)^p - b(x)$. Then

$$a(x) = \prod_{k=0}^{p-1} \gcd(a(x), b(x) - k) \tag{8.6}$$

is a non-trivial factorization of $a(x)$.

Proof. From (8.5) it follows that

$$a(x) \mid \prod_{k=0}^{p-1} \gcd(a(x), b(x) - k).$$

Conversely, by the definition of the greatest common divisor, we see that for all $k = 0, \ldots, p-1$,

$$\gcd(a(x), b(x) - k) \mid a(x).$$

Since the polynomials $b(x) - k$ are pairwise relatively prime for different values $k = 0, \ldots, p-1$, we finally get

$$\prod_{k=0}^{p-1} \gcd(a(x), b(x) - k) \mid a(x).$$

This yields (8.6).

Every factor on the right-hand side of (8.6) has degree $\leq m$ and therefore $< n$. The polynomial on the left-hand side, however, has degree n. Therefore at least two factors on the right-hand side cannot be constant, which proves that the factorization is non-trivial. $\qquad \square$

Hence, we have transformed the factorization problem in $\mathbb{Z}_p[x]$: now we search for a polynomial $b(x)$ with $a(x) \mid b(x)^p - b(x)$ whose degree is $< n$. For this purpose we can write $b(x) = b_{n-1} x^{n-1} + \cdots + b_1 x + b_0$, where we allow $b_{n-1} = 0$. Then, by the arithmetic in $\mathbb{Z}_p$, we obtain

$$
\begin{aligned}
b(x)^p &= (b_{n-1} x^{n-1} + \cdots + b_1 x + b_0)^p \\
&= b_{n-1}^p x^{(n-1)p} + \cdots + b_1^p x^p + b_0^p \\
&= b_{n-1} x^{(n-1)p} + \cdots + b_1 x^p + b_0,
\end{aligned}
$$

where we have used Fermat's Little Theorem in the last step.

Next we use n divisions with remainder and write all occurring powers of x in the form

$$
x^{jp} = a(x) q_j(x) + r_j(x) \qquad (j = 0, \ldots, n-1),
$$

which finally gives us

$$
b(x)^p = b_{n-1} r_{n-1}(x) + \cdots + b_1 r_1(x) + b_0 + a(x) q(x)
$$

for some $q(x) \in \mathbb{Z}_p[x]$. Therefore $a(x) \mid b(x)^p - b(x)$ if and only if $a(x)$ is a divisor of

$$
b_{n-1} r_{n-1}(x) + \cdots + b_1 r_1(x) + b_0 - b(x).
$$

This polynomial has degree $< n$, by construction, and hence it is divisible by $a(x)$ if and only if it is the zero polynomial. Therefore we have reduced the search for $b(x)$ to the polynomial equation

$$
b_{n-1} r_{n-1}(x) + \cdots + b_1 r_1(x) + b_0 - b(x) = 0. \tag{8.7}
$$

This equation can be solved by equating the coefficients. We get n linear equations in the n indeterminates $b_0, \ldots, b_{n-1}$. If we find a non-constant solution, this yields a non-trivial factorization of $a(x)$ by Lemma 8.9. Furthermore it is clear that the considered linear system has a solution if and only if (8.7) can be satisfied, i.e., if $a(x)$ is a divisor of $b(x)^p - b(x)$.

We have proved the following theorem.

Theorem 8.10. (Berlekamp Algorithm) Let $p \in \mathbb{P}$ and let $a(x) \in \mathbb{Z}_p[x]$ be of degree n. Then the algorithm described above finds a non-trivial factorization of $a(x)$ if such a factorization exists. This algorithm can be applied recursively to find the full factorization of $a(x)$. $\qquad\square$

Session 8.11. *Let us implement Berlekamp's algorithm in Mathematica. For this, let the polynomial* $A(x) := 4x^6 + 3x^5 + x^4 + 7x^3 + 6x^2 + 2x + 4 \in \mathbb{Z}_{13}[x]$ *be given:*

```
In[1]:= A = 4x^6 + 3x^5 + x^4 + 7x^3 + 6x^2 + 2x + 4
Out[1]= 4 x^6 + 3 x^5 + x^4 + 7 x^3 + 6 x^2 + 2 x + 4
```

```
In[2]:= p = 13
Out[2]= 13
```

We compute the degree of $A(x)$ *by*

```
In[3]:= n = Exponent[A, x]
Out[3]= 6
```

and its leading coefficient by

```
In[4]:= lcoeff = Coefficient[A, x, n]
```

Out[4]= 4

We divide A(x) by its leading coefficient and get the monic associated polynomial a(x):

In[5]:= **a = PolynomialMod[A PowerMod[lcoeff,-1,p],p]**

Out[5]= $x^6 + 4x^5 + 10x^4 + 5x^3 + 8x^2 + 7x + 1$

By n-fold division with remainder, we get the remainders $r_j(x)$ ($j = 0, \ldots, n-1$):

In[6]:= **rests = Table[PolynomialMod[**
           PolynomialRemainder[x^{jp},a,x],p],{j,0,n-1}]

Out[6]= $\{1, x^5 + 12x^4 + 5x^3 + 11x^2 + 7x + 3,$
          $x^5 + 6x^4 + 11x^3 + 4x^2 + 4x + 10,$
          $8x^5 + 4x^4 + 9x^3 + 10x^2 + 7x + 2,$
          $10x^5 + 8x^4 + 10x^3 + 12x^2 + 2x + 4,$
          $11x^5 + 9x^4 + 10x^3 + 5x^2 + 5x + 12\}$

Next, we compute the coefficients of the polynomial on the left-hand side of (8.7):

In[7]:= **list = CoefficientList$\left[\sum_{k=0}^{n-1}\beta_k\,(\text{rests[[k+1]]} - x^k), x\right]$**

Out[7]= $\{3\beta_1 + 10\beta_2 + 2\beta_3 + 4\beta_4 + 12\beta_5, 6\beta_1 + 4\beta_2 + 7\beta_3 + 2\beta_4 + 5\beta_5,$
          $11\beta_1 + 3\beta_2 + 10\beta_3 + 12\beta_4 + 5\beta_5, 5\beta_1 + 11\beta_2 + 8\beta_3 + 10\beta_4 + 10\beta_5,$
          $12\beta_1 + 6\beta_2 + 4\beta_3 + 7\beta_4 + 9\beta_5, \beta_1 + \beta_2 + 8\beta_3 + 10\beta_4 + 10\beta_5\}$

Finally we solve the resulting linear system in $\mathbb{Z}_p$:

In[8]:= **sol = Solve[**
        Union[Map[# == 0&, list]],
        Intersection[Variables[list], Table[β$_k$,{k,0,n-1}]],
    Modulus → 13]

Out[8]= $\{\{\beta_5 \to \beta_1 + 6\beta_3, \beta_4 \to 7\beta_1 + \beta_3, \beta_2 \to 10\beta_1\}\}$

Therefore we get the polynomial b(x)

In[9]:= **b = PolynomialMod$\left[\sum_{k=0}^{n-1}\beta_k x^k\text{/.sol[[1]]/.}\beta_{1\_} \to 1, p\right]$**

Out[9]= $7x^5 + 8x^4 + x^3 + 10x^2 + x + 1$

with $a(x) \mid b(x)^p - b(x)$.[1]

The computation

In[10]:= **tab = Table[PolynomialGCD[a, b - k, Modulus → p],**
        {k,0,p-1}]

Out[10]= $\{1, 1, 1, 1, 1, 1, x + 7, x^2 + 9x + 10, 1, 1, x^3 + x^2 + 12x + 8, 1, 1\}$

therefore yields the non-trivial factors according to Lemma 8.9, and we finally get the factorization of a(x) as follows:

In[11]:= **Apply[Times, tab]**

Out[11]= $(x+7)(x^2 + 9x + 10)(x^3 + x^2 + 12x + 8)$

The computation

In[12]:= **Factor[A, Modulus → 13]**

---

[1] There are several such polynomials which form a vector space over $\mathbb{Z}_p$. Since for our further considerations a single element is sufficient, we set without loss of generality $\beta_l = 1$ for the remaining variables.

$Out[12] = 4\,(x+7)\,\left(x^2 + 9\,x + 10\right)\left(x^3 + x^2 + 12\,x + 8\right)$

using the built-in command Factor *confirms that the resulting factors are irreducible. This is not automatically the case as we will see in the next example. Whether there are further factors can be checked by the same algorithm.*

The full algorithm is therefore given by the program

```
In[13] := Clear[Berlekamp]
          Berlekamp[A_, x_, p_] :=
            Module[{n, lcoeff, a, rests, j, k, sol, β, b, tab},
              n = Exponent[A, x];
              lcoeff = Coefficient[A, x, n];
              a = PolynomialMod[A PowerMod[lcoeff, -1, p], p];
              rests =
                Table[PolynomialMod[
                  PolynomialRemainder[x^(j p), a, x], p],
                  {j, 0, n - 1}];
              list =
                Union[Map[# == 0&,
                  CoefficientList[
                    Sum[β_k (rests[[k + 1]] - x^k), {k, 0, n - 1}],
                    x]]];
              sol =
                Solve[list, Intersection[Variables[list],
                    Table[β_k, {k, 0, n - 1}]], Modulus → p];
              b = PolynomialMod[Sum[β_k x^k /. sol[[1]] /. β_1_ → 1,
                  {k, 0, n - 1}], p];
              tab = Table[PolynomialGCD[a, b - k, Modulus → p],
                  {k, 0, p - 1}];
              lcoeff Apply[Times, tab]
            ]
```

and we obtain again

$In[14] := $ **Berlekamp[A, x, p]**
$Out[14] = 4\,(x+7)\,\left(x^2 + 9\,x + 10\right)\left(x^3 + x^2 + 12\,x + 8\right)$

For the example

$In[15] := $ **Berlekamp$\left[\displaystyle\sum_{k=0}^{7} x^k, x, 13\right]$**
$Out[15] = (x+1)\,(x+5)\,(x+8)\,\left(x^4 + 1\right)$

it remains to check whether $x^4 + 1$ is irreducible. Using the same algorithm, we get

$In[16] := $ **Berlekamp[$x^4 + 1$, x, 13]**
$Out[16] = \left(x^2 + 5\right)\left(x^2 + 8\right)$

in agreement with

$In[17] := $ **Factor$\left[\displaystyle\sum_{k=0}^{7} x^k, \text{Modulus} → 13\right]$**
$Out[17] = (x+1)\,(x+5)\,(x+8)\,\left(x^2 + 5\right)\left(x^2 + 8\right)$

Note that the previous computations have short timings, which clearly shows the efficiency of the method.

The complexity, however, is linear with respect to p and therefore exponential in the length of p. For large $p \in \mathbb{P}$ there are variants of Berlekamp's algorithm which are much more efficient. This is illustrated by the example

$In[18] :=$ **Timing** $\left[\textbf{Berlekamp} \left[\sum_{k=0}^{7} \textbf{x}^{\textbf{k}}, \textbf{x}, \textbf{1000003} \right] \right]$

$Out[18] = \{54.75 \text{ Second}, (x+1)(x^2+1)(x^2+410588\,x+1000002)(x^2+589415\,x+1000002)\}$

$In[19] :=$ **Timing** $\left[\textbf{Factor} \left[\sum_{k=0}^{7} \textbf{x}^{\textbf{k}}, \textbf{Modulus} \rightarrow \textbf{1000003} \right] \right]$

$Out[19] = \{0. \text{ Second}, (x+1)(x^2+1)(x^2+410588\,x+1000002)(x^2+589415\,x+1000002)\}$

In the sequel we will show that Berlekamp's algorithm provides even more, namely it additionally computes the exact number of different irreducible factors of $a(x)$ as a side effect.

For this purpose, we first rearrange the problem in matrix form. Let

$$R = \begin{pmatrix} r_{00} & r_{01} & \cdots & r_{0,n-1} \\ r_{10} & r_{11} & \cdots & r_{1,n-1} \\ \vdots & \vdots & \ddots & \vdots \\ r_{n-1,0} & r_{n-1,1} & \cdots & r_{n-1,n-1} \end{pmatrix}$$

be the $n \times n$ matrix which is constructed from the remainder polynomials

$$r_k(x) := \sum_{j=0}^{n-1} r_{jk} x^j .$$

The coefficients of the polynomials $r_k(x)$ therefore build the columns of R. Under these circumstances $(b_0, b_1, \ldots, b_{n-1})$ are the coefficients of a solution polynomial $b(x) = \sum_{j=0}^{n-1} b_j x^j$ of (8.7) if and only if the homogeneous linear system

$$(R-E) \cdot \begin{pmatrix} b_0 \\ b_1 \\ \vdots \\ b_{n-1} \end{pmatrix} = \begin{pmatrix} 0 \\ 0 \\ \vdots \\ 0 \end{pmatrix}$$

has a solution, where E denotes the $n \times n$ unit matrix.

Let $\dim \text{kern} A$ denote the dimension of the kernel of a matrix A. Then we get the following theorem.

Theorem 8.12. (Berlekamp Algorithm Part II) Let $p \in \mathbb{P}$ and let $a(x) \in \mathbb{Z}_p[x]$ be of degree n. Then the dimension of the kernel of the matrix $R - E$ gives the number of *different* irreducible factors of $a(x)$. In particular, $a(x)$ is irreducible if and only if $a(x)$ is squarefree and $\dim \text{kern}(R - E) = 1$.

Proof. Let us first assume that $a(x)$ has exactly J different factors. Then $a(x) = \prod_{j=1}^{J} p_j(x)^{e_j}$, such that the polynomials $p_j(x) \in \mathbb{Z}_p[x]$ are irreducible. Now, if for some $b(x) \in \mathbb{Z}_p[x]$ we have

$a(x) \mid b(x)^p - b(x) = \prod\limits_{k=0}^{p-1} (b(x) - k)$, then every $p_j(x) \mid b(x) - k$ for some $k = 0, \ldots, p-1$. Since $b(x) - k$ and $b(x) - l$ are relatively prime for $k \neq l$, the number k for which $p_j(x) \mid b(x) - k$ is *uniquely determined* and we therefore denote it by k_j. This defines a function φ which maps the set of polynomials $b(x) \in \mathbb{Z}_p[x]$, for which $a(x) \mid b(x)^p - b(x)$ is valid, onto the set of vectors $(k_1, \ldots, k_J) \in \mathbb{Z}_p^J$.

Next we construct the inverse function φ^{-1}. By the Chinese remainder theorem for polynomials (Theorem 7.6), for every J-tuple $(k_1, \ldots, k_J) \in \mathbb{Z}_p^J$ there is a uniquely determined polynomial $b(x) \in \mathbb{Z}_p[x]$ of degree $< n$ with

$$b(x) \equiv k_j \ (\mathrm{mod}\ p_j(x)^{e_j}),$$

in other words $p_j(x)^{e_j} \mid b(x) - k_j$. For such a $b(x)$ we get furthermore that

$$p_j(x)^{e_j} \mid \prod_{k=0}^{p-1} (b(x) - k) = b(x)^p - b(x), \qquad \text{for all } j = 1, \ldots, J,$$

so that we finally see that $a(x)$ is a divisor of $b(x)^p - b(x)$. The given construction therefore provides the inverse function φ^{-1}.

By construction φ is one-to-one, therefore there exists a unique correspondence between the polynomials $b(x)$ of degree $< n$ with $a(x) \mid b(x)^p - b(x)$ and the J-tuples $(k_1, \ldots, k_J) \in \mathbb{Z}_p^J$ for which J is the number of different irreducible factors of $a(x)$. Since there are p^J such vectors, we have p^J possible polynomials $b(x)$.

The possible polynomials $b(x) \in \mathbb{Z}_p[x]$ therefore form a vector space over $\mathbb{Z}_p$ of dimension J. This, however, is equivalent to the statement $\dim \mathrm{kern}(R - E) = J$. $\square$

Session 8.13. *We define the function*

```
In[1]:= Clear[NumberOfFactors]
        NumberOfFactors[a_,x_,p_] :=
          Module[{n, rests, R, k, j, ns},
            n = Exponent[a,x];
            rests =
              Table[PolynomialMod[
                PolynomialRemainder[x^(p*k),a,x],
                p],{k,0,n-1}];
            R = Table[Coefficient[rests[[k]],x,j],
                {j,0,Length[rests]-1},
                {k,Length[rests]}];
            ns = NullSpace[
                R - IdentityMatrix[Length[rests]],
                Modulus → p];
            Length[ns]
          ]
```

which implements the newly introduced method. The function NumberOfFactors *determines the number of* different irreducible factors *of a polynomial* $a(x)$.[2] *The call*

---

[2] If we had designed our function for highest efficiency, then of course we would have combined the functionalities of NumberOfFactors and Berlekamp. In this case, the kernel must be computed only once.

In[2]:= **NumberOfFactors [x^6 +x^5 +x^4 +x^3 +x^2 +x+1, x, 2]**

Out[2]= 2

shows that the computation

In[3]:= **Berlekamp [x^6 +x^5 +x^4 +x^3 +x^2 +x+1, x, 2]**

Out[3]= $\left(x^3 +x+1\right)\left(x^3 +x^2 +1\right)$

has already computed the complete factorization. The computation

In[4]:= **NumberOfFactors [x^6 +x^5 +x^4 +x^3 +x^2 +x+1, x, 3]**

Out[4]= 1

shows the irreducibility of $x^6 +x^5 +x^4 +x^3 +x^3 +x+1 \in \mathbb{Z}_3[x]$ (under the premise of squarefreeness). In the following example, the computation

In[5]:= **NumberOfFactors [x^{10} +x^9 +x^7 +x^3 +x^2 +1, x, 2]**

Out[5]= 3

shows that the factorization

In[6]:= **factors = Berlekamp [x^{10} +x^9 +x^7 +x^3 +x^2 +1, x, 2]**

Out[6]= $\left(x^4 +x^3 +x^2 +1\right)\left(x^6 +x^4 +1\right)$

is still incomplete. We must therefore apply Berlekamp's algorithm to the generated factors again, and we obtain

In[7]:= **NumberOfFactors [factors [[1]], x, 2]**

Out[7]= 2

Therefore we must get

In[8]:= **NumberOfFactors [factors [[2]], x, 2]**

Out[8]= 1

With the calculation

In[9]:= **Berlekamp [factors [[1]], x, 2]**

Out[9]= $(x+1)\left(x^3 +x+1\right)$

we have finally found a factorization in different factors, which, however, still contains squares:

In[10]:= **Factor [x^{10} +x^9 +x^7 +x^3 +x^2 +1, Modulus → 2]**

Out[10]= $(x+1)\left(x^3 +x+1\right)\left(x^3 +x^2 +1\right)^2$

We thus see that a squarefree factorization is necessary before Berlekamp's algorithm can be applied.

8.3 Squarefree Factorization of Polynomials over Finite Fields

Since Berlekamp's algorithm computes the number of *different* factors, one must find a squarefree factorization for a given $a(x) \in \mathbb{Z}_p[x]$ before its application. However, Theorem 6.42 is not valid over $\mathbb{Z}_p$, hence we must proceed differently. For this purpose, we consider the following example.

Example 8.14. Let $a(x) = x^{17} + 1 \in \mathbb{Z}_{17}[x]$. Then

$$a'(x) = 17 \cdot x^{16} = 0,$$

since $\mathbb{Z}_{17}$ has characteristic 17. In this example, however, the relation (4.7) yields the squarefree factorization

$$(1 + x)^{17} = x^{17} + 1.$$

This shows that $x = \text{mod}(-1, 17) = 16$ is a zero of order 17 of $a(x)$. $\triangle$

Next we show that the idea of the above example works for arbitrary polynomials $a(x) \in \text{GF}(q)[x]$ with $a'(x) = 0$.

Theorem 8.15. Let $a(x) \in \text{GF}(p^n)[x]$ with $a'(x) = 0$. Then $a(x) = b(x)^p$ for some polynomial $b(x) \in \text{GF}(p^n)[x]$ which can be computed algorithmically.

Proof. Let $a(x) \in \text{GF}(p^n)[x]$. Since $a'(x) = 0$ and since the characteristic of $\text{GF}(p^n)[x]$ equals p, the powers occurring in $a(x)$ must be divisible by p. Hence we get

$$a(x) = a_0 + a_p x^p + \cdots + a_{kp} x^{kp}$$

for some $k \in \mathbb{N}$. Now we set

$$b(x) = b_0 + b_1 x + \cdots + b_k x^k,$$

where the coefficients b_j should be chosen according to the relation $b_j^p = a_{jp}$. Since for $a \in \text{GF}(q)$ we have $a^q = a$, see (7.9), and since

$$\left(a_{jp}^{p^{n-1}} \right)^p = a_{jp}^{p^n} = a_{jp}^q = a_{jp},$$

we can choose the coefficients b_j as follows:

$$b_j = a_{jp}^{p^{n-1}}.$$

This choice finally implies

$$\begin{aligned}
b(x)^p &= \left(b_0 + b_1 x + \cdots + b_k x^k \right)^p \\
&= b_0^p + b_1^p x^p + \cdots + b_k^p x^{kp} \\
&= a_0 + a_p x^p + \cdots + a_{kp} x^{kp} = a(x),
\end{aligned}$$

which completes the proof. $\qquad\square$

Theorem 8.15 yields an algorithm for the squarefree factorization of a polynomial $a(x) \in \text{GF}(q)[x]$ with $a'(x) = 0$. If, however, $a'(x) \neq 0$, then the algorithm given in Theorem 6.43 is applicable. Joining both methods yields a complete algorithm for squarefree factorization in $\text{GF}(q)[x]$.

Session 8.16. FactorSquareFree *yields squarefree factorizations in* $\mathbb{Z}_p[x]$, *but not in finite fields* $\text{GF}(p^n)[x]$ *for* $n > 1$. *Therefore we can factor*

```
In[1]:= FactorSquareFree[x^17 + 1, Modulus → 17]
Out[1]= (x + 1)^17
```

An example without a vanishing derivative follows:

$In[2]:=$ **a = x^5 + 4x^2 + 9x + 2**
$Out[2]=$ $x^5 + 4x^2 + 9x + 2$

$In[3]:=$ **FactorSquareFree[a, Modulus → 17]**
$Out[3]=$ $(x+10)^3 (x^2 + 4x + 5)$

This example has a nontrivial squarefree factorization over $\mathbb{Z}_{17}$.

8.4 Efficient Factorization in $\mathbb{Q}[x]$

In order to factorize a polynomial $a(x) \in \mathbb{Q}[x]$ efficiently, we first compute the primitive part of $a(x)$. Therefore, we can assume that $a(x) \in \mathbb{Z}[x]$ has integer coefficients. The next step would be a squarefree factorization. Hence in the sequel we will assume that $a(x)$ is squarefree.

We also assume that the polynomial $a(x) \in \mathbb{Z}[x]$ of degree n is monic. Otherwise, if $a_n \neq 0$ is the leading coefficient of $a(x)$, then we can substitute x by x/a_n and multiply the resulting polynomial by a_n^{n-1} to get a monic polynomial whose factorization is equivalent to the original problem.

Let for example $a(x) = 7x^2 + 4x - 3$. Then

$$b(x) = 7\left(7\left(\frac{x}{7}\right)^2 + 4\left(\frac{x}{7}\right) - 3\right) = x^2 + 4x - 21 = (x-3)(x+7).$$

Back substitution yields the factorization of $a(x)$ as $a(x) = \frac{1}{7}(7x-3)(7x+7) = (7x-3)(x+1)$.

Hence our aim is to factorize squarefree monic polynomials $a(x) \in \mathbb{Z}[x]$. One possibility for this purpose is the following: we first choose one (or several) $p \in \mathbb{P}$ and compute—using Berlekamp's or a more advanced algorithm—the factorization(s) of $a(x) \in \mathbb{Z}_p[x]$, interpreted as elements of $\mathbb{Z}_p[x]$. If $a(x)$ is irreducible for one of those p, then $a(x)$ is also irreducible as element of $\mathbb{Z}[x]$. Otherwise, one can choose a $p \in \mathbb{P}$ such that there are only few factors.[3]

The following lemma shows that this method leads to a valid factorization of $a(x) \in \mathbb{Z}[x]$ if p is large enough. If the values of the modulo function $\mathrm{mod}(x, p) \to (-p/2, p/2]$ are chosen symmetrically around the origin, then the modulo function always takes the smallest possible absolute value. This is essential for our purposes, and we will take this convention for granted throughout this chapter. Hence, if we write $a(x) \in \mathbb{Z}_p[x]$, this means in particular that the coefficients of $a(x)$ lie in the interval $(-p/2, p/2]$ and are therefore bounded in absolute value by $p/2$.

Lemma 8.17. Let $a(x) \in \mathbb{Z}[x]$ be monic with $a(x) \equiv b(x)c(x) \,(\mathrm{mod}\ p)$ and $b(x), c(x) \in \mathbb{Z}_p[x]$. Furthermore, let the absolute values of all coefficients of all possible factors of $a(x)$ be smaller than $p/2$. Then the relation $a(x) = B(x)C(x)$ in $\mathbb{Z}[x]$ with $B(x) \equiv b(x) \,(\mathrm{mod}\ p)$ and $C(x) \equiv c(x) \,(\mathrm{mod}\ p)$ implies that $B(x) = b(x)$ and $C(x) = c(x)$.

Proof. Since $B(x) \equiv b(x) \,(\mathrm{mod}\ p)$, either $B(x) = b(x)$ or there is a coefficient of $B(x)$ which differs from the corresponding coefficient of $b(x)$ by a multiple of p. In such a case this coefficient

---

[3] Unfortunately, even for irreducible $a(x) \in \mathbb{Z}[x]$, it is possible that $a(x) \in \mathbb{Z}_p[x]$ has many factors for every $p \in \mathbb{P}$, see Exercise 8.10.

would lie outside the interval $(-p/2, p/2]$ and therefore $B(x)$ could not be a factor of $a(x)$, by assumption. Similarly we get $C(x) = c(x)$.   □

The following lemma yields a common bound for the coefficients of all potential factors [Zas1969].

Lemma 8.18. (Zassenhaus Bound) Let $a(x) = \sum\limits_{k=0}^{n} a_k x^k \in \mathbb{Z}[x]$ be monic and of degree n. Then the following statements are valid:

(a) Every complex root $x_0 \in \mathbb{C}$ of $a(x)$ satisfies the inequality

$$|x_0| \leq R_0 := \frac{1}{\sqrt[n]{2}-1} \cdot \max_{k=1,\ldots,n} \sqrt[k]{\frac{|a_{n-k}|}{\binom{n}{k}}} \,. \tag{8.8}$$

(b) The coefficients of every proper monic factor $b(x) = \sum\limits_{j=0}^{m} b_j x^j \mid a(x)$ of $a(x)$ satisfy the inequality $|b_{m-k}| \leq \binom{m}{k} R_0^k$, in particular

$$|b_j| \leq \max_{k=1,\ldots,m} \binom{m}{k} R_0^k \,.$$

Proof. (a) We choose

$$M := \max_{k=1,\ldots,n} \sqrt[k]{\frac{|a_{n-k}|}{\binom{n}{k}}}$$

according to (8.8). Then we get for all $k = 1, \ldots, n$ the inequality

$$M \geq \sqrt[k]{\frac{|a_{n-k}|}{\binom{n}{k}}}$$

or, equivalently,

$$|a_{n-k}| \leq \binom{n}{k} M^k \,.$$

Now let $x_0 \in \mathbb{C}$ be a zero of $a(x)$. Then from $x_0^n + \sum\limits_{k=0}^{n-1} a_k x_0^k = 0$ we obtain the relation

$$|x_0|^n = \left| \sum_{k=0}^{n-1} a_k x_0^k \right| \,.$$

The triangle inequality yields

$$|x_0|^n \leq \sum_{k=0}^{n-1} |a_k| \cdot |x_0|^k \leq \sum_{k=1}^{n} |a_{n-k}| \cdot |x_0|^{n-k} \leq \sum_{k=1}^{n} \binom{n}{k} \cdot M^k \cdot |x_0|^{n-k} = \left(|x_0| + M \right)^n - |x_0|^n \,.$$

Hence we have $2|x_0|^n \leq (|x_0| + M)^n$ or, equivalently,

$$|x_0| \leq \frac{M}{\sqrt[n]{2}-1} \,,$$

as announced.

(b) Let

$$b(x) = \sum_{j=0}^{m} b_j x^j = \prod_{k=1}^{m} (x + x_k)$$

be a divisor of $a(x)$, therefore let $-x_k$ ($k = 1, \ldots, m$) be the m complex zeros of $b(x)$. Because of $b(x) \mid a(x)$ those zeros are automatically zeros of $a(x)$ as well, so that $|x_k| \leq R_0$. Because of[4]

$$
\begin{aligned}
b_{m-1} &= x_1 + x_2 + \cdots + x_m \\
b_{m-2} &= x_1 x_2 + x_1 x_3 + x_1 x_4 + \ldots + x_2 x_3 + \ldots = \sum_{j<k} x_j x_k \\
b_{m-3} &= \sum_{i<j<k} x_i x_j x_k \\
&\vdots \\
b_0 &= \prod_{k=1}^{m} x_k
\end{aligned}
$$

we get, using the triangle inequality,

$$
\begin{aligned}
|b_{m-1}| &\leq |x_1| + |x_2| + \cdots + |x_m| \leq m R_0 \\
|b_{m-2}| &\leq \sum_{j<k} |x_j||x_k| \leq \binom{m}{2} R_0^2 \\
|b_{m-3}| &\leq \sum_{i<j<k} |x_i||x_j||x_k| \leq \binom{m}{3} R_0^3 \\
&\vdots \\
|b_0| &\leq \prod_{k=1}^{m} |x_k| \leq R_0^m
\end{aligned}
$$

and induction yields the result. $\qquad\square$

Example 8.19. We consider the rather special example $a(x) = x^n - 1 \in \mathbb{C}[x]$. The complex zeros of this polynomial are the nth roots of unity, $x_k = e^{2\pi i k/n}$ ($k = 0, \ldots, n-1$), i.e.,

$$a(x) = x^n - 1 = \prod_{k=0}^{n-1} (x - x_k),$$

and all roots have modulus 1. Therefore for every $n \in \mathbb{N}$ the optimal zero bound is $R = 1$. For this example, Lemma 8.18 (a) states

$$|x_k| \leq R_n = \frac{1}{\sqrt[n]{2} - 1}.$$

For small n this is still rather accurate, e.g., $R_2 = 1 + \sqrt{2} \approx 2.41421356237$. But when $n \to \infty$, we have $\lim_{n\to\infty} R_n = \infty$.

---

[4] These functions that connect the zeros of a polynomial with its coefficients are called *elementary symmetric polynomials*. In *Mathematica* they can be called by `SymmetricPolynomial`.

Since we know the true zero bound $R = 1$ for this example, we can deduce from Lemma 8.18 (b) that the coefficients of a factor of $a(x)$ of degree n are bounded by the formula given in Lemma 8.18 (b). Unfortunately this bound grows exponentially (compare Section 6.2):

$$\binom{m}{\lfloor m/2 \rfloor} \sim \frac{2^{m-1/2}}{\sqrt{\pi m}} .$$

One can show that for this particular example arbitrary large integer coefficients can indeed occur for $m \to \infty$, see also Exercise 8.8.

Session 8.20. *We would like to use the given algorithm for the factorization of the following integer polynomial $a(x) \in \mathbb{Z}[x]$:*

```
In[1]:= a = x^8 + x^6 + 3x^5 - 9x^4 + 6x^3 - 13x^2 + 11x - 2
```
$Out[1]= x^8 + x^6 + 3x^5 - 9x^4 + 6x^3 - 13x^2 + 11x - 2$

For this purpose, we first implement the symmetric modulo function for polynomials:

```
In[2]:= Clear[SymmetricPolynomialMod];
        SymmetricPolynomialMod[a_, x_, p_] :=
          Module[{n},
            n = Exponent[a, x];
```
$$\sum_{k=0}^{n} \text{Mod}\Big[\text{Coefficient}[a, x, k], p, -\frac{p}{2}\Big] x^k$$
```
          ]
```

Using this function, we get the representation

```
In[3]:= SymmetricPolynomialMod[a, x, 9]
```
$Out[3]= x^8 + x^6 + 3x^5 - 3x^3 - 4x^2 + 2x - 2$

for our polynomial $a(x)$, considered as a member of $\mathbb{Z}_9[x]$. The Zassenhaus bound given in Lemma 8.18 can be computed by

```
In[4]:= Clear[ZassenhausBound];
        ZassenhausBound[a_, x_, m_] :=
          Module[{n, M, R},
            n = Exponent[a, x];
```
$$M = \text{Max}\Big[\text{Table}\Big[\sqrt[k]{\frac{\text{Abs}[\text{Coefficient}[a, x, n-k]]}{\text{Binomial}[n, k]}},$$
```
              {k, n}]];
```
$$R = \frac{M}{\sqrt[n]{2} - 1};$$
```
            Max[Table[Binomial[m, k] R^k, {k, m}]]
          ]
```

The coefficients of all quadratic factors of our polynomial $a(x)$ are therefore bounded by

```
In[5]:= z = ZassenhausBound[a, x, 2]
```
$$Out[5]= \frac{\sqrt[4]{2}}{\left(\sqrt[8]{2} - 1\right)^2}$$

or

```
In[6]:= N[z]
Out[6]= 145.173
```

The polynomial $a(x)$ has degree 8, therefore it is sufficient to search for factors of degree ≤ 4. For such factors we get the coefficient bounds

```
In[7]:= list = Table[N[ZassenhausBound[a, x, m]], {m, 1, 4}]
Out[7]= {12.0488, 145.173, 1749.16, 21075.2}
```

Hence no coefficient of such a factor can be larger than

```
In[8]:= Max[list]
Out[8]= 21075.2
```

To find factors of degree ≤ 4, by Lemma 8.17 it is therefore sufficient to compute modulo p with[5]

```
In[9]:= p = NextPrime[Ceiling[2 Max[list]]]
Out[9]= 42157
```

Factorization of $a(x)$ modulo p yields

```
In[10]:= fac = Factor[PolynomialMod[a, p], Modulus → p]
```
$$Out[10]= (x+4915)(x+16724)(x^3+x+42156)(x^3+20518\,x^2+16212\,x+8859)$$

Next, we convert all factors towards their symmetric equivalents:

```
In[11]:= b = Map[SymmetricPolynomialMod[#, x, p]&, fac]
```
$$Out[11]= (x+4915)(x+16724)(x^3+x-1)(x^3+20518\,x^2+16212\,x+8859)$$

Only these candidates are potential irreducible divisors of $a(x)$. By division it can be checked which of these factors are indeed divisors of $a(s)$:

```
In[12]:= Map[Together[a/#]&, Apply[List, b]]
```
$$Out[12]= \left\{ \frac{x^8+x^6+3\,x^5-9\,x^4+6\,x^3-13\,x^2+11\,x-2}{x+4915}, \right.$$
$$\frac{x^8+x^6+3\,x^5-9\,x^4+6\,x^3-13\,x^2+11\,x-2}{x+16724}, x^5+4\,x^2-9\,x+2,$$
$$\left. \frac{x^8+x^6+3\,x^5-9\,x^4+6\,x^3-13\,x^2+11\,x-2}{x^3+20518\,x^2+16212\,x+8859} \right\}$$

This is the case only for the third factor, therefore the complete factorization of $a(x) \in \mathbb{Z}[x]$ is given by

```
In[13]:= Factor[a]
```
$$Out[13]= (x^3+x-1)(x^5+4\,x^2-9\,x+2)$$

Note that this result is also deduced by Mathematica's `Factor` *command.*

Finally, let us mention that there are many other bounds available for the coefficients of factors, all of which can be incorporated in an implementation.

---

[5] The `Ceiling` function rounds up. We also use the `NextPrime` function considered in Chapter 1.

8.5 Hensel Lifting

In this section, we introduce another algorithm for the factorization in $\mathbb{Z}[x]$ which avoids the use of large primes, so that Berlekamp's algorithm can be applied for small $p \in \mathbb{P}$. This *Hensel lifting* uses the factorization modulo a small prime p and "lifts" it towards a factorization modulo p^2 and iteratively modulo $p^4, p^8, p^{16}, \ldots$. For this purpose, we will show how—starting with a factorization of $a(x)$ modulo q—we get a factorization of $a(x)$ modulo q^2. Applying this algorithm iteratively then yields factorizations of $a(x)$ modulo fast growing modules q. This is again a typical divide-and-conquer approach which goes back to Zassenhaus [Zas1969].

Theorem 8.21. (Hensel Lifting) Let $a(x)$ be a monic polynomial in $\mathbb{Z}[x]$ and $a(x) \equiv b(x)c(x)$ (mod q), where $b(x), c(x)$ are also monic and relatively prime modulo q. Then there are uniquely determined monic polynomials $\tilde{b}(x), \tilde{c}(x)$ in $\mathbb{Z}[x]$ with $\tilde{b}(x) \equiv b(x)$ (mod q) and $\tilde{c}(x) \equiv c(x)$ (mod q) such that $a(x) \equiv \tilde{b}(x)\tilde{c}(x)$ (mod q^2) and $\tilde{b}(x), \tilde{c}(x)$ are relatively prime modulo q^2. This factorization can be found algorithmically.

For the proof of this theorem we need the following lemma.

Lemma 8.22. Let $a(x), b(x) \in \mathbb{Z}[x]$, $b(x) \neq 0$ with $a(x) = q(x)b(x) + r(x)$ and $\deg(r(x), x) < \deg(b(x), x)$. If $a(x) \equiv 0$ (mod m), then it follows that $q(x) \equiv r(x) \equiv 0$ (mod m).

Proof. Let $a(x) \equiv 0$ (mod m). Then we have $a(x) = m\tilde{a}(x)$ for some $\tilde{a}(x) \in \mathbb{Z}[x]$. Polynomial division yields $\tilde{q}(x), \tilde{r}(x) \in \mathbb{Z}[x]$ with $\tilde{a}(x) = \tilde{q}(x)b(x) + \tilde{r}(x)$ and $\deg(\tilde{r}(x), x) < \deg(b(x), x)$. Therefore $a(x) = m\tilde{a}(x) = (m\tilde{q}(x))b(x) + (m\tilde{r}(x))$. The uniqueness of division with remainder yields the result. □

Proof. Proof of theorem: Since $b(x)$ and $c(x)$ are relatively prime modulo q, the extended Euclidean algorithm yields $s(x), t(x) \in \mathbb{Z}[x]$ with $s(x)b(x) + t(x)c(x) \equiv 1$ (mod q), where the Bézout polynomials $s(x)$ and $t(x)$ have the properties $\deg(s(x), x) < \deg(c(x), x)$ and $\deg(t(x), x) < \deg(b(x), x)$, see Exercise 8.9.

Our aim is to show that under these conditions there are monic polynomials $\tilde{b}(x), \tilde{c}(x) \in \mathbb{Z}[x]$ that are uniquely determined modulo q^2, with the properties $\tilde{b}(x) \equiv b(x)$ (mod q), $\tilde{c}(x) \equiv c(x)$ (mod q) and $a(x) \equiv \tilde{b}(x)\tilde{c}(x)$ (mod q^2). We further assume that the polynomials $\tilde{b}(x), \tilde{c}(x)$ are relatively prime modulo q^2, i.e., there are $\tilde{s}(x), \tilde{t}(x) \in \mathbb{Z}[x]$ with $\tilde{s}(x)\tilde{b}(x) + \tilde{t}(x)\tilde{c}(x) \equiv 1$ (mod q^2). Additionally, $\deg(\tilde{s}(x), x) < \deg(\tilde{c}(x), x), \deg(\tilde{t}(x), x) < \deg(\tilde{b}(x), x), \tilde{s}(x) \equiv s(x)$ (mod q) and $\tilde{t}(x) \equiv t(x)$ (mod q). We will compute all these polynomials algorithmically.

Because $a(x) \equiv b(x)c(x)$ (mod q), there is $k(x) \in \mathbb{Z}[x]$ with $a(x) = b(x)c(x) + qk(x)$. Since $a(x), b(x)$ and $c(x)$ are monic, we have $\deg(k(x), x) < \deg(a(x), x)$. In particular, $a(x) \equiv b(x)c(x) + qk(x)$ (mod q^2).

We search for $\tilde{b}(x), \tilde{c}(x)$ of the form $\tilde{b}(x) = b(x) + qB(x)$, $\tilde{c}(x) = c(x) + qC(x)$ with $B(x), C(x) \in \mathbb{Z}[x]$. By the computation

$$
\begin{aligned}
a(x) &\equiv \tilde{b}(x)\tilde{c}(x) \\
&\equiv (b(x) + qB(x))(c(x) + qC(x)) \\
&\equiv b(x)c(x) + q\Big(b(x)C(x) + B(x)c(x)\Big) \\
&\equiv b(x)c(x) + qk(x) \qquad (\text{mod } q^2)
\end{aligned}
$$

we therefore set

$$k(x) \equiv b(x)C(x) + B(x)c(x) \,(\text{mod } q)\,. \tag{8.9}$$

Since $b(x)$ and $c(x)$ are relatively prime modulo q, equation (8.9) has solutions $B(x)$ and $C(x)$. From $s(x)b(x) + t(x)c(x) \equiv 1 \,(\text{mod } q)$ we find that $C(x) = k(x)s(x)$ and $B(x) = k(x)t(x)$ are such solutions.

However, $B(x)$ and $C(x)$ are not yet uniquely determined. Starting from the polynomials just given, we would like to select $B(x)$ and $C(x)$ such that $\deg(B(x), x) < \deg(b(x), x)$ and $\deg(C(x), x) < \deg(c(x), x)$. This assures that the degrees of $b(x)$ and $\tilde{b}(x)$ as well as those of $c(x)$ and $\tilde{c}(x)$ agree. For this purpose, we choose the new $C(x)$ as the remainder of $k(x)s(x)$ when divided by $c(x)$, and the new $B(x)$ as the remainder of $k(x)t(x)$ when divided by $b(x)$. This implies that

$$k(x)s(x) = c(x)p(x) + C(x) \quad \text{and} \quad k(x)t(x) = b(x)q(x) + B(x)$$

with $\deg(C(x), x) < \deg(c(x), x)$ and $\deg(B(x), x) < \deg(b(x), x)$. After substitution we get

$$\begin{aligned}
k(x) &\equiv k(x)\Big(s(x)b(x) + t(x)c(x)\Big) \equiv k(x)s(x)b(x) + k(x)t(x)c(x) \\
&\equiv c(x)p(x)b(x) + C(x)b(x) + b(x)q(x)c(x) + c(x)B(x) \\
&\equiv c(x)b(x)\Big(p(x) + q(x)\Big) + C(x)b(x) + c(x)B(x) \quad (\text{mod } q)
\end{aligned}$$

and therefore

$$c(x)b(x)\Big(p(x) + q(x)\Big) \equiv k(x) - C(x)b(x) - c(x)B(x) \,(\text{mod } q)\,.$$

The term on the left-hand side is a multiple of the monic polynomial $c(x)b(x)$ of degree $\deg(c(x)b(x), x) = \deg(a(x), x)$. Under the assumption that $p(x) + q(x) \not\equiv 0 \,(\text{mod } q)$, the degree of the left-hand side would be $\geq \deg(a(x), x)$. However, since $\deg(k(x), x) < \deg(a(x), x)$, the polynomial on the right-hand side has degree $< \deg(a(x), x)$. Therefore we deduce that $p(x) + q(x) \equiv 0 \,(\text{mod } q)$, and the remainders $B(x)$ and $C(x)$ solve equation (8.9) as announced. This completes the construction of the required polynomials $\tilde{b}(x) := b(x) + qB(x)$ and $\tilde{c}(x) := c(x) + qC(x)$.

The next step is to show that these polynomials $\tilde{b}(x)$ and $\tilde{c}(x)$ are relatively prime modulo q^2. In order to minimize the degrees of the associated polynomials $\tilde{s}(x), \tilde{t}(x)$, we again use the division algorithm. First, we set $d(x) := s(x)\tilde{b}(x) + t(x)\tilde{c}(x) - 1$. Since $b(x)$ and $c(x)$ are relatively prime modulo q, we have $s(x)\tilde{b}(x) + t(x)\tilde{c}(x) \equiv s(x)b(x) + t(x)c(x) \equiv 1 \,(\text{mod } q)$ and therefore $d(x) \equiv 0 \,(\text{mod } q)$. Hence the sought $\tilde{s}(x), \tilde{t}(x)$ are given by

$$\tilde{s}(x) := s(x) - S(x) \quad \text{and} \quad \tilde{t}(x) := t(x) - T(x)\,,$$

where $S(x)$ and $T(x)$ are the remainders in the division of $s(x)d(x)$ by $\tilde{c}(x)$ and of $t(x)d(x)$ by $\tilde{b}(x)$, respectively. Hence we have the equations

$$s(x)d(x) = P(x)\tilde{c}(x) + S(x) \quad \text{and} \quad t(x)d(x) = Q(x)\tilde{b}(x) + T(x)\,. \tag{8.10}$$

The inequality $\deg(S(x), x) < \deg(\tilde{c}(x), x)$ implies that $\deg(\tilde{s}(x), x) < \deg(\tilde{c}(x), x)$, and analogously we get $\deg(\tilde{t}(x), x) < \deg(\tilde{b}(x), x)$. Furthermore, using Lemma 8.22, we get from (8.10) that $S(x) \equiv T(x) \equiv 0 \,(\text{mod } q)$ and therefore $\tilde{s}(x) \equiv s(x) \,(\text{mod } q)$ and $\tilde{t}(x) \equiv t(x) \,(\text{mod } q)$. Hence

$$\begin{aligned}
&\tilde{s}(x)\tilde{b}(x) + \tilde{t}(x)\tilde{c}(x) - 1 \\
&\equiv (s(x) - S(x))\tilde{b}(x) + (t(x) - T(x))\tilde{c}(x) - 1 \\
&\equiv \Big(s(x) - s(x)d(x) + P(x)\tilde{c}(x)\Big)\tilde{b}(x) + \Big(t(x) - t(x)d(x) + Q(x)\tilde{b}(x)\Big)\tilde{c}(x) - 1 \\
&\equiv s(x)\tilde{b}(x) + t(x)\tilde{c}(x) - 1 - d(x)\Big(s(x)\tilde{b}(x) + t(x)\tilde{c}(x)\Big) + \Big(P(x) + Q(x)\Big)\tilde{b}(x)\tilde{c}(x)
\end{aligned}$$

$$\equiv\; d(x) - d(x)(d(x)+1) + \big(P(x) + Q(x)\big)\tilde{b}(x)\tilde{c}(x)$$

$$\equiv\; -d(x)^2 + \big(P(x) + Q(x)\big)\tilde{b}(x)\tilde{c}(x)$$

$$\equiv\; \big(P(x) + Q(x)\big)\tilde{b}(x)\tilde{c}(x) \quad (\mathrm{mod}\; q^2)$$

Using a similar argument as above we obtain $P(x) + Q(x) \equiv 0 \;(\mathrm{mod}\; q^2)$, and therefore $\tilde{b}(x), \tilde{c}(x)$ are relatively prime modulo q^2. We have thus constructed a solution pair $\tilde{b}(x), \tilde{c}(x)$ that has the desired properties.

To show uniqueness, we assume that $\tilde{b}(x) \equiv b(x) \;(\mathrm{mod}\; q)$ and $\hat{b}(x) \equiv b(x) \;(\mathrm{mod}\; q)$, as well as $\tilde{c}(x) \equiv c(x) \;(\mathrm{mod}\; q)$ und $\hat{c}(x) \equiv c(x) \;(\mathrm{mod}\; q)$, where $\tilde{b}(x), \hat{b}(x), \tilde{c}(x), \hat{c}(x)$ are monic and $a(x) \equiv \tilde{b}(x)\tilde{c}(x) \equiv \hat{b}(x)\hat{c}(x) \;(\mathrm{mod}\; q^2)$. Then

$$\hat{b}(x) \;=\; \tilde{b}(x) + q\,m(x) \quad \text{with} \quad \deg(m(x), x) < \deg(\tilde{b}(x), x),$$
$$\hat{c}(x) \;=\; \tilde{c}(x) + q\,n(x) \quad \text{with} \quad \deg(n(x), x) < \deg(\tilde{c}(x), x).$$

We will show that $m(x) \equiv n(x) \equiv 0 \;(\mathrm{mod}\; q)$. For this purpose we assume that $n(x) \not\equiv 0 \;(\mathrm{mod}\; q)$. Since $\tilde{b}(x)\tilde{c}(x) \equiv \hat{b}(x)\hat{c}(x) \;(\mathrm{mod}\; q^2)$, it follows that

$$\begin{aligned}
0 \;&\equiv\; \hat{b}(x)\hat{c}(x) - \tilde{b}(x)\tilde{c}(x) \\
&\equiv\; (\tilde{b}(x) + q\,m(x))(\tilde{c}(x) + q\,n(x)) - \tilde{b}(x)\tilde{c}(x) \\
&\equiv\; q\big(\tilde{b}(x)n(x) + m(x)\tilde{c}(x)\big) \quad (\mathrm{mod}\; q^2)
\end{aligned}$$

and hence

$$\tilde{b}(x)n(x) + m(x)\tilde{c}(x) \equiv 0 \;(\mathrm{mod}\; q). \tag{8.11}$$

We further know that

$$\tilde{b}(x)\tilde{s}(x) + \tilde{t}(x)\tilde{c}(x) \equiv 1 \;(\mathrm{mod}\; q). \tag{8.12}$$

Next, we multiply (8.11) by $\tilde{s}(x)$ and replace $\tilde{s}(x)\tilde{b}(x)$ according to (8.12). This yields

$$\big(1 - \tilde{c}(x)\tilde{t}(x)\big)n(x) + \tilde{s}(x)\tilde{c}(x)m(x) \equiv 0 \;(\mathrm{mod}\; q),$$

and therefore

$$n(x) \equiv \tilde{c}(x)\big(\tilde{t}(x)n(x) - \tilde{s}(x)m(x)\big) \;(\mathrm{mod}\; q).$$

Since $\tilde{c}(x)$ is monic and $n \not\equiv 0$, we therefore get $\tilde{t}(x)n(x) - \tilde{s}(x)m(x) \not\equiv 0 \;(\mathrm{mod}\; q)$, and hence $\deg(n(x), x) \geq \deg(\tilde{c}(x), x)$. But this is a contradiction.

Therefore $n(x) \equiv 0 \;(\mathrm{mod}\; q)$. Analogously one shows that $m(x) \equiv 0 \;(\mathrm{mod}\; q)$. $\square$

Session 8.23. *Now we implement Hensel lifting. For this purpose we load the functions of the last Mathematica session and consider the example polynomial*

$In[1]:=$ **a = x^8 – 46 x^7 – 1062 x^6 + 4028 x^5 – 4944 x^4 + 2104 x^3 –**
              66 x^2 – 35 x + 2

$Out[1]=$ $x^8 - 46\,x^7 - 1062\,x^6 + 4028\,x^5 - 4944\,x^4 + 2104\,x^3 - 66\,x^2 - 35\,x + 2$

We factorize a(x) modulo small primes and get

$In[2]:=$ **list = Table[Factor[a, Modulus → Prime[n]], {n, 10}]**

$Out[2]=$ $\{x\,(x+1)\,(x^3+x+1)\,(x^3+x^2+1), (x+1)\,(x+2)^5\,(x^2+1),$
              $(x+1)\,(x^3+x^2+x+4)\,(x^4+2\,x^3+2\,x^2+3),$
              $(x+3)\,(x+5)\,(x^2+2\,x+5)\,(x^4+x^2+6),$
              $(x^4+3\,x^3+2\,x^2+10)\,(x^4+6\,x^3+7\,x^2+2\,x+9),$
              $(x+4)\,(x+6)\,(x^2+3\,x+6)\,(x^2+7\,x+1)\,(x^2+12\,x+2),$
              $(x+13)\,(x^3+4\,x^2+2\,x+9)\,(x^4+5\,x^3+6\,x^2+16),$
              $(x+12)\,(x^3+x^2+7\,x+11)\,(x^4+17\,x^3+9\,x^2+16\,x+17),$
              $(x^2+10\,x+20)\,(x^2+19\,x+8)\,(x^4+17\,x^3+21\,x^2+12\,x+21),$
              $(x+11)\,(x+14)\,(x+27)\,(x^2+12\,x+28)\,(x^3+6\,x^2+2\,x+13)\}$

$In[3]:=$ **Map[Length, list]**

$Out[3]=$ $\{4, 3, 3, 4, 2, 5, 3, 3, 3, 5\}$

For the considered primes we therefore get between 2 and 5 factors. The smallest prime q with (only) two factors is given by

$In[4]:=$ **q = Prime[5]**

$Out[4]=$ 11

and a(x) has the following factorization modulo q = 11:

$In[5]:=$ **fac = Factor[a, Modulus → q]**

$Out[5]=$ $(x^4+3\,x^3+2\,x^2+10)\,(x^4+6\,x^3+7\,x^2+2\,x+9)$

The factors b(x) and c(x) are given by

$In[6]:=$ **b = SymmetricPolynomialMod[fac[[1]], x, q]**

$Out[6]=$ $x^4+3\,x^3+2\,x^2-1$

and

$In[7]:=$ **c = SymmetricPolynomialMod[fac[[2]], x, q]**

$Out[7]=$ $x^4-5\,x^3-4\,x^2+2\,x-2$

respectively. The extended Euclidean algorithms computes the polynomials s(x) and t(x)

$In[8]:=$ **{g, {s, t}} = PolynomialMod[**
              Expand[PolynomialExtendedGCD[b, c, Modulus → q]], q]

$Out[8]=$ $\{1, \{8\,x^3+8\,x^2+x+6, 3\,x^3+5\,x^2+7\,x+2\}\}$

As a side effect, it has checked that the greatest common divisor of b(x) and c(x) modulo q equals 1 as expected.

The procedure HenselLifting *executes the step* $q \mapsto q^2$:

```
In[9]:= Clear[HenselLifting]
        HenselLifting[a_,b_,c_,s_,t_,x_,q_] :=
          Module[{k,B,C,d,S,T},
            k = Expand[ (a - b c) / q ];
            B = SymmetricPolynomialMod[
                PolynomialRemainder[k t,b,x],x,q];
            C = SymmetricPolynomialMod[
                PolynomialRemainder[k s,c,x],x,q];
            d = s (b + q B) + t (c + q C) - 1;
            S = SymmetricPolynomialMod[
                PolynomialRemainder[s d,c + q C,x],x,q^2];
            T = SymmetricPolynomialMod[
                PolynomialRemainder[t d,b + q B,x],x,q^2];
            {Expand[b + q B], Expand[c + q C], Expand[s - S],
              Expand[t - T], q^2}
          ]
```

This procedure takes as input the polynomials $a(x), b(x), c(x), s(x)$ and $t(x)$ as well as q, and the outputs are $\tilde{b}(x), \tilde{c}(x), \tilde{s}(x)$ and $\tilde{t}(x)$ as well as q^2, hence the data for the next iteration step of the lifting process.

We start with $q = 11$ and get the first two iteration steps

```
In[10]:= {b,c,s,t,q} = HenselLifting[a,b,c,s,t,x,q]
Out[10]= {x^4 + 58 x^3 + 57 x^2 - 1, x^4 + 17 x^3 - 48 x^2 + 35 x - 2,
            -47 x^3 - 3 x^2 + 12 x + 61, 47 x^3 - 6 x^2 - 4 x - 31, 121}
```

```
In[11]:= {b,c,s,t,q} = HenselLifting[a,b,c,s,t,x,q]
Out[11]= {x^4 - 63 x^3 + 57 x^2 - 1, x^4 + 17 x^3 - 48 x^2 + 35 x - 2,
            316 x^3 + 3869 x^2 + 3400 x + 3086,
            -316 x^3 + 6770 x^2 + 4231 x + 5777, 14641}
```

The module $q = 14641$ already yields the valid factorization in $\mathbb{Z}[x]$:

```
In[12]:= Factor[a]
Out[12]= (x^4 - 63 x^3 + 57 x^2 - 1) (x^4 + 17 x^3 - 48 x^2 + 35 x - 2)
```

This can be checked by trial division:

```
In[13]:= {PolynomialRemainder[a,b,x],
          PolynomialRemainder[a,c,x]}
Out[13]= {0,0}
```

Please note that the Zassenhaus bound

```
In[14]:= Ceiling[ZassenhausBound[a,x,4]]
Out[14]= 21438295
```

has not yet been reached. However, the next Hensel iteration step surpassed this bound:

```
In[15]:= {b,c,s,t,q} = HenselLifting[a,b,c,s,t,x,q]
Out[15]= {x^4 - 63 x^3 + 57 x^2 - 1, x^4 + 17 x^3 - 48 x^2 + 35 x - 2,
            -3572088 x^3 - 64723992 x^2 + 39065588 x - 71444994,
            3572088 x^3 - 6684167 x^2 + 16123972 x - 71456944, 214358881}
```

This finalizes the lifting algorithm, even without trial division.

8.6 Multivariate Factorization

In this section we want to briefly discuss how to obtain a multivariate factorization, for example in $\mathbb{Z}[x, y]$, using the univariate algorithms. For this purpose, polynomial interpolation can be used again. We present this method by an example.

Session 8.24. *We would like to factorize the polynomial $a(x, y) \in \mathbb{Z}[x, y]$ given by*

$In[1] := $ **a = Expand[(x - 3y + x^3) (2x + 45y^2 - xy)]**
$Out[1] = $ $45\,x^3\,y^2 + x^4\,(-y) - x^2\,y + 2\,x^4 + 2\,x^2 + 48\,x\,y^2 - 6\,x\,y - 135\,y^3$

We see that $a(x, y)$ has degree 3 with respect to y. Therefore, every proper factorization in $\mathbb{Z}[x, y]$ must have at least one linear factor with respect to y which we want to compute.

We first compute the polynomial $a(x, y_0)$ for two particular values $y_0 \in \mathbb{Z}$. From the univariate factorizations of $a(x, y_0)$ we can then determine potential factors $b(x, y)$ by polynomial interpolation. For $y_0 = 1$ we get $a(x, 1) \in \mathbb{Z}[x]$,

$In[2] := $ **a/.y→1**
$Out[2] = $ $x^4 + 45\,x^3 + x^2 + 42\,x - 135$

and a univariate factorization yields

$In[3] := $ **res1 = Factor[a/.y→1]**
$Out[3] = $ $(x + 45)\,(x^3 + x - 3)$

At random we choose the first factor of the above result,

$In[4] := $ **value[1] = res1[[1]]**
$Out[4] = $ $x + 45$

and try to match it with a factor of $a(x, 2) \in \mathbb{Z}[x]$,

$In[5] := $ **a/.y→2**
$Out[5] = $ $180\,x^3 + 180\,x - 1080$

$In[6] := $ **res2 = Factor[a/.y→2]**
$Out[6] = $ $180\,(x^3 + x - 6)$

For this purpose we choose the factor

$In[7] := $ **value[2] = res2[[2]]**
$Out[7] = $ $x^3 + x - 6$

Therefore, the interpolating polynomial

$In[8] := $ **b = Expand[InterpolatingPolynomial[**
      {{1, value[1]}, {2, value[2]}}, y]]
$Out[8] = $ $x^3\,y - x^3 + x - 51\,y + 96$

is a candidate for our factorization problem. Trial division, however, shows that this does not lead to a factor of $a(x, y)$:

$In[9] := $ **Together[PolynomialRemainder[a, b, y]]**
$Out[9] = $ $\dfrac{1}{(x^3 - 51)^3}$

$\quad (x^{13} + 45\,x^{12} + 2\,x^{11} - 156\,x^{10} - 11069\,x^9 - 255\,x^8 + 6372\,x^7 + $
$\qquad 894186\,x^6 + 16614\,x^5 + 304020\,x^4 - 24882912\,x^3 - $
$\qquad 431136\,x^2 - 17330112\,x + 119439360)$

Therefore we try to combine the second factor of a(x, 1) with the considered factor of a(x, 2):

```
In[10]:= value[1] = res2[[2]]
```
$Out[10]= x^3 + x - 3$

Now we get the interpolating polynomial

```
In[11]:= b = Expand[InterpolatingPolynomial[
              {{1, value[1]}, {2, value[2]}}, y]]
```
$Out[11]= x^3 + x - 3y$

and this time trial division is successful:

```
In[12]:= Together[PolynomialRemainder[a, b, y]]
```
$Out[12]= 0$

We have thus found the required factor b(x, y) of a(x, y). The second factor of a(x, y) is given by the computation

```
In[13]:= PolynomialQuotient[a, b, y]
```
$Out[13]= -xy + 2x + 45y^2$

This result is confirmed by Mathematica's built-in Factor *routine:*

```
In[14]:= Factor[a]
```
$Out[14]= -((x^3 + x - 3y)(xy - 2x - 45y^2))$

Of course, in case of polynomials of higher degree, the number of support points for the interpolation process must be enlarged. Furthermore, the process must be iterated in order to obtain all factors. The number of variables can be recursively increased.

As a second example, we consider the same factorization problem as above, but interchange the role of the two variables.

Since a(x, y) is of degree 4 with respect to x, every proper factorization must have a factor of degree ≤ 2. Therefore this time we interpolate with three support points. First we factorize a(1, y),

```
In[15]:= a/.x → 1
```
$Out[15]= -135y^3 + 93y^2 - 8y + 4$

```
In[16]:= res1 = Factor[a/.x → 1]
```
$Out[16]= -(3y - 2)(45y^2 - y + 2)$

and choose one factor:

```
In[17]:= value[1] = res1[[3]]
```
$Out[17]= 45y^2 - y + 2$

Then we factorize a(2, y),

```
In[18]:= a/.x → 2
```
$Out[18]= -135y^3 + 456y^2 - 32y + 40$

```
In[19]:= res2 = Factor[a/.x → 2]
```
$Out[19]= -(3y - 10)(45y^2 - 2y + 4)$

and again choose one factor:

```
In[20]:= value[2] = res2[[3]]
```
$Out[20]= 45y^2 - 2y + 4$

Finally, we factorize a(3, y):

```
In[21]:= a/.x → 3
```
$Out[21] = -135\,y^3 + 1359\,y^2 - 108\,y + 180$

```
In[22]:= res3 = Factor[a/.x → 3]
```
$Out[22] = -9\,(y-10)\,(15\,y^2 - y + 2)$

The last factor is the one we search, but does not yet have the correct leading coefficient. In order to generate the summand $45y^2$ as before, we therefore choose the factor

```
In[23]:= value[3] = 3 res3[[3]]
```
$Out[23] = 3\,(15\,y^2 - y + 2)$

How the correct leading coefficient can be found, was investigated by Wang [Wan1978], see also [GCL1992], page 377.

Now we can compute the interpolating polynomial:

```
In[24]:= b = Expand[InterpolatingPolynomial[
              {{1, value[1]}, {2, value[2]}, {3, value[3]}}, x]]
```
$Out[24] = -x\,y + 2\,x + 45\,y^2$

Trial division then yields

```
In[25]:= Together[PolynomialRemainder[a, b, x]]
```
$Out[25] = 0$

The computation

```
In[26]:= Together[PolynomialQuotient[a, b, x]]
```
$Out[26] = x^3 + x - 3\,y$

generates the second factor, again.

8.7 Additional Remarks

Up to these days factorization is a rather important research domain since it has many applications and since the complexity can increase immensely with the number of variables. Highly efficient methods are therefore extremely important.

In 1982, Lenstra, Lenstra und Lovász published their famous LLL lattice reduction algorithm [LLL1982], with the help of which factorization of polynomials is possible in polynomial run time. All previous algorithms were exponential in the lifting phase if there are many modular factors.

However, it turned out that the original algorithm given in [LLL1982] was not efficient enough for realistic examples. In [Hoe2002] Mark van Hoeij eliminated this shortage by using the LLL lattice reduction algorithm in a completely different way. His implementation was very efficient for practical purposes, and his algorithm is the current state of art for univariate polynomial factorization. In [BHKS2005] this algorithm was further simplified and it was shown that van Hoeij's algorithm has polynomial time complexity, too.

All modern computer algebra systems are able to execute very complicated factorizations (even without van Hoeij's approach) so that polynomial factorization clearly is one of the highlights of computer algebra.

8.8 Exercises

Exercise 8.1. *Implement a function* `RationalZeros` *which computes all rational zeros of a polynomial* $a(x) \in \mathbb{Q}[x]$.

Exercise 8.2. *Implement a function* `Zeros` *which computes all zeros* $x_k \in \mathbb{Z}_p$ *of a polynomial* $a(x) \in \mathbb{Z}_p[x]$.

Exercise 8.3. *Prove Lemma 8.5 about the primitivity of the product of two primitive polynomials by using the fact that for a primitive polynomial* $a(x) \in \mathbb{Z}[x]$ *and every prime* $p \in \mathbb{P}$ *the relation* $a(x) \not\equiv 0 \pmod{p}$ *is valid.*

Exercise 8.4. (Berlekamp Algorithm) *Check the size of the prime* $p \in \mathbb{P}$ *such that Berlekamp's algorithm still works in "reasonable" time using suitable example polynomials.*

Exercise 8.5. (Berlekamp Algorithm) *Apply Berlekamp's algorithm for a full factorization over* $\mathbb{Z}_p$ *of the following polynomials:*

(a) $a(x) = x^5 + 3x^3 + 2x + 4$ $(p = 7, 13, 37, 97)$;
(b) $b(x) = x^{14} + 5x^{10} + 7x^6 + 6x^2 + 1$ $(p = 7, 13, 17, 37, 97)$;
(c) $c(x) = x^{105} - 1$ $(p = 7, 1009)$.

Exercise 8.6. (Hensel Lifting) *Apply Hensel lifting to* $a(x) = x^4 - 11$ *with respect to the prime* $p = 5$. *The polynomial* $a(x)$ *is irreducible in* $\mathbb{Z}[x]$. *However, it splits into four linear factors modulo* $p = 5$.

To determine via Hensel lifting that $a(x)$ *is irreducible, all combinations of potential divisors of* $a(x)$ *must be checked. In our case it must be checked concretely whether one of the 4 linear factors is a divisor of* $a(x)$. *Moreover, it must be checked whether one of the 6 quadratic polynomials given as products of two linear ones is a divisor of* $a(x)$. *Execute all these computations in our example case.*[6]

Exercise 8.7. *(Factorization in* $\mathbb{Q}[x]$*) Factorize*

$$
\begin{aligned}
a(x) \; = \; & 45x^{13} + 216x^{12} + \frac{1776x^{11}}{5} + \frac{5827x^{10}}{30} - \frac{649x^9}{15} - \frac{3341x^8}{30} - \frac{724x^7}{5} \\
& - \frac{97x^6}{3} + \frac{2117x^5}{15} + \frac{253x^4}{5} - \frac{284x^3}{5} - \frac{83x^2}{10} + \frac{161x}{15} - \frac{49}{30} \in \mathbb{Q}[x]
\end{aligned}
$$

by applying the method presented in Section 8.4. For this purpose, first compute an equivalent polynomial $\in \mathbb{Z}[x]$ *and afterwards an equivalent monic polynomial* $\in \mathbb{Z}[x]$. *Next, apply a squarefree factorization. Finally, apply Berlekamp's algorithm to every squarefree factor for a suitable* $p \in \mathbb{P}$, *followed by Hensel lifting. Compare your results with the ones of* `Factor`.

Exercise 8.8. (Cyclotomic Polynomials) *The cyclotomic polynomials are defined by*

$$
\Phi_n(x) = \prod_{\substack{1 \le k < n \\ \gcd(k, n) = 1}} \left(x - e^{\frac{2\pi i k}{n}} \right) \in \mathbb{C}[x].
$$

---

[6] Obviously such behavior can lead—with increasing degree of the input polynomial and with an unsuitable choice of $p \in \mathbb{P}$—to combinatorial explosion which results in exponential run time. Unfortunately there are polynomials for which every $p \in \mathbb{P}$ is "unsuitable".

(a) Show that $x^n - 1 = \prod_{k \mid n} \Phi_k(x)$.

(b) Show that $\Phi_n(x) = \sum_{k=0}^{n-1} x^k$, if $n \in \mathbb{P}$.

(c) Show that $\Phi_{2n}(x) = \Phi_n(-x)$, if n is odd.

(d) Show that $\Phi_{kn}(x) \cdot \Phi_n(x) = \Phi_n(x^k)$, if $\gcd(k, n) = 1$.

(e) Show that $\Phi_{kn}(x) = \Phi_n(x^k)$, if every prime divisor of k is a divisor of n.

(f) Show that $\Phi_n(x) \in \mathbb{Z}[x]$.

(g) Explain how (b)–(e) can be used as an algorithm to compute $\Phi_n(x)$.

(h) Implement the algorithm given in (g) for the computation of $\Phi_n(x)$ and compare your result with the built-in function `Cyclotomic`.

(i) For small $n \in \mathbb{N}$ the coefficients a_k of $\Phi_n(x) = \sum a_k x^k$ are all bounded by 1: $a_k \in \{-1, 0, 1\}$. For every $a = 2, \dots, 10$ determine the smallest $n \in \mathbb{N}$, for which $\Phi_n(x)$ has a coefficient a_k with $|a_k| = a$.[7] Which conjecture do you get from your results?

Exercise 8.9. Let $a(x), b(x) \in \mathbb{K}[x]$, let $g(x) = \gcd(a(x), b(x))$ and let $s(x), t(x) \in \mathbb{K}[x]$ be the polynomials created by the extended Euclidean algorithm so that $s(x) a(x) + t(x) b(x) = g(x)$ is valid. Show that $\deg(t(x), x) < \deg(a(x), x)$ and $\deg(s(x), x) < \deg(b(x), x)$. Show by an example that the equation $s(x) a(x) + t(x) b(x) = g(x)$ can also be satisfied by polynomials $s(x), t(x) \in \mathbb{K}[x]$ that violate the relation $\deg(t(x), x) < \deg(a(x), x)$.

Exercise 8.10. (Swinnerton-Dyer Polynomials) Let

$$\text{SD}_n(x) := \prod \left(x \pm \sqrt{2} \pm \sqrt{3} \pm \sqrt{5} \cdots \pm \sqrt{p_n} \right),$$

where p_n denotes the nth prime and the product is taken over all 2^n possible combinations of plus and minus signs. These polynomials $\text{SD}_n(x)$ are called the Swinnerton-Dyer polynomials. $\text{SD}_n(x)$ is a polynomial of degree 2^n.

One can further show that

(a) $\text{SD}_n(x) \in \mathbb{Z}[x]$;

(b) $\text{SD}_n(x)$ is irreducible;

(c) however, if $\text{SD}_n(x) \in \mathbb{Z}_p[x]$ is interpreted as an element of $\mathbb{Z}_p[x]$, then for every $p \in \mathbb{P}$ it splits into 2^{n-1} factors.

Implement the computation of the Swinnerton-Dyer polynomials and compute $\text{SD}_n(x)$ for $n = 2, \dots, 4$. Check the above properties (a)–(c).

Exercise 8.11. (Multivariate Factorization) Factorize the polynomial

$$
\begin{aligned}
a(x) \;=\; & -xy^2z^7 - 2x^2yz^6 + y^3z^5 + 2x^2yz^5 + xyz^5 + 2xy^4z^4 + 4x^3z^4 + 2x^2z^4 - 3x^2y^2z^4 \\
& + xy^2z^4 + 4x^2y^3z^3 - 2xy^2z^3 - y^2z^3 - 6x^3yz^3 + 2x^2yz^3 - 2y^5z^2 - 4x^2y^3z^2 \\
& + xy^3z^2 - y^3z^2 + 6x^3yz^2 + x^2yz^2 - xyz^2 + 12x^4z + 2x^3z - 2x^2z - 8x^3y^2z \\
& - 4x^2y^2z + 4xy^4 + 2y^4 - 6x^2y^2 - xy^2 + y^2 \in \mathbb{Z}[x, y, z]
\end{aligned}
$$

with the method introduced in Section 8.6.

For univariate factorization you can use the built-in function `Factor`.

---

[7] By (a) the divisors of the polynomial $x^n - 1$ which has only coefficients in $\{-1, 0, 1\}$ are the cyclotomic polynomials. The hope that the coefficient bounds are inherited by the divisors is therefore destroyed. One can show that the cyclotomic polynomials $\Phi_n(x)$ have arbitrary large coefficients if only n is chosen large enough. Please note that the computation times for this exercise are quite high.

Chapter 9
Simplification and Normal Forms

9.1 Normal Forms and Canonical Forms

If we call a simplification command like `Simplify`, `FullSimplify`, `Expand` or `Together` in *Mathematica*, then algebraic expressions are replaced by (hopefully) mathematically equivalent ones. Therefore the general question arises under which circumstances such transformations are possible and which types of simplifications can be executed.

The questions in which form an expression is simplest, is not always easy to decide. For example, for the polynomial

$$p(x) = (1+x)^{100} \in \mathbb{Z}[x]$$

obviously the factorized representation is preferrable over the expanded form

$$p(x) = 1 + 100x + 4950x^2 + 161700x^3 + 3921225x^4 + 75287520x^5 + \cdots + x^{100},$$

whereas the polynomial

$$q(x) = 1 - x^{10} \in \mathbb{Z}[x]$$

possesses the relatively complicated factorization

$$q(x) = (1-x)(1+x)\left(1-x+x^2-x^3+x^4\right)\left(1+x+x^2+x^3+x^4\right).$$

However, we will not pursue this aspect of "simplicity of expressions". Much rather, we are interested in the aspect of uniqueness of such transformations.

An important problem in connection with the simplification of expressions, i.e. elements of a set M of syntactic expressions (see Definition 9.2), is the question whether and how we can assure that two given expressions are mathematically equivalent, for example[1]

$$\frac{1+\tan x}{1-\tan x} = e^{2\,\text{arctanh}\,\frac{\sin(2x)}{1+\cos(2x)}}. \tag{9.1}$$

---

[1] You may check (9.1) (for $x \in \mathbb{R}$) with *Mathematica* by a graphical representation of both sides and by `FullSimplify[(1+Tan[x])/(1-Tan[x])-E^(2*ArcTanh[Sin[2*x]/(1+Cos[2*x])])]`. Please note that in [Teg2020] a normal form was defined with the help of which this statement can be algorithmically proved.

© Springer Nature Switzerland AG 2021
W. Koepf, *Computer Algebra*, Springer Undergraduate Texts in Mathematics and Technology,
https://doi.org/10.1007/978-3-030-78017-3_9

This mathematical equivalence is generated by an equivalence relation $\sim$, and expressions that are equivalent, e.g., $a \sim b$, define equivalence classes which—similarly as for the residue classes—are elements of the *quotient set $M/\sim$*.

The equivalence expressed by the equal sign, for example in (9.1) or if we write

$$1 - x^{10} = (1-x)(1+x)\left(1-x+x^2-x^3+x^4\right)\left(1+x+x^2+x^3+x^4\right),$$

does not mean that these two expressions are *identical*, but that they are equal in the sense of the equivalence relation $\sim$ of mathematically equivalent expressions.[2] In contrast, if two expressions a and b are syntactically equivalent—and therefore identical —then (for comparison purposes) we write $a \equiv b$ in this chapter.[3]

Session 9.1. *Mathematica knows this difference. Whereas with the usual equal sign* == (Equal) *arbitrary equations (whether true or false) can be formulated*

```
In[1]:= (1+x)^2 == 1+2x+x^2
Out[1]= (x+1)^2 == x^2+2x+1
```

which, however, remain unevaluated (except for numbers), the command === (SameQ) *checks syntactical equality:*

```
In[2]:= (1+x)^2 === 1+2x+x^2
Out[2]= False
```

```
In[3]:= Expand[(1+x)^2] === 1+2x+x^2
Out[3]= True
```

The following code generates an equation between numbers and is therefore evaluated by ==:

```
In[4]:= Expand[(1+x)^2 - (1+2x+x^2)] == 0
Out[4]= True
```

This distinction between == *and* === *must be carefully taken into consideration in the implementation process. If, for example, the symbol* Equal *is used in an* If *instruction, then this statement is not generally recognized to be true, since the equation remains unevaluated.*

The two representations of $p(x)$ and $q(x)$ given at the beginning of this chapter are equivalent. Since both $p(x)$ and $q(x)$ are members of the polynomial ring $\mathbb{Z}[x]$, they can be subtracted from each other. This fact was used in Session 9.1. Subtraction results in an expression which is equivalent to 0. Sometimes it is easier to prove that an expression is equivalent to 0.

This leads to the following concept of normal and canonical forms.

Definition 9.2. (Normal Forms and Canonical Forms) Let $(M, +)$ be an Abelian group of syntactical expressions with an addition + and a zero element $0 \in M$, and let $\sim$ denote an equivalence relation on M which stands for mathematical equivalence and which is *compatible* with addition, i.e.

$(N_0)\ a \sim b \quad \Rightarrow \quad a + c \sim b + c \quad$ for all $a, b, c \in M$.

---

[2] Hence, technically we should write

$$1 - x^{10} \sim (1-x)(1+x)\left(1-x+x^2-x^3+x^4\right)\left(1+x+x^2+x^3+x^4\right).$$

[3] Note that contrary to the notation in residue class rings, the sign $\equiv$ now stands for equality.

A *normal function* is a function, given by an algorithm[4] $f : M \to M$ with the properties

(N_1) $f(a) \sim a$ for all $a \in M$;
(N_2) $a \sim 0$ $\Rightarrow$ $f(a) \equiv f(0)$ for all $a \in M$.

If instead of (N_2) even the property

(N_3) $a \sim b$ $\Rightarrow$ $f(a) \equiv f(b)$ for all $a, b \in M$

is valid, then f is called a *canonical function*.

If f is a normal function, then the expression $\tilde{a} \in M$ is called a *normal form* if $f(\tilde{a}) \equiv \tilde{a}$ is valid. If $\tilde{a}$ is a normal form and $a \sim \tilde{a}$, then $\tilde{a}$ is called a normal form of a.

Analogously an expression $\tilde{a} \in M$ with $f(\tilde{a}) \equiv \tilde{a}$ is called a *canonical form* or the canonical form of a, respectively, if f is a canonical function and $f(a) \equiv \tilde{a}$. $\triangle$

Example 9.3. Let $M = \mathbb{Z}[x]$ be the polynomial ring with integer coefficients. Consider the simplification function f_1, given by the following algorithm:

(i) Expand all products using the distributive law.
(ii) Collect all terms of the same degree.

Then this algorithm obviously yields a normal function f_1 since the algorithm can safely decide whether a polynomial is equivalent to the zero polynomial. (This is the case if and only if all resulting coefficients equal 0.) Here, f_1 is no canonical form, since the order of the powers of x is not defined.[5] However, if we issue the step

(iii) Sort the powers in x in decreasing degree.

after (i) and (ii), then we get a canonical simplification function f_2. This function transforms every $a(x) \in \mathbb{Z}[x]$ into its canonical form, namely as an expanded polynomial sorted by its powers of x. Note that in Section 6.1 the resulting representation was called the standard representation of a polynomial. $\triangle$

Session 9.4. *Mathematica's function* Expand *is a (multivariate) version of* f_2 *which sorts powers in rising order. We get for example*

$In[1]:=$ **Expand[$(1+x)^5$]**
$Out[1]=$ $x^5 + 5x^4 + 10x^3 + 10x^2 + 5x + 1$

It seems that Mathematica has sorted for decreasing degree. However, this is a property of the notebook frontend[6], as the internal representation shows:

$In[2]:=$ **InputForm[Expand[$(1+x)^5$]]**
$Out[2]=$ `1 + 5*x + 10*x^2 + 10*x^3 + 5*x^4 + x^5`

Next we show that, having a normal form at hand, we can solve the *identification problem* of the equivalence of two expressions by checking whether the difference of the expression is equivalent to 0. The latter is called the *zero equivalence problem*.

---

[4] One should imagine a *simplification function*. Generally one can assume that the function is *computable*. However, we will not discuss computability in detail.

[5] In *Maple* polynomials generally are *not* sorted (for efficiency reasons). Powers are shown in the order of their appearance in the computer memory, and therefore may be sorted differently in every session. Therefore this output is the normal form considered.

[6] We use the frontend in the TraditionalForm mode.

Theorem 9.5. Let f be a normal function on M. Then we have for all $a, b \in M$:

$$a \sim b \qquad \Leftrightarrow \qquad f(a-b) \equiv f(0).$$

Proof. "⇒": From (N_0) and $a \sim b$ we get the equivalence $a - b \sim 0$, hence with (N_2) it follows that $f(a-b) \equiv f(0)$.
"⇐": Since $\sim$ is an equivalence relation, we have $x \sim x$. Therefore, $x \equiv y$ implies $x \sim y$. From $f(a-b) \equiv f(0)$ it follows in particular that $f(a-b) \sim f(0)$. Using (N_1) we get the relation

$$a-b \sim f(a-b) \sim f(0) \sim 0 \,,$$

and because of the transitivity of $\sim$ also $a - b \sim 0$. The rule (N_0) then implies that $a \sim b$. □

Canonical forms go even further: On top of the properties of a normal form they create for every equivalence class in $M/\sim$ a unique representative, as the following theorem shows.

Theorem 9.6. Let f be a canonical function on M. Then the following properties are valid:

(a) **(Idempotence)** $f \circ f = f$;
(b) **(Characterization)** $f(a) \equiv f(b)$ if and only if $a \sim b$;
(c) **(Existence and Uniqueness)** Every equivalence class in $M/\sim$ contains exactly one canonical form.

Proof. (a) Property (N_1) implies that $f(a) \sim a$ for all $a \in M$. Therefore, using property (N_3), we get $f(f(a)) \equiv f(a)$.
(b) One direction follows directly from the definition, namely property (N_3). Therefore, if $f(a) \equiv f(b)$, then in particular $f(a) \sim f(b)$, and property (N_1) takes us to

$$a \sim f(a) \sim f(b) \sim b \,.$$

Transitivity of the equivalence relation implies $a \sim b$.
(c) Existence: Let a be an element of one of the equivalence classes. We define $\tilde{a} := f(a)$. Then

$$f(\tilde{a}) \equiv f(f(a)) \equiv f(a) \equiv \tilde{a} \tag{9.2}$$

by using the idempotence. Hence $\tilde{a}$ is a canonical form of a.
Uniqueness: Let $\tilde{a}$ and $\tilde{b}$ be two canonical forms of a in the same equivalence class. Then $\tilde{a} \sim \tilde{b}$, and hence $f(\tilde{a}) \equiv f(\tilde{b})$ due to property (N_3). This implies by definition $\tilde{a} \equiv f(\tilde{a}) \equiv f(\tilde{b}) \equiv \tilde{b}$. □

Property (a) of Theorem 9.6 is a highly desirable property of every simplification function. It tells that renewed simplification does not change a simplified result any more. Nevertheless, not every simplification function in every computer algebra system has this desired property. Generally, simplification routines in *Mathematica* should have this property since it is the philosophy behind *Mathematica* to apply every transformation rule until changes do no longer occur. This, however, implies idempotence.

9.2 Normal Forms and Canonical Forms for Polynomials

As we have already seen in Example 9.3, *Mathematica*'s command `Expand` is a canonical form in $\mathbb{Z}[x]$. Analogously, this is valid if we replace $\mathbb{Z}$ by another ring R. We must, however, demand that R itself has a canonical form. In particular, this creates a canonical form for the recursive representation of polynomials in several variables. For example,

$$x^6 + (-y^2 - 2)x^5 + (-y^2 - 4)x^4 + 3x^3 + (y^4 + 2y^2)x^2 + (y^4 + 4y^2)x - y^4 - 3y^2$$

is the canonical form of

$$p(x, y) := (x + x^2 - 1)(y^2 - x + 3)(y^2 - x^3)$$

considered as an element of $\mathbb{Z}[y][x]$.

Session 9.7. *The recursive canonical form of a multivariate polynomial is created by* `Collect`. *For example, let* $p(x, y)$ *be given as above:*

$In[1] :=$ `p = (x+x^2 - 1) * (y^2 - x + 3) * (y^2 - x^3);`

Then the command

$In[2] :=$ `Collect[p,x]`
$Out[2] =$ $x^5 (-y^2 - 2) + x^4 (-y^2 - 4) + x^2 (y^4 + 2 y^2) + x^6 + 3 x^3 + x (y^4 + 4 y^2) - y^4 - 3 y^2$

yields a variant of the representation as shown above.

Next, we would like to compute the canonical form with respect to y, hence we consider $p(x, y) \in \mathbb{Z}[x][y]$:

$In[3] :=$ `Collect[p,y]`
$Out[3] =$ $(x^2 + x - 1) y^4 + (x^2 + x - 1)(-x^3 - x + 3) y^2 + (x^2 + x - 1)(x^4 - 3 x^3)$

In this case, `Collect` *has collected all coefficients, but has not simplified them further. This is due to the fact that our input was in factorized form. For efficiency reasons* `Collect` *does not expand those factors and therefore yields only a normal form.*

However, if we expand first, then `Collect` *yields the canonical form:*

$In[4] :=$ `s = Collect[Expand[p],y]`
$Out[4] =$ $(x^2 + x - 1) y^4 + (-x^5 - x^4 + 2 x^2 + 4 x - 3) y^2 + x^6 - 2 x^5 - 4 x^4 + 3 x^3$

The fact that the powers with respect to y are sorted internally in increasing order, can be seen by

$In[5] :=$ `InputForm[s]`
```
Out[5]= 3*x^3 - 4*x^4 - 2*x^5 + x^6 +
        (-3 + 4*x + 2*x^2 - x^4 - x^5)*y^2 +
        (-1 + x + x^2)*y^4
```

Analogously we proceed in case of several variables by a recursive application of `Collect`.

It is easy to see that the distributive representation of multivariate polynomials represents a canonical form. For this purpose, we need a suitable order to sort the monomials. The distributive form of a polynomial $p \in R[x_1, x_2, \ldots, x_n]$ in the variables $X := (x_1, x_2, \ldots, x_n)$ has the form

$$p = \sum_{\alpha} c_\alpha X^\alpha \tag{9.3}$$

where $\alpha = (\alpha_1, \ldots, \alpha_n) \in \mathbb{N}_{\geq 0}^n$ is a *multi-index*, the coefficients c_α lie in R and the notation $X^\alpha = x_1^{\alpha_1} \cdots x_n^{\alpha_n}$ is a shortcut for the *multivariate monomials*.

We create a *distributive canonical form* if the monomials X^α are sorted in a suitable manner.[7] In the distributive representation, addition of polynomials is especially simple. If we want to make the multiplication efficient as well, then we choose an order > which is compatible with multiplication, i.e.,

$$X^s > X^t \quad \Rightarrow \quad X^{s+u} > X^{t+u} \qquad \text{for all} \quad s, t, u \in \mathbb{N}_{\geq 0}^n.$$

A *well-order*[8] with this property is called *monomial order* or *term order*.

Session 9.8. *Mathematica sorts all multivariate polynomials with respect to the* lexicographic *order, which is a monomial order. The lexicographic order first sorts the variables in alphabetical manner (by their ASCII code). For two monomials we have $X^s > X^t$, if for some $m \in \mathbb{N}$*

$$s_k = t_k \text{ for all } k = 1, \ldots, m-1 \qquad \text{and} \quad s_m > t_m.$$

The distributive canonical form with respect to the lexicographic order is generated by Expand:

```
In[1]:= p = (x + x^2 - 1) * (y^2 - x + 3) * (y^2 - x^3);
```

```
In[2]:= InputForm[Expand[p]]
Out[2]= 3*x^3 - 4*x^4 - 2*x^5 + x^6 - 3*y^2 + 4*x*y^2 +
        2*x^2*y^2 - x^4*y^2 - x^5*y^2 - y^4 + x*y^4 + x^2*y^4
```

The terms are given in increasing order. Hence $y > x$, and $x^2 y^4$ is the overall largest monomial.

Note that the notebook interface in $\boxed{\textbf{TraditionalForm}}$ *mode resorts the polynomial*

```
In[3]:= Expand[p]
```
$$Out[3]= -x^5 y^2 - x^4 y^2 + x^2 y^4 + 2 x^2 y^2 + x^6 - 2 x^5 - 4 x^4 + 3 x^3 + x y^4 + 4 x y^2 - y^4 - 3 y^2$$

so that the output looks more "traditional".

Finally one can generate a canonical form in $R[x_1, \ldots, x_n]$ by factorization, if R does not have zero divisors, see for example [GCL1992], Section 3.4. However, this form is more difficult to compute and therefore in general not very efficient.[9]

9.3 Normal Forms for Rational Functions

Rational functions are quotients of polynomials. These are unique only after canceling common divisors, and even then numerators and denominators are only unique up to a common factor.

If we cancel common factors (using the Euclidean algorithm), make the denominator monic[10] and expand both the numerator and the denominator, we clearly get a canonical form.

---

[7] Without sorting we already have a normal form.

[8] An order > on a set M is called well-order, if every non-empty subset of M possesses a smallest element with respect to >.

[9] In contrast to *Mathematica*, the Computer Algebra System *Reduce* always uses one of these canonical forms, and allows switching between those forms using the commands `on exp;` or `on factor;`, respectively.

[10] One should clarify what "monic" means in the multivariate case.

Session 9.9. *For efficiency reasons the above program is typically not executed completely, and we are satisfied with a normal form. For this purpose, Mathematica's function* `Together` *essentially carries out the Euclidean algorithm step. Potential factorizations that have been carried out should not get lost and the computations should be done in the simplest possible way.*

Let us choose again

$In[1]:=$ **p = (x + x^2 − 1) * (y^2 − x + 3) * (y^2 − x^3);**

and compute

$In[2]:=$ **t = Together** $\left[\dfrac{p}{x^3 - yx^2 + x^2 - yx - x + y} \right]$

$Out[2]=$ $\dfrac{(x - y^2 - 3)(x^3 - y^2)}{x - y}$

We can see that common factors were canceled, but the numerator and the denominator were not brought into canonical form for efficiency reasons. Nevertheless, it is clear that `Together` *easily enables the detection of the rational function zero:*

$In[3]:=$ **Together** $\left[\dfrac{p}{x^3 - yx^2 + x^2 - yx - x + y} - t \right]$

$Out[3]=$ 0

Hence `Together` *yields a normal form for rational functions.*

Since we have now successfully presented normal forms for important classes of functions, the question remains whether such a normal form exists for *all* mathematical expressions. Unfortunately this is not the case for sufficiently comprehensive classes, as Richardson's Theorem [Ric1968] shows.

Theorem 9.10. (Undecidability) There are expressions containing the symbols +, *, Exp, Sin and Abs as well as the constants π and log 2 for which the decision whether they are identically zero is undecidable. ☐

However, in the next section and in Section 10.3.1 we will show that normal forms are available for suitable subsets of transcendental functions.

9.4 Normal Forms for Trigonometric Polynomials

Let R be an integral domain with 1. In this section we consider polynomials $p(x, y) \in R[x, y]$ in the variables $x = \cos t$ and $y = \sin t$. In this case, as a consequence of the *Pythagorean identity* $\cos^2 t + \sin^2 t = 1$, the variables x and y are not independent of each other. This actually means that we must consider the quotient ring $R[x, y]/\langle x^2 + y^2 - 1 \rangle$.[11]

Theorem 9.11. Let R be an integral domain with 1 and with a canonical form. Let $p(\cos t, \sin t) \in R[x, y]/\langle x^2 + y^2 - 1 \rangle$ be a trigonometric polynomial. Then there are two canonical forms which are linear with respect to either $x = \cos t$ or $y = \sin t$:

$$p(x, y) = p_1(y) + x p_2(y) \quad (p_1(y), p_2(y) \in R[y])$$

---

[11] This quotient ring consists of equivalence classes modulo $x^2 + y^2 - 1$.

and

$$p(x, y) = p_3(x) + y\, p_4(x) \quad (p_3(x), p_4(x) \in R[x]),$$

respectively.

Proof. By the substitution $\cos^2 t \mapsto 1 - \sin^2 t$ all powers of $\cos t$ with exponents > 1 are recursively reduced by smaller powers of $\cos t$ or powers of $\sin t$ until the power of $\cos t$ has reached 0 or 1. An analogous statement is valid for the replacement $\sin^2 t \mapsto 1 - \cos^2 t$. These reductions correspond in $R[x, y]/\langle x^2 + y^2 - 1\rangle$ to the computation of the remainder in the polynomial division by $x^2 + y^2 - 1$ with respect to x and with respect to y, respectively, and are therefore uniquely determined. $\qquad\qquad\qquad\qquad\qquad\qquad\qquad\qquad\qquad\qquad\qquad\qquad\square$

Session 9.12. *We give an example using Mathematica. The expression*

```
In[1]:= p = (Cos[t] + Sin[t])^10
```
$Out[1] = (\sin(t) + \cos(t))^{10}$

is fully expanded as element of $\mathbb{Z}[\cos t, \sin t]$ *by* Expand:

```
In[2]:= q = Expand[p]
```
$Out[2] = \sin^{10}(t) + \cos^{10}(t) + 45 \sin^2(t) \cos^8(t) + 120 \sin^3(t) \cos^7(t) +$
$\qquad\quad 210 \sin^4(t) \cos^6(t) + 252 \sin^5(t) \cos^5(t) + 210 \sin^6(t) \cos^4(t) + 120 \sin^7(t) \cos^3(t) +$
$\qquad\quad 45 \sin^8(t) \cos^2(t) + 10 \sin(t) \cos^9(t) + 10 \sin^9(t) \cos(t)$

In this computation the Pythagorean identity is not used at all, so that the canonical forms of Theorem 9.11 cannot be generated. The factorization of the expanded expression is also just carried out in $\mathbb{Z}[x, y]$.

```
In[3]:= Factor[q]
```
$Out[3] = (\sin(t) + \cos(t))^{10}$

The fact that Mathematica considers our expression as a polynomial in the variables $x = \cos t$ *and* $y = \sin t$ *can also be seen by the computation*

```
In[4]:= Variables[p]
```
$Out[4] = \{\cos(t), \sin(t)\}$

Now, we reduce p modulo $\cos^2 t + \sin^2 t - 1$ *by the recursive reduction* $\cos^2 t \mapsto 1 - \sin^2 t$ *and obtain the canonical representation of Theorem 9.11:*

```
In[5]:= Collect[(Expand[p] //. {Cos[t]^k -/; k > 1 →
                    Expand[Cos[t]^(k-2) * (1 - Sin[t]^2)]}) //Expand,
         Cos[t]]
```
$Out[5] = 80 \sin^8(t) - 160 \sin^6(t) + 40 \sin^4(t) + 40 \sin^2(t) + (32 \sin^9(t) - 64 \sin^7(t) -$
$\qquad\quad 48 \sin^5(t) + 80 \sin^3(t) + 10 \sin(t)) \cos(t) + 1$

Note that this can be also computed in a simpler way using polynomial division by $x^2 + y^2 - 1$:

```
In[6]:= PolynomialRemainder[(x+y)^10, x^2 + y^2 - 1, x]
```
$Out[6] = x(32 y^9 - 64 y^7 - 48 y^5 + 80 y^3 + 10 y) + 80 y^8 - 160 y^6 + 40 y^4 + 40 y^2 + 1$

Mathematica contains several built-in functions for trigonometric simplification, namely TrigExpand, TrigFactor, TrigFactorList, TrigReduce *and* TrigToExp. *For example, the function* TrigReduce *yields the representation*

```
In[7]:= TrigReduce[p]
```
$Out[7] = \dfrac{1}{16}\, (210 \sin(2t) - 45 \sin(6t) + \sin(10t) - 120 \cos(4t) + 10 \cos(8t) + 126)$

for our trigonometric polynomial p. We will soon see that this result also represents a canonical form. The other mentioned Mathematica functions, however, do not constitute normal forms.

The *trigonometric addition theorems*

$$
\begin{aligned}
\cos(t+u) &= \cos t \cos u - \sin t \sin u \\
\cos(-t) &= \cos t \\
\sin(t+u) &= \sin t \cos u + \cos t \sin u \\
\sin(-t) &= -\sin t
\end{aligned}
\tag{9.4}
$$

allow the reduction of all trigonometric functions with argument sums. In particular, (9.4) implies the recursive reduction of multiple angles:

$$
\begin{aligned}
\cos(kt) &= \cos((k-1)t)\cos t - \sin((k-1)t)\sin t \quad &(k \in \mathbb{N}) \\
\sin(kt) &= \sin((k-1)t)\cos t + \cos((k-1)t)\sin t \quad &(k \in \mathbb{N}),
\end{aligned}
$$

Therefore, all functions of the form

$$
a_0 + \sum_{k=1}^{N} \left(a_k \cos(kt) + b_k \sin(kt) \right) \qquad (a_k, b_k \in R, \ N \in \mathbb{N})
\tag{9.5}
$$

with finitely many summands can be represented as trigonometric polynomials.

If we read the replacement rules (9.4) from left to right, however, then every trigonometric polynomial can be written in the form (9.5). This is accomplished by the substitution rules

$$
\begin{aligned}
\cos t \cos u &= \frac{1}{2}\cos(t-u) + \frac{1}{2}\cos(t+u) \\
\sin t \cos u &= \frac{1}{2}\sin(t-u) + \frac{1}{2}\sin(t+u) \\
\sin t \sin u &= \frac{1}{2}\cos(t-u) - \frac{1}{2}\cos(t+u)
\end{aligned}
\tag{9.6}
$$

which can be derived by addition and subtraction of the rules in (9.4). With the help of these substitution rules, products of trigonometric functions are transformed into sums. In particular, we obtain the following recursive formulas for powers:

$$
\begin{aligned}
\cos^k t &= \left(\frac{1}{2} + \frac{1}{2}\cos(2t) \right) \cos^{k-2} t \quad &(k \geq 2) \\
\sin^k t &= \left(\frac{1}{2} - \frac{1}{2}\cos(2t) \right) \sin^{k-2} t \quad &(k \geq 2)
\end{aligned}
\tag{9.7}
$$

Please note that for this process we need to divide by 2, therefore we assume that $R = \mathbb{K}$ is a field of characteristic $\neq 2$ for these transmissions.

To show the uniqueness of the representation in (9.5), we assume further that $\mathbb{K} \subset \mathbb{R}$. This takes us to the following theorem.

Theorem 9.13. Let $\mathbb{K} \subset \mathbb{R}$ be a field with a canonical form. Then every polynomial $p(\cos t, \sin t) \in \mathbb{K}[x,y]/\langle x^2 + y^2 - 1 \rangle$ has a canonical form of the type (9.5).

Proof. The reductions (9.7) obviously transform every trigonometric polynomial into a term of the type (9.5). We have to show that the coefficients $a_k, b_k \in \mathbb{K}$ are unique.

These coefficients can be generated in the same way as in Fourier theory. Since for $k \geq 1$ we have

$$\int_0^{2\pi} \cos(kt)\,dt = \int_0^{2\pi} \sin(kt)\,dt = 0\,,$$

it follows for

$$p(t) = a_0 + \sum_{k=1}^{N} \left(a_k \cos(kt) + b_k \sin(kt) \right)$$

and $n \geq 1$ that

$$
\begin{aligned}
\int_0^{2\pi} p(t)\sin(nt)\,dt &= \sum_{k=1}^{N} a_k \int_0^{2\pi} \cos(kt)\sin(nt)\,dt + \sum_{k=1}^{N} b_k \int_0^{2\pi} \sin(kt)\sin(nt)\,dt \\
&= \sum_{k=1}^{N} a_k \int_0^{2\pi} \left(\frac{1}{2}\sin((n-k)t) + \frac{1}{2}\sin((n+k)t) \right) dt \\
&\quad + \sum_{k=1}^{N} b_k \int_0^{2\pi} \left(\frac{1}{2}\cos((n-k)t) - \frac{1}{2}\cos((n+k)t) \right) dt \\
&= \pi \cdot b_n
\end{aligned}
$$

by using (9.6), since only the integrand $\cos((n-k)t)$ for $k = n$ generates an integral value which differs from zero.

Similarly we get for $n \geq 1$

$$\int_0^{2\pi} p(t)\cos(nt)\,dt = \pi \cdot a_n$$

as well as

$$\int_0^{2\pi} p(t)\,dt = 2\pi a_0\,.$$

Hence we see that the coefficients a_k and b_k of $p(t)$ are uniquely determined.   □

Session 9.14. *We consider again the trigonometric polynomial*

```
In[1]:= p = (Cos[t] + Sin[t])^10
Out[1]= (sin(t) + cos(t))^10
```

and its canonical form using `TrigReduce`*:*

```
In[2]:= q = TrigReduce[p]
Out[2]= 1/16 (210 sin(2t) - 45 sin(6t) + sin(10t) - 120 cos(4t) + 10 cos(8t) + 126)
```

```
In[3]:= Expand[q]
Out[3]= 105/8 sin(2t) - 45/16 sin(6t) + 1/16 sin(10t) - 15/2 cos(4t) + 5/8 cos(8t) + 63/8
```

The arising coefficients can be computed by the integrals considered in the proof of Theorem 9.13:

```
In[4]:= 1/(2π) ∫_0^{2π} p dt
Out[4]= 63/8
```

generates a_0,

$$In[5]:= \text{Table}\Big[\frac{1}{\pi}\int_0^{2\pi} \text{p Cos[k t] dt}, \{k,1,10\}\Big]$$

$$Out[5]= \Big\{0,0,0,-\frac{15}{2},0,0,0,\frac{5}{8},0,0\Big\}$$

generates the values a_k *(*$k = 1, \ldots, 10$*) and finally*

$$In[6]:= \text{Table}\Big[\frac{1}{\pi}\int_0^{2\pi} \text{p Sin[k t] dt}, \{k,1,10\}\Big]$$

$$Out[6]= \Big\{0,\frac{105}{8},0,0,0,-\frac{45}{16},0,0,0,\frac{1}{16}\Big\}$$

generates the values b_k *(*$k = 1, \ldots, 10$*). Of course, it is much more efficient to compute those coefficients by the reduction via (9.6).*

We have seen that there are canonical forms for trigonometric polynomials which are also valid for terms with multiple angles. On the other hand, the ring $\mathbb{K}[x,y]/\langle x^2+y^2-1\rangle$ is no *unique factorization domain* whose elements possess a unique factorization. For example, for $x = \cos t$ and $y = \sin t$ we have

$$(1+x)y+1-x^2 = y(1+x+y) = (1+x)(1-x+y),$$

therefore it is difficult to define greatest common divisors. This, however, is an essential ingredient for the simplification of quotients of trigonometric polynomials. In [MM2001] this problem is discussed in detail.

9.5 Additional Remarks

Simplification is an important mathematical concept, and therefore canonical and normal forms are very important in computer algebra. Unfortunately, computer algebra systems in practice do not consider this concept carefully enough. In particular the user information about the functionality of the built-in simplification functions leaves a lot to be desired.

While in *Mathematica* the simplification functions Expand and TrigReduce result in uniquely identifiable results (hence canonical forms) for particular types of input, and Together represents at least a normal form, such statements are not valid for most other simplification commands. In particular, the popular commands Simplify and FullSimplify lead to results which may be "simple" (which is not always the case), but whose forms are not predictable in advance.

As mentioned in Section 9.4 trigonometric factorization of trigonometric polynomials does not constitute a canonical form. Therefore it would be desirable to know some details about the functionality of the command TrigFactor.[12] The documentation, however, does not provide such information.

Similar remarks are valid for the simplification commands in other computer algebra systems like *Maple*, *Maxima* and *Reduce*. Note, however, that *Reduce* contains all normal functions mentioned in this section [KBM1995].

---

[12] In [KBM1995] and [MM2001] potential factorization algorithms are treated.

9.6 Exercises

Exercise 9.1. *Represent the function*

$$\frac{1 + \tan x}{1 - \tan x} - e^{2 \operatorname{arctanh} \frac{\sin(2x)}{1+\cos(2x)}} \, ,$$

which equals zero by (9.1), graphically in a suitable interval of $\mathbb{R}$. *Why does the resulting graph not seem to correspond to the zero function?*

Exercise 9.2. *The function* $\frac{1+\tan x}{1-\tan x} - e^{2 \operatorname{arctanh} \frac{\sin 2x}{1+\cos 2x}}$ *is simplified by* `FullSimplify` *to 0.*

Use the identities

$$
\begin{aligned}
\tanh x &= \frac{e^x - e^{-x}}{e^x + e^{-x}} \\
\sin(2x) &= 2 \sin(x) \cos(x) \\
\cos(2x) &= \cos^2(x) - \sin^2(x) \\
\tan(x) &= \frac{\sin(x)}{\cos(x)} \\
\sin(x)^2 + \cos(x)^2 &= 1
\end{aligned}
$$

(a) *to simplify* $\frac{\sin(2x)}{1+\cos(2x)}$ *towards* $\tan x$;
(b) *to represent the inverse function* $\operatorname{arctanh}(x)$ *by logarithms and exponential functions only;*
(c) *to comprehend Mathematica's simplification.*

Finally, discuss the equivalent expressions $\frac{1+\tan x}{1-\tan x}$ *and* $e^{\operatorname{arctanh} \frac{\sin 2x}{1+\cos 2x}}$ *with respect to their potential domains of definition in* $\mathbb{R}$.

Exercise 9.3. (Partial Fraction Decomposition) *Another normal form for rational functions is given by partial fraction decomposition which is computed by Mathematica's command* `Apart`.

Let $\mathbb{K}$ *be a field and* $r(x) = \frac{p(x)}{q(x)} \in \mathbb{K}(x)$ *a rational function with coprime numerators* $p(x) \in \mathbb{K}[x]$ *and denominators* $q(x) \in \mathbb{K}[x]$. *Without loss of generality let* $q(x)$ *be monic. Assume that*

$$q(x) = \prod_{i=1}^{m} q_i(x) \tag{9.8}$$

is a partial factorization of the denominator with $\gcd(q_i(x), q_j(x)) = 1$ *for* $i \neq j$. *Then there is a representation of the form*

$$r(x) = P(x) + \sum_{i=1}^{m} \frac{p_i(x)}{q_i(x)}$$

with polynomials $P(x) \in \mathbb{K}[x]$ *and* $p_i(x) \in \mathbb{K}[x]$ *with* $\deg(p_i(x), x) < \deg(q_i(x), x)$. *This representation is called partial fraction decomposition with respect to the factorization (9.8).*

If in particular $q(x) = \prod_{i=1}^{m} q_i(x)^{e_i}$ *is the complete factorization of the denominator over* $\mathbb{K}$, *then the partial fraction decomposition has the form*

$$r(x) = P(x) + \sum_{i=1}^{m} \sum_{k=1}^{e_i} \frac{A_{ik}}{q_i(x)^k} \, , \tag{9.9}$$

where $P(x) \in \mathbb{K}[x]$ is the polynomial part and $A_{ik} \in \mathbb{K}$ are constants.

For $\mathbb{K} = \mathbb{C}$ we get in particular

$$r(x) = P(x) + \sum_{i=1}^{m} \sum_{k=1}^{e_i} \frac{A_{ik}}{(x - a_i)^k},$$

where $a_i \in \mathbb{C}$ are the zeros of $q(x)$ and e_i their multiplicities.

Show these statements and prove that the partial fraction decomposition (9.9) yields a canonical form.

Exercise 9.4. (Morley's Theorem) *Morley's Theorem of plane geometry states: If all angles of a triangle are divided into three equal parts, then the three adjacent rays meet each other in three points which always form an equilateral triangle.*

Formulate this theorem in trigonometric terms and prove it with Mathematica.

Exercise 9.5. (Trigonometric Simplification)

(a) *Implement the generation of the trigonometric canonical forms presented in Theorem 9.11.*

(b) *Simplify* $\dfrac{\sin(x+a) + \sin(x-a)}{\cos(x+a) + \cos(x-a)}$.

(c) *A geometrical theorem which is related to Morley's theorem, and which was detected experimentally by Douglas Hofstader, leads to the determinant condition*

$$\begin{vmatrix} \dfrac{\sin(r\alpha)}{\sin((1-r)\alpha)} & \dfrac{\sin(2\alpha)}{\sin(-\alpha)} & \dfrac{\sin((2-r)\alpha)}{\sin((r-1)\alpha)} \\[2ex] \dfrac{\sin(r\beta)}{\sin((1-r)\beta)} & \dfrac{\sin(2\beta)}{\sin(-\beta)} & \dfrac{\sin((2-r)\beta)}{\sin((r-1)\beta)} \\[2ex] \dfrac{\sin(r\gamma)}{\sin((1-r)\gamma)} & \dfrac{\sin(2\gamma)}{\sin(-\gamma)} & \dfrac{\sin((2-r)\gamma)}{\sin((r-1)\gamma)} \end{vmatrix} = 0$$

under the condition that α, β and γ denote the angles in a plane triangle and $r > 0$, see e.g. [Eng1995]. Simplify the Hofstadter determinant. Is the condition $r > 0$ necessary?

(d) *Simplify the Hofstadter determinant if α, β and γ are assumed to be independent of each other.*

Exercise 9.6. (Trigonometric Simplification)

(a) *Implement the simplification rules (9.4) and (9.6), respectively, to obtain similar results as* `TrigExpand` *or* `TrigReduce`. *Check your implementation with* $\sin^{10} t$ *and* $\sin(10t)$ *and more trigonometric example polynomials.*

(b) *Describe a divide-and-conquer algorithm for the expansion of $\sin(nt)$, $\cos(nt)$ and $\sin^n t$, $\cos^n t$ for large $n \in \mathbb{N}$.*

Exercise 9.7. (Trigonometric Polynomial Ring)

(a) *Implement the computation of the normal forms of Theorem 9.11 for $\mathbb{Q}[x, y]/\langle x^2 + y^2 - 1 \rangle$. Compute both normal forms of x^k and $x^k - y^k$ for $k = 1, \ldots, 10$.*

(b) *Show that if y is a divisor of $p(x) \in \mathbb{Q}[x, y]/\langle x^2 + y^2 - 1 \rangle$, then the same is true for $x + 1$ and $x - 1$.*

(c) *Show that—using the definition of a greatest common divisor given in Definition 3.10 on page 51—in general no greatest common divisor exists in* $\mathbb{Q}[x, y]/\langle x^2 + y^2 - 1 \rangle$.

Exercise 9.8. (Simplification of Transcendental Functions)

(a) *Simplify*[13]

$$\frac{1 + \tan^2(\frac{x}{2})}{1 - \tan^2(\frac{x}{2})}.$$

The result should be very simple!

(b) *Show with the help of a suitable simplification mechanism that the identity [GCL1992, (3.7)]*

$$\ln\left(\tan\frac{x}{2} + \sec\frac{x}{2}\right) = \operatorname{arcsinh}\left(\frac{\sin x}{1 + \cos x}\right)$$

is valid for $x \in \mathbb{R}$.

Hint: Take the maximal domains of definition of the expressions into consideration and note that Simplify *has an option to enter its data! If necessary, use suitable replacement rules, but please justify them.*

---

[13] Because of the simplification $\frac{\sin x}{\cos x} \mapsto \tan x$, which is automatically performed by *Mathematica*, the replacement Tan[x] → Sin[x]/Cos[x] cannot be accomplished in *Mathematica*. This is clearly a design bug in *Mathematica*. However, TrigFactor works!

Chapter 10

Power Series

10.1 Formal Power Series

If R is an integral domain, then the same is true for the set $R[x]$ of polynomials. Addition and multiplication in $R[x]$ were defined in Section 6.1. Analogously one can define the integral domain of formal power series

$$a(x) = a_0 + a_1 x + \cdots = \sum_{k=0}^{\infty} a_k x^k$$

in the variable x over R where $(a_k)_{k \in \mathbb{N}_{\geq 0}}$ is an arbitrary sequence of elements of R. The *order* $\mathrm{ord}(a(x), x)$ of a formal power series $a(x)$ is the smallest integer k for which $a_k \neq 0$. The power series whose coefficients a_k are all zero is called the *zero power series*. The zero power series has order ∞. The power series with $a_k = 0$ $(k > 0)$ are called *constant power series*, and the coefficient a_0 is called the *constant term* of the series.

Addition of two formal power series $a(x)$ and

$$b(x) = b_0 + b_1 x + \cdots = \sum_{k=0}^{\infty} b_k x^k$$

is defined term-wise as for polynomials, and for the product

$$a(x) \cdot b(x) = \sum_{k=0}^{\infty} c_k x^k$$

we use again the Cauchy product formula:

$$c_k = \sum_{j=0}^{k} a_j b_{k-j} . \tag{10.1}$$

In the present situation this formula is valid for all $k \in \mathbb{N}_{\geq 0}$. In the above way, we have defined the *ring of (formal) power series* $R[[x]]$, which forms an integral domain. The zero power series is the neutral element of addition in $R[[x]]$, and the constant 1 is the neutral element of multiplication in the ring $R[[x]]$.

As an example of a multiplication, we consider the power series of the exponential function e^x,

© Springer Nature Switzerland AG 2021
W. Koepf, *Computer Algebra*, Springer Undergraduate Texts in Mathematics and Technology,
https://doi.org/10.1007/978-3-030-78017-3_10

$$e(x) := \sum_{k=0}^{\infty} \frac{1}{k!} x^k ,$$

which we want to multiply with itself. On the side of the *generating functions* this multiplication is easy: $e^x \cdot e^x = e^{2x}$, hence we should get the series

$$e^x \cdot e^x = e^{2x} = e(2x) = \sum_{k=0}^{\infty} \frac{1}{k!} (2x)^k = \sum_{k=0}^{\infty} \frac{2^k}{k!} x^k . \qquad (10.2)$$

On the side of the formal power series, however, the computation is more involved. To multiply $e(x)$ with itself, we use formula (10.1) and get

$$c_k = \sum_{j=0}^{k} \frac{1}{j!(k-j)!} = \frac{1}{k!} \sum_{j=0}^{k} \binom{k}{j} ,$$

and we finally have to compute this sum. Such and related questions will be considered in a more general context (and from an algorithmic point of view) in Chapter 11. But from the binomial formula, we already know the result for our special case:

$$c_k = \frac{1}{k!} \sum_{j=0}^{k} \binom{k}{j} = \frac{2^k}{k!} .$$

This, however, yields (10.2) again.

We would like to point out that the formal power series

$$\sum_{k=0}^{\infty} k! x^k ,$$

whose radius of convergence is 0, is an element of $\mathbb{C}[[x]]$ and $\mathbb{Z}[[x]]$. This formal power series does not converge (in $\mathbb{C}$) anywhere besides at the origin and is therefore not a Taylor series of real analysis or of an analytic function in function theory.[1]

Convergence does not play a role in the theory of formal power series. As long as we consider those series as elements of a power series ring, we are therefore not allowed to substitute any values $x \in R$ for the indeterminate x. For this purpose, convergence is needed.

Obviously $R[x]$ is that subring of $R[[x]]$ whose elements $(a_k)_{k \in \mathbb{N}_{\geq 0}}$ have only finitely many components different from zero.[2]

While the element $a(x) = 1 - x \in \mathbb{Z}[x]$ does not have a multiplicative inverse in $\mathbb{Z}[x]$, $a(x) = 1 - x \in \mathbb{Z}[[x]]$, considered as an element of the power series ring $\mathbb{Z}[[x]]$, possesses such an inverse: since

$$(1-x) \cdot (1 + x + x^2 + \cdots) = (1-x) + (x - x^2) + (x^2 - x^3) + \cdots = 1 ,$$

the relation

$$(1-x)^{-1} = \sum_{k=0}^{\infty} x^k$$

---

[1] However, such divergent series play a role as *asymptotic series*.

[2] We can represent the elements of $R[[x]]$ by $\sum_{k=0}^{\infty} a_k x^k$, or similarly by the defining sequence $(a_k)_{k \in \mathbb{N}_{\geq 0}}$.

is valid in $\mathbb{Z}[[x]]$.

We know that in $\mathbb{Z}$ there are only two units 1 and -1. This fact was inherited by $\mathbb{Z}[x]$. Now we just found that $\mathbb{Z}[[x]]$ possesses further units. The following theorem throws light on this finding.

Theorem 10.1. Let R be an integral domain. The units of the power series ring $R[[x]]$ are exactly those power series whose constant term a_0 is a unit in the coefficient ring R.

If R is a field, then every element $a(x) \in R[[x]]$ with $\mathrm{ord}(a(x), x) = 0$, i.e. with $a_0 \neq 0$, is a unit.

Proof. Let

$$a(x) = a_0 + a_1 x + \cdots - \sum_{k=0}^{\infty} a_k x^k$$

be a unit in $R[[x]]$. Then there exists a multiplicative inverse series

$$b(x) = b_0 + b_1 x + \cdots = \sum_{k=0}^{\infty} b_k x^k .$$

The identity $a(x) \cdot b(x) = 1$ implies by definition of multiplication the equations

$$
\begin{aligned}
1 &= a_0 b_0 \\
0 &= a_0 b_1 + a_1 b_0 \\
&\vdots \\
0 &= a_0 b_k + a_1 b_{k-1} + \cdots + a_{k-1} b_1 + a_k b_0 \\
&\vdots
\end{aligned}
\tag{10.3}
$$

The first equation shows that a_0 must be a unit in R with $a_0^{-1} = b_0$.

Let now conversely a_0 be a unit in R. Then the equations (10.3) can be iteratively solved for the coefficients b_k ($k \in \mathbb{N}_{\geq 0}$) according to the scheme

$$
\begin{aligned}
b_0 &= \frac{1}{a_0} \\
b_1 &= -\frac{1}{a_0}(a_1 b_0) \\
&\vdots \\
b_k &= -\frac{1}{a_0}(a_1 b_{k-1} + \cdots + a_{k-1} b_1 + a_k b_0) \\
&\vdots
\end{aligned}
$$

In this way we construct the coefficients of a series $b(x) \in R[[x]]$ with the property $a(x) \cdot b(x) = 1$. Hence $a(x)$ is a unit in $R[[x]]$.

Since every element in a field, besides 0, possesses a multiplicative inverse, the second proposition follows. □

Now that addition, subtraction, multiplication and division of power series are theoretically settled, we can ask ourselves how the *composition* of power series can be defined and accomplished. The power series $(a \circ b)(x)$ is defined by substitution of the power series $b(x)$ in the power series $a(x)$:

$$(a \circ b)(x) = a(b(x)) := \sum_{k=0}^{\infty} a_k \left(\sum_{j=0}^{\infty} b_j x^j \right)^k.$$

Expanding the above formula transforms it towards a power series again—we do not give a general formula here— if the second series has at least order 1. This will be generally assumed in the sequel.

Finally we will prove that formal power series of order 1 can be inverted in general.

Theorem 10.2. Let R be an integral domain and let $a(x) \in R[[x]]$ be given with $\mathrm{ord}(a(x), x) = 1$. Then there is a uniquely determined *inverse* $a^{-1}(x) \in R[[x]]$ if a_1 is a unit.[3]

Proof. We give a constructive proof by searching for $i(x) = i_0 + i_1 x + i_2 x^2 + \cdots \in R[[x]]$ for which the relation $i(a(x)) = x$ is valid. In order to find the coefficients of $i(x)$, we substitute $a(x)$ in $i(x)$ and get

$$i(a(x)) = i_0 + a_1 i_1 x + (a_2 i_1 + a_1^2 i_2) x^2 + (a_3 i_1 + 2 a_1 a_2 i_2 + a_1^3 i_3) x^3 + \cdots,$$

which follows from the computations

$In[7] :=$ **series** $= \displaystyle\sum_{k=1}^{3} \mathbf{a_k x^k}$

$Out[7] = a_3 x^3 + a_2 x^2 + a_1 x$

$In[8] :=$ **inverse** $= \displaystyle\sum_{k=0}^{3} \mathbf{i_k x^k}$

$Out[8] = i_3 x^3 + i_2 x^2 + i_1 x + i_0$

$In[9] :=$ **Collect[Expand[inverse/.x $\to$ series]/.{x^n--/;n > 3 $\to$ 0},x]**

$Out[9] = x^3 (i_3 a_1^3 + 2 a_2 i_2 a_1 + a_3 i_1) + x^2 (i_2 a_1^2 + a_2 i_1) + a_1 i_1 x + i_0$

This expression must be equal to x. Equating the coefficients implies first $i_0 = 0$ as well as $i_1 = a_1^{-1}$ and then yields successive equations which can be solved uniquely for the coefficients $i_2, i_3, \ldots$ if a_1 is a unit. This process computes a uniquely determined sequence of coefficients $(i_k)_{k \in \mathbb{N}_{\geq 0}}$ and therefore $i(x)$, which, as a result of the equation $i(a(x)) = x$, is the inverse of $a(x)$. $\square$

Session 10.3. *We can work with formal power series in Mathematica. However, they do not exist as (potentially) infinite sums (although such a mechanism exists and is implemented, for example, in the CAS Axiom), but as truncated power series. To declare such a series, one can use the Big-O-notation which refers to the order of truncation.*

By[4]

$In[1] :=$ $\mathbf{s_1} = \displaystyle\sum_{k=0}^{10} \mathbf{x^k} + \mathbf{O[x]^{11}}$

$Out[1] = 1 + x + x^2 + x^3 + x^4 + x^5 + x^6 + x^7 + x^8 + x^9 + x^{10} + O(x^{11})$

---

[3] $a^{-1}(x)$ therefore corresponds to the inverse function, i.e. the inverse with respect to composition, which is not to be confused with the inverse with respect to multiplication, hence the reciprocal $a(x)^{-1}$.

[4] Note that we must enter a power of $O(x)$ although in *Mathematica's* output (in $\boxed{\textbf{TraditionalForm}}$) the exponent appears in the argument of the Big-O function.

$In[2]:= $ **s$_2$** $= \sum\limits_{k=0}^{10}$ **k! x^k + O[x]^{11}**

$Out[2]= 1+x+2\,x^2+6\,x^3+24\,x^4+120\,x^5+720\,x^6+$
$\qquad\qquad 5040\,x^7+40320\,x^8+362880\,x^9+3628800\,x^{10}+O\!\left(x^{11}\right)$

we have declared the two series

$$s_1 := \sum_{k=0}^{\infty} x^k \qquad and \qquad s_2 := \sum_{k=0}^{\infty} k!\,x^k$$

up to order 10, considered as elements of $\mathbb{Q}[[x]]$. Now we can perform computations with those truncated series:

$In[3]:= $ **s$_1$ * s$_2$**

$Out[3]= 1+2\,x+4\,x^2+10\,x^3+34\,x^4+154\,x^5+874\,x^6+$
$\qquad\qquad 5914\,x^7+46234\,x^8+409114\,x^9+4037914\,x^{10}+O\!\left(x^{11}\right)$

$In[4]:= \dfrac{\mathbf{s_1}}{\mathbf{s_2}}$

$Out[4]= 1-x^2-4\,x^3-17\,x^4-88\,x^5-549\,x^6-$
$\qquad\qquad 3996\,x^7-33089\,x^8-306432\,x^9-3135757\,x^{10}+O\!\left(x^{11}\right)$

$In[5]:= $ **s$_1$^{2}**

$Out[5]= 1+2\,x+3\,x^2+4\,x^3+5\,x^4+6\,x^5+7\,x^6+8\,x^7+9\,x^8+10\,x^9+11\,x^{10}+O\!\left(x^{11}\right)$

The complete arithmetic of (truncated) power series is accessible and can be used. For example, we can compute the derivative of a power series:

$In[6]:= $ **D[s$_1$, x]**

$Out[6]= 1+2\,x+3\,x^2+4\,x^3+5\,x^4+6\,x^5+7\,x^6+8\,x^7+9\,x^8+10\,x^9+O\!\left(x^{10}\right)$

Note, however, that this—as well as some other operations—necessarily decreases the order of truncation.

As another example for arithmetic operations, division of s_1 by x^3 yields

$In[7]:= \dfrac{\mathbf{s_1}}{\mathbf{x^3}}$

$Out[7]= \dfrac{1}{x^3}+\dfrac{1}{x^2}+\dfrac{1}{x}+1+x+x^2+x^3+x^4+x^5+x^6+x^7+O\!\left(x^8\right)$

Note that in this example the scope had to be enlarged. The result is a formal Laurent series

$$a(x) = \sum_{k=k_0}^{\infty} a_k x^k \qquad (k_0 \in \mathbb{Z})\,,$$

for which a finite number of powers with negative exponents is allowed. If $R = \mathbb{K}$ is a field, then the formal Laurent series build again a field $\mathbb{K}((x))$, namely the quotient field of $\mathbb{K}[[x]]$.

If we divide s_1 by $(1-x^2)$, then the term $\frac{1}{1-x^2}$ is converted into a power series and the corresponding product is computed:

$In[8]:= \dfrac{\mathbf{s_1}}{\mathbf{1-x^2}}$

$Out[8]= 1+x+2\,x^2+2\,x^3+3\,x^4+3\,x^5+4\,x^6+4\,x^7+5\,x^8+5\,x^9+6\,x^{10}+O(x^{11})$

We can also compute the square root of a truncated series:[5]

---

[5] Reason how this is accomplished!

$In[9]:=$ $\sqrt{s_1}$

$Out[9]=$ $1+\dfrac{x}{2}+\dfrac{3\,x^2}{8}+\dfrac{5\,x^3}{16}+\dfrac{35\,x^4}{128}+\dfrac{63\,x^5}{256}+\dfrac{231\,x^6}{1024}+$

$\quad\dfrac{429\,x^7}{2048}+\dfrac{6435\,x^8}{32768}+\dfrac{12155\,x^9}{65536}+\dfrac{46189\,x^{10}}{262144}+O(x^{11})$

If we build, however, the square root of s_1-1, *then this computation leaves the power series ring* $\mathbb{Q}[[x]]$ *again:*

$In[10]:=$ $\sqrt{s_1-1}$

$Out[10]=$ $\sqrt{x}+\dfrac{x^{3/2}}{2}+\dfrac{3\,x^{5/2}}{8}+\dfrac{5\,x^{7/2}}{16}+\dfrac{35\,x^{9/2}}{128}+\dfrac{63\,x^{11/2}}{256}+$

$\quad\dfrac{231\,x^{13/2}}{1024}+\dfrac{429\,x^{15/2}}{2048}+\dfrac{6435\,x^{17/2}}{32768}+\dfrac{12155\,x^{19/2}}{65536}+O(x^{21/2})$

A series of the form

$$a(x)=\sum_{k=k_0}^{\infty} a_k x^{\frac{k}{q}} \qquad (k_0\in\mathbb{Z},q\in\mathbb{N})$$

with rational exponents is called formal Puiseux series.[6] Mathematica can execute computations with Puiseux series.

The above result indicates that the series of $\sqrt{\frac{s_1-1}{x}}$ *is a power series, again:*

$In[11]:=$ $\sqrt{\dfrac{s_1-1}{x}}$

$Out[11]=$ $1+\dfrac{x}{2}+\dfrac{3\,x^2}{8}+\dfrac{5\,x^3}{16}+\dfrac{35\,x^4}{128}+\dfrac{63\,x^5}{256}+\dfrac{231\,x^6}{1024}+\dfrac{429\,x^7}{2048}+\dfrac{6435\,x^8}{32768}+\dfrac{12155\,x^9}{65536}+O(x^{10})$

Next, we compute the composition $(s_2\circ(s_1-1))(x)$. *To understand the composition better, we first substitute only, without simplification. For this purpose, we convert both series (using* Normal*) into polynomials.[7]*

$In[12]:=$ **comp = Normal[s$_2$]/.x → Normal[s$_1$ - 1]**

$Out[12]=$ $x^{10}+x^9+x^8+x^7+x^6+x^5+x^4+x^3+x^2+$

$\quad 3628800\,(x^{10}+x^9+x^8+x^7+x^6+x^5+x^4+x^3+x^2+x)^{10}+$

$\quad 362880\,(x^{10}+x^9+x^8+x^7+x^6+x^5+x^4+x^3+x^2+x)^{9}+$

$\quad 40320\,(x^{10}+x^9+x^8+x^7+x^6+x^5+x^4+x^3+x^2+x)^{8}+$

$\quad 5040\,(x^{10}+x^9+x^8+x^7+x^6+x^5+x^4+x^3+x^2+x)^{7}+$

$\quad 720\,(x^{10}+x^9+x^8+x^7+x^6+x^5+x^4+x^3+x^2+x)^{6}+$

$\quad 120\,(x^{10}+x^9+x^8+x^7+x^6+x^5+x^4+x^3+x^2+x)^{5}+$

$\quad 24\,(x^{10}+x^9+x^8+x^7+x^6+x^5+x^4+x^3+x^2+x)^{4}+$

$\quad 6\,(x^{10}+x^9+x^8+x^7+x^6+x^5+x^4+x^3+x^2+x)^{3}+$

$\quad 2\,(x^{10}+x^9+x^8+x^7+x^6+x^5+x^4+x^3+x^2+x)^{2}+x+1$

After back-conversion we again get a series modulo $O(x^{11})$

$In[13]:=$ **comp + O[x]^{11}**

$Out[13]=$ $1+x+3\,x^2+11\,x^3+49\,x^4+261\,x^5+1631\,x^6+$

$\quad 11743\,x^7+95901\,x^8+876809\,x^9+8877691\,x^{10}+O(x^{11})$

---

[6] Note that a Puiseux series does *not* allow arbitrary rational exponents, but the exponents must have a common denominator q. This makes the Puiseux series a ring again.

[7] The data structures of polynomials and truncated series are completely different in *Mathematica*.

The above computation can be also reached by direct substitution of the series,

$In[14]:=$ **$s_2/.x \to s_1 - 1$**
$Out[14]=$ $1 + x + 3x^2 + 11x^3 + 49x^4 + 261x^5 + 1631x^6 +$
$11743x^7 + 95901x^8 + 876809x^9 + 8877691x^{10} + O(x^{11})$

or by the command ComposeSeries

$In[15]:=$ **ComposeSeries[$s_2, s_1 - 1$]**
$Out[15]=$ $1 + x + 3x^2 + 11x^3 + 49x^4 + 261x^5 + 1631x^6 +$
$11743x^7 + 95901x^8 + 876809x^9 + 8877691x^{10} + O(x^{11})$

The command InverseSeries *computes inverse series:*

$In[16]:=$ **inv = InverseSeries[$s_2 - 1$]**
$Out[16]=$ $x - 2x^2 + 2x^3 - 4x^4 - 4x^5 - 48x^6 - 336x^7 - 2928x^8 - 28144x^9 - 298528x^{10} + O(x^{11})$

We check the result:

$In[17]:=$ **inv/.x $\to s_2 - 1$**
$Out[17]=$ $x + O(x^{11})$

As a last piece of information for this session, note that Mathematica internally represents formal series by SeriesData *objects:*

$In[18]:=$ **FullForm[s_2]**
$Out[18]=$ SeriesData[x, 0, List[1, 1, 2, 6, 24, 120, 720, 5040, 40320, 362880, 3628800], 0, 11, 1]

Detailed information about this data structure can be found in the help pages.

10.2 Taylor Polynomials

In the sequel we consider functions which we want to expand as power series (or, if necessary, as Laurent or Puiseux series). The expansion of a C^∞-function as a power series is principally possible using Taylor's theorem

$$f(x) = \sum_{k=0}^{\infty} \frac{f^{(k)}(x_0)}{k!} (x - x_0)^k .$$

For $x_0 = 0$, this leads to a formal power series as considered in the last section. Now, however, we start with a function which we want to represent by a Taylor series (or at least approximate by a Taylor polynomial). Therefore, the resulting series generally have a positive radius of convergence. However, they do not necessarily lead to good approximations.

Session 10.4. *The approximating nth Taylor polynomial can be implemented, e.g., by:*

```
In[1]:= Clear[Taylor]
       Taylor[f_, {x_, x0_, n_}] :=
         Module[{derivatives, k},
           derivatives = NestList[D[#, x]&, f, n];
           derivatives = derivatives /. {x → x0};
```

$$\sum_{k=0}^{n} \frac{derivatives[[k+1]]}{k!} (x-x0)^k +$$

```
             O[x-x0]^{n+1}
         ]
```

We get for example

```
In[2]:= Taylor[ArcSin[x], {x, 0, 10}]
```
$$Out[2]= x + \frac{x^3}{6} + \frac{3 x^5}{40} + \frac{5 x^7}{112} + \frac{35 x^9}{1152} + O(x^{11})$$

Note that this implementation does not yet allow the computation of

```
In[3]:= Taylor[ ArcSin[x]/x , {x, 0, 10}]
Out[3]= Indeterminate
```

since already the value $\left.\frac{\arcsin x}{x}\right|_{x=0}$ *at the origin is not well defined. However, the value at $x = 0$ can be computed by the limit* $\lim_{x\to 0} \frac{\arcsin x}{x} = 1$ *since this is a removable singularity.*[8]

Therefore the adapted version

```
In[4]:= Clear[Taylor]
       Taylor[f_, {x_, x0_, n_}] :=
         Module[{derivatives, k},
           derivatives = NestList[D[#, x]&, f, n];
           derivatives = Map[Limit[#, x → x0]&, derivatives];
```

$$\sum_{k=0}^{n} \frac{derivatives[[k+1]]}{k!} (x-x0)^k + O[x-x0]^{n+1}$$

```
         ]
```

yields a fully functional version (for analytic expressions) for the computation of the nth Taylor polynomial. Now we indeed get

```
In[5]:= Taylor[ ArcSin[x]/x , {x, 0, 10}]
```
$$Out[5]= 1 + \frac{x^2}{6} + \frac{3 x^4}{40} + \frac{5 x^6}{112} + \frac{35 x^8}{1152} + \frac{63 x^{10}}{2816} + O(x^{11})$$

The following example shows that the computed Taylor series does not have to represent the input function, see e.g. [Koe1993a]:

```
In[6]:= Taylor[ Exp[ -1/x^2 ], {x, 0, 100}]
```
$$Out[6]= O(x^{101})$$

Independent of the chosen order, the function e^{-1/x^2} is represented by the Taylor polynomial 0. The fact that this function approaches the x-axis more intensely than every power, can also be seen in the following graph:

```
In[7]:= Plot[ Exp[ -1/x^2 ], {x, -1, 1}]
```

---

[8] This completion makes the current function continuous and analytic at the origin.

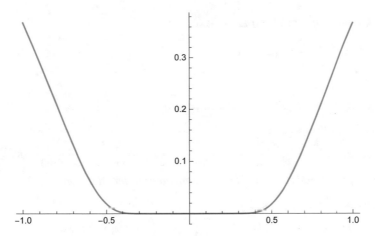

The above implementation of `Taylor` is not very efficient.[9] Much better is the following recursive approach. For the computation of the nth Taylor polynomial of a decomposed expression, one computes the nth Taylor approximations of all subexpressions and combines these appropriately. Most current computer algebra systems compute Taylor approximations like this. However, we do not know the details of the internal *Mathematica* implementation.

Session 10.5. *The* `Series` *command in Mathematica computes power, Laurent and Puiseux approximations and thus generates a* `SeriesData` *object.*

We compute again the series of $\frac{\arcsin x}{x}$ and subsequently work with it:

$In[1]:= \mathbf{s_1} = \mathbf{Series}\left[\mathbf{f} = \dfrac{\mathbf{ArcSin[x]}}{\mathbf{x}}, \{\mathbf{x}, \mathbf{0}, \mathbf{10}\}\right]$

$Out[1]= 1 + \dfrac{x^2}{6} + \dfrac{3x^4}{40} + \dfrac{5x^6}{112} + \dfrac{35x^8}{1152} + \dfrac{63x^{10}}{2816} + O(x^{11})$

$In[2]:= \dfrac{\mathbf{s_1 - 1}}{\mathbf{x^2}}$

$Out[2]= \dfrac{1}{6} + \dfrac{3x^2}{40} + \dfrac{5x^4}{112} + \dfrac{35x^6}{1152} + \dfrac{63x^8}{2816} + O(x^9)$

$In[3]:= \mathbf{Series}\left[\dfrac{\mathbf{f - 1}}{\mathbf{x^2}}, \{\mathbf{x}, \mathbf{0}, \mathbf{10}\}\right]$

$Out[3]= \dfrac{1}{6} + \dfrac{3x^2}{40} + \dfrac{5x^4}{112} + \dfrac{35x^6}{1152} + \dfrac{63x^8}{2816} + \dfrac{231x^{10}}{13312} + O(x^{11})$

The last two results show the loss of information due to the division process.[10]

Now we would like to show the identity $\sin^2 x + \cos^2 x = 1$ on the level of truncated power series:

$In[4]:= \mathbf{s_2} = \mathbf{Sin[x]} + \mathbf{O[x]}^{15}$

$Out[4]= x - \dfrac{x^3}{6} + \dfrac{x^5}{120} - \dfrac{x^7}{5040} + \dfrac{x^9}{362880} - \dfrac{x^{11}}{39916800} + \dfrac{x^{13}}{6227020800} + O(x^{15})$

$In[5]:= \mathbf{s_3} = \mathbf{Cos[x]} + \mathbf{O[x]}^{15}$

$Out[5]= 1 - \dfrac{x^2}{2} + \dfrac{x^4}{24} - \dfrac{x^6}{720} + \dfrac{x^8}{40320} - \dfrac{x^{10}}{3628800} + \dfrac{x^{12}}{479001600} - \dfrac{x^{14}}{87178291200} + O(x^{15})$

$In[6]:= \mathbf{s_2}^2 + \mathbf{s_3}^2$

$Out[6]= 1 + O(x^{15})$

---

[9] The computation of limits in *Mathematica* is not efficient, and in fact it is not even algorithmic.

[10] However, `Series` always computes—other than the `series` command in *Maple*—the requested order.

The following computation yields a Laurent series:

$In[7]:=$ **Series$\left[\dfrac{\text{Sin}[x^5]}{x^{10}}, \{x, 0, 15\}\right]$**

$Out[7]=\ \dfrac{1}{x^5} - \dfrac{x^5}{6} + \dfrac{x^{15}}{120} + O(x^{16})$

and the next computation takes us to a Puiseux series:

$In[8]:=$ **Series$\left[e^{\sqrt{x}}, \{x, 0, 5\}\right]$**

$Out[8]=\ 1 + \sqrt{x} + \dfrac{x}{2} + \dfrac{x^{3/2}}{6} + \dfrac{x^2}{24} + \dfrac{x^{5/2}}{120} + \dfrac{x^3}{720} + \dfrac{x^{7/2}}{5040} + \dfrac{x^4}{40320} + \dfrac{x^{9/2}}{362880} + \dfrac{x^5}{3628800} + O(x^{11/2})$

Finally Taylor approximations can also be computed at different centers:

$In[9]:=$ **Series$[\text{Sin}[x], \{x, \pi, 10\}]$**

$Out[9]=\ -(x-\pi) + \dfrac{1}{6}(x-\pi)^3 - \dfrac{1}{120}(x-\pi)^5 + \dfrac{(x-\pi)^7}{5040} - \dfrac{(x-\pi)^9}{362880} + O((x-\pi)^{11})$

Can you see how this series illustrates the relation $\sin x = \sin(\pi - x)$*?*

10.3 Computation of Formal Power Series

In the last section we investigated how truncated Taylor series can be computed. In this section we would like to learn under which conditions the truncation can be avoided and even infinite series can be generated. In other words, we would like to find a formula for the kth coefficient a_k for

$$f(x) = \sum_{k=0}^{\infty} a_k x^k .$$

Examples for such series are known from calculus, e.g.

$$e^x = \sum_{k=0}^{\infty} \frac{1}{k!} x^k , \qquad \sin x = \sum_{k=0}^{\infty} \frac{(-1)^k}{(2k+1)!} x^{2k+1} , \qquad \cos x = \sum_{k=0}^{\infty} \frac{(-1)^k}{(2k)!} x^{2k} .$$

We would like to demonstrate a method which allows to compute those series in closed form. However, it is clear that such a method cannot solve every problem of this type. For example, a series like

$$\tan x = \sum_{k=1}^{\infty} (-1)^{k+1} \frac{2^{2k}(2^{2k}-1)}{(2k)!} B_{2k} x^{2k-1} , \tag{10.4}$$

for whose representation the *Bernoulli numbers* B_k are needed, is algorithmically not feasible.[11] We would like, however, to cover all series whose coefficients can be represented by factorials or binomial coefficients.

Session 10.6. *First we would like to check the representation (10.4). Mathematica knows the Bernoulli numbers and computes*

$In[1]:=$ **Series$[\text{Tan}[x], \{x, 0, 20\}]$**

---

[11] Later on, we will prove that the tangent function is not covered by the approach given in this section. However, in his thesis [Teg2020], B. Teguia Tabuguia has given an extended variant of our algorithm which covers the tangent function.

$Out[1]= x + \dfrac{x^3}{3} + \dfrac{2\,x^5}{15} + \dfrac{17\,x^7}{315} + \dfrac{62\,x^9}{2835} + \dfrac{1382\,x^{11}}{155925} +$

$\dfrac{21844\,x^{13}}{6081075} + \dfrac{929569\,x^{15}}{638512875} + \dfrac{6404582\,x^{17}}{10854718875} + \dfrac{443861162\,x^{19}}{1856156927625} + O\!\left(x^{21}\right)$

$In[2]:= \displaystyle\sum_{k=1}^{10} \dfrac{(-1)^{k+1}\,2^{2k}\,(2^{2k}-1)}{(2k)!}\,\text{BernoulliB}[2k]\,x^{2k-1} + O[x]^{21}$

$Out[2]= x + \dfrac{x^3}{3} + \dfrac{2\,x^5}{15} + \dfrac{17\,x^7}{315} + \dfrac{62\,x^9}{2835} + \dfrac{1382\,x^{11}}{155925} +$

$\dfrac{21844\,x^{13}}{6081075} + \dfrac{929569\,x^{15}}{638512875} + \dfrac{6404582\,x^{17}}{10854718875} + \dfrac{443861162\,x^{19}}{1856156927625} + O\!\left(x^{21}\right)$

This shows (10.4) up to degree 20. The Bernoulli numbers are generated by the function[12]

$$\frac{x}{e^x - 1} = \sum_{k=0}^{\infty} \frac{B_k}{k!}\,x^k.$$ (10.5)

Let Mathematica confirm this formula up to degree 20:

$In[3]:= \text{Series}\left[\dfrac{x}{e^x - 1}, \{x, 0, 20\}\right]$

$Out[3]= 1 - \dfrac{x}{2} + \dfrac{x^2}{12} - \dfrac{x^4}{720} + \dfrac{x^6}{30240} - \dfrac{x^8}{1209600} +$

$\dfrac{x^{10}}{47900160} - \dfrac{691\,x^{12}}{1307674368000} + \dfrac{x^{14}}{74724249600} -$

$\dfrac{3617\,x^{16}}{10670622842880000} + \dfrac{43867\,x^{18}}{5109094217170944000} -$

$\dfrac{174611\,x^{20}}{802857662698291200000} + O(x^{21})$

$In[4]:= \displaystyle\sum_{k=0}^{20} \dfrac{\text{BernoulliB}[k]}{k!}\,x^k + O[x]^{21}$

$Out[4]= 1 - \dfrac{x}{2} + \dfrac{x^2}{12} - \dfrac{x^4}{720} + \dfrac{x^6}{30240} - \dfrac{x^8}{1209600} +$

$\dfrac{x^{10}}{47900160} - \dfrac{691\,x^{12}}{1307674368000} + \dfrac{x^{14}}{74724249600} -$

$\dfrac{3617\,x^{16}}{10670622842880000} + \dfrac{43867\,x^{18}}{5109094217170944000} -$

$\dfrac{174611\,x^{20}}{802857662698291200000} + O(x^{21})$

Mathematica can solve many integrals, and it can also simplify many sums:

$In[5]:= \displaystyle\sum_{k=0}^{\infty} \dfrac{x^k}{k!}$

$Out[5]= e^x$

$In[6]:= \displaystyle\sum_{k=0}^{\infty} \dfrac{(-1)^k}{(2k+1)!}\,x^{2k+1}$

$Out[6]= \sin(x)$

In this section, we are interested in the inverse question, namely how to convert functions into their representing series. For this purpose we would first like to demonstrate some results of the

---

[12] The radius of convergence of the *generating function* $\sum_{k=0}^{\infty} B_k x^k$ of the Bernoulli numbers B_k is 0. Therefore one considers the *exponential generating function* (10.5) which has a finite radius of convergence.

package SpecialFunctions *which contains an algorithm for the computation of infinite series.*[13]

The output of FPS *(FormalPowerSeries) in the* SpecialFunctions *package does not use the built-in function* Sum *since Mathematica typically simplifies such results instantly. Instead, it uses the function name* sum.

To get started, we load the package

In[7]:= **Needs["SpecialFunctions`"]**

SpecialFunctions, (C) Wolfram Koepf, version 2.2, 2020, with English command names

and then compute several series:

In[8]:= **FPS[e^x,x]**

Out[8]= $\text{sum}\left(\dfrac{x^k}{k!}, \{k, 0, \infty\}\right)$

In[9]:= **FPS[Sin[x],x]**

Out[9]= $\text{sum}\left(\dfrac{(-1)^k x^{2k+1}}{(2k+1)!}, \{k, 0, \infty\}\right)$

In[10]:= **FPS[Cos[x],x]**

Out[10]= $\text{sum}\left(\dfrac{(-1)^k x^{2k}}{(2k)!}, \{k, 0, \infty\}\right)$

In[11]:= **s = FPS[Cos[$\sqrt[4]{x}$],x]**

Out[11]= $\text{sum}\left(\dfrac{x^k}{(4k)!}, \{k, 0, \infty\}\right) + \text{sum}\left(-\dfrac{x^{k+\frac{1}{2}}}{2(2k+1)(4k+1)!}, \{k, 0, \infty\}\right)$

These series were not found in a lookup table, but were computed algorithmically. Note that the command

In[12]:= **specfunprint**

yields intermediate results.

Now that the command specfunprint *is activated, let us repeat two of the above computations:*

In[13]:= **FPS[Cos[x],x]**

SpecialFunctions, (C) Wolfram Koepf, version 2.2, 2020, with English command names

```
specfun - info : DE :
    f(x) + f''(x) == 0
specfun - info : RE for all k >= 0 :
    a[k+2] = a[k]/((1 + k) * (2 + k))
specfun - info : function of hypergeometric type
specfun - info : a[0] = 1
specfun - info : a[1] = 0
specfun - info : 2 - fold symmetric function (e.g. even)
```

Out[13]= $\text{sum}\left(\dfrac{(-1)^k x^{2k}}{(2k)!}, \{k, 0, \infty\}\right)$

---

[13] The package SpecialFunctions was started in 1992 by an implementation of Axel Rennoch, and was further developed by the author in the last three decades. It can be downloaded from the web page www.mathematik.uni-kassel.de/~koepf/Publikationen. As we shall see, it contains many further algorithms. It is not possible to describe the complete package here. Furthermore, the package is more like a test package to check algorithms, therefore false input may not be detected and can generate strange error messages. This should not irritate the user.

$In[14]:=$ **FPS$\left[\text{Cos}\left[\sqrt[4]{x}\right],x\right]$**

SpecialFunctions, (C) Wolfram Koepf, version 2.2, 2020, with English command names

specfun - info : DE :

$\quad$ 256 f'''' (x) x^3 + 1152 f''' (x) x^2 + 816 f'' (x) x + 24 f' (x) – f (x) == 0

specfun - info : RE for all k >= 3/4 :

$\quad$ a[k+1] = a[k]/(8 * (1 + k) * (1 + 2 * k) * (1 + 4 * k) * (3 + 4 * k))

specfun - info : RE modified to $(k \to \dfrac{k}{4})$

specfun - info : RE for all k >= 0 :

$\quad$ a[k+4] = a[k]/((1 + k) * (2 + k) * (3 + k) * (4 + k))

specfun - info : function of hypergeometric type

specfun - info : a[0] = 1

specfun - info : a[1] = 0

specfun - info : a[2] = $-\dfrac{1}{2}$

specfun - info : a[3] = 0

$Out[14]=$ $\text{sum}\left(\dfrac{x^k}{(4\,k)!},\{k,0,\infty\}\right)+\text{sum}\left(-\dfrac{x^{k+\frac{1}{2}}}{2(2\,k+1)(4\,k+1)!},\{k,0,\infty\}\right)$

$In[15]:=$ **specfunprintoff**

The last call stops the verbose mode.

The above computations show that for the conversion of $f(x)$

(a) a differential equation for $f(x)$ is computed;

(b) this differential equation is converted into a recurrence equation for the coefficients a_k;

(c) finally this recurrence equation is solved.

By pattern matching one could easily detect that $\cos \sqrt[4]{x}$ possesses the same series as $\cos x$, where x is replaced by $\sqrt[4]{x}$. However, it is not possible to detect all series by pattern matching. Therefore a completely different approach is used which also yields different intermediate results. For example, $\cos \sqrt[4]{x}$ does *not* satisfy a differential equation of second order (of the type considered), in clear contrast to $\cos x$. Therefore instead of

$$\cos \sqrt[4]{x} = \sum_{k=0}^{\infty}(-1)^k \frac{1}{(2k)!}\,x^{k/2}$$

our procedure computes an equivalent series which consists of two summands, namely a regular power series and a (pure) Puiseux series:

$$\cos \sqrt[4]{x} = \sum_{k=0}^{\infty} \frac{1}{(4k)!}\,x^k - \sum_{k=0}^{\infty} \frac{1}{2(2k+1)(4k+1)!}\,x^{k+\frac{1}{2}}\,.$$

10.3.1 Holonomic Differential Equations

Now we describe the three steps of the above algorithm in detail. Let us start with the first step, namely the search for an ordinary differential equation which is valid for given $f(x)$.

More exactly we search for a *homogeneous linear differential equation* which has *polynomial coefficients* $\in \mathbb{K}[x]$, where $\mathbb{K}$ is a field. (In most cases we assume that $\mathbb{K} = \mathbb{Q}$.) Such a differential equation is called *holonomic*. A function satisfying a holonomic differential equation is also called holonomic. The smallest order of a holonomic differential equation valid for a holonomic function $f(x)$ is called the *holonomic degree* of $f(x)$ and is denoted by holdeg$(f(x), x)$.

Example 10.7. (a) The exponential function is obviously holonomic with holdeg$(e^x, x) = 1$.
(b) The sine function $f(x) = \sin x$ is holonomic since it satisfies the differential equation $f''(x) + f(x) = 0$. It has the degree holdeg$(\sin x, x) = 2$ because[14]

$$\frac{f'(x)}{f(x)} = \cot x \notin \mathbb{Q}(x)$$

and therefore no holonomic differential equation of first order is valid. △

Session 10.8. *Let us search for a holonomic differential equation for the exponential function* $f_0 = e^{x^2}$:

$In[1]:=$ $\mathbf{f_0 = e^{x^2}}$
$Out[1]=$ e^{x^2}

For this purpose, we compute the derivative f_1 *of* f_0:

$In[2]:=$ $\mathbf{f_1 = D[f_0, x]}$
$Out[2]=$ $2\, e^{x^2} x$

A holonomic differential equation of first order obviously exists if and only if the quotient f_1/f_0 *is an element of* $\mathbb{Q}(x)$:

$In[3]:=$ $\mathbf{Together\left[\frac{f_1}{f_0}\right]}$
$Out[3]=$ $2\,x$

This is the case for our current example, so that we have found the differential equation

$In[4]:=$ $\mathbf{DE = F'[x] - 2x\, F[x] == 0}$
$Out[4]=$ $F'(x) - 2\,x\,F(x) == 0$

Therefore the function $f_0(x) = e^{x^2}$ *is holonomic of order* 1.

Next, we consider a more complicated example:

$In[5]:=$ $\mathbf{f_0 = ArcSin[x]}$
$Out[5]=$ $\sin^{-1}(x)$

$In[6]:=$ $\mathbf{f_1 = D[f_0, x]}$
$Out[6]=$ $\dfrac{1}{\sqrt{1-x^2}}$

Obviously f_1/f_0 *is not rational since* $\arcsin x$ *and* $\frac{1}{\sqrt{1-x^2}}$ *are linearly independent over* $\mathbb{Q}(x)$[15]; *hence* f_0 *does not have holonomic degree* 1. *As a next step, we search for a differential equation of order* 2 *by considering*

$In[7]:=$ $\mathbf{f_2 = D[f_1, x]}$
$Out[7]=$ $\dfrac{x}{\left(1-x^2\right)^{3/2}}$

and trying (for yet undetermined $A_k \in \mathbb{Q}(x)$*) the ansatz* $\sum_{k=0}^{2} A_k f^{(k)}(x) = 0$, *where we can set* $A_2 = 1$ *since we already know that no differential equation of first order exists.*

---

[14] Rational functions have (in $\mathbb{C}$) only finitely many zeros.
[15] Reason why!

$In[8]:=$ **ansatz** $= \displaystyle\sum_{k=0}^{2} \mathbf{A_k} \, \mathbf{f_k} \, / \, . \, \mathbf{A_2} \to \mathbf{1}$

$Out[8]= \dfrac{A_1}{\sqrt{1-x^2}} + A_0 \sin^{-1}(x) + \dfrac{x}{\left(1-x^2\right)^{3/2}}$

Next, we check which of the summands differ only by a rational factor and are thus linearly dependent over $\mathbb{Q}(x)$. These can be collected. In our case the first and the third summand are the only such candidates. We get

$In[9]:=$ **Together** $\Big[\dfrac{\texttt{ansatz[[1]]}}{\texttt{ansatz[[3]]}} \Big]$

$Out[9]= -\dfrac{x}{A_1 \left(x^2-1\right)}$

Therefore these summands actually are linearly dependent over $\mathbb{Q}(x)$ and we can merge them.

Our ansatz can only be identically zero if the coefficients of the linearly independent summands (over $\mathbb{Q}(x)$) vanish. This leads to

$In[10]:=$ **sol = Solve[{ansatz[[1]] + ansatz[[3]] == 0,**
                 ansatz[[2]] == 0}, {A$_0$, A$_1$}]

$Out[10]= \left\{\left\{ A_0 \to 0, A_1 \to \dfrac{x}{x^2-1} \right\}\right\}$

Therefore the left-hand side of the differential equation is given by

$In[11]:=$ **DE** $= \displaystyle\sum_{k=0}^{2} \mathbf{A_k} \, \mathbf{D[F[x], \{x,k\}]} \, / \, . \, \mathbf{sol[[1]]} \, / \, . \, \mathbf{A_2} \to \mathbf{1}$

$Out[11]= \dfrac{x\,\mathrm{F}'(x)}{x^2-1} + \mathrm{F}''(x)$

After multiplication by the common denominator, we finally get the following holonomic differential equation for $F(x) = \arcsin x$:

$In[12]:=$ **Collect[Numerator[Together[DE]],**
                 Table[D[F[x], {x,k}], {k, 0, 2}]] == 0

$Out[12]= \left(x^2-1\right) \mathrm{F}''(x) + x\,\mathrm{F}'(x) == 0$

Hence this function is holonomic of degree 2.

The package `SpecialFunctions` *contains the function* `holonomicDE` *for the computation of holonomic differential equations using the described approach.*

We load the package:

$In[13]:=$ **Remove[F]**
          Needs["SpecialFunctions`"]

Note that before loading we have removed the (already used) global variable F from the global context to avoid conflicts with the variable F used in the `SpecialFunctions` *package.*

Now we can compute several differential equations:

$In[14]:=$ **holonomicDE[e$^{\mathbf{x}}$, F[x]]**

$Out[14]= \mathrm{F}'(x) - \mathrm{F}(x) == 0$

$In[15]:=$ **holonomicDE[ArcSin[x], F[x]]**

$Out[15]= x\,F'(x) + (x-1)\,(x+1)\,F''(x) == 0$

We also find a holonomic differential equation for every power of the inverse sine function:

In[16]:= **Table[holonomicDE[ArcSin[x]^k,F[x]],{k,1,4}]**

Out[16]= $\{x\,F'(x)+(x-1)\,(x+1)\,F''(x)==0,$

$\qquad F'(x)+3\,x\,F''(x)+(x-1)\,(x+1)\,F^{(3)}(x)==0,$

$\qquad (x-1)^2\,F^{(4)}(x)\,(x+1)^2+6\,(x-1)\,x\,F^{(3)}(x)\,(x+1)+$

$\qquad\qquad x\,F'(x)+(7\,x^2-4)\,F''(x)==0,$

$\qquad (x-1)^2\,F^{(5)}(x)\,(x+1)^2+10\,(x-1)\,x\,F^{(4)}(x)\,(x+1)+$

$\qquad\qquad F'(x)+15\,x\,F''(x)+5\,(5\,x^2-2)\,F^{(3)}(x)==0\}$

Also sums of holonomic function satisfy a holonomic differential equation again:

In[17]:= **holonomicDE[e$^{\alpha\,x}$+Sin[β x],F[x]]**

Out[17]= $-\alpha\,F''(x)+\beta^2\,F'(x)+F'''(x)-\alpha\,\beta^2\,F(x)==0$

In[18]:= **DE = holonomicDE[ArcSin[x]^2+Sin[x]^2,F[x]]**

Out[18]= $4\,(2\,x^4+7\,x^2+13)\,F'(x)+4\,x\,(6\,x^4+13\,x^2+37)\,F''(x)+$

$\qquad (8\,x^6-2\,x^4+23\,x^2-7)\,F^{(3)}(x)+$

$\qquad x\,(6\,x^4+13\,x^2+37)\,F^{(4)}(x)+$

$\qquad (x-1)\,(x+1)\,(2\,x^4+x^2+5)\,F^{(5)}(x)==0$

We check the last result:

In[19]:= **DE[[1]]/.Table[Derivative[k][F][x] →**

$\qquad\qquad$ **D[ArcSin[x]^2+Sin[x]^2,{x,k}],{k,0,5}]**

Out[19]= $(8\,x^6-2\,x^4+23\,x^2-7)\left(\dfrac{6\,\sin^{-1}(x)\,x^2}{(1-x^2)^{5/2}}+\right.$

$\qquad\qquad \left.\dfrac{6\,x}{(1-x^2)^2}+\dfrac{2\,\sin^{-1}(x)}{(1-x^2)^{3/2}}-8\,\cos(x)\,\sin(x)\right)+$

$\qquad 4\,(2\,x^4+7\,x^2+13)\left(\dfrac{2\,\sin^{-1}(x)}{\sqrt{1-x^2}}+2\,\cos(x)\,\sin(x)\right)+$

$\qquad (x-1)\,(x+1)\,(2\,x^4+x^2+5)$

$\qquad \left(\dfrac{210\,\sin^{-1}(x)\,x^4}{(1-x^2)^{9/2}}+\dfrac{210\,x^3}{(1-x^2)^4}+\dfrac{180\,\sin^{-1}(x)\,x^2}{(1-x^2)^{7/2}}+\right.$

$\qquad\qquad \left.\dfrac{110\,x}{(1-x^2)^3}+\dfrac{18\,\sin^{-1}(x)}{(1-x^2)^{5/2}}+32\,\cos(x)\,\sin(x)\right)+$

$\qquad 4\,x\,(6\,x^4+13\,x^2+37)$

$\qquad \left(2\,\cos^2(x)-2\,\sin^2(x)+\dfrac{2\,x\,\sin^{-1}(x)}{(1-x^2)^{3/2}}+\dfrac{2}{1-x^2}\right)+$

$\qquad x\,(6\,x^4+13\,x^2+37)\left(\dfrac{30\,\sin^{-1}(x)\,x^3}{(1-x^2)^{7/2}}+\dfrac{30\,x^2}{(1-x^2)^3}+\right.$

$\qquad\qquad \left.\dfrac{18\,\sin^{-1}(x)\,x}{(1-x^2)^{5/2}}-8\,\cos^2(x)+8\,\sin^2(x)+\dfrac{8}{(1-x^2)^2}\right)$

In[20]:= **%//Together**

Out[20]= 0

In the last example it was not at all simple to prove the validity of the differential equation. We were lucky that Together *was successful since we know that there is no normal form for arbitrary transcendental expressions.*[16]

---

[16] Theorem 10.12 will show that for the functions considered in this section a normal form exists. However, *Mathematica* does not support this normal form.

The result of the last session suggests the holonomy of sums and products of holonomic functions. Indeed the following theorem is valid.

Theorem 10.9. The sum and the product of holonomic functions are also holonomic. Therefore we get the *ring of holonomic functions*. If $f(x)$ and $g(x)$ are holonomic of degree m and n, respectively, then $f(x) + g(x)$ has degree $\leq m + n$, and $f(x) \cdot g(x)$ has degree $\leq m \cdot n$.

Proof. Let the functions $f(x)$ and $g(x)$ be holonomic of degree m and n, respectively. We first consider the vector space $V(f) = \langle f(x), f'(x), f''(x), \ldots \rangle$ over the field of rational functions $\mathbb{K}(x)$, generated by all the derivatives of $f(x)$.

Because $f(x)$ is holonomic of degree m, it satifies a holonomic differential equation of order m, but none of order $m - 1$. Successive differentiation shows that all higher derivatives can be represented as linear combinations (over $\mathbb{K}(x)$) of the functions $f(x), f'(x), \ldots, f^{(m-1)}$. Hence $\{f(x), f'(x), f''(x), \ldots, f^{(m-1)}(x)\}$ forms a basis of $V(f)$ which therefore has dimension $\dim(V(f)) = m$. Analogously we construct the vector space $V(g)$ of all derivatives of $g(x)$ over $\mathbb{K}(x)$ for which $\dim(V(g)) = n$ is valid.

Now we build the sum $V(f) + V(g)$, which is a vector space of dimension $\leq m + n$. Since $h := f + g$, $h' = (f + g)'$, $\ldots$, $h^{(k)} = (f + g)^{(k)}$, $\ldots$ are elements of $V(f) + V(g)$, any $m + n + 1$ of these functions are linearly dependent (over $\mathbb{K}(x)$). In other words, there is a differential equation of order $\leq m + n$ with coefficients in $\mathbb{K}(x)$ for h. After multiplication by the common denominator, this yields the holonomic differential equation sought.

This nice linear algebra proof has one disadvantage, namely does not show (directly) how the sought differential equation can be found. How can this be accomplished? For this purpose the given holonomic differential equations for f and g are brought into the explicit form

$$f^{(m)} = \sum_{j=0}^{m-1} p_j f^{(j)} \qquad \text{and} \qquad g^{(n)} = \sum_{k=0}^{n-1} q_k g^{(k)},$$

with rational coefficient functions $p_j, q_k \in \mathbb{K}(x)$. Successive differentiation and recursive substitution of those explicit representations for $f^{(m)}$ and $g^{(n)}$ yield for all derivatives of higher order representations of the same form

$$f^{(l)} = \sum_{j=0}^{m-1} p_j^l f^{(j)} \ (l \geq m) \qquad \text{and} \qquad g^{(l)} = \sum_{k=0}^{n-1} q_k^l g^{(k)} \ (l \geq n), \tag{10.6}$$

of course with other coefficient functions $p_j^l, q_k^l \in \mathbb{K}(x)$.

The linearity of the derivative implies the following equations:

$$\begin{aligned}
h &= f + g \\
h' &= f' + g' \\
h'' &= f'' + g'' \\
&\vdots \qquad \vdots \\
h^{(n+m)} &= f^{(n+m)} + g^{(n+m)}.
\end{aligned} \tag{10.7}$$

Now we search for a holonomic differential equation, starting with the possible order $J :=$ $\max\{m, n\}$.[17] If this is not successful, we increase J by 1 and start searching again.

Next we choose the $\max\{m, n\}$ equations (10.7) and use the replacement rules (10.6) to eliminate the higher derivatives of f ($l \geqq m$) and g ($l \geqq n$). On the right-hand side the $m + n$ variables $f^{(l)}$ ($l = 0, ..., m - 1$) and $g^{(l)}$ ($l = 0, ..., n - 1$) remain. We solve the resulting linear system for the variables $h^{(l)}$ ($l = 0, ..., \max\{m, n\}$) and try to eliminate the variables $f^{(l)}$ ($l = 0, ..., m - 1$) and $g^{(l)}$ ($l = 0, ..., n - 1$). This is possible by a variant of Gaussian elimination which is implemented in the function `Eliminate` in *Mathematica*.[18] If successful, this yields the holonomic differential equation sought. It is at $J = m + n$ at the latest that the search must be successful.

The construction of the "product differential equation" can be handled similarly. The only difference is that now the product $h := f \cdot g$ is differentiated using the Leibniz rule[19], and the variables to be eliminated are the $m \cdot n$ products $f^{(j)} g^{(k)}$ ($j = 0, ..., m - 1$; $k = 0, ..., n - 1$). This procedure then yields a holonomic differential equation of order $\leqq m \cdot n$.

The termination of the algorithm is assured by the fact that we eventually arrive at a homogeneous linear system with $m \cdot n + 1$ equations for which $m \cdot n$ variables must be eliminated. $\square$

Session 10.10. *In the* `SpecialFunctions` *package the algorithms of Theorem 10.9 are implemented. The function* `HolonomicDE` *computes a holonomic differential equation again, this time, however, using the algorithms of Theorem 10.9.*

We first consider the example

$In[1]:=$ **Needs["SpecialFunctions`"]**

$In[2]:=$ **HolonomicDE$\left[\sqrt{1+x} + \dfrac{1}{\sqrt{1+x}}, F[x]\right]$**

$Out[2]=$ $4(x+1)F'[(x) + 4(x+1)^2 F''(x) - F(x) == 0$

As we will soon see, the resulting differential equation is not of lowest order. This is due to the fact that the described sum algorithm only "knows" the differential equations of the summands, but not the summands themselves. Therefore, this algorithm cannot make sure that the two summands are linearly dependent over $\mathbb{Q}(x)$*. This becomes particularly clear in the computation*

$In[3]:=$ **HolonomicDE$\left[A\sqrt{1+x} + \dfrac{B}{\sqrt{1+x}}, F[x]\right]$**

$Out[3]=$ $4(x+1)F'[(x) + 4(x+1)^2 F''(x) - F(x) == 0$

which shows that the computed differential equation is the differential equation of the linear hull of the two functions $\sqrt{1+x}$ *and* $\frac{1}{\sqrt{1+x}}$*. Therefore these two functions form a solution basis of the computed differential equation of second order.*

For our original function we can do better with the call

$In[4]:=$ **HolonomicDE$\left[\text{Simplify}\left[\sqrt{1+x} + \dfrac{1}{\sqrt{1+x}}\right], F[x]\right]$**

$Out[4]=$ $2(x+1)(x+2)F'(x) - xF(x) == 0$

where `Simplify` *converts the sum into a product. Similarly we can ignore the use of the sum and product algorithms using the method described in Session 10.8, which we can call by*

---

[17] Of course we cannot expect to get a differential equation for $f + g$ or $f \cdot g$ whose order is lower than that of f and g. In certain cases this might lead to a holonomic differential equation which is not of lowest possible order, for example for $f(x) = \sin x$ and $g(x) = -\sin x$.

[18] Of course, one can also use `Solve`.

[19] Obviously *Mathematica* can handle this task for us!

$In[5] :=$ **holonomicDE** $\left[\sqrt{1+x} + \dfrac{1}{\sqrt{1+x}}, F[x] \right]$

$Out[5] = 2(x+1)(x+2) F'(x) - x F(x) == 0$

Let us check the single substeps of the computation using HolonomicDE. *In the first call the sum algorithm is called using* SumDE:

$In[6] :=$ **DE$_1$ = HolonomicDE** $\left[\sqrt{1+x}, F[x] \right]$

$Out[6] = 2(x+1) F'(x) - F(x) == 0$

$In[7] :=$ **DE$_2$ = HolonomicDE** $\left[\dfrac{1}{\sqrt{1+x}}, F[x] \right]$

$Out[7] = 2(x+1) F'(x) + F(x) == 0$

$In[8] :=$ **SumDE[DE$_1$, DE$_2$, F[x]]**

$Out[8] = 4(x+1) F'](x) + 4(x+1)^2 F''(x) - F(x) == 0$

The second call yields a product:[20]

$In[9] :=$ **Simplify** $\left[\sqrt{1+x} + \dfrac{1}{\sqrt{1+x}} \right]$

$Out[9] = \dfrac{x+2}{\sqrt{x+1}}$

Therefore the product algorithm is used, executed by the procedure ProductDE:

$In[10] :=$ **DE$_3$ = HolonomicDE[x + 2, F[x]]**

$Out[10] = (-x-2) F'(x) + F(x) == 0$

$In[11] :=$ **ProductDE[DE$_2$, DE$_3$, F[x]]**

$Out[11] = 2(x+1)(x+2) F'(x) - x F(x) == 0$

A further example is given by the computation

$In[12] :=$ **ProductDE[F$'$[x] - F[x] == 0, F$''$[x] + F[x] == 0, F[x]]**

$Out[12] = -2 F'(x) + F''(x) + 2 F(x) == 0$

which obviously computes a differential equation for the product $e^x \sin x$ (as well as for every linear combination of $e^x \sin x$ and $e^x \cos x$):

$In[13] :=$ **HolonomicDE[e^x * Sin[x], F[x]]**

$Out[13] = -2 F'(x) + F''(x) + 2 F(x) == 0$

Now we can also work with much more complicated functions. The solutions of the most simple holonomic differential equation of second order,

$$F''(x) = x F(x),$$

whose coefficients are not constant, are called Airy functions. The computation

$In[14] :=$ **DE = ProductDE[F$'$[x] - F[x] == 0, F$''$[x] - x F[x] == 0, F[x]]**

$Out[14] = -2 F'(x) + F''(x) + (1-x) F(x) == 0$

therefore yields a holonomic differential equation for the product of an Airy function with the exponential function. This can be checked by Mathematica's built-in differential equation solver DSolve:

$In[15] :=$ **DSolve[{DE, F[0] == 1}, F[x], x]**

$Out[15] = \left\{ \left\{ F(x) \to \dfrac{1}{3} e^x \left(3 c_1 \, \text{Ai}(x) + 3 \sqrt[6]{3} \, \Gamma\!\left(\dfrac{2}{3}\right) \text{Bi}(x) - \sqrt{3} \, c_1 \, \text{Bi}(x) \right) \right\} \right\}$

---

[20] The internal representation of a/b is a product Times[a, Power[b,-1]]!

Note that Mathematica uses for the presentation of the result the Gamma function $\Gamma(x) = (x-1)!$, *see Exercise 10.4, and the Airy functions* $\mathrm{Ai}(x)$ *and* $\mathrm{Bi}(x)$.

When we discussed the algebraic numbers, it turned out that division does not yield something new, so that we arrived at an algebraic extension field. Unfortunately, this is different in the current situation, since the following theorem is valid.

Theorem 10.11. The function $\tan x = \frac{\sin x}{\cos x}$ is not holonomic.

Proof. It is easy to verify that $f(x) = \tan x$ satisfies the following non-linear differential equation of first order:

$$f' = 1 + f^2. \tag{10.8}$$

Taking the derivative in equation (10.8) and applying the chain rule yields

$$f'' = (1 + f^2)' = 2f f' = 2f(1 + f^2),$$

where we have substituted (10.8) again in the last step.

By successive differentiation we therefore get, for all $k \in \mathbb{N}$, representations of the form $f^{(k)} = P_k(f)$ for polynomials $P_k(y) \in \mathbb{Z}[y]$.

Now if f was a solution of the holonomic differential equation

$$\sum_{k=0}^{n} p_k(x) f^{(k)} = 0 \qquad (p_k(x) \in \mathbb{Q}[x]),$$

then we could substitute the formulas derived above for $f^{(k)}$ and would get an algebraic equation

$$G(x, f) = \sum_{j=0}^{J} \sum_{k=0}^{K} c_{jk} x^j f^k = 0, \qquad (G(x, y) \in \mathbb{Q}[x, y]) \tag{10.9}$$

for the tangent function. Therefore $\tan x$ would be an *algebraic function.*[21] An algebraic function, however, cannot possess infinitely many zeros because solving the implicit equation (10.9) for f yields at most K different solution branches. Therefore that algebraic function can also have at most K zeros. This contradicts the fact that $\tan x$ has infinitely many zeros (in $\mathbb{R}$ and in $\mathbb{C}$). Therefore $f(x) = \tan x$ cannot satisfy a holonomic differential equation. $\square$

Since both $\sin x$ and $\cos x$ satisfy the holonomic differential equation $F''(x) + F(x) = 0$ and hence are holonomic functions, Theorem 10.11 shows that the quotient of holonomic functions is generally not holonomic.

However, in the same way as an algebraic number is best represented by its minimal polynomial, holonomic functions are best represented by their corresponding differential equation, as the following theorem indicates.

Theorem 10.12. In the ring of holonomic functions the holonomic differential equation of degree n together with n suitably chosen initial values provides a normal form.

Proof. This result follows from the theory of ordinary differential equations which guarantees that such a differential equation of order n together with n initial values $y^{(k)}(x_0)$ ($k = 0, \ldots, n-1$)

---

[21] Algebraic functions are considered in more detail in Section 10.4.

has a unique solution in a suitably chosen real interval. Thereby, an initial value x_0 is suitable if it is not a zero of the polynomial coefficient of the highest derivative $f^{(n)}(x)$ of the differential equation, see e.g. [Wal2000].

However, the proof that this creates a normal form requires some more thoughts. If we used algorithms which guarantee that the generated holonomic differential equations are of lowest possible order, then this approach would clearly yield a canonical form. However, the algorithms considered here cannot ensure this. To prove an identity about two holonomic functions, it is sufficient to show that both sides satisfy the same holonomic differential equation and have the same initial values. If the orders n and m of the computed differential equations D_1 and D_2 are different, however, we can still prove such an identity. Then we must show that the validity of the differential equation D_1 of lower order implies the validity of the differential equation D_2 of higher order. For this purpose, we solve D_1 for the highest derivative $f^{(n)}(x)$ and substitute this term and all resulting higher-order derivatives in D_2. If after rational simplification we then get the identity $0 = 0$, then D_1 and D_2 are compatible (and we only have to check m initial values), otherwise not. □

Obviously the theorem tells us that the identification problem can be solved for holonomic functions by computing the underlying differential equation and suitable initial values.

Session 10.13. *We use Theorem 10.12 for the proof of two transcendental identities.*

First we load the SpecialFunctions *package:*

In[1]:= **Needs["SpecialFunctions`"]**

The equation

$$\sin^2 x = 1 - \cos^2 x$$

is proved by the computations

In[2]:= **f$_1$[x_] := Sin[x]^2**

In[3]:= **{HolonomicDE[f$_1$[x],F[x]],**
        Table[Derivative[k][f$_1$][0],{k,0,2}]}
Out[3]= $\left\{4\,F'(x) + F'''(x) == 0, \{0, 0, 2\}\right\}$

and

In[4]:= **f$_2$[x_] := 1 - Cos[x]^2**

In[5]:= **{HolonomicDE[f$_2$[x],F[x]],**
        Table[Derivative[k][f$_2$][0],{k,0,2}]}
Out[5]= $\left\{4\,F'(x) + F'''(x) == 0, \{0, 0, 2\}\right\}$

Note that this relation is proved without prior knowledge of the Pythagorean identity $\sin^2 x + \cos^2 x = 1$. *Conversely, the executed computations yield (using Theorem 10.12) an algebraic proof of the Pythagorean identity.*

As another example, the computations

In[6]:= **f$_3$[x_] := ArcTanh[x]**

In[7]:= **{HolonomicDE[f$_3$[x],F[x]],**
        Table[Derivative[k][f$_3$][0],{k,0,1}]}
Out[7]= $\left\{\left(x^2 - 1\right) F''(x) + 2\,x\,F'(x) == 0, \{0, 1\}\right\}$

and

$In[8]:=$ $\mathbf{f_4}\,[\mathbf{x\_}] := \dfrac{1}{2}\,\mathbf{Log}\left[\dfrac{1+\mathbf{x}}{1-\mathbf{x}}\right]$

$In[9]:=$ $\{\mathbf{HolonomicDE}\,[\mathbf{f_4}\,[\mathbf{x}]\,,\mathbf{F}\,[\mathbf{x}]]\,,$
$\qquad\quad\mathbf{Table}\,[\mathbf{Derivative}\,[\mathbf{k}]\,[\mathbf{f_4}]\,[\mathbf{0}]\,,\{\mathbf{k},\mathbf{0},\mathbf{1}\}]\}$

$Out[9]=$ $\left\{-2\,x\,(x-1)^2\,F'(x)-((x+1)\,(x-1)^3\,F''(x))==0,\{0,1\}\right\}$

show the identity

$$\operatorname{arctanh} x = \frac{1}{2}\ln\frac{1+x}{1-x}. \tag{10.10}$$

To realize this, the last differential equation must be divided by the common factor $-(x-1)^2$.
Note that Mathematica knows this identity in a suitable real interval:

$In[10]:=$ $\mathbf{FullSimplify}\,[\mathbf{f_3}\,[\mathbf{x}] - \mathbf{f_4}\,[\mathbf{x}]\,,\mathbf{x} > -\mathbf{1}\,\&\&\,\mathbf{x} < \mathbf{1}]$

$Out[10]=$ 0

The following graph confirms this computation:

$In[11]:=$ $\mathbf{Plot}\,[\mathbf{f_3}\,[\mathbf{x}] - \mathbf{f_4}\,[\mathbf{x}]\,,\{\mathbf{x},-\mathbf{1},\mathbf{1}\}]$

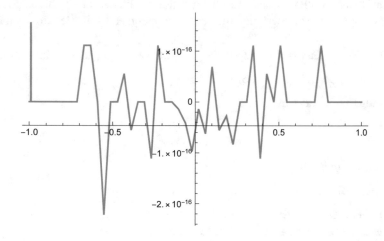

The numerical computation yields values that are different from zero but of order 10^{-16}.

10.3.2 Holonomic Recurrence Equations

In this section we describe the second step of the conversion process of an expression towards its representing power series. This step converts the derived holonomic differential equation into a recurrence equation for the associated coefficients a_k. It will turn out that this recurrence equation is holonomic if and only if the starting differential equation is holonomic.

Definition 10.14. (Holonomic Sequences and Recurrence Equations) A recurrence equation for a_k is *holonomic* if it is homogeneous, linear and has polynomial coefficients $\in \mathbb{K}[k]$. A sequence a_k satisfying a holonomic recurrence equation is also called holonomic. The difference between the largest and lowest shift j of the terms a_{k+j} occurring in the recurrence equation is called the *degree* of the recurrence equation. The smallest degree of a holonomic recurrence equation valid for a_k is called the *holonomic degree of* a_k and is denoted by holdeg(a_k, k).

Example 10.15. (a) The factorial $a_k = k!$ is holonomic of degree $\text{holdeg}(k!, k) = 1$ since

$$a_{k+1} - (k+1)a_k = 0.$$

(b) The sequence $F(n, k) = \binom{n}{k}$ is holonomic with respect to both n and k since it satisfies the two recurrence equations

$$\frac{F(n, k+1)}{F(n, k)} = \frac{\binom{n}{k+1}}{\binom{n}{k}} = \frac{n-k}{k+1} \quad \text{and} \quad \frac{F(n+1, k)}{F(n, k)} = \frac{\binom{n+1}{k}}{\binom{n}{k}} = \frac{n+1}{n+1-k}$$

or

$$(k+1)F(n, k+1) + (k-n)F(n, k) = 0 \quad \text{and} \quad (n+1-k)F(n+1, k) - (n+1)F(n, k) = 0,$$

respectively. It has degrees $\text{holdeg}(\binom{n}{k}, k) = \text{holdeg}(\binom{n}{k}, n) = 1$. $\triangle$

How can the holonomic recurrence equations of a holonomic sequence be determined? This is handled in the same way as for holonomic functions, using the following theorem.

Theorem 10.16. The sum and the product of holonomic sequences are holonomic. Therefore we get the *ring of holonomic sequences*. If a_k and b_k have degree m and n, respectively, then $a_k + b_k$ is of degree $\leq m + n$, and $a_k \cdot b_k$ is of degree $\leq m \cdot n$.

Proof. The proof of this theorem is analogous to Theorem 10.9 and is left to the reader. Note that the operation corresponding to the *differential operator* $D_x : f(x) \mapsto f'(x)$ is now the *forward shift operator* $S_k : a_k \mapsto a_{k+1}$. $\square$

Session 10.17. *The package* `SpecialFunctions` *contains the algorithms of Theorem 10.16 as well as the recursive algorithm invoking those. We get for example*

```
In[1]:= Needs["SpecialFunctions`"]
```

```
In[2]:= HolonomicRE[k!,a[k]]
Out[2]= (k+1) a(k) - a(k+1) == 0
```

```
In[3]:= RE₁ = HolonomicRE[Binomial[n,k],a[k]]
Out[3]= a(k) (k-n) + (k+1) a(k+1) == 0
```

```
In[4]:= RE₂ = HolonomicRE[Binomial[k,n],a[k]]
Out[4]= a(k+1) (k-n+1) + (-k-1) a(k) == 0
```

Next, we compute the sum recurrence equation of RE_1 *and* RE_2 *by*

```
In[5]:= SumRE[RE₁, RE₂, a[k]]
Out[5]= (k+1) (k-n) (2 k² - 2 n k + 7 k + n² - 3 n + 6) a(k)+
        (-n⁴ + 4 k n³ + 4 n³ - 6 k² n² - 12 k n² - 5 n² + 4 k³ n+
            12 k² n + 10 k n + 2 n + 2 k³ + 8 k² + 10 k + 4) a(k+1)-
        (k+2) (k-n+2) (2 k² - 2 n k + 3 k + n² - n + 1) a(k+2) == 0
```

This result can also be directly computed by the call

```
In[6]:= HolonomicRE[Binomial[n,k]+Binomial[k,n],a[k]]
Out[6]= (k+1) (k-n) (2 k² - 2 n k + 7 k + n² - 3 n + 6) a(k)+
        (-n⁴ + 4 k n³ + 4 n³ - 6 k² n² - 12 k n² - 5 n² + 4 k³ n+
            12 k² n + 10 k n + 2 n + 2 k³ + 8 k² + 10 k + 4) a(k+1)-
        (k+2) (k-n+2) (2 k² - 2 n k + 3 k + n² - n + 1) a(k+2) == 0
```

We give another example:

```
In[7]:= HolonomicRE[(1-k!)Binomial[n,k],a[k]]
```
$Out[7]= (k+1)(k-n)(k-n+1)\,a(k)+\left(k^2+3\,k+1\right)(k-n+1)\,a(k+1)+k\,(k+2)\,a(k+2)==0$

In this example, the sum algorithm is called first (for the difference $1-k!$) followed by the product algorithm.

Again, we get a normal form, as the following theorem reveals:

Theorem 10.18. In the ring of holonomic sequences the holonomic recurrence equation of degree n together with n suitable initial values provides a normal form.   □

Hence for holonomic sequences, one can solve the identification problem again by computing the corresponding recurrence equations and suitable initial values.

Session 10.19. *We use Theorem 10.18 for the proof of two discrete transcendental identities.*

By

$$(a)_k := \underbrace{a\cdot(a+1)\cdots(a+k-1)}_{k\ factors}$$

we denote the Pochhammer symbol (also called shifted factorial).

We first consider the following identity:[22]

$$\left(\frac{1}{2}\right)_k = \frac{(2k)!}{4^k\,k!}\ . \tag{10.11}$$

Mathematica knows the `Pochhammer` *function and represents it (in* $\boxed{\textbf{TraditionalForm}}$ *) in the usual way:*

```
In[1]:= Pochhammer[1/2,k]
```
$Out[1]= \left(\frac{1}{2}\right)_k$

Let us start again by loading the `SpecialFunctions` *package:*

```
In[2]:= Needs["SpecialFunctions`"]
```

Then we compute the holonomic recurrence equations of both sides of (10.11) and their initial values:

```
In[3]:= {HolonomicRE[Pochhammer[1/2,k],a[k]],

            Pochhammer[1/2,k]/.k→0}
```
$Out[3]= \{(-2\,k-1)\,a(k)+2\,a(k+1)==0,\ 1\}$

```
In[4]:= {HolonomicRE[(2k)!/(4^k k!),a[k]], (2k)!/(4^k k!)/.k→0}
```
$Out[4]= \{(-2\,k-1)\,a(k)+2\,a(k+1)==0,\ 1\}$

This seems to prove identity (10.11). But attention please: after loading the package the automatic simplification

```
In[5]:= Pochhammer[1/2,k]
```

---

[22] Your first step should be to prove this identity by hand computation!

$$Out[5] = \frac{4^{-k}\,(2\,k)!}{k!}$$

is valid! This shows that in the current case our computations were not successful in proving this identity, but the package did this conversion itself! However, it remains an easy exercise to show the recurrence $(2k+1)a_k - 2a_{k+1} = 0$ for $\left(\frac{1}{2}\right)_k$ and therefore to finish this proof.

Since we do not yet know many discrete holonomic sequences, we consider another simple bivariate example, namely the binomial coefficients. They satisfy the well-known recurrence equation of the Pascal triangle:

$$\binom{n}{k} = \binom{n-1}{k} + \binom{n-1}{k-1}. \tag{10.12}$$

We would like to check this identity. For this purpose, we first compute the holonomic recurrence equation with respect to k of the left-hand side of (10.12),

```
In[6]:= RE = HolonomicRE[Binomial[n,k],a[k]]
```
$$Out[6] = (k-n)\,a(k) + (k+1)\,a(k+1) == 0$$

and then the holonomic recurrence equation with respect to k of the right-hand side of (10.12):

```
In[7]:= RE₁ = HolonomicRE[
            Binomial[n-1,k] + Binomial[n-1,k-1],a[k]]
```
$$Out[7] = (k-n)\,(k-n+1)\,a(k) + 2\,(k+1)\,(k-n+1)\,a(k+1) + (k+1)\,(k+2)\,a(k+2) == 0$$

We see that the sum algorithm—by the same reason that was thoroughly discussed in the respective differential equation situation—has computed a recurrence equation for the right-hand side whose order is too high. One could easily show by the use of another shift $k \mapsto k+1$ for the recurrence equation for the left-hand side that this recurrence implies the recurrence equation for the right-hand side, see Exercise 10.18.

However the current case is easier since we know the left-hand side in "closed form". Therefore we can just substitute this term in RE_1 to get

```
In[8]:= RE₁/.a[k_]->Binomial[n,k]
```
$$Out[8] = (k-n)\,(k-n+1)\binom{n}{k} + 2\,(k+1)\,(k-n+1)\binom{n}{k+1} + (k+1)\,(k+2)\binom{n}{k+2} == 0$$

The question is now whether this is an identity. This, however, can be easily checked by FunctionExpand*:*

```
In[9]:= RE₁/.a[k_] → Binomial[n,k]//FunctionExpand
```
$$Out[9] = \text{True}$$

Now we have seen that the left- and right-hand side of (10.12) satisfy the same recurrence equation of second order. It therefore remains to check two initial values:

```
In[10]:= {Binomial[n,k],
            Binomial[n-1,k] + Binomial[n-1,k-1]}/.k → 0//FunctionExpand
```
$$Out[10] = \{1, 1\}$$

```
In[11]:= {Binomial[n,k],
            Binomial[n-1,k] + Binomial[n-1,k-1]}/.k → 1
```
$$Out[11] = \{n, n\}$$

This completes the proof of (10.12)!

In the next chapter we will consider more complicated (and more interesting) discrete transcendental identities.

Now we come back to our original question: how do we convert a holonomic differential equation into a corresponding recurrence equation for the associated series coefficients?

Theorem 10.20. Let $f(x)$ be a function satisfying the holonomic differential equation DE. If we substitute the series representation $f(x) = \sum\limits_{n=0}^{\infty} a_n x^n$ in DE, then equating the coefficients yields a holonomic recurrence equation RE for a_n. If DE is fully expanded, then we get RE by the formal substitution

$$x^j f^{(k)} \mapsto (n+1-j)_k \cdot a_{n+k-j} \qquad (10.13)$$

in the differential equation DE.

Proof. Let us assume that $f(x)$ has the representation

$$f(x) = \sum_{n=0}^{\infty} a_n x^n = \sum_{n=-\infty}^{\infty} a_n x^n$$

with $a_n = 0$ for $n < 0$. This avoids case distinctions when index shifts are necessary.

We further assume that the differential equation in fully expanded form is given by

$$\sum_{j=0}^{J} \sum_{k=0}^{K} c_{jk} x^j f^{(k)}(x) = 0 \qquad (c_{jk} \in \mathbb{K},\, J, K \in \mathbb{N}_{\geq 0}). \qquad (10.14)$$

By induction, we see that

$$f^{(k)}(x) = \sum_{n=-\infty}^{\infty} (n+1-k)_k a_n x^{n-k} \qquad (10.15)$$

is valid. Hence (10.15) can be substituted in (10.14), and we get

$$
\begin{aligned}
0 &= \sum_{j=0}^{J} \sum_{k=0}^{K} c_{jk} x^j \sum_{n=-\infty}^{\infty} (n+1-k)_k a_n x^{n-k} \\
&= \sum_{n=-\infty}^{\infty} \sum_{j=0}^{J} \sum_{k=0}^{K} c_{jk} (n+1-k)_k a_n x^{n-k+j} \\
&= \sum_{n=-\infty}^{\infty} \sum_{j=0}^{J} \sum_{k=0}^{K} c_{jk} (n+1-j)_k a_{n+k-j} x^n,
\end{aligned}
$$

where we carried out an index shift with respect to n in the last step. Equating the coefficients yields our claim.

Since for fixed $k \in \mathbb{N}_{\geq 0}$ the term $(n+1-j)_k$ is a polynomial in n (of degree k), it follows that the resulting recurrence equation is holonomic and vice versa. $\qquad \Box$

Session 10.21. *The described conversion algorithm is implemented in the* Special-Functions *package as function* DEtoRE. *We start again by loading the package:*

In[1]:= **Needs["SpecialFunctions`"]**

Next, we compute the coefficient recurrence equation which belongs to the differential equation of the exponential function:

In[2]:= **DE = F'[x] - F[x] == 0**

Out[2]= $F'(x) - F(x) == 0$

In[3]:= **RE = DEtoRE[DE, F[x], a[k]]**
Out[3]= $(k+1) a(k+1) - a(k) == 0$

We just mention that this recurrence equation can be converted back by REtoDE*:*

In[4]:= **REtoDE[RE, a[k], F[x]]**

InverseFunction :: "if un": *Inverse functions are being used. Values may be lost for multivalued inverses.*
Out[4]= $F'(x) - F(x) == 0$

Here is a more complicated example:

In[5]:= **DE = HolonomicDE$\left[\left(\dfrac{\text{ArcSin}\left[\sqrt{\text{x}}\right]}{\sqrt{\text{x}}}\right)^2, \text{F}[\text{x}]\right]$**
Out[5]= $2(x-1)x^2 F'''(x) + 3(4x-3)x F''(x) + 2(7x-3) F'(x) + 2 F(x) == 0$

In[6]:= **DEtoRE$\left[\text{DE}, \text{F}[\text{x}], \text{a}[\text{k}]\right]$**
Out[6]= $2(k+1)^3 a(k) - (k+1)(k+2)(2k+3) a(k+1) == 0$

The fact that the resulting recurrence equation is so simple (since it has degree 1), makes it possible to solve it explicitly. This will be considered in detail in the next section.

10.3.3 Hypergeometric Functions

The exponential function as well as the function $\left(\frac{\arcsin\sqrt{x}}{\sqrt{x}}\right)^2$ just considered have in common that both are represented by a series $\sum\limits_{k=0}^{\infty} A_k$ whose coefficients A_k satisfy a holonomic recurrence equation of first order so that

$$\frac{A_{k+1}}{A_k} \in \mathbb{K}(k), \tag{10.16}$$

i.e. the quotient of successive terms is a rational function. Such series are called *hypergeometric series*, and we will now consider this type of series in detail.

In the last section we introduced the Pochhammer symbol as

$$a_k = (a)_k = \underbrace{a \cdot (a+1)\cdots(a+k-1)}_{k \text{ factors}} .$$

In particular $(1)_k = k!$. The Pochhammer symbol satisfies the exceptionally simple recurrence equation

$$\frac{a_{k+1}}{a_k} = \frac{(a)_{k+1}}{(a)_k} = a + k .$$

Therefore the quotient

$$A_k = \frac{(\alpha_1)_k \cdot (\alpha_2)_k \cdots (\alpha_p)_k}{(\beta_1)_k \cdot (\beta_2)_k \cdots (\beta_q)_k} \frac{x^k}{k!} A_0 \tag{10.17}$$

of p Pochhammer symbols α_k ($k = 1, \ldots, p$) in the numerator and $q+1$ Pochhammer symbols β_k ($k = 0, \ldots, q$, $\beta_0 = 1$) in the denominator has the term ratio

$$\frac{A_{k+1}}{A_k} = \frac{(k+\alpha_1)\cdot(k+\alpha_2)\cdots(k+\alpha_p)}{(k+\beta_1)\cdot(k+\beta_2)\cdots(k+\beta_q)\cdot(k+1)}x \quad (k \in \mathbb{N}_{\geq 0}). \tag{10.18}$$

This is equivalent to the following recurrence equation of first order for A_k:

$$(k+\beta_1)\cdot(k+\beta_2)\cdots(k+\beta_q)\cdot(k+1)A_{k+1} - (k+\alpha_1)\cdot(k+\alpha_2)\cdots(k+\alpha_p)xA_k = 0. \tag{10.19}$$

Note that every rational function $r(k) \in \mathbb{C}(k)$ in fully factored form has the representation (10.18). This leads to the following definition.

Definition 10.22. (Hypergeometric Series) The *generalized hypergeometric function* or *series* $_pF_q$ is defined by

$$_pF_q\left(\begin{array}{c}\alpha_1,\alpha_2,\cdots,\alpha_p\\\beta_1,\beta_2,\cdots,\beta_q\end{array}\bigg| x\right) := \sum_{k=0}^{\infty} A_k = \sum_{k=0}^{\infty} \frac{(\alpha_1)_k\cdot(\alpha_2)_k\cdots(\alpha_p)_k}{(\beta_1)_k\cdot(\beta_2)_k\cdots(\beta_q)_k}\frac{x^k}{k!}. \tag{10.20}$$

All series $\sum_{k=0}^{\infty} A_k$ whose summand A_k has a rational term ratio according to (10.16) can therefore (after complete factorization) be represented by a generalized hypergeometric series. The coefficients A_k are uniquely determined by (10.18) or (10.19), respectively, and $A_0 = 1$. The summand A_k of a generalized hypergeometric series is called a *hypergeometric term*.

Therefore a hypergeometric series is a sum of hypergeometric terms, and the ratio of successive hypergeometric terms is rational. In other words, hypergeometric terms are exactly those summands that satisfy a holonomic recurrence equation of first order.

The numbers $\alpha_k \in \mathbb{K}$ are called the *upper* and $\beta_k \in \mathbb{K}$ are called the *lower parameters* of $_pF_q$. Note that $_pF_q$ is well defined if no lower parameter is a negative integer (or zero), and the series converges by the quotient criterion for $p \leq q$ and for $p = q+1$ and $|x| < 1$. Further details about convergence can be found in [Bai1935]. △

The function $_2F_1(a,b;c;x)$ is called *Gaussian hypergeometric series*, $_1F_1(a;b;x)$ is called *Kummer's hypergeometric series* and $_3F_2(a,b,c;d,e;x)$ is called *Clausen's hypergeometric series*. These are the most important hypergeometric series.

Example 10.23. Most functions of calculus are hypergeometric. We give some examples.

(a) For the exponential series

$$e^x = \sum_{k=0}^{\infty} A_k = \sum_{k=0}^{\infty} \frac{1}{k!}x^k$$

we have

$$\frac{A_{k+1}}{A_k} = \frac{k!x^{k+1}}{(k+1)!x^k} = \frac{1}{k+1}x.$$

Therefore from (10.18) it follows that

$$e^x = \sum_{k=0}^{\infty} \frac{1}{k!}x^k = {}_0F_0\left(\begin{array}{c}-\\-\end{array}\bigg| x\right).$$

Hence the exponential function is the simplest hypergeometric function. This can also be read off directly from the hypergeometric coefficient formula (10.17).

(b) For the sine series

$$\sin x = \sum_{k=0}^{\infty} A_k = \sum_{k=0}^{\infty} \frac{(-1)^k}{(2k+1)!} x^{2k+1}$$

we get

$$\frac{A_{k+1}}{A_k} = \frac{(-1)^{k+1}(2k+1)! \, x^{2k+3}}{(-1)^k (2k+3)! \, x^{2k+1}} = \frac{-x^2}{(2k+3)(2k+2)} = \frac{1}{(k+\frac{3}{2})(k+1)} \cdot \left(\frac{-x^2}{4}\right),$$

so that with $A_0 = x$ we obtain according to (10.17) and (10.18) that

$$\sin x = \sum_{k=0}^{\infty} \frac{(-1)^k}{(2k+1)!} x^{2k+1} = x \cdot {}_0F_1\left(\begin{array}{c} \\ \frac{3}{2} \end{array} \middle| \frac{-x^2}{4} \right).$$

In the sequel, we will consider further examples using *Mathematica*. △

Session 10.24. *Mathematica knows the generalized hypergeometric series under the name* HypergeometricPFQ:[23]

In[1]:= **HypergeometricPFQ[{a,b},{c},x]**
Out[1]= ${}_2F_1(a,b;c;x)$

Mathematica can identify many hypergeometric series, for example the exponential series:

In[2]:= **HypergeometricPFQ[{},{},x]**
Out[2]= e^x

The binomial series is evoked as follows:

In[3]:= **HypergeometricPFQ[{-α},{},-x]**
Out[3]= $(x+1)^{\alpha}$

The logarithmic series is evoked by

In[4]:= **x∗HypergeometricPFQ[{1,1},{2},x]**
Out[4]= $-\log(1-x)$

For the sine series, we find

In[5]:= **x∗HypergeometricPFQ$\left[\{\},\left\{\frac{3}{2}\right\},-\frac{x^2}{4}\right]$**
Out[5]= $\sin(x)$

Similarly, for the cosine series, we find

In[6]:= **HypergeometricPFQ$\left[\{\},\left\{\frac{1}{2}\right\},-\frac{x^2}{4}\right]$**
Out[6]= $\cos(x)$

And as a last example, here is the inverse tangent series:

In[7]:= **x∗HypergeometricPFQ$\left[\left\{\frac{1}{2},1\right\},\left\{\frac{3}{2}\right\},-x^2\right]$**
Out[7]= $\tan^{-1}(x)$

Mathematica also recognizes when a hypergeometric series is terminating:

---

[23] The Gaussian series ${}_2F_1$ can also be called via Hypergeometric2F1, and Kummer's series ${}_1F_1$ via Hypergeometric1F1.

$In[8]:=$ **HypergeometricPFQ[{-5,a},{b},x]**

$Out[8]=$ $-\dfrac{a\,(a+1)\,(a+2)\,(a+3)\,(a+4)\,x^5}{b\,(b+1)\,(b+2)\,(b+3)\,(b+4)}+\dfrac{5\,a\,(a+1)\,(a+2)\,(a+3)\,x^4}{b\,(b+1)\,(b+2)\,(b+3)}-$

$\dfrac{10\,a\,(a+1)\,(a+2)\,x^3}{b\,(b+1)\,(b+2)}+\dfrac{10\,a\,(a+1)\,x^2}{b\,(b+1)}-\dfrac{5\,a\,x}{b}+1$

These simplifications are executed completely automatically, without having to call Simplify, *and therefore cannot be avoided by the user.*

The inverse question, i.e., the reconstruction of the hypergeometric representation from the sum representation, is accomplished by the function SumToHypergeometric *from the package* SpecialFunctions. *First, we load the package again:*

$In[9]:=$ **Needs["SpecialFunctions`"]**

As an example, we compute the hypergeometric representation of the exponential series,[24]

$In[10]:=$ **SumToHypergeometric** $\left[\text{sum}\left[\dfrac{1}{k!}\,x^k,\{k,0,\infty\}\right]\right]$

$Out[10]=$ hypergeometricPFQ({}, {}, x)

and of the sine series:

$In[11]:=$ **SumToHypergeometric** $\left[\text{sum}\left[\dfrac{(-1)^k}{(2k+1)!}\,x^{2k+1},\{k,0,\infty\}\right]\right]$

$Out[11]=$ x hypergeometricPFQ $\left(\{\},\{\dfrac{3}{2}\},-\dfrac{x^2}{4}\right)$

As another example, let us consider the series

$$\sum_{k=0}^{\infty}\binom{n}{k}=2^n,\tag{10.21}$$

whose evaluation we know from the binomial formula. This series is converted into

$In[12]:=$ **SumToHypergeometric[sum[Binomial[n,k],{k,0,∞}]]**

$Out[12]=$ hypergeometricPFQ({$-n$}, {}, -1)

Therefore we have deduced the hypergeometric identity

$${}_1F_0\left(\begin{array}{c}-n\\ -\end{array}\middle|-1\right)=2^n\quad(n\in\mathbb{N}_{\geq0}).$$

It is easy to observe that a hypergeometric series has only finitely many summands if and only if one of the upper parameters is a negative integer (or zero). If $n\in\mathbb{N}_{\geq0}$, then this applies to the above example and the sum has the range $k=0,\dots,n$. Because of the appearing binomial coefficient this can of course be seen from (10.21) as well. The appearing bounds are called the natural bounds of the sum (10.21).

The natural bounds cannot always be easily detected from the given representation. However, after successful conversion these can be read off. As an example we consider the Hermite polynomials ($n\in\mathbb{N}_{\geq0}$):

$$H_n(x)=n!\sum_{k=0}^{\infty}\frac{(-1)^k}{k!\,(n-2k)!}\,(2x)^{n-2k}.\tag{10.22}$$

---

[24] To avoid automatic simplification, we use the unevaluated forms sum and hypergeometricPFQ.

We get the hypergeometric representation

$In[13]:=$ **SumToHypergeometric** $\left[\text{sum}\left[\frac{n!\,(-1)^k}{k!\,(n-2k)!}\,(2x)^{n-2k}, \{k,0,\infty\}\right]\right]$

$Out[13]=$ $2^n\,x^n$ hypergeometricPFQ$\left(\left\{-\dfrac{n}{2}, \dfrac{1-n}{2}\right\}, \{\}, -\dfrac{1}{x^2}\right)$

with the natural bounds $k = 0, \ldots, \lfloor n/2 \rfloor.$

Now we continue our considerations with respect to the computation of the power series coefficients of a given holonomic function.

In Session 10.21 we computed the recurrence equation

$$(k+1)a_{k+1} - a_k = 0$$

for the power series coefficients a_k of the exponential function. This equation is of first order and yields—together with one initial value $a_0 = e^0 = 1$—the solution as hypergeometric term $a_k = \frac{1}{k!}$.

In the same session the recurrence equation

$$2(k+1)^3 a_k - (k+1)(k+2)(2k+3)a_{k+1} = 0$$

was deduced for the power series coefficients a_k of the function $f(x) = \left(\dfrac{\arcsin\sqrt{x}}{\sqrt{x}}\right)^2$. Since the coefficients of this first-order recurrence equation are already completely factorized, we can read off these coefficients explicitly using the hypergeometric coefficient formula

$$a_k = \frac{(1)_k (1)_k (1)_k}{(2)_k \left(\frac{3}{2}\right)_k k!} = \frac{k!}{(1+k)\left(\frac{3}{2}\right)_k} = \frac{4^k (k!)^2}{(1+k)(1+2k)!}$$

where we used $a_0 = \lim_{x\to 0} f(x) = 1$. Therefore we get

$$\left(\frac{\arcsin\sqrt{x}}{\sqrt{x}}\right)^2 = \sum_{k=0}^{\infty} \frac{4^k (k!)^2}{(1+k)(1+2k)!} x^k = {}_3F_2\left(\begin{array}{c}1,1,1\\2,\frac{3}{2}\end{array}\middle|\,x\right).$$

This completes the computation of the power series, and it is consistent with the result of the calls

$In[14]:=$ **s = FPS** $\left[\left(\dfrac{\text{ArcSin}\left[\sqrt{x}\right]}{\sqrt{x}}\right)^2, x\right]$

$Out[14]=$ sum$\left(\dfrac{4^k\,x^k\,(k!)^2}{(k+1)\,(2\,k+1)!}, \{k,0,\infty\}\right)$

and

$In[15]:=$ **SumToHypergeometric[s]**

$Out[15]=$ hypergeometricPFQ$\left(\{1,1,1\}, \left\{2, \dfrac{3}{2}\right\}, x\right)$

Hence the function $\arcsin^2 x$ is represented by a Clausen-type hypergeometric series. If we use $\arcsin^2 x$ directly as the input for our algorithm, then we analogously get:

$In[16]:=$ **s = FPS[ArcSin[x]^2, x]**

$Out[16]=$ sum$\left(\dfrac{4^k\,x^{2\,k+2}\,(k!)^2}{(k+1)\,(2\,k+1)!}, \{k,0,\infty\}\right)$

This case yields the recurrence equation

In[17]:= **DEtoRE[HolonomicDE[ArcSin[x]², F[x]], F[x], a[k]]**

Out[17]= $k^3 a(k) - k(k+1)(k+2)a(k+2) == 0$

as intermediate result. Next, we discuss this special situation of a recurrence equation of order larger than 1 with two summands in detail.

If the resulting recurrence equation is of order $m \in \mathbb{N}$, but has the special form

$$\frac{a_{k+m}}{a_k} = R(k) \in \mathbb{K}(k),$$

then we can use a similar approach.[25] We call the functions which are associated with such power series representations *functions of hypergeometric type* and m their *symmetry number*. This completes the algorithm for the computation of power series representations of hypergeometric type. Of course this procedure gives only a "closed form" power series result of $f(x)$ in the hypergeometric, but not in the general holonomic case.

Next we consider some examples.

Example 10.25. (a) Let us consider $f(x) = \arcsin x$. For this function we get according to Section 10.3.1 the differential equation

$$(1 - x^2)f''(x) - xf'(x) = 0,$$

and the transformation discussed in Section 10.3.2 yields the recurrence equation

$$(n+2)(n+1)a_{n+2} - n^2 a_n = 0 \tag{10.23}$$

for the coefficients a_n of the solution $f(x) = \sum_{n=0}^{\infty} a_n x^n$, which is therefore of hypergeometric type with symmetry number 2. Since $\arcsin 0 = 0$, the even part vanishes by (10.23). Therefore we set

$$h(x) = \sum_{k=0}^{\infty} c_k x^k$$

such that $f(x) = xh(x^2)$ or $c_k = a_{2k+1}$, respectively. The substitution $n = 2k+1$ in (10.23) then gives the following recurrence equation for c_k.

$$c_{k+1} = \frac{(k + \frac{1}{2})^2}{(k + \frac{3}{2})(k+1)} c_k.$$

The initial value is $c_0 = a_1 = \arcsin'(0) = 1$, so that finally it follows from the hypergeometric coefficient formula (10.18) that

$$c_k = \frac{\left(\frac{1}{2}\right)_k \cdot \left(\frac{1}{2}\right)_k}{\left(\frac{3}{2}\right)_k k!} = \frac{(2k)!}{(2k+1)4^k(k!)^2},$$

or

$$\arcsin x = \sum_{k=0}^{\infty} \frac{(2k)!}{(2k+1)4^k(k!)^2} x^{2k+1} = x \, {}_2F_1\left(\begin{array}{c} \frac{1}{2}, \frac{1}{2} \\ \frac{3}{2} \end{array} \Bigg| x^2\right).$$

---

[25] We will not give all details here. More details can be found in Example 10.25 and in Theorem 10.26.

The function $\arcsin x$ is therefore represented by a Gaussian type hypergeometric series.

(b) Next, we discuss the *error function*:

$$f(x) = \operatorname{erf} x := \frac{2}{\sqrt{\pi}} \int_0^x e^{-t^2} dt .$$

We find the holonomic differential equation[26]

$$f''(x) + 2xf'(x) = 0$$

and the corresponding recurrence equation

$$(n+2)(n+1)a_{n+2} + 2na_n = 0 .$$

Hence the error function is of hypergeometric type with symmetry number 2. As above it follows that $f(x)$ is odd since $\operatorname{erf} 0 = 0$. Finally, using the initial value $\operatorname{erf}'(0) = \frac{2}{\sqrt{\pi}}$, we get the hypergeometric power series representation

$$\operatorname{erf} x = \frac{2}{\sqrt{\pi}} \sum_{k=0}^{\infty} \frac{(-1)^k}{(2k+1)k!} x^{2k+1} = \frac{2x}{\sqrt{\pi}} {}_1F_1 \left(\left. \begin{array}{c} \frac{1}{2} \\ \frac{3}{2} \end{array} \right| -x^2 \right).$$

Therefore the error function is an example of a Kummer-type hypergeometric series. △

We collect the above algorithm in the next theorem.

Theorem 10.26. (Computation of Hypergeometric Power Series Representations) Let a function $f(x)$ of hypergeometric type with symmetry number m be given as an expression. Then $f(x) = \sum_{k=0}^{\infty} a_k x^k$ has a power series representation of the form

$$f(x) = \sum_{j=1}^{m} f_j(x) = \sum_{j=0}^{m-1} \sum_{k=0}^{\infty} a_{jk} x^{mk+j} , \tag{10.24}$$

whose m subseries $f_j(x)$ $(j = 0, \ldots, m-1)$ are called *m-fold symmetric series*.

The following algorithm computes the coefficients a_{jk} of these power series representations:

(a) Compute a holonomic differential equation for $f(x)$ according to Section 10.3.1.
(b) Convert the differential equation towards a holonomic recurrence equation for a_k according to Section 10.3.2.
(c) If this recurrence equation is of hypergeometric type with symmetry number m, then according to (10.24) the series can be decomposed into m-fold symmetric series $f_j(x)$ $(j = 0, \ldots, m-1)$ whose coefficients a_{jm} satisfy a recurrence equation of order 1 for every $j = 0, \ldots, m-1$.
(d) These m recurrence equations can be solved using m initial values according to the hypergeometric coefficient formula.

---

[26] By definition, the derivatives of the error function are elementary and do not contain any integrals.

Proof. One can show [Koe1992] that every function $f(x)$ of hypergeometric type is holonomic.[27] Therefore step (a) will be successful. The conversion in step (b) works for all holonomic differential equations. Steps (c) and (d) were justified in this section.

If the conversion in (b) does not yield a recurrence equation of hypergeometric type,[28] although $f(x)$ has this property, then an algorithm given by Teguia [Teg2020] finds those solutions of hypergeometric type.   □

Note that the given algorithm can be extended to the computation of Laurent and Puiseux series. Furthermore, note that the symmetry number of a function of hypergeometric type is not unique. For example, every multiple of a symmetry number is also a symmetry number. Therefore the output of the given algorithm is generally not unique.

10.3.4 Efficient Computation of Taylor Polynomials of Holonomic Functions

Of course not every holonomic function $f(x)$ is of hypergeometric type. In this case the algorithm of Theorem 10.26 will not yield an explicit formula for the power series representation of $f(x)$. Nevertheless, the algorithm yields a particularly efficient method for the computation of a truncated Taylor series of high order:[29] to compute the nth Taylor polynomial of a holonomic function for whose Taylor coefficients we have computed a holonomic recurrence equation RE of order $m \ll n$ in step (b), we compute the first m coefficients by Taylor's formula. All further Taylor coefficients can then be iteratively computed (very fast!) using the recurrence equation RE.

Session 10.27. *The function* `Taylor` *which is part of the* `SpecialFunctions` *package uses this algorithm for the computation of Taylor polynomials (in contrast to* `Series`*). Of course this command is only applicable to holonomic functions! Note that the result is a polynomial, not a* `SeriesData` *object. We consider some examples:*

$In[1]:=$ **Series[ArcTan[x]^2, {x, 0, 10}]**

$Out[1]=$ $x^2 - \dfrac{2x^4}{3} + \dfrac{23x^6}{45} - \dfrac{44x^8}{105} + \dfrac{563x^{10}}{1575} + O(x^{11})$

$In[2]:=$ **Taylor[ArcTan[x]^2, {x, 0, 10}]**

$Out[2]=$ $\dfrac{563x^{10}}{1575} - \dfrac{44x^8}{105} + \dfrac{23x^6}{45} - \dfrac{2x^4}{3} + x^2$

Of course both results agree. Now let us check the efficiency of both approaches by choosing a higher order:

$In[3]:=$ **Series[ArcTan[x]^2, {x, 0, 2000}]; //Timing**

$Out[3]=$ {0.703125 Second, Null}

$In[4]:=$ **Taylor[ArcTan[x]^2, {x, 0, 2000}]; //Timing**

$Out[4]=$ {0.125 Second, Null}

---

[27] The argument is that the generalized hypergeometric function satisfies a holonomic differential equation, see Exercise 10.16. Hence $f_j(x)$ given by (10.24) are—as compositions with rational functions—also holonomic. As a sum of holonomic functions, $f(x)$ is then holonomic, too.

[28] This happens very rarely!

[29] This is the asymptotically fastest algorithm for this purpose.

The timing difference is visible, although it is still modest for this order. In essence, the efficiency advantage depends on the complexity of the given input function. For a product as, e.g.,

$In[5] := $ **Series[e^{x}Cos[x],{x,0,20}]**

$Out[5] = 1 + x - \dfrac{x^3}{3} - \dfrac{x^4}{6} - \dfrac{x^5}{30} + \dfrac{x^7}{630} + \dfrac{x^8}{2520} + \dfrac{x^9}{22680} -$

$\dfrac{x^{11}}{1247400} - \dfrac{x^{12}}{7484400} - \dfrac{x^{13}}{97297200} + \dfrac{x^{15}}{10216206000} + \dfrac{x^{16}}{81729648000} +$

$\dfrac{x^{17}}{1389404016000} - \dfrac{x^{19}}{237588086736000} - \dfrac{x^{20}}{2375880867360000} + O(x^{21})$

$In[6] := $ **Taylor[e^{x}Cos[x],{x,0,20}]**

$Out[6] = -\dfrac{x^{20}}{2375880867360000} - \dfrac{x^{19}}{237588086736000} + \dfrac{x^{17}}{1389404016000} +$

$\dfrac{x^{16}}{81729648000} + \dfrac{x^{15}}{10216206000} - \dfrac{x^{13}}{97297200} - \dfrac{x^{12}}{7484400} -$

$\dfrac{x^{11}}{1247400} + \dfrac{x^9}{22680} + \dfrac{x^8}{2520} + \dfrac{x^7}{630} - \dfrac{x^5}{30} - \dfrac{x^4}{6} - \dfrac{x^3}{3} + x + 1$

the efficiency difference is much higher:

$In[7] := $ **Series[e^{x}Cos[x],{x,0,2000}]; //Timing**

$Out[7] = $ {22.4063 Second, Null}

$In[8] := $ **Taylor[e^{x}Cos[x],{x,0,2000}]; //Timing**

$Out[8] = $ {0.0625 Second, Null}

For this example not even the computation using an "explicit" formula is faster:[30]

$In[9] := $ **ps = FPS[e^{x}Cos[x],x]//Timing**

$Out[9] = \left\{0.015625 \text{ Second}, \text{sum}\left(\dfrac{2^{k/2} x^k \cos\left(\frac{k\pi}{4}\right)}{k!}, \{k, 0, \infty\}\right)\right\}$

$In[10] := $ **ps/.{sum → Sum, ∞ → 2000}; //Timing**

$Out[10] = $ {0.0625 Second, Null}

If the order is rather high, then for this example Taylor *is as efficient as the explicit computation.*

10.4 Algebraic Functions

In this section we consider a further class of holonomic functions.

Definition 10.28. (Algebraic Functions) A function $y(x)$ which is implicitly given as the zeros of an irreducible bivariate polynomial,

$$F(x, y(x)) = 0, \qquad F(x, y) \in \mathbb{K}[x, y], \qquad (10.25)$$

is called an *algebraic function* over $\mathbb{K}$. The highest degree of F with respect to the variable y in (10.25) is called the *degree of the algebraic function* and is denoted by $\text{algdeg}(y(x), x)$. Every algebraic function over $\mathbb{R}$ geometrically represents a subset of the x-y-plane $\mathbb{R}^2$, the *graph of the algebraic function*.

---

[30] The formula is found since the associated recurrence equation has constant coefficients.

Since an algebraic function is not necessarily unique, we might locally consider one of its branches. △

Before we show that every algebraic function is also holonomic, we would like to consider some examples. For the sake of simplicity we will write y instead of $y(x)$. However, we must pay attention to the dependency of y on x, for example in the differentiation process.

Example 10.29. (a) We consider the implicit equation

$$F(x, y) = x^2 + y^2 - r^2 = 0.$$

                                                                                    (10.26)

This is obviously an algebraic function of degree 2 over $\mathbb{Q}$ (or $\mathbb{R}$). Of course we know that the graph of this equation is a circle:

```
In[1]:= F = x² + y² - r² == 0
Out[1]= -r² + x² + y² == 0

In[2]:= ContourPlot[Evaluate[F/.{r → 1}], {x, -1, 1}, {y, -1, 1}]
```

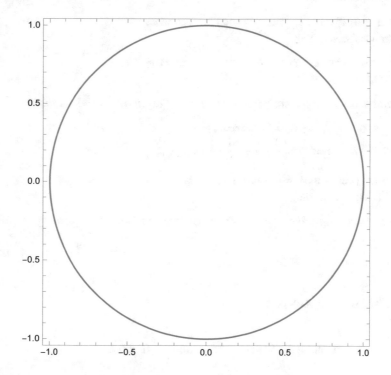

If we solve the equation $F(x, y) = 0$ for y, then we get the two branches

$$y_+(x) = \sqrt{r^2 - x^2} \quad \text{and} \quad y_-(x) = -\sqrt{r^2 - x^2}$$

of the algebraic function which refer to the upper and the lower half-circle, respectively.

To find a holonomic differential equation for $y(x)$, we initially solve the defining equation for the highest power in $y(x)$ to get

$$y^2 = r^2 - x^2.$$

                                                                                    (10.27)

Differentiation of (10.26) further yields

$$2x + 2yy'(x) = 0.\tag{10.28}$$

We solve for $y'(x)$, and obtain the explicit form

$$y'(x) = -\frac{x}{y}.$$

This is an *explicit differential equation of first order* which, however, is not holonomic. But the ratio $\frac{y'(x)}{y(x)}$ can be simplified using (10.27):

$$\frac{y'(x)}{y(x)} = -\frac{x}{y(x)^2} = -\frac{x}{r^2 - x^2}.$$

Since the right term is independent of y, we have discovered the holonomic differential equation of first order:

$$(r^2 - x^2)y'(x) + xy(x) = 0.$$

Hence, $y(x)$ is holonomic of degree 1.

(b) As our next example we consider the similar equation of a circle centered at $(0, r)$:

$$(y(x) - r)^2 + x^2 - r^2 = 0.$$

In this case, simplification of $\frac{y'(x)}{y(x)}$ yields the expression

$$\frac{y'(x)}{y(x)} = \frac{x}{x^2 - ry},$$

which obviously is *not* independent of y. This proves that a holonomic differential equation of first order *does not exist*.

Hence we iterate the process and differentiate the equation

$$y'(x) = \frac{xy}{x^2 - ry}$$

a second time, continue[31] similarly as in Session 10.8 and finally get the holonomic differential equation

$$x(r^2 - x^2)y''(x) - r^2 y'(x) = 0.$$

Therefore $y(x)$ is holonomic of degree 2. △

Theorem 10.30. (Holonomicity of Algebraic Functions) An algebraic function $y(x)$ of degree n is holonomic of degree $\mathrm{holdeg}(y(x), x) \leqq n$.

Proof. We prove this theorem by presenting an algorithm for the computation of a holonomic differential equation of order n. Of course, this algorithm can again be invoked iteratively, for $m = 1, \ldots, n$, in order to find a holonomic differential equation of lowest order m.

For this purpose, let the algebraic function be given by the equation (10.25):

$$F(x, y(x)) = \sum_{k=0}^{n} p_k(x)y(x)^k = 0 \qquad (p_k(x) \in \mathbb{K}[x]).\tag{10.29}$$

---

[31] The details can be found in Theorem 10.30.

Let further

$$y^{(m)}(x) + \sum_{k=0}^{m-1} A_k y^{(k)}(x) = 0 \qquad (10.30)$$

be an ansatz for a holonomic differential equation of order $m \leq n$ for $y(x)$ with yet undetermined coefficients $A_k \in \mathbb{K}(x)$.

If we differentiate (10.29), then the chain rule generates a linear factor $y'(x)$, so that we can solve the differentiated equation for $y'(x)$, and we obtain

$$y'(x) = R(x, y) \in \mathbb{K}(x, y).$$

Iterative differentiation further yields, for all $k = 1, \ldots, n$,

$$y^{(k)}(x) = R_k(x, y) \in \mathbb{K}(x, y).$$

In the next step we substitute these derivatives in (10.30). Next we substitute the powers $y(x)^j$ of $y(x)$ with $j \geq n$ recursively according to (10.25). For efficiency reasons in every iteration step, one should multiply by the respective common denominator. This procedure results in a polynomial equation (where $z_0, z_1, \ldots, z_{m-1}$ denote m formal variables)

$$P(x, y(x), A_0, A_1, \ldots, A_{m-1}) = 0 \quad (P \in \mathbb{K}(x, y, z_0, z_1, \ldots, z_{m-1}))$$

whose degree with respect to the second variable y is at most $n - 1$. This polynomial can only be identically zero if all of its coefficients with respect to y vanish. Therefore we can equate the coefficients, which yields a linear system for the variables $A_0, A_1, \ldots, A_{m-1}$. For $m = n$ the existence of a solution can be shown by an algebraic argument similar as in Theorem 10.9. The resulting solution has generated $A_j \in \mathbb{K}(x)$. If we substitute these variables in the ansatz (10.30) and multiply by their common denominator, we finally get the wanted holonomic differential equation for $y(x)$.   □

Session 10.31. *The given algorithm is implemented in the package* `SpecialFunctions` *as the command* `AlgebraicDE`. *Let us consider Example 10.29 again:*

`In[1]:= Needs["SpecialFunctions`"]`

`In[2]:= DE1 = AlgebraicDE[x² + y² - r², y[x]]`
`Out[2]=` $\left(x^2 - r^2\right) y'(x) - x\, y(x) == 0$

`In[3]:= DE2 = AlgebraicDE[(y - r)² + x² - r², y[x]]`
`No differential equation of order 1 found`
`Out[3]=` $r^2 y'(x) - x (r - x) (r + x) y''(x) == 0$

These computations confirm the results of Example 10.29. As another example, the algebraic function of degree 8 which is given by[32]

$$(y^2 - x^2)^4 - x^2 y^2 = 0$$

has the holonomic degree 3:

`In[4]:= DE3 = AlgebraicDE[ (y² - x²)⁴ - x²y², y[x]]`
`No differential equation of order 1 found`

---

[32] Actually the defining polynomial can be factorized over $\mathbb{Q}$ into two factors of degree 4, so that over $\mathbb{Q}$ this is the union of two algebraic functions of order 4.

```
No differential equation of order 2 found
```
$Out[4] = 2(512x^4 - 27)x^2 y''(x) + (256x^4 + 27)x^3 y'''(x) + 69 x y'(x) - 45 y(x) == 0$

This example again shows that the holonomic degree can be smaller that the algebraic degree.

Now we again consider the question how an algebraic function can be represented by a power series. Without loss of generality we choose the origin as center. To compute the value of an algebraic function at the origin, we substitute $x = 0$ in the defining polynomial equation. Therefore an algebraic function of degree n has up to n different values which belong to the n branches of the algebraic function. The power series expansions of the branches of an algebraic function turn out to be Puiseux series in general. Of course these Puiseux series can be determined by conversion of the holonomic differential equation into a holonomic recurrence equation for its coefficients.

$In[5] :=$ **DEtoRE[DE1,y[x],a[k]]**
$Out[5] = (k-1)a(k) - (k+2)r^2 a(k+2) == 0$

$In[6] :=$ **DEtoRE[DE2,y[x],a[k]]**
$Out[6] = (k-1)k a(k) - k(k+2)r^2 a(k+2) == 0$

$In[7] :=$ **DEtoRE[DE3,y[x],a[k]]**
$Out[7] = 256(k-1)k(k+2)a(k) + 3(k+1)(3k+7)(3k+11)a(k+4) == 0$

The above computations show that all three algebraic functions which we considered are of the hypergeometric type.

Example 10.32. We continue our considerations from Example 10.29. For the algebraic function $x^2 + y^2 - r^2 = 0$ we had computed the holonomic differential equation $(r^2 - x^2)y'(x) + xy(x) = 0$, and for the coefficients a_k of the corresponding power series $y(x) = \sum_{k=0}^{\infty} a_k x^k$ the recurrence equation $(k-1)a_k - (k+2)r^2 a_{k+2} = 0$ was obtained.

Since $y(0)^2 - r^2 = 0$, we have $y(0) = \pm|r|$, and the differentiated equation (10.28) implies $y'(0) = 0$. Therefore we get

$$y(x) = \pm \sum_{k=0}^{\infty} \frac{|r|(2k)!}{4^k r^{2k}(2k-1)k!^2} x^{2k}.$$

In a similar way, $(y(x) - r)^2 + x^2 - r^2 = 0$ leads to the two representations

$$y(x) = \sum_{k=0}^{\infty} \frac{(2k)!}{2^{2k+1} r^{2k+1}(1+k)k!^2} x^{2+2k}$$

and

$$y(x) = 2r - \sum_{k=0}^{\infty} \frac{(2k)!}{2^{2k+1} r^{2k+1}(1+k)k!^2} x^{2+2k}.$$

The third example can be treated in the same way. In this case, however, we need four initial values. △

10.5 Implicit Functions

The method for the computation of Taylor polynomials as described in Section 10.3.4 can also be applied to algebraic functions. Algebraic functions are special cases of *implicit functions*.

In the sequel, we shall study how Taylor polynomials of implicit functions can be computed in general. To this end, let a function $F(x, y) : \mathbb{R}^2 \to \mathbb{R}$ of two variables be given and let $F(x, y) = 0$, where we consider $y(x)$ as a function of the variable x. The Implicit Function Theorem, see e.g. [Koe1994], guarantees that (under weak conditions) such solution branches exist in a suitable neighborhood of a given point (x_0, y_0), for which $F(x_0, y_0) = 0$ is satisfied.

Example 10.33. A typical example is again the equation of the unit circle,

$$F(x, y) = x^2 + y^2 - 1 = 0,$$

which moreover is an algebraic function. For every (x_0, y_0) with $x_0^2 + y_0^2 = 1$ and $y_0 > 0$ this yields the branch

$$y_+(x) = \sqrt{1 - x^2},$$

whereas for every (x_0, y_0), with $x_0^2 + y_0^2 = 1$ and $y_0 < 0$, we get the branch

$$y_-(x) = -\sqrt{1 - x^2}.$$

The case $y_0 = 0$ is degenerated, because in this case the moderate conditions of the Implicit Function Theorem are not satisfied.[33] △

Now we present an iterative procedure for the computation of the Taylor coefficients of a branch of an implicit function. For this purpose we differentiate the defining equation $F(x, y(x)) = 0$ with the multidimensional chain rule to get

$$\frac{\partial F}{\partial x}(x, y) + \frac{\partial F}{\partial y}(x, y) \cdot y'(x) = 0.$$

This equation can be solved for $y'(x)$ with the result

$$y'(x) = -\frac{\frac{\partial F}{\partial x}(x, y)}{\frac{\partial F}{\partial y}(x, y)} =: F_1(x, y). \tag{10.31}$$

To obtain the higher derivatives iteratively, we differentiate F_1 and get

$$y''(x) = \frac{\partial F_1}{\partial x}(x, y) + \frac{\partial F_1}{\partial y}(x, y) \cdot y'(x) = \frac{\partial F_1}{\partial x}(x, y) + \frac{\partial F_1}{\partial y}(x, y) \cdot F_1(x, y) =: F_2(x, y)$$

and subsequently

$$y'''(x) = \frac{\partial F_2}{\partial x}(x, y) + \frac{\partial F_2}{\partial y}(x, y) \cdot F_1(x, y) =: F_3(x, y).$$

This procedure can obviously be continued and therefore represents an iterative method for the successive computation of the higher derivatives of $y(x)$. An application of Taylor's theorem then yields the Taylor coefficients.

---

[33] At these points the slope is unbounded.

Session 10.34. *The following function* ImplicitTaylor *is an implementation of the itera-tive algorithm presented:*

```
In[1]:= Clear[ImplicitTaylor]
        ImplicitTaylor[F_,y_[x_],y0_,n_] :=
          Module[{yprime,k},
            yprime = -D[F,x]/D[F,y];
            derivatives =
              NestList[
                Together[D[#,x]+D[#,y]*yprime]&,
                yprime,n-1];
            y0+
              Sum[
                Limit[Limit[derivatives[[k]],y→y0],
                      x→0]/k!*x^k,{k,1,n}]
          ]
```

Based on this algorithm, let us compute the Taylor polynomial for the equation of the unit circle with the initial value $y(0) = 1$*:*

```
In[2]:= ImplicitTaylor[x^2+y^2-1,y[x],1,8]
```
$$Out[2]= -\frac{5\,x^8}{128} - \frac{x^6}{16} - \frac{x^4}{8} - \frac{x^2}{2} + 1$$

Since we have not declared the variable derivatives—*which contains the list of the first derivatives of the considered function—as* local, *this variable is now globally accessible. There-fore the following call shows the first 8 derivatives of* $y(x)$*:*

```
In[3]:= derivatives
```
$$Out[3]= \Big\{ -\frac{x}{y},\ \frac{-x^2-y^2}{y^3},\ -\frac{3\,(x^3+y^2\,x)}{y^5},$$

$$-\frac{3\,(5\,x^4+6\,y^2\,x^2+y^4)}{y^7},\ -\frac{15\,(7\,x^5+10\,y^2\,x^3+3\,y^4\,x)}{y^9},$$

$$-\frac{45\,(21\,x^6+35\,y^2\,x^4+15\,y^4\,x^2+y^6)}{y^{11}},$$

$$-\frac{315\,(33\,x^7+63\,y^2\,x^5+35\,y^4\,x^3+5\,y^6\,x)}{y^{13}},$$

$$-\frac{315\,(429\,x^8+924\,y^2\,x^6+630\,y^4\,x^4+140\,y^6\,x^2+5\,y^8)}{y^{15}} \Big\}$$

In particular we have computed the differential equation $y' = -\frac{x}{y}$ *for the equation of the unit circle. Note that these computations yield an* explicit *differential equation of first order (10.31) for every implicit function. However, this differential equation is generally not linear.*

Our next example is the Taylor polynomial of the function $y(x)$*, given by the equation* $y\,e^y = x$ *and the initial value* $y(0) = 0$*:*

```
In[4]:= imp = ImplicitTaylor[y e^y-x,y[x],0,10]
```
$$Out[4]= -\frac{156250\,x^{10}}{567} + \frac{531441\,x^9}{4480} - \frac{16384\,x^8}{315} +$$

$$\frac{16807\,x^7}{720} - \frac{54\,x^6}{5} + \frac{125\,x^5}{24} - \frac{8\,x^4}{3} + \frac{3\,x^3}{2} - x^2 + x$$

Mathematica knows this function under the name ProductLog*:*[34]

---

[34] This function is well known as *Lambert W function*. In TraditionalForm *Mathematica* also uses the abbre-viation *W*.

```
In[5]:= Series[ProductLog[x],{x,0,10}]
```
$$Out[5]= x - x^2 + \frac{3\,x^3}{2} - \frac{8\,x^4}{3} + \frac{125\,x^5}{24} - \frac{54\,x^6}{5} +$$
$$\frac{16807\,x^7}{720} - \frac{16384\,x^8}{315} + \frac{531441\,x^9}{4480} - \frac{156250\,x^{10}}{567} + O(x^{11})$$

```
In[6]:= Solve[y e^y - x == 0]
```
$$Out[6]= \{\{y \to W(x)\}\}$$

Arguments from function theory show that the radius of convergence of the power series of the ProductLog function equals $1/e$, since the singularity in $\mathbb{C}$ nearest to the origin is at the point $x = -\frac{1}{e}$. This follows from the computation

```
In[7]:= x e^x /. Solve[D[x e^x, x] == 0, x][[1]]
```
$$Out[7]= -\frac{1}{e}$$

Next comes the graph of the ProductLog function together with the computed Taylor approximation of order 10 in a reasonable interval:

```
In[8]:= Plot[{ProductLog[x], imp}, {x, -\frac{1}{e}, \frac{1}{2}}]
```

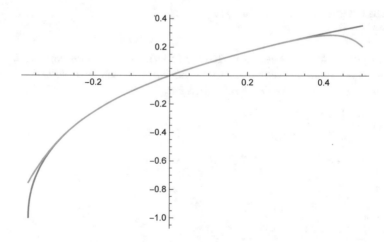

Finally we consider the implicit function $x^2 \ln y + 1 = 0$, $y(0) = 0$:

```
In[9]:= ImplicitTaylor[x^2 Log[y] + 1, y[x], 0, 10]
```
$$Out[9]= 0$$

If we solve the equation $x^2 \ln y + 1 = 0$ for y, then we get $y = e^{-1/x^2}$. Therefore the Taylor polynomial of every order equals 0. This example shows that the order of the limit computation is important, since in the given case a two-dimensional limit does not exist.

Note that the presented algorithm is rather efficient for small orders. Taylor polynomials of high order, however, should not be computed using this method:

```
In[10]:= ImplicitTaylor[x^2 + y^2 - 1, y[x], 1, 255]; //Timing
```
$$Out[10]= \{1.53125\ \text{Second}, \text{Null}\}$$

In the sequel we will consider a much more efficient method for this purpose.

To accelerate the computation we have a divide-and-conquer approach in mind. For this purpose, we will apply a variant of *Newton's method*[35] which is used in numerical analysis for the approximation of the zeros of a real function. Its nice property is the *quadratic convergence*: under rather weak conditions Newton's method approximately doubles the number of correct digits in each iteration step (see [Koe1993a], Chapter 10, and [Koe2010]).

For the computation of a solution of the equation $f(x) = 0$, hence solving for the variable x, one starts with an approximation x_0 and iterates according to the scheme

$$x_{n+1} = x_n - \frac{f(x_n)}{f'(x_n)} \qquad (n \in \mathbb{N}_{\geq 0}) \ . \tag{10.32}$$

Session 10.35. *Before we adapt this method to the computation of truncated power series, we shortly want to show how the ordinary Newton method can be implemented. Under the assumption that the initial value x_0 is a decimal number, the continued application of iteration (10.32) stops with a repeated value since there are only finitely many decimal numbers of fixed length.*[36]

```
In[1]:= Clear[NewtonList, NewtonMethod]

    NewtonList[f_, {x_, x0_Real}] := Module[{G, z},
        G[z_] := z - (f/D[f, x] /. x → z);
        FixedPointList[G, x0]
    ]

    NewtonMethod[f_, {x_, x0_Real}] := Module[{G, z},
        G[z_] := z - (f/D[f, x] /. x → z);
        FixedPoint[G, x0]
    ]
```

As an example, we compute an approximation of π as zero of the sine function:

```
In[2]:= NewtonMethod[f = Sin[x], {x, 3.0}]
Out[2]= 3.14159
```

NewtonList *returns the iteration process:*

```
In[3]:= NewtonList[f, {x, 3.0}]
Out[3]= {3., 3.14255, 3.14159, 3.14159, 3.14159}
```

If we call NewtonMethod *with a non-numerical initial value, no computation is executed since the symbolic computation prevents convergence in such a case:*

```
In[4]:= NewtonMethod[f, {x, 3}]
Out[4]= NewtonMethod(sin(x), {x, 3})
```

This effect can be seen by the 4-fold iteration:

```
In[5]:= FixedPointList[# - (f/D[f, x] /. x → #) &, 3, 4]
Out[5]= {3, 3 - tan(3), 3 - tan(3) - tan(3 - tan(3)),
         3 - tan(3) - tan(3 - tan(3)) - tan(3 - tan(3) - tan(3 - tan(3))),
         3 - tan(3) - tan(3 - tan(3)) - tan(3 - tan(3) - tan(3 - tan(3))) -
             tan(3 - tan(3) - tan(3 - tan(3)) - tan(3 - tan(3) - tan(3 - tan(3))))}
```

Our last example computes the solution of the equation $x = e^{-x}$:

---

[35] This method is also known under the name *Newton-Raphson method.*

[36] This is true whenever the convergence is *not* towards 0. Why?

In[6]:= **NewtonMethod[x - e^{-x}, {x, 0.0}]**

Out[6]= 0.567143

Note that the result can be written in terms of the ProductLog function:

In[7]:= **sol = Solve[x - e^{-x} == 0, x]**

InverseFunction :: **"ifun"**: *Inverse functions are being used. Values may be lost for multivalued inverses.*

Solve :: **"ifun"**: *Inverse functions are being used by* Solve, *so some solutions may not be found*

Out[7]= $\{\{x \rightarrow W[1]\}\}$

In[8]:= **N[sol]**

Out[8]= $\{\{x \rightarrow 0.567143\}\}$

Obviously the two decimal numbers agree.

Note that the given implementation did not yet use the complete efficiency of the method. In fact, since every iteration doubles the number of correct digits, we could have started with lower precision. We will use this effect soon.

To solve the implicit equation $F(x, y) = 0$ for y, we apply Newton's method to $f(y) := F(x, y)$. This yields iteratively *approximating functions* in the variable x, which can be expanded as Taylor series so that we get power series approximations for $y(x)$. By the quadratic convergence of the method we expect that each iteration step will double the number of correct power series coefficients. This is the typical divide-and-conquer effect. It can be proved that this assertion is true, see [GG1999], Algorithm 9.22.

Session 10.36. *The function* ImplicitTaylor2 *is an implementation of the above method:*

```
In[1]:= Clear[ImplicitTaylor2]
        ImplicitTaylor2[f_, y_[x_], y0_, n_] :=
          Module[{F, z},
            F[z_] := (f /. y -> z);
            approx = Nest[# - F[#]/F'[#]&, y0,
              Log[2, n + 1]];
            approx + O[x]^(n + 1)
          ] /; IntegerQ[Log[2, n + 1]]
```

Without loss of generality the program assumes that the order n is of the form $n = 2^k - 1$ for some $k \in \mathbb{N}$ since then the number of computed coefficients is a power of 2. The number of iterations is $\log_2(n + 1)$, and we hope that this is enough. This implementation builds the Taylor polynomial in the last step.

As a first application of the algorithm, we again get

In[2]:= **ImplicitTaylor2[x^2 + y^2 - 1, y[x], 1, 7]**

Out[2]= $1 - \dfrac{x^2}{2} - \dfrac{x^4}{8} - \dfrac{x^6}{16} + O(x^8)$

Since approx *is a global variable, we can check which approximation function was computed:*

In[3]:= **approx**

$$Out[3]= \; -\frac{x^2}{2}-\frac{x^2+\left(1-\frac{x^2}{2}\right)^2-1}{2\left(1-\frac{x^2}{2}\right)}-\frac{x^2+\left(-\frac{x^2}{2}-\frac{x^2+\left(1-\frac{x^2}{2}\right)^2-1}{2\left(1-\frac{x^2}{2}\right)}+1\right)^2-1}{2\left(-\frac{x^2}{2}-\frac{x^2+\left(1-\frac{x^2}{2}\right)^2-1}{2\left(1-\frac{x^2}{2}\right)}+1\right)}+1$$

We see that the generated expression is quite complex. This can be improved by the following implementation:

```
In[4]:= Clear[ImplicitTaylor3]
        ImplicitTaylor3[f_,y_[x_],y0_,n_] :=
          Module[{F,z},
            F[z_] := (f/.y→z);
            approx = Nest[Together[#-F[#]/F'[#]]&,
              y0,Log[2,n+1]];
            approx+O[x]^(n+1)
          ]/; IntegerQ[Log[2,n+1]]
```

The above example now yields

```
In[5]:= ImplicitTaylor3[x^2+y^2-1,y[x],1,7]
```
$$Out[5]= \; 1-\frac{x^2}{2}-\frac{x^4}{8}-\frac{x^6}{16}+O(x^8)$$

```
In[6]:= approx
```
$$Out[6]= \; \frac{-x^8+32\,x^6-160\,x^4+256\,x^2-128}{8\,(x^2-2)\,(x^4-8\,x^2+8)}$$

Next we consider several other examples. The exponential function can be considered as the inverse of the logarithm function:

```
In[7]:= ImplicitTaylor3[Log[y]-x,y[x],1,7]
```
$$Out[7]= \; 1+x+\frac{x^2}{2}+\frac{x^3}{6}+\frac{x^4}{24}+\frac{x^5}{120}+\frac{x^6}{720}+\frac{x^7}{5040}+O(x^8)$$

And the ProducLog function is defined implicitly by

```
In[8]:= ImplicitTaylor3[y e^y-x,y[x],0,7]
```
$$Out[8]= \; x-x^2+\frac{3\,x^3}{2}-\frac{8\,x^4}{3}+\frac{125\,x^5}{24}-\frac{54\,x^6}{5}+\frac{16807\,x^7}{720}+O(x^8)$$

If we compare the examples with those of `ImplicitTaylor`, *we see that our hope was confirmed: in all the above cases we have computed* $8 = 2^3$ *coefficients with only three iterations.*

However, the following example does not work:

```
In[9]:= ImplicitTaylor3[x^2 Log[y]+1,y[x],0,7]
```

Power :: **"infy"**: *Infinite expression* $\frac{1}{0}$ *encountered*

$Out[9]=$ Indeterminate

The reason is that the input function $F(x,y) = x^2 \ln y + 1$ *does not possess a power series representation with respect to* y. *Such examples are out of the scope of the current method.*

Unfortunately, the present implementation is not much faster than the previous one, despite the supposed divide-and-conquer property. Therefore we would like to repair these shortcomings by the following measures:

1. *we use a truncated series already at the very beginning so that no large terms are generated;*

2.   *we increase the computation length of the resulting truncated series step-wise.*

In the current situation these measures are quite successful.

```
In[10]:= Clear[FastImplicitTaylor]
         FastImplicitTaylor[f_,y_[x_],y0_,n_] :=
           Module[{F,z,approx},
             F[z_] :=
               Normal[(f/.y→z)+O[x]^(n+1)];
             approx = y0;
             Do[
               approx =
                 Normal[#-F[#]/F'[#]&[approx]+
                     O[x]^(2^k)],
               {k,1,Log[2,n+1]}];
             approx+O[x]^(n+1)
           ]/; IntegerQ[Log[2,n+1]]
```

Let us compare the timings of the four different implementations of ImplicitTaylor:

```
In[11]:= Timing[imp1 = ImplicitTaylor[x² +y² -1,y[x],1,511];]
Out[11]= {6.3125 Second, Null}
```

```
In[12]:= Timing[imp2 = ImplicitTaylor2[x² +y² -1,y[x],1,511];]
Out[12]= {3.125 Second, Null}
```

```
In[13]:= Timing[imp3 = ImplicitTaylor3[x² +y² -1,y[x],1,511];]
Out[13]= {2. Second, Null}
```

```
In[14]:= Timing[
           imp4 = FastImplicitTaylor[x² +y² -1,y[x],1,511];]
Out[14]= {0.375 Second, Null}
```

We also compare the results of the different implementations:

```
In[15]:= {imp1 - imp2, imp1 - imp3, imp1 - imp4}
Out[15]= {O(x^{512}), O(x^{512}), O(x^{512})}
```

Now we can compute high-order Taylor approximations which are out of reach for Implicit-Taylor:

```
In[16]:= Timing[FastImplicitTaylor[x² +y² -1,y[x],1,2047];]
Out[16]= {15.625 Second, Null}
```

Of course in most cases this algorithm cannot compete against the computation by Series, *using an explicit representation of y:*

```
In[17]:= Timing[Series[√(1-x²), {x,0,2047}];]
Out[17]= {0.015625 Second, Null}
```

For counter-examples, see [Koe2010]. Nevertheless, the results show that Newton's method is obviously suitable to compute Taylor polynomials of high order for implicitly given functions.

10.6 Additional Remarks

A disadvantage of the recursive computation of truncated power series is the fact that some operations, such as differentiation and division, result in a loss of the truncation order. In order to avoid this, one can work with *strings*[37] and *lazy evaluation*. This means that the series is computed up to a given truncation order together with an instruction how to compute the next elements. Those elements can then be computed on demand.[38] Further details can be found in [Nor1975] and [JS1992].

The computation of Taylor series of functions of hypergeometric type was presented in [Koe1992]. This algorithm was extended and linearized in [Teg2020]. The special case of algebraic functions was considered in [CC1986], [CC1987] and [Koe1996].

10.7 Exercises

Exercise 10.1. (SeriesData)

(a) Enter the following formal power series up to order 10.

$$a(x) = \sum_{k=0}^{\infty} x^k, \quad b(x) = \sum_{k=0}^{\infty} \frac{1}{k+1} x^k \quad and \quad c(x) = \sum_{k=0}^{\infty} k! x^k.$$

(b) Compute all possible sums and products of the three series.

(c) Compute the reciprocals $a(x)^{-1}, b(x)^{-1}$ and $c(x)^{-1}$.

(d) Describe the meaning of the components of a `SeriesData` object in detail.

(e) Use the results of (d) to implement the function `PowerSeriesQ[a]`, which returns whether a is a power series or not (in contrast to a Laurent or Puiseux series), and the functions `Variable[a]`, `TruncationOrder[a]`, which return the variable and the truncation order of the series a. These should be very short programs!

(f) Implement the algorithm of Theorem 10.1 as procedure `Reciprocal[a]`, which returns the reciprocal of a.[39] Based on this procedure, compute $a(x)^{-1}, b(x)^{-1}$ and $c(x)^{-1}$ again, and compare your results with those of (c).

Exercise 10.2. (Truncated Power Series)

(a) Generate the truncated power series of the exponential function $f(x) = e^x - 1$ with truncation order 20. What is the general form of the infinite series?

(b) Use the result of (a) to compute the series of the inverse function of f, using `Inverse Series`. Which function corresponds to $f^{-1}(x)$? What is the general form of the infinite series $f^{-1}(x)$?

(c) Execute (a) and (b) with the functions $g(x) = \sin x$ and $h(x) = \arctan x$. For which of these functions is it difficult to obtain the general term of the series?

(d) Use `ComposeSeries` to compute the series for $k(x) := e^{\sin x}$ with truncation order 20, and compare your result with the result of `Series`.

---

[37] Strings are sequences with potentially arbitrary elements.

[38] In contrast to *Axiom*, the work with strings is not supported in *Mathematica*, however.

[39] You can use the functions `Solve`, `Table`, `Sum`, `CoefficientList` and `Join`.

(e) *Implement a procedure* `SeriesSqrt[a]` *which computes the square root of a series* a.[40] *Note that, in general, there are two square roots. Therefore assume that* $a_0 > 0$ *and choose the square root such that it has the property*

$$\left.\sqrt{a(x)}\right|_{x=0} = +\sqrt{a_0} \,.$$

Compute the square root of the (diverging) series

$$c(x) = \sum_{k=0}^{\infty} k! \, x^k$$

and compare your result with the result of `Sqrt`.

Exercise 10.3. (Inverse Series)

(a) *Implement the algorithm of Theorem 10.2 as* `inverseSeries[a]`.
(b) *Compute the 10th Taylor polynomial of the inverse of* $f(x)$ *in each of the following cases:*
   (i) $f(x) = e^x - 1$;
   (ii) $f(x) = \sqrt{x}$
   (iii) $f(x) = x^2$;
   (iv) $f(x) = \ln(1 + x)$;
   (v) $f(x) = xe^x$.
(c) *Compare your results in (b) with those of* `InverseSeries`.

Exercise 10.4. (Gamma Function) *We define* $\Gamma(x)$ *by the improper integral*

$$\Gamma(x) := \int_{0}^{\infty} t^{x-1} e^{-t} \, dt \,,$$

which exists for $x > 0$ *(or more general: for* $\mathrm{Re}\, x > 0$*). This function is called Euler's Gamma function.*

(a) *Show that the Gamma function interpolates the factorials. In detail, we have*

$$\Gamma(x + 1) = x\Gamma(x) \tag{10.33}$$

and therefore for all $n \in \mathbb{N}_{\geq 0}$

$$\Gamma(n + 1) = n! \,.$$

Using (10.33), the Gamma function can be continued as meromorphic function in $\mathbb{C}$*, see e.g. [Koe2014], Chapter 1.*
(b) *Plot the Gamma function in the interval* $[-3, 6]$.

Exercise 10.5. (Pascal Triangle) *Solve the following questions with Mathematica.*

(a) *Compute the row sums in Pascal's triangle:*

$$\sum_{k=0}^{n} \binom{n}{k} \,.$$

---

[40] For this purpose you can adapt the function `Reciprocal` of Exercise 10.1.

(b) *Compute the column partial sums in Pascal's triangle:*

$$\sum_{m=0}^{n} \binom{m}{k}.$$

(c) *Compute the diagonal partial sums in Pascal's triangle:*

$$\sum_{m=0}^{n} \binom{m+k}{k}.$$

(d) *Compute the alternating row partial sums in Pascal's triangle:*

$$\sum_{k=0}^{m} (-1)^k \binom{n}{k}.$$

(e) *Compute the Vandermonde sum*

$$\sum_{k=0}^{n} \binom{x}{k}\binom{y}{n-k}.$$

(f) *If one expresses the exponential function by its power series,*

$$e^x = \sum_{k=0}^{\infty} \frac{1}{k!} x^k,$$

then, due to the Cauchy product, the addition theorem of the exponential function

$$e^x \cdot e^y = e^{x+y}$$

can be expressed by an equivalent summation formula. Which of the above formulas emerges?

(g) *If one expresses the power function by its power series,*

$$(1+x)^\alpha = \sum_{k=0}^{\infty} \binom{\alpha}{k} x^k,$$

then, due to the Cauchy product, the addition theorem of the power function

$$(1+x)^\alpha \cdot (1+x)^\beta = (1+x)^{\alpha+\beta}$$

can be expressed by an equivalent summation formula. Which of the above formulas emerges?

Exercise 10.6. (Generation of Differential Equations) *With the method of Session 10.8 (without the package* `SpecialFunctions`*), determine holonomic differential equations for*

(a) $\left(\frac{1+x}{1-x}\right)^\alpha$;

(b) e^{-1/x^2};

(c) $\arcsin^2 x$ *(smallest order is 3)*;

(d) $\sin x + \arctan x$ *(smallest order is 4)*.

What does the holonomic differential equation look like for the rational function $\frac{p(x)}{q(x)}$ with $p(x), q(x) \in \mathbb{K}[x]$?

Exercise 10.7. (Differential Vector Spaces) *Compute a list of the first derivatives for the following functions $f(x)$ and compute a basis of the vector space $V = \langle f(x), f'(x), f''(x), \ldots \rangle$ over $\mathbb{Q}(x)$, respectively. What is the dimension of V? Which holonomic degree does $f(x)$ therefore have? Check your findings using* `holonomicDE`.

(a) $f(x) = \arcsin x$;
(b) $f(x) = \arctan x$;
(c) $f(x) = e^{\operatorname{arcsinh} x}$;
(d) $f(x) = e^{\arcsin x}$;
(e) $f(x) = \sin^3 x$.

Exercise 10.8. *In Theorem 10.11 it was shown that the function $f(x) = \tan x$ is not holonomic. Specify a (simplest possible) generator of the (infinite-dimensional) vector space $V = \langle f(x), f'(x), f''(x), \ldots \rangle$.*

Exercise 10.9. (Holonomic Degree)

(a) *Compute* $\operatorname{holdeg}(e^{\frac{1}{x}} + x)$ *manually, and determine the holonomic differential equation of the function.*
(b) *Compute* $\operatorname{holdeg}\left(\sin(x^k)\right), k \in \mathbb{N}_{\geq 1}$.
(c) *For which $t \in \mathbb{R}$ does e^{tx} not satisfy a holonomic differential equation with polynomial coefficients in $\mathbb{Q}[x]$?*

Exercise 10.10. (Non-Holonomic Functions and Sequences)

(a) *Show that $\sec x = \frac{1}{\cos x}$ is not holonomic.*
(b) *Show that the Benoulli numbers B_k are not holonomic. Hint: Show that the exponential generating function $f(x) = \sum\limits_{k=0}^{\infty} \frac{B_k}{k!} x^k$ satisfies the non-linear differential equation*

$$x f'(x) = (1 - x) f(x) - f(x)^2$$

and apply a reasoning similar to $f(x) = \tan x$.

Exercise 10.11. (Holonomic Products)

(a) *Determine the holonomic differential equation of $\arcsin^8 x$ in the following three ways:*
   (i)   *by a 7-fold application of* `ProductDE` *using* `Nest`;
   (ii) *by a threefold application of* `ProductDE` *(as divide-and-conquer power!) using* `Nest`;
   (iii) *with* `HolonomicDE`.

   Compare the respective computing times. Check your result. What is the holonomic degree of $\arcsin^8 x$?
(b) *Implement the computation of the differential equation of an integer power using the above divide-and-conquer algorithm. Check your implementation with the above and two more examples.*
(c) *This power function is available in the package* `SpecialFunctions` *with the syntax* `PowerDE[f(x),f,x,n]`. *However, this functionality, is not exported since this function*

lies in the "private part" of the package. Nevertheless, one can use it under the full name
`SpecialFunctions'Private'PowerDE`. *Check this function with the three examples executed in (b).*

(d) Let $\text{holdeg}(f(x), x) = n$. *Explain why* $\text{holdeg}(f(x)^2, x) \leqq \frac{n(n+1)}{2}$.

Exercise 10.12. (Solution Bases of Holonomic Differential Equations)

(a) *The two functions* $f(x) = \sqrt{1+x}$ *and* $g(x) = xe^{x^2}$ *both are holonomic of degree 1. Which differential equations do they satisfy? Which differential equation does their sum satisfy? The sum differential equation has order 2 and therefore possesses a solution basis of two functions that are linearly independent over* $\mathbb{Q}(x)$. *Which are these functions? Is* `DSolve` *able to find this basis?* [41]

(b) *Compute the holonomic differential equation of* $h(x) = \sin x + \arctan x$. *What is its order? Determine a solution basis of this differential equation. Check your result.*

(c) *Compute the holonomic differential equation of* $k(x) = \sin x \cdot \arctan x$. *What is its order? Determine a solution basis of this differential equation. Check your result.*

Exercise 10.13. (Hypergeometric Terms)

(a) *Show that* $a_k = k!$ *and* $b_k = (-1)^k$ *are hypergeometric terms.*

(b) *Show that* $c_k = a_k + b_k = k! + (-1)^k$ *is not a hypergeometric term. What is its holonomic degree?*

(c) *Show: If* $F(n, k)$ *is a hypergeometric term with respect to both* n *and* k, *then* $F(n+m, k) - F(n, k)$ *is a hypergeometric term with respect to both* n *and* k *for every* $m \in \mathbb{N}$.

(d) *Show: If* a_k *and* b_k *are hypergeometric terms, then the same is true for* $a_k \cdot b_k$.

(e) *Compute a holonomic recurrence equation with symmetry number* m *for* $a_k = k!$.

Exercise 10.14. (Hypergeometric Functions and Series) *Convert the following functions into hypergeometric form using* `FPS` *and* `SumToHypergeometric`:

(a) $f(x) = \arcsin^2 x$;

(b) $f(x) = e^{\text{arcsinh} x}$;

(c) $f(x) = \text{Ai}(x)$;

(d) $f(x) = \text{Bi}(x)$;

(e) $f(x) = L_n^{(\alpha)}(x)$. *These are the (generalized) Laguerre polynomials with the Mathematica name* `LaguerreL`.

(f) *The two functions* $\text{Ai}(x)$ *and* $\text{Bi}(x)$ *form a solution basis of the holonomic differential equation* $f''(x) - x f(x) = 0$. *Mathematica knows these two functions under the names* `AiryAi` *and* `AiryBi`. *Give a complete holonomic representation of the two functions* $\text{Ai}(x)$ *and* $\text{Bi}(x)$.

(g) *Determine a second hypergeometric representation of the Laguerre polynomials by reversing the summation order.*

Convert the following series into hypergeometric form, if applicable:

(h) $s_n = \sum_{k=0}^{\infty} \binom{n}{k}^2$;

(i) $s_n = \sum_{k=0}^{\infty} \binom{n-k}{k}$;

(j) $s_n = \sum_{k=0}^{\infty} \binom{n}{k}^2 \cdot \binom{n+k}{k}^2$. *These numbers are called Apéry's numbers.*

---

[41] If necessary, interrupt this computation after a minute!

(k) $s = \sum\limits_{k=0}^{\infty} \frac{1}{F_k}$, where F_k are the Fibonacci numbers.

(l) For (h)–(k) determine the natural bounds, respectively.

(m) Compute the sum s_n given in (h) for several values $n \in \mathbb{N}_{\geq 0}$ and guess a sum-free formula. Is the result a hypergeometric term?

(n) Compute the sum s_n given in (i) for several values $n \in \mathbb{N}_{\geq 0}$ and guess a sum-free formula. Is the result a hypergeometric term?

Exercise 10.15. (Hypergeometric Terms)

(a) Implement a Mathematica function hyperterm[plist,qlist,x,k], which returns the kth summand of the hypergeometric function $_pF_q\left(\begin{array}{c|c} plist \\ qlist \end{array} x\right)$, where plist and qlist express the lists of upper and lower parameters, respectively.

(b) Compare for some examples the results of your procedure with the procedure HyperTerm of the package SpecialFunctions.

Exercise 10.16. (Hypergeometric Differential Equation) Let $F(x) = \sum\limits_{k=0}^{\infty} a_k x^k$ be a power series and let θ denote the differential operator

$$\theta F(x) = xF'(x).$$

(a) Show that

$$\theta F(x) = \sum_{k=0}^{\infty} k a_k x^k.$$

(b) Show by induction that for all $j \in \mathbb{N}$

$$\theta^j F(x) = \sum_{k=0}^{\infty} k^j a_k x^k.$$

(c) Show that from (b) for every polynomial $P \in \mathbb{Q}(x)$ the operator equation

$$P(\theta)F(x) = \sum_{k=0}^{\infty} P(k) a_k x^k$$

follows.

(d) Show that the hypergeometric function $F(x) := {}_pF_q\left(\begin{array}{c|c} \alpha_1, \alpha_2, \ldots, \alpha_p \\ \beta_1, \beta_2, \ldots, \beta_q \end{array} x\right)$ satisfies the holonomic differential equation

$$\theta(\theta + \beta_1 - 1)\cdots(\theta + \beta_q - 1)F(x) - x(\theta + \alpha_1)(\theta + \alpha_2)\cdots(\theta + \alpha_p)F(x) = 0.$$

Hint: Substitute the series in the differential equation and equate the coefficients.

(e) Implement the Mathematica function hypDE[alist,blist,f[x]], which returns the hypergeometric differential equation.

(f) Check hypDE for $_2F_1\left(\begin{array}{c|c} a,b \\ c \end{array} x\right)$, $_1F_1\left(\begin{array}{c|c} a \\ b \end{array} x\right)$ and $_3F_2\left(\begin{array}{c|c} a,b,c \\ d,e \end{array} x\right)$. Compare the results with the ones of HolonomicDE.

Exercise 10.17. (Conversion Differential Equation ⇆ Recurrence Equation)

(a) Implement the conversion `detore[DE,f[x],a[k]]` of a holonomic differential equation DE for $f(x)$ into the corresponding holonomic recurrence equation of the series coefficients a_k.

(b) Check your procedure with the following examples:

  (i) $f''(x) + f(x) = 0$;

  (ii) $(1 - x^2) f''(x) - 2x f'(x) + n^2 f(x) = 0$;

  (iii) `HolonomicDE[Sin[x]+ArcTan[x],F[x]]`.

  Compare your results with those of DEtoRE.

(c) Use the properties of the differential operator θ as defined in Exercise 10.16 for the implementation of the conversion `retode[RE,a[k],f[x]]` of a holonomic recurrence equation RE for a_k into the corresponding holonomic differential equation of its generating function $f(x)$.

(d) Check your procedure with suitable examples and compare your results with those of RTEtoDE.

Exercise 10.18. Show that the recurrence equation

$$(k - n)a_k + (k + 1)a_{k+1} = 0 \tag{10.34}$$

of $\binom{n}{k}$ is compatible with the recurrence equation of second order

$$(k - n)(k - n + 1)a_k + 2(k + 1)(k - n + 1)a_{k+1} + (k + 1)(k + 2)a_{k+2} = 0 \tag{10.35}$$

of $\binom{n-1}{k} + \binom{n-1}{k-1}$, i.e., every solution of (10.34) is also solution of (10.35), see Session 10.19.

Exercise 10.19. Explain the concept of "suitable initial values of a holonomic recurrence equation", see Theorem 10.18.

Exercise 10.20. (Algebraic Functions) Determine the holonomic differential equations of the following algebraic functions:

(a) $(y(x) - r)^2 + x^2 - r^2 = 0$;

(b) $y(x)^3 + xy(x)^2 + x^2 = 0$.

After the above computations, load `SpecialFunctions`.

(c) Check (a) and (b) with `AlgebraicDE`. Convert these differential equations into their corresponding recurrence equations and determine which of the given algebraic functions is of hypergeometric type.

Now, compute the holonomic differential equations of the following algebraic functions as well as their algebraic and holonomic degrees, and again determine which of these algebraic functions are of hypergeometric type:

(d) $(x^2 + y(x)^2)^4 - x^2 y(x)^2 = 0$;

(e) $y(x)^5 + 2xy(x)^4 - xy(x)^2 - 2x^2 y(x) + x^4 - x^3 = 0$.

Exercise 10.21. (Algebraic Curves) We consider graphs given by algebraic functions. In each of the following cases, plot the graph (for suitable parameters), determine the algebraic and the holonomic degree of $y(x)$ and determine whether the representing power series is hypergeometric:

(a) **(Folium of Descartes)** $x^3 + y^3 - axy = 0$;
(b) **(Conchoid)** $(x^2 + y^2)(x - p)^2 - s^2 x^2 = 0$;
(c) **(Lemniscate)** $(x^2 + y^2)^2 - 2a^2(x^2 - y^2) = 0$;
(d) **(Cissoid)** $y^2(a - x) = x^3$;
(e) **(Semicubical Parabola)** $(y - b)^2 = (x - a)^3$.

Explain what happens if we have $a = b = 0$ in (e).

Exercise 10.22. (Power Series Solutions of Explicit Differential Equations) *Let an initial value problem be given in the form*

$$y' = F(x, y) \qquad \text{with} \qquad y(x_0) = y_0 .$$

(a) *Implement a function* DSolveTaylor[F,$y[x]$,x_0,y_0,n] *which computes the Taylor approximation of order n for the solution of the initial value problem, if applicable. Proceed as in Section 10.5.*
(b) *Determine the 10th Taylor approximation for the following initial value problems, respectively:*
  (i) $y' = \frac{x}{y}$, $y(0) = 1$;
  (ii) $y' = x^2 + y^3$, $y(0) = 1$;
  (iii) $y' = y$, $y(0) = 1$;
  (iv) $y' = e^{xy}$, $y(0) = 1$;
  (v) $y' = \sqrt{(1 - y^2)(1 - k^2 y^2)}$, $y(0) = 0$.
  Compare your results with the explicit solutions if these can be found.

Exercise 10.23. (Newton's Method)

(a) *Use Newton's method to approximate $\sqrt[3]{7}$ (hence to solve the equation $x^3 - 7 = 0$). Specify the first 5 approximations with starting value 2 in a suitable numerical precision. Compute the difference to the exact value with 20 decimal places. How can you realize the quadratic convergence?*
(b) *Consider the function $F : \mathbb{R}[[x]] \to \mathbb{R}[[x]]$, given by $F(y) = x^2 y + y^2 x - y - 1$. Let further $y_0 = -1$. Then we can compute the Taylor series of $y(x)$ from the given implicit representation $F(y) = 0$.*
  (i)   *Compute the explicit solution $y(x)$.*
  (ii) *Use (i) to find the Taylor polynomial of degree 15 using* Series.
  (iii) *Compute the holonomic differential equation from the explicit solution $y(x)$.*
  (iv) *Use the method of Section 10.4 to compute this differential equation again.*
  (v)   *Use the programs* ImplicitTaylor *and* FastImplicitTaylor *considered in Section 10.5 for the computation of the truncated Taylor series of degree 255. Compare the computing times.*

Chapter 11
Algorithmic Summation

11.1 Definite Summation

Since a rather simple (although not very efficient) algorithm exists for them we would first like
to discuss bivariate sums of the form

$$s_n = \sum_{k=-\infty}^{\infty} F(n,k) \tag{11.1}$$

whose summand $F(n,k)$ depends on the summation variable $k \in \mathbb{Z}$ and on another discrete
variable $n \in \mathbb{Z}$. We will generally assume that those sums actually have only finitely many
summands, hence that for every $n \in \mathbb{Z}$ (or $n \in \mathbb{N}_{\geq 0}$, respectively) the natural bounds of the
summation with respect to k are finite. For example, this is the case if $F(n,k)$ contains a factor
like $\binom{n}{k}$.

What would we like to know about s_n? Of course we would prefer a "simple" formula for s_n, in
particular, if s_n is a hypergeometric term. We consider the general case that s_n is a holonomic
sequence, since holonomic sequences have a normal form consisting of a holonomic recurrence
equation together with suitable initial values. We will see that s_n is indeed holonomic if the
summand $F(n,k)$ satisfies a recurrence equation of a special form, namely a *k-free recurrence*.

Theorem 11.1. Let $F(n,k)$ satisfy a recurrence equation of the form

$$\sum_{i=0}^{I} \sum_{j=0}^{J} a_{ij} F(n+j,k+i) = 0 \qquad (a_{ij} \in \mathbb{K}[n], I, J \in \mathbb{N}), \tag{11.2}$$

whose polynomial coefficients a_{ij} do not contain the summation index. Such a recurrence equation is called *k-free*. Then s_n as defined in (11.1) is holonomic (over $\mathbb{K}$).

Proof. An index shift shows that for all $i, j \in \mathbb{Z}$

$$\sum_{k=-\infty}^{\infty} F(n+j,k+i) = s_{n+j}$$

© Springer Nature Switzerland AG 2021
W. Koepf, *Computer Algebra*, Springer Undergraduate Texts in Mathematics and Technology,
https://doi.org/10.1007/978-3-030-78017-3_11

is valid.[1] If we sum the given k-free recurrence equation of $F(n, k)$ for $k = -\infty, \ldots, \infty$, then this obviously yields a holonomic recurrence equation over $\mathbb{K}$ for s_n. $\square$

Zeilberger [Zei1990b] gave an algorithm for the computation of a recurrence equation for s_n if $F(n, k)$ is holonomic with respect to n and k. We will, however, restrict our considerations to the special case that $F(n, k)$ is a hypergeometric term with respect to both variables n and k. Therefore we assume that

$$\frac{F(n+1, k)}{F(n, k)} \in \mathbb{K}(n, k) \quad \text{and} \quad \frac{F(n, k+1)}{F(n, k)} \in \mathbb{K}(n, k). \tag{11.3}$$

Example 11.2. The binomial coefficients

$$F(n, k) = \binom{n}{k}$$

satisfy the *recurrence equation of Pascal's triangle,*

$$\binom{n+1}{k+1} = \binom{n}{k} + \binom{n}{k+1},$$

or equivalently

$$F(n+1, k+1) = F(n, k) + F(n, k+1). \tag{11.4}$$

Now we apply Theorem 11.1 and find for

$$s_n = \sum_{k=0}^{n} \binom{n}{k} = \sum_{k=-\infty}^{\infty} F(n, k),$$

by summation of (11.4) from $k = -\infty, \ldots, \infty$ that

$$s_{n+1} = s_n + s_n = 2 s_n$$

with the obvious solution $s_n = 2^n s_0 = 2^n$. Of course this is a special case of the binomial formula. $\triangle$

The algorithm presented in the following *Mathematica* session is due to Celine Fasenmyer [Fas1945].

Session 11.3. *We consider a slightly more complicated problem, namely let*

$$s_n = \sum_{k=-\infty}^{\infty} F(n, k)$$

with $F(n, k) = k \binom{n}{k}$:

```
In[1]:= f = k Binomial[n, k]
```
$$Out[1]= k \binom{n}{k}$$

---

[1] This step shows again that it is smart to work with bilateral sums!

When searching for a recurrence equation of the form (11.2) for the summand $F(n,k)$, we try the choice $I = J = 1$. Therefore, we have the ansatz

$$In[2]:= \text{ansatz} = \sum_{j=0}^{1}\sum_{i=0}^{1} a[i,j]\, F[n+j,k+i] == 0$$

$Out[2]= a(0,0)\,F(n,k) + a(1,0)\,F(n,k+1) + a(0,1)\,F(n+1,k) + a(1,1)\,F(n+1,k+1) == 0$

Division by $F(n,k)$ gives

$$In[3]:= \sum_{j=0}^{1}\sum_{i=0}^{1} a[i,j]\,\frac{F[n+j,k+i]}{F[n,k]} == 0$$

$Out[3]= \dfrac{a(1,0)\,F(n,k+1)}{F(n,k)} + \dfrac{a(0,1)\,F(n+1,k)}{F(n,k)} + \dfrac{a(1,1)\,F(n+1,k+1)}{F(n,k)} + a(0,0) == 0$

Since with $\frac{F(n+1,k)}{F(n,k)}, \frac{F(n,k+1)}{F(n,k)} \in \mathbb{K}(n,k)$ all terms of the form $\frac{F(n\,|\,j,k\,|\,i)}{F(n,k)} \in \mathbb{K}(n,k)$ are rational by induction for $i,j \in \mathbb{N}$, this always yields a rational identity, provided that (11.3) is satisfied.

In our case, we have $\mathbb{K} = \mathbb{Q}$, and we substitute $F(n,k) = k\binom{n}{k}$ to get

$$In[4]:= \text{sum} = \sum_{j=0}^{1}\sum_{i=0}^{1} a[i,j]\,\text{FunctionExpand}\Big[$$

$$\frac{F[n+j,k+i]}{F[n,k]}\, /.\,\{F[n\_,k\_] \to f\}\Big] == 0$$

$Out[4]= \dfrac{(n+1)\,a(0,1)}{-k+n+1} + \dfrac{(n-k)\,a(1,0)}{k} + \dfrac{(n+1)\,a(1,1)}{k} + a(0,0) == 0$

Note that the Mathematica function FunctionExpand *expands the occurring factorials and binomial coefficients so that the considered ratios are identified as rational functions.*

Hence, after multiplying with the common denominator, the following term must equal zero:

$In[5]:= \text{sum} = \text{Numerator}[\text{Together}[\text{sum}[[1]]]]$

$Out[5]= -a(0,0)\,k^2 + a(1,0)\,k^2 + n\,a(0,0)\,k + a(0,0)\,k + n\,a(0,1)\,k +$
$\qquad a(0,1)\,k - 2n\,a(1,0)\,k - a(1,0)\,k - n\,a(1,1)\,k - a(1,1)\,k +$
$\qquad n^2\,a(1,0) + n\,a(1,0) + n^2\,a(1,1) + 2n\,a(1,1) + a(1,1)$

To find the k-free coefficients a_{ij}, we consider this term as a polynomial with respect to k, hence as an element of $\mathbb{Q}(n)[k]$, and equate the coefficients. This yields a system of linear equations which we can solve for the variables $\{a_{00}, a_{01}, a_{10}, a_{11}\}$. If there is a solution, then we get rational functions $\in \mathbb{Q}(n)$ for the sought variables. This computation can be executed by Mathematica:

$In[6]:= \text{list} = \text{CoefficientList}[\text{sum}, k]$

$Out[6]= \{a(1,0)\,n^2 + a(1,1)\,n^2 + a(1,0)\,n + 2\,a(1,1)\,n + a(1,1),$
$\qquad n\,a(0,0) + a(0,0) + n\,a(0,1) + a(0,1) - 2n\,a(1,0) -$
$\qquad a(1,0) - n\,a(1,1) - a(1,1), a(1,0) - a(0,0)\}$

$In[7]:= \text{solution} = \text{Solve}[\text{list} == 0,$
$\qquad\qquad \text{Flatten}[\text{Table}[a[i,j], \{i,0,1\}, \{j,0,1\}]]]$

Solve :: **"svars"**: *Equations may not give solutions for all "solve" variables.*

$Out[7]= \left\{\left\{a(0,0) \to a(1,0), a(0,1) \to 0, a(1,1) \to -\dfrac{n\,a(1,0)}{n+1}\right\}\right\}$

Therefore we obtain the recurrence equation

$In[8]:= \text{RE} = \text{Numerator}[\text{Together}[\text{ansatz}[[1]]\,/.\,\text{solution}[[1]]]] == 0$

$Out[8]= n\,a(1,0)\,F(n,k) + a(1,0)\,F(n,k) + n\,a(1,0)\,F(n,k+1) +$
$\qquad a(1,0)\,F(n,k+1) - n\,a(1,0)\,F(n+1,k+1) == 0$

or, in sorted form,

$$(n+1)F(n,k+1)+(n+1)F(n,k)-nF(n+1,k+1)=0$$

for $F(n,k)$.

The whole procedure can be implemented by the following algorithm

```
In[9]:= Clear[kfreeRE]
        kfreeRE[f_, {k_, kmax_}, {n_, nmax_}] :=
          kfreeRE[f, {k, kmax}, {n, nmax}] =
            Module[{variables, SUM, i, j, ansatz, list,
                solution, RE},
              variables =
                Flatten[Table[a[i, j], {i, 0, kmax},
                    {j, 0, nmax}]];
              ansatz = Sum[a[i, j] * F[n+j, k+i],
                  {j, 0, nmax}, {i, 0, kmax}];
              SUM =
                Sum[a[i, j] * FunctionExpand[
                      (f/.{k→k+i, n→n+j})/f], {j, 0, nmax},
                  {i, 0, kmax}];
              SUM = Numerator[Together[SUM]];
              list = CoefficientList[SUM, k];
              Off[Solve :: "svars"];
              solution = Solve[list == 0, variables];
              On[Solve :: "svars"];
              solution = Simplify[solution];
              If[
                Union[Flatten[Table[a[i, j]/.solution[[1]],
                      {i, 0, kmax}, {j, 0, nmax}]]] === {0},
                Return["Such a recurrence equation does not exist"]];
              RE = Numerator[Together[ansatz/.lösung[[1]]]];
              RE = Collect[RE/.a[__]→1, F[__]];
              RE = Map[Factor, RE];
              RE == 0
            ]
```

Thereby, k_{max} and n_{max} correspond to the numbers I and J and hence bound the orders of the sought recurrence equation with respect to k and n. Since the linear system is homogeneous, all coefficients of the resulting recurrence equation still have a common factor depending on a_{ij}. This factor is canceled by setting all remaining coefficients a_{ij} to 1 in the output. In order to get the output as simple as possible, we factorize the polynomial coefficients in a final step.

Now let us try to deduce some recurrence equations. The recurrence equation from our above example is directly computed by the call

```
In[10]:= RE = kfreeRE[k Binomial[n, k], {k, 1}, {n, 1}]
Out[10]= (n+1) F(n,k)+(n+1) F(n,k+1)-n F(n+1,k+1) == 0
```

The next computation yields the recurrence equation of Pascal's triangle which is valid for the binomial coefficients:

```
In[11]:= kfreeRE[Binomial[n, k], {k, 1}, {n, 1}]
```

Out[11]= $F(n,k)+F(n,k+1)-F(n+1,k+1)==0$

Finally we compute a recurrence equation which is valid for the squares of the binomial coefficients:

In[12]:= **kfreeRE[Binomial[n,k]^2,{k,2},{n,2}]**

Out[12]= $(n+1)F(n,k)-2(n+1)F(n,k+1)+$
$\quad (n+1)F(n,k+2)-(2n+3)F(n+1,k+1)-$
$\quad (2n+3)F(n+1,k+2)+(n+2)F(n+2,k+2)==0$

This recurrence is quite complicated and obviously cannot be computed manually so easily.[2] Mathematica needs a little more computing time.[3] One reason is that the present algorithm is not very efficient.[4] For this example a rather complicated linear system with polynomial coefficients had to be solved for 9 variables.

Next, we consider the second part of the algorithm given in Theorem 11.1 by computing a holonomic recurrence equation for s_n.

For our example function $F(n,k) = k\binom{n}{k}$ we get

In[13]:= **RE = RE[[1]]/.{F[n+j_.,k+i_.] → s[n+j]}**

Out[13]= $2(n+1)s(n)-ns(n+1)==0$

and hence the recurrence equation

$$n s_{n+1} - 2(n+1)s_n = 0$$

for $s_n = \sum_{k=-\infty}^{\infty} F(n,k)$. Therefore s_n is a hypergeometric term with

$$\frac{s_{n+1}}{s_n} = 2\frac{n+1}{n}.$$

This equation shows that an index shift of n by 1 is suitable to generate the term $(n+1)$ in the denominator, thus converting the term into the coefficient of a general hypergeometric series. Hence we set $t_n := s_{n+1}$ and get

$$\frac{t_{n+1}}{t_n} = 2\frac{n+2}{n+1},$$

so that together with the initial value $t_0 = s_1 = \sum_{k=0}^{1} k\binom{1}{k} = 1$ the coefficient formula (10.18) of the general hypergeometric series implies

$$t_n = 2^n \frac{(2)_n}{n!} = (n+1)2^n \quad (n \geq 0).$$

Therefore we obtain the following final result:

$$s_n = t_{n-1} = \sum_{k=-\infty}^{\infty} k\binom{n}{k} = n2^{n-1} \quad (n \geq 1).$$

---

[2] Try it!

[3] This was the case for the first (German language) edition of this book in 2006. Today's computers are much faster. Therefore you might try kfreeRE[Binomial[n,k]^3,{k,3},{n,3}] instead to generate a larger computing time.

[4] It can be made more efficient by suitable measures, see [Ver1976] and [Weg1997].

Let us now summarize the complete algorithm for the computation of a holonomic recurrence equation for s_n in the following Mathematica function:[5]

```
In[14]:= Clear[FasenmyerRE]
         FasenmyerRE[f_, {k_, kmax_}, {n_, nmax_}] :=
           Module[{RE, tmp},
             RE = kfreeRE[f, {k, kmax}, {n, nmax}];
             If[RE === "Such a recurrence equation does not exist",
               Return["Such a recurrence equation does not exist"],
               RE = RE[[1]]];
             RE = RE/.{F[n+j_., k+i_.] -> S[n+j]};
             tmp = S[n+nmax]/.Solve[RE == 0, S[n+nmax]][[1]];
             RE = Denominator[tmp] * S[n+nmax] - Numerator[tmp];
             RE = Collect[RE, S[___]];
             Map[Factor, RE] == 0
           ]
```

For our example we get again

```
In[15]:= FasenmyerRE[k Binomial[n, k], {k, 1}, {n, 1}]
```
$Out[15]= n S(n+1) - 2(n+1) S(n) == 0$

Similarly as in Example 11.2, for the sum of the binomial coefficients we get the hypergeometric recurrence equation

```
In[16]:= FasenmyerRE[Binomial[n, k], {k, 1}, {n, 1}]
```
$Out[16]= S(n+1) - 2 S(n) == 0$

It turns out that for the sum of squares of the binomial coefficients we also get a hypergeometric term:

```
In[17]:= FasenmyerRE[Binomial[n, k]^2, {k, 2}, {n, 2}]
```
$Out[17]= (n+2) S(n+2) - 2(2n+3) S(n+1) == 0$

Note, however, that the algorithm was rather inefficient to obtain this result: although a holonomic recurrence equation of first order exists for s_n, we needed to find a second order recurrence for the summand. The reason is that for the summand $F(n,k)$ no k-free recurrence equation with $J = 1$ exists.

The alternating sum of squares of the binomial coefficients is also of hypergeometric type:

```
In[18]:= FasenmyerRE[(-1)^k Binomial[n, k]^2, {k, 2}, {n, 2}]
```
$Out[18]= 4(n+1) S(n) + (n+2) S(n+2) == 0$

We consider another example, namely the Legendre polynomials which are orthogonal polynomials defined by

$$P_n(x) = \sum_{k=0}^{\infty} \binom{n}{k} \binom{-n-1}{k} \left(\frac{1-x}{2}\right)^k. \tag{11.5}$$

Therefore

$$P_n(x) = \sum_{k=-\infty}^{\infty} F(n,k) \qquad with \qquad F(n,k) = \binom{n}{k} \binom{-n-1}{k} \left(\frac{1-x}{2}\right)^k.$$

---

[5] Please note that for the sake of simplicity the variables a, F and S, which are used for the output of kfreeRE and FasenmyerRE, respectively, are defined as global variables. As a consequence, these variables should not have a value.

We obtain a three-term recurrence equation for the sum

$In[19]:=$ **FasenmyerRE$\left[$Binomial[n,k] Binomial[-n-1,k]$\right.$**

$$\left(\frac{1-x}{2}\right)^k, \{k,1\}, \{n,2\}\Big]$$

$Out[19]=$ $-(2n+3) x\, S(n+1) + (n+1)\, S(n) + (n+2)\, S(n+2) == 0$

It is well known that all orthogonal polynomial families satisfy a three-term recurrence equation.

There are further hypergeometric representations of the Legendre polynomials, for example

$$P_n(x) = \frac{1}{2^n} \sum_{k=0}^{\infty} (-1)^k \binom{n}{k}\binom{2n-2k}{n} x^{n-2k}, \tag{11.6}$$

with the result

$In[20]:=$ **FasenmyerRE$\left[\dfrac{1}{2^n}(-1)^k$Binomial[n,k]$\right.$**

$$\text{Binomial[2n-2k,n]}\,x^{n-2k}, \{k,1\}, \{n,2\}\Big]$$

$Out[20]=$ $-(2n+3) x\, S(n+1) + (n+1)\, S(n) + (n+2)\, S(n+2) == 0$

The resulting two recurrence equations are identical. Therefore by checking two initial values our computations have proved that (11.5) and (11.6) represent the same family of functions indeed!

The technique that we used in our examples can be summarized in the following algorithm.

Theorem 11.4. (Fasenmyer Algorithm)   The following algorithm computes a holonomic recurrence equation for series of the form (11.1):

1. (kfreeRE)
    Choose suitable bounds $I, J \in \mathbb{N}$. Then the following procedure computes a k-free recurrence equation of order (I, J) with polynomial coefficients for the summand $F(n, k)$ if such a recurrence equation is valid.
    a. Input: $F(n, k)$ with property (11.3).
    b. Define (11.2) with initially unknown coefficients a_{ij} and substitute the given term $F(n, k)$.
    c. Divide by $F(n, k)$ and simplify the occurring quotients so that the final result is a rational identity.
    d. Bring the rational function in normal form and multiply by its common denominator.
    e. Equate the coefficients with respect to k and solve the resulting homogeneous linear system for the variables a_{ij} $(i = 0, \ldots, I, j = 0, \ldots, J)$.
    f. If there is only the trivial solution $a_{ij} \equiv 0$, then there is no non-trivial k-free recurrence equation of this order for $F(n, k)$. In this case, increase I or J suitably and restart.
    g. If there is a non-trivial solution, then substitute it into the ansatz and multiply by the common denominator.
    h. Output: The resulting k-free recurrence equation for $F(n, k)$.

2. (FasenmyerRE)
    Choose suitable bounds $I, J \in \mathbb{N}$ so that J is as low as possible. Then the following procedure searches for a holonomic recurrence equation or order J for s_n given by (11.1).
    a. Input: The summand $F(n, k)$ with property (11.3).
    b. Apply procedure kfreeRE to $F(n, k)$.

c.   If this is successful, then take the resulting recurrence equation for $F(n, k)$ and replace the pattern $F(n+j, k+i)$ by the pattern s_{n+j}. This generates the holonomic recurrence equation for s_n sought.

d.   Output: The resulting holonomic recurrence equation for s_n.

Proof. (kfreeRE): As soon as the rational function in step (c) is established, the algorithm is pure rational arithmetic. The resulting expression in (d) is a polynomial in $\mathbb{K}[a_{ij}][k]$ which is the zero polynomial if and only if all its coefficients vanish. Therefore the existence of a k-free recurrence equation of the considered type is equivalent to the existence of a non-trivial solution of the linear system considered in (e). All remaining parts of kfreeRE are pure linear algebra.

(FasenmyerRE): If the k-free recurrence equation

$$\sum_{i=0}^{I}\sum_{j=0}^{J} a_{ij} F(n+j, k+i) = 0$$

is summed from $k = -\infty, \ldots, \infty$, then we get

$$
\begin{aligned}
0 &= \sum_{k=-\infty}^{\infty}\sum_{i=0}^{I}\sum_{j=0}^{J} a_{ij}(n) F(n+j, k+i) \\
&= \sum_{i=0}^{I}\sum_{j=0}^{J} a_{ij}(n)\left(\sum_{k=-\infty}^{\infty} F(n+j, k+i)\right) \\
&= \sum_{i=0}^{I}\sum_{j=0}^{J} a_{ij}(n) s_{n+j} = \sum_{j=0}^{J}\left(\sum_{i=0}^{I} a_{ij}(n)\right) s_{n+j}
\end{aligned}
$$

since the coefficients $a_{ij}(n)$ do not depend on k. Therefore the given substitution generates the wanted holonomic recurrence equation, provided that the first step kfreeRE was successful. □

11.2 Difference Calculus

In the preceding section we considered bivariate infinite sums of hypergeometric terms and computed holonomic recurrence equations for them. However, in many cases we wish to simplify sums with finite bounds. This is our next topic.

As a motivation, we first consider the problem of integration. The Fundamental Theorem of Calculus allows us to evaluate every definite integral over $f(x)$ once we know an *antiderivative*, i.e., a function $F(x)$ with the property

$$F'(x) = f(x),$$

using the simple rule[6]

---

[6] Of course f must be, for example, continuous. Otherwise the fundamental theorem may not be applicable.

$$\int_a^b f(x)\,dx = F(b) - F(a).$$

If we express this rule using the *differential operator* $D_x : f(x) \mapsto f'(x)$ and the *integral operator* $I_a^b : f(x) \mapsto \int_a^b f(x)\,dx$, then the fundamental theorem reads

$$I_a^b D_x F(x) = F(b) - F(a). \tag{11.7}$$

If we denote by $\int_x : f(x) \mapsto \int f(x)\,dx$ the operator which creates the antiderivative, then obviously this is the inverse mapping of D_x:

$$D_x \int_x f(x) = f(x).$$

Now we introduce the corresponding operators for sums. Corresponding to the differential operator we define the *forward difference operator* $\Delta_k : a_k \mapsto a_{k+1} - a_k$ and the *backward difference operator* $\nabla_k : a_k \mapsto a_k - a_{k-1}$.[7] If the variable is clear, we simply write Δ and ∇.

Using the operator Δ, it follows for arbitrary bounds $a \leq b$ that

$$\sum_{k=a}^b a_k = (s_{b+1} - s_b) + (s_b - s_{b-1}) + \cdots + (s_{a+1} - s_a) = s_{b+1} - s_a, \tag{11.8}$$

if s_k is a sequence for which the relation

$$\Delta s_k = s_{k+1} - s_k = a_k$$

is valid. We call s_k an *antidifference* and (11.8) a *telescoping sum*. Analogously we get for the operator ∇

$$\sum_{k=a}^b a_k = t_b - t_{a-1},$$

if $\nabla t_k = a_k$. The theory of the operators Δ and ∇ is called *difference calculus*.

For the sum operator $\sum_{k=a}^b$ it therefore follows from (11.8) that

$$\sum_{k=a}^b \Delta s_k = s_{b+1} - s_a,$$

analogously to (11.7). The inverse mapping of Δ is denoted by $\sum_k$. Obviously the *indefinite sum* $\sum_k a_k = s_k$ is an antidifference. It is easy to see that two antidifferences s_k of a_k differ from each other by a constant, as in the continuous case.

Due to the existence of the two difference operators Δ and ∇, there are two equivalent but different summation theories.[8] In this chapter, we will consider the theory of the operator Δ.

---

[7] The two operators are called Delta and Nabla, respectively.

[8] For example, in *Maple* the command `sum(a,k)` refers to the inverse of the operator Δ, whereas *Reduce* refers to ∇.

Example 11.5. We would like to compute $\sum_k k$, hence the antidifference of $a_k = k$. Because of $\int x \, dx = \frac{x^2}{2}$ we try $s_k = ak^2 + bk + c$ as a quadratic polynomial. This yields

$$a_k = k = s_{k+1} - s_k = (a(k+1)^2 + b(k+1) + c) - (ak^2 + bk + c) = 2ak + (a+b).$$

By equating the coefficients we obtain the linear system

$$a + b = 0 \qquad \text{and} \qquad 2a = 1$$

for the unknowns a, b, c. Therefore c is arbitrary and we get $a = \frac{1}{2}$, as well as $b = -\frac{1}{2}$, and therefore $\sum_k k = \frac{1}{2}k(k-1) + c.$ $\triangle$

In the next step, we would like to find out how the operator Δ acts on polynomials. The differential operator satisfies the power rule

$$D_x x^n = n x^{n-1}. \tag{11.9}$$

Unfortunately the operator Δ acts much more complicated on polynomials, for example

$$\Delta_k k^3 = (k+1)^3 - k^3 = 3k^2 + 3k + 1. \tag{11.10}$$

Whereas by (11.9) the powers x^n are *eigenfunctions* of the operator $\theta_x = x D_x$ (with eigenvalues n), and therefore essentially reproduce themselves under D_x, this is not the case for the operator Δ. Therefore we still have to find those reproducing functions. But this is not very difficult, as you will see in the following steps. Since $\Delta a_k = a_{k+1} - a_k$, it would be helpful if a_{k+1} had many factors in common with a_k. The Pochhammer symbol $(k)_n = k(k+1) \cdots (k+n-1)$ is a polynomial in k of degree n which has many factors in common with both $(k+1)_n$ and $(k-1)_n$. For the operator ∇_k we get

$$
\begin{aligned}
\nabla_k (k)_n &= (k)_n - (k-1)_n \\
&= k(k+1) \cdots (k+n-1) - (k-1)k(k+1) \cdots (k+n-2) \\
&= k(k+1) \cdots (k+n-2) \cdot \Big(k+n-1 - (k-1) \Big) \\
&= n \cdot (k)_{n-1}.
\end{aligned}
$$

The Pochhammer symbol $(k)_n$ therefore reproduces itself under the operator ∇_k. Analogously we define the *falling factorials*,[9] first for $n \in \mathbb{N}_{\geq 0}$

$$k^{\underline{n}} := k(k-1) \cdots (k-n+1).$$

The falling factorials satisfy

$$\Delta k^{\underline{n}} = n \cdot k^{\underline{n-1}}, \tag{11.11}$$

analogously to (11.9). Initially this relation is valid for $n \in \mathbb{N}_{\geq 0}$, but in order to make sure that this relation—as in the continuous case—remains valid for all $n \in \mathbb{Z}$, one sets for $n \in \mathbb{N}$

$$k^{\underline{-n}} = \frac{1}{(k+1)(k+2) \cdots (k+n)}.$$

Hence we have found the reproducing functions under Δ_k. In the next section we will study how to use this knowledge for the summation of arbitrary polynomials.

---

[9] The Pochhammer symbol is also called *rising factorial*, and the alternative notation $(k)_n = k^{\overline{n}}$ is used.

11.3 Indefinite Summation

Because of

$$\Delta k^{\underline{n}} = n \cdot k^{\underline{n-1}},$$

we have for $n \in \mathbb{Z}$, $n \neq -1$, the summation formula

$$\sum_k k^{\underline{n}} = \frac{1}{n+1} k^{\underline{n+1}}. \tag{11.12}$$

For $n = -1$ this summation formula is complemented by

$$\sum_k k^{\underline{-1}} = \sum_k \frac{1}{k+1} = H_k,$$

where

$$H_k := \sum_{j=1}^{k} \frac{1}{j}$$

are the *harmonic numbers* that obviously form the discrete analog of the logarithm function.

Using the summation formula (11.12) and the linearity of $\sum_k$, one can sum every polynomial $p(k) \in \mathbb{K}[k]$ and therefore compute every polynomial sum with arbitrary lower and upper bounds.

Session 11.6. *We define the operator Δ in Mathematica:*

```
In[1]:= Clear[Delta]
        Delta[f_,k_] := (f/.k→k+1) - f
```

For example, we get

```
In[2]:= Delta[k³,k]
Out[2]= (k+1)³ - k³
```

and if we simplify, then the result is again (11.10).

```
In[3]:= Delta[k³,k]//Expand
Out[3]= 3 k² + 3 k + 1
```

Furthermore, we define the falling factorial as

```
In[4]:= Clear[fallingFactorial]
        fallingFactorial[k_,j_] := Pochhammer[k-j+1,j]
```

The following computation confirms (11.11) for $n = 3$:

```
In[5]:= Delta[fallingFactorial[k,3],k]
Out[5]= (k-1) k (k+1) - (k-2) (k-1) k
```

or in factorized form

```
In[6]:= Delta[[k,3],k]//Factor
Out[6]= 3 (k-1) k
```

Next, we compute $\sum k^3$. Since

$$k^3 = k^{\underline{3}} + 3k^{\underline{2}} - 2k = k^{\underline{3}} + 3k^{\underline{2}} + k = k^{\underline{3}} + 3k^{\underline{2}} + k^{\underline{1}} \tag{11.13}$$

it follows by indefinite summation that

$$\sum k^3 = \frac{k^{\underline{4}}}{4} + 3\frac{k^{\underline{3}}}{3} + \frac{k^{\underline{2}}}{2}. \tag{11.14}$$

The computation (11.13) is therefore confirmed by

```
In[7]:= fallingFactorial[k,3]+
           3fallingFactorial[k,2]+
           fallingFactorial[k,1]//Expand
Out[7]= k³
```

Now we implement the computation of the antidifference for an arbitrary polynomial. We use the linearity of summation and apply (11.12) recursively:

```
In[8]:= Clear[Summation]
        Summation[c_,k_]:=c k/;FreeQ[c,k];
        Summation[c_ f_,k_]:=c Summation[f,k]/;FreeQ[c,k]
        Summation[f_+g_,k_]:=Summation[f,k]+Summation[g,k]
        Summation[FallingFactorial[k_,n_],k_]:=
          FallingFactorial[k,n+1]
          ──────────────────────── /;n≥0
                  n+1
        Summation[k_^n_.,k_]:=
          Module[{j},
            Expand[Summation[FallingFactorial[k,n],k]+
                                   n-1
              Summation[Expand[kⁿ-∏(k-j)],k]]]]
                                   j=0
```

In order to represent the results by falling factorials instead of powers, we use the function FallingFactorial. This function is undefined in Mathematica and remains therefore unevaluated.

Using our implementation Summation, *we again get (11.14),*

```
In[9]:= Summation[k³,k]
Out[9]= ½ FallingFactorial(k,2)+FallingFactorial(k,3)+¼ FallingFactorial(k,4)
```

and after simplification

```
In[10]:= s = Summation[k³,k]/.
               FallingFactorial → fallingFactorial//Factor
Out[10]= ¼ (k-1)² k²
```

If we would like to compute the definite sum $\sum_{k=1}^{n} k^3$, *we get*

```
In[11]:= (s/.k→n+1) - (s/.k→1)
Out[11]= ¼ n² (n+1)²
```

hence we have discovered the summation formula

$$\sum_{k=1}^{n} k^3 = \sum_{k} k^3 \Bigg|_{k=1}^{k=n+1} = \frac{1}{4} n^2 (n+1)^2.$$

Please note that this was done completely algorithmically and not like a proof by mathematical induction.

Whereas we can now sum all polynomials, the indefinite summation for rational functions, hypergeometric terms or other transcendental expressions is completely open. Similarly as with differentiation and integration, the list of differences also yields a list of antidifferences.

Certain special rational functions of the form $k^{-n} = \frac{1}{(k+1)\cdots(k+n)}$ are (for $n \in \mathbb{N}$) captured by the rule (11.12).

The exponential function a^k has the difference

$$\Delta a^k = a^{k+1} - a^k = (a-1)a^k,$$

so that for $a \neq 1$ it follows

$$\sum_k a^k = \frac{a^k}{a-1}.$$

For the factorial, we get

$$\Delta k! = (k+1)! - k! = k \cdot k!,$$

so that

$$\sum_k k \cdot k! = k!.$$

However, it remains unclear in which form we can sum the factorial $k!$ itself.

To sum more complicated expressions, it is possible to use similar techniques as for integration.

The operator Δ satisfies the product rule

$$\begin{aligned}
\Delta(a_k \cdot b_k) &= a_{k+1} \cdot b_{k+1} - a_k \cdot b_k \\
&= (a_{k+1} \cdot b_{k+1} - a_k \cdot b_{k+1}) + (a_k \cdot b_{k+1} - a_k \cdot b_k) \\
&= \Delta a_k \cdot b_{k+1} + a_k \cdot \Delta b_k.
\end{aligned}$$

By summation the rule of *partial summation* follows:

$$\sum_k a_k \cdot \Delta b_k = a_k \cdot b_k - \sum_k \Delta a_k \cdot b_{k+1}.$$

As an example for the use of partial summation, we consider the sum $\sum_k H_k$ of the harmonic numbers H_k and set $a_k = H_k$ and $\Delta b_k = 1$, hence $\Delta a_k = \frac{1}{k+1}$ and $b_k = \sum_k 1 = k$. Hence we get

$$\sum_k H_k = k H_k - \sum_k \frac{1}{k+1}(k+1) = k H_k - \sum_k 1 = k(H_k - 1).$$

Another example is the sum

$$\sum_k k a^k = k \frac{a^k}{a-1} - \sum_k \frac{a^{k+1}}{a-1} = k \frac{a^k}{a-1} - \frac{a}{a-1} \sum_k a^k = \frac{(k(a-1)-a)a^k}{(a-1)^2}.$$

Although with the methods of this section we can compute rather difficult indefinite sums in certain cases, an algorithmic aspect is missing: if one fails to find a definite sum, then one does not know whether this is unskillfulness or a principal problem.

For example, with these methods it is not possible to decide if $\sum_k k!$ or $\sum_k \frac{1}{k+1} = H_k$ are (again) hypergeometric terms. This question will be settled in the next section, where we will show that both these antidifferences are *not* hypergeometric terms.

We finish this section by a table of the deduced antidifferences.

Table 11.1 Table of antidifferences

a_k	$\sum_k a_k$	a_k	$\sum_k a_k$
$k^{\underline{n}}$	$\frac{1}{n+1} k^{\underline{n+1}}$	1	k
$k\,k!$	$k!$	k	$\frac{k(k-1)}{2}$
H_k	$k(H_k - 1)$	k^2	$\frac{k(2k-1)(k-1)}{6}$
a^k	$\frac{a^k}{a-1}$	k^3	$\frac{k^2(k-1)^2}{4}$
$k a^k$	$\frac{(k(a-1)-a)\,a^k}{(a-1)^2}$	k^4	$\frac{k(2k-1)(k-1)(3k^2-3k-1)}{30}$

11.4 Indefinite Summation of Hypergeometric Terms

In the previous section we have developed some heuristics for indefinite summation and an algorithm for the summation of polynomials. In this section we will consider the specific situation when an antidifference is a hypergeometric term.

Hence, we would like to find an antidifference s_k for a given term a_k with

$$s_{k+1} - s_k = a_k,\tag{11.15}$$

which is a hypergeometric term, i.e. $\frac{s_{k+1}}{s_k} \in \mathbb{K}(k)$.

From $\frac{s_{k+1}}{s_k} \in \mathbb{K}(k)$ it follows that

$$\frac{a_{k+1}}{a_k} = \frac{s_{k+2} - s_{k+1}}{s_{k+1} - s_k} = \frac{s_{k+1}}{s_k} \frac{\frac{s_{k+2}}{s_{k+1}} - 1}{\frac{s_{k+1}}{s_k} - 1} = \frac{u_k}{v_k} \in \mathbb{K}(k),$$

where we can choose $u_k, v_k \in \mathbb{K}[k]$ with $\gcd(u_k, v_k) = 1$. Hence, if we demand that the output s_k is a hypergeometric term, then the same is true for the input a_k. Therefore this section is about the indefinite summation of hypergeometric terms. If such a hypergeometric term a_k has a hypergeometric term antidifference s_k, then we call it *Gosper summable* since the underlying algorithm was found by Gosper [Gos1978]. The input for Gosper summation is the hypergeometric term a_k or else the two representing coprime polynomials u_k and v_k.

If we divide (11.15) by s_k, we get

$$\frac{s_{k+1}}{s_k} - 1 = \frac{a_k}{s_k} \in \mathbb{K}(k),$$

which implies that the quotient s_k/a_k is always rational,

$$\frac{s_k}{a_k} = \frac{g_k}{h_k} \in \mathbb{K}(k)$$

with $g_k, h_k \in \mathbb{K}[k]$ and $\gcd(g_k, h_k) = 1$. In such a way we have iteratively replaced the appearing hypergeometric terms by polynomials which are easier to deal with of course. In particular, we recall that the output term s_k can be written in terms of polynomials,

$$s_k = \frac{g_k}{h_k} \cdot a_k, \tag{11.16}$$

and is therefore a rational multiple of the input term a_k.

Finally, we divide (11.15) by a_k to get

$$\frac{s_{k+1}}{a_k} - \frac{s_k}{a_k} = \frac{a_{k+1}}{a_k} \cdot \frac{s_{k+1}}{a_{k+1}} - \frac{s_k}{a_k} = 1$$

or

$$\frac{u_k}{v_k} \cdot \frac{g_{k+1}}{h_{k+1}} - \frac{g_k}{h_k} = 1,$$

respectively. By multiplication with the common denominator we can write this equation as a polynomial equation:

$$h_k u_k g_{k+1} - h_{k+1} v_k g_k = h_k h_{k+1} v_k. \tag{11.17}$$

Here, u_k and v_k are given and g_k and h_k are sought polynomials. Assume for the moment that the denominator h_k of the rational function s_k/a_k is known. Then (11.17) defines an inhomogeneous recurrence equation of first order with polynomial coefficients for the numerator polynomial g_k. We will see that it is easy to compute the polynomial solutions of such a recurrence equation.

Therefore, in the next step we would like to find h_k. After division by h_{k+1} and h_k, using $\gcd(g_k, h_k) = 1$, the following divisibility conditions follow immediately from (11.17):

$$h_{k+1} \mid h_k \cdot u_k \qquad \text{as well as} \qquad h_k \mid h_{k+1} \cdot v_k$$

and

$$h_k \mid h_{k-1} \cdot u_{k-1} \qquad \text{as well as} \qquad h_k \mid h_{k+1} \cdot v_k. \tag{11.18}$$

This directly implies the following theorem.

Theorem 11.7. (Computation of Denominator) Let u_k, v_k and $h_k \in \mathbb{K}[k]$ and assume (11.18). Then it follows that

$$h_k \mid \gcd\left(\prod_{j=0}^{N} u_{k-1-j}, \prod_{j=0}^{N} v_{k+j}\right), \tag{11.19}$$

where[10]

$$N := \max\{j \in \mathbb{N}_{\geq 0} \mid \gcd(u_{k-1}, v_{k+j}) \neq 1\}. \tag{11.20}$$

If the set in (11.20) is empty[11], then $h_k = 1$.

---

[10] The maximum of the empty set is defined as $-\infty$.

[11] The empty product is declared as 1.

Proof. The given divisibility conditions imply by mathematical induction that for all $n \in \mathbb{N}_{\geq 0}$ we have

$$h_k \mid h_{k-1-n} \cdot \prod_{j=0}^{n} u_{k-1-j} \qquad \text{as well as} \qquad h_k \mid h_{k+1+n} \cdot \prod_{j=0}^{n} v_{k+j} \, .$$

Clearly, for suitably large $n \in \mathbb{N}$ the polynomial h_k must be coprime with h_{k-1-n} and with h_{k+1+n}. Hence for such an n the denominator h_k is a divisor of both $\prod_{j=0}^{n} u_{k-1-j}$ and $\prod_{j=0}^{n} v_{k+j}$ and therefore of $\gcd\left(\prod_{j=0}^{n} u_{k-1-j}, \prod_{j=0}^{n} v_{k+j} \right)$.

To find the smallest such $n \in \mathbb{N}_{\geq 0}$, we assume that t_k is a non-constant irreducible divisor of h_k. Then the irreducibility of t_k implies that there are integers $i, j \in \mathbb{N}_{\geq 0}$ such that $t_k \mid u_{k-1-i}$ as well as $t_k \mid v_{k+j}$. Since t_k is not constant, we get

$$\gcd(u_{k-1-i}, v_{k+j}) \neq 1 \Leftrightarrow \gcd(u_{k-1}, v_{k+(i+j)}) \neq 1 \, .$$

If we finally choose N as the largest positive index shift which, applied to v_k, generates a common divisor with u_{k-1} according to (11.20), then it is guaranteed that (11.19) is satisfied.

If the set in (11.20) is empty, then obviously $h_k = 1$. $\square$

The number N defined in (11.20) is called the *dispersion* of u_{k-1} and v_k. We saw in the proof of Theorem 11.7 that it pops up very naturally in hypergeometric summation. Later we will discuss how the dispersion can be computed algorithmically.

If we now *define*

$$h_k := \gcd\left(\prod_{j=0}^{N} u_{k-1-j}, \prod_{j=0}^{N} v_{k+j} \right) , \tag{11.21}$$

—if the set in (11.20) is non-empty, otherwise we set $h_k := 1$—then by Theorem 11.7 this polynomial is always a multiple of the denominator of s_k/a_k; it may, however, have common factors with the numerator g_k of s_k/a_k. If we do so, then the property $\gcd(g_k, h_k) = 1$ might no longer be valid, and therefore the degree of the sought numerator g_k might be higher than necessary. When we want to avoid this, we are led to the famous *Gosper algorithm*.[12] However, we would like to content ourselves with h_k according to (11.21) and will now consider some examples.

Session 11.8. *(a) We begin with the example*

$$a_k = (-1)^k \binom{n}{k}$$

which is entered into Mathematica as follows:

```
In[1]:= a = (-1)^k Binomial[n, k]
```
$$Out[1]= (-1)^k \binom{n}{k}$$

In this example, we consider n as an indeterminate variable, hence we work over the ground field $\mathbb{Q}(n)$.

---

[12] Gosper's algorithm [Gos1978] was probably the first algorithm which would not have been found without computer algebra. Gosper writes in his paper: "Without the support of MACSYMA and its developers, I could not have collected the experiences necessary to provoke the conjectures that led to this algorithm."

We obviously have

$$\frac{a_{k+1}}{a_k} = \frac{k-n}{k+1}$$

which Mathematica happily confirms:

```
In[2]:= rat = FunctionExpand[ (a/.k→k+1) / a ]
```

$$Out[2]= \frac{k-n}{k+1}$$

Therefore the two polynomials $u_k = k - n$ and $v_k = k + 1$

```
In[3]:= u = Numerator[rat]; v = Denominator[rat];
```

are our input polynomials $\in \mathbb{Q}(n)[k]$. In this example it is obvious that no index shift of the summation variable k applied to v_k can generate a common divisor with u_{k-1} [13], therefore we have $h_k = 1$ by Theorem 11.7:

```
In[4]:= h = 1;
```

The recurrence equation (11.17) for g_k is therefore given by

```
In[5]:= RE = h*u*g[k+1] - (h/.k→k+1)*v*g[k] == h*(h/.k→k+1)*v
Out[5]= g(k+1)(k-n)-(k+1)g(k) == k+1
```

hence

$$(k-n)\,g_{k+1} - (k+1)\,g_k = k+1\,. \tag{11.22}$$

We will soon study how to solve equation (11.17) systematically. For the moment we search for a polynomial solution for g_k of degree 1. Therefore we substitute the ansatz $g_k = A + Bk$ in (11.22) and get the following polynomial identity:

```
In[6]:= RE2 = RE/.{g[k_] → A+B k}//ExpandAll
Out[6]= A(-n)-A-B k n-B n == k+1
```

By equating the coefficients we get the linear system

```
In[7]:= eqs = CoefficientList[RE2[[1]] - RE2[[2]], k] == 0
Out[7]= {A(-n)-A-B n-1, -B n-1} == 0
```

with the solution

```
In[8]:= solution = Solve[eqs, {A, B}]
```

$$Out[8]= \left\{\left\{A \to 0, B \to -\frac{1}{n}\right\}\right\}$$

```
In[9]:= g = A+B k/.solution[[1]]
```

$$Out[9]= -\frac{k}{n}$$

hence

$$g_k = -\frac{k}{n}\,.$$

For the antidifference s_k we get according to (11.16)

```
In[10]:= s = g/h a
```

$$Out[10]= -\frac{(-1)^k\, k\, \binom{n}{k}}{n}$$

Therefore we have completely solved our summation problem! We thus have

---

[13] Of course, this would be different if $n \in \mathbb{Z}$ was given.

$$\sum_k (-1)^k \binom{n}{k} = \sum_k a_k = s_k = \frac{g_k}{h_k} a_k = -\frac{k}{n}(-1)^k \binom{n}{k}.$$

In particular, we have proved that a_k is Gosper summable.

Using our result we can compute every definite sum $\sum_{k=a}^{b} a_k = s_{b+1} - s_a$ with arbitrary bounds. As a special case, we consider $\sum_{k=0}^{n} a_k$ and get

```
In[11]:= (s/.k→n+1) - (s/.k→0)//FunctionExpand
Out[11]= 0
```

hence

$$\sum_{k=0}^{n} (-1)^k \binom{n}{k} = 0.$$

Of course, this is a well-known special case of the binomial formula. Note that this result is only valid for $n \in \mathbb{N}$, but not for $n = 0$. That the case $n = 0$ should be treated separately can be observed by our derivation, since the antidifference s_k has a pole at $n = 0$ and is therefore not defined for this particular value.

Using the function `AntiDifference` *of the* `SpecialFunctions` *package, the above result can be called directly:*

```
In[12]:= Needs["SpecialFunctions`"]
SpecialFunctions, (C) Wolfram Koepf, version 2.2, 2020, with English command names

In[13]:= AntiDifference[a,k]
```
$$Out[13]= \frac{(-1)^{k+1} k \binom{n}{k}}{n}$$

(b) In our next example we consider $a_k = k\,k!$, which was already treated in Section 11.3. Here we have

```
In[14]:= a = k k!
Out[14]= k k!
```

$$In[15]:= \text{rat} = \text{FunctionExpand}\left[\frac{a/.k \to k+1}{a}\right]$$
$$Out[15]= \frac{(k+1)^2}{k}$$

```
In[16]:= u = Numerator[rat]; v = Denominator[rat];
```

hence

$$\frac{a_{k+1}}{a_k} = \frac{(k+1)^2}{k}$$

and $u_k = (k+1)^2$ or $u_{k-1} = k^2$ as well as $v_k = k$. In this case the dispersion obviously equals $N = 0$. Since

```
In[17]:= h = PolynomialGCD[(u/.k→k-1),v]
Out[17]= k
```

we can work with $h_k = \gcd(u_{k-1}, v_k) = k$ according to (11.21). The recurrence equation (11.17) for g_k then reads as

```
In[18]:= RE = h*u*g[k+1] - (h/.k→k+1)*v*g[k] == h*(h/.k→k+1)*v
```

Out[18]= $k\,(k+1)^2\,g(k+1) - k\,(k+1)\,g(k) == k^2\,(k+1)$

After substituting $g_k = A$ for a constant polynomial g_k, we finally get

In[19]:= **RE2 = RE/.{g[k_] → A}//ExpandAll**
Out[19]= $A\,k^3 + A\,k^2 == k^3 + k^2$

In[20]:= **eqs = CoefficientList[RE2[[1]] - RE2[[2]], k] == 0**
Out[20]= $\{0, 0, A-1, A-1\} == 0$

In[21]:= **Solve[eqs, {A}]**
Out[21]= $\{\{A \to 1\}\}$

hence

$$g_k = 1 \,.$$

Therefore we have derived the result

$$s_k = \sum_k a_k = \sum_k k\,k! = \frac{g_k}{h_k}\,a_k = \frac{1}{k}\,k\,k! = k! \,,$$

as expected. Accordingly, the built-in function AntiDifference *generates the output*

In[22]:= **AntiDifference[a,k]**
Out[22]= $k!$

(c) Finally we consider $a_k = k!$:

In[23]:= **a = k!**
Out[23]= $k!$

We get

In[24]:= **rat = FunctionExpand$\left[\dfrac{\text{a/.k → k+1}}{\text{a}}\right]$**
Out[24]= $k+1$

In[25]:= **u = Numerator[rat]; v = Denominator[rat];**

therefore $u_k = k+1$ and $v_k = 1$, and hence obviously $h_k = 1$:

In[26]:= **h = 1;**

The recurrence equation for g_k is given by

In[27]:= **RE = h * u * g[k+1] - (h/.k → k+1) * v * g[k] == h * (h/.k → k+1) * v**
Out[27]= $(k+1)\,g(k+1) - g(k) == 1$

In the current case we can prove ad hoc that this recurrence equation cannot have a polynomial solution. In fact, if g_k has degree m, then $(k+1)\,g_{k+1}$ has degree $m+1$. This means, however, that—since g_k is not the zero polynomial—the polynomial $(k+1)\,g_{k+1} - g_k$ has at least degree 1 and therefore cannot be constant. This contradicts the right-hand side of the recurrence equation. Therefore the sought $g_k \in \mathbb{K}[k]$ does not exist, and a_k is not Gosper summable. This is the reason for the answer

In[28]:= **AntiDifference[a,k]**
There is no hypergeometric term antidifference

The above computation answers the question earlier posed: there is no antidifference of $k!$ which is a hypergeometric term!

Our method still has the disadvantage that we had to guess the potential degree of the numerator polynomial g_k or to prove ad hoc that such a polynomial does not exist. In particular, in case of failure we still cannot decide whether a solution exists or not. In our next step we discuss this question systematically.

Gosper designed the following algorithm which detects the potential degree of a polynomial solution of recurrence equation (11.17) a priori. As we have seen, once we know such a degree bound, g_k can be computed algorithmically by solving a linear system.

Theorem 11.9. (Degree Bound) For every polynomial solution g_k of the recurrence equation

$$A_k g_{k+1} + B_k g_k = C_k \tag{11.23}$$

with the polynomials $A_k, B_k, C_k \in \mathbb{K}[k]$, the following algorithm finds a *degree bound*:

1. If $\deg(A_k - B_k, k) \leqq \deg(A_k + B_k, k)$, then we have

$$\deg(g_k, k) = \deg(C_k, k) - \deg(A_k + B_k, k).$$

2. If $n := \deg(A_k - B_k, k) > \deg(A_k + B_k, k)$, then let a and b be the coefficients $a := \mathrm{coeff}(A_k - B_k, k, n) \neq 0$ and $b := \mathrm{coeff}(A_k + B_k, k, n-1)$.
    a. If $-2b/a \notin \mathbb{N}_{\geq 0}$, then we have

$$\deg(g_k, k) = \deg(C_k, k) - n + 1.$$

    b. If, however, $-2b/a \in \mathbb{N}_{\geq 0}$, then

$$\deg(g_k, k) \in \{-2b/a, \deg(C_k, k) - n + 1\}.$$

Proof. We rewrite the recurrence equation (11.23) in the form

$$C_k = (A_k + B_k)\frac{g_{k+1} + g_k}{2} + (A_k - B_k)\frac{g_{k+1} - g_k}{2}. \tag{11.24}$$

For every polynomial $g_k \neq 0$ the relation

$$\deg(g_{k+1} - g_k, k) = \deg(g_{k+1} + g_k, k) - 1 \tag{11.25}$$

is valid since the highest coefficient of $g_{k+1} - g_k$ cancels, in contrast to $g_{k+1} + g_k$. (This can be seen by using the binomial formula.) If we furthermore define $\deg(0) := -1$, then (11.25) is valid for all $g_k \in \mathbb{K}[k] \setminus \{0\}$.

Now, if $\deg(A_k - B_k, k) \leqq \deg(A_k + B_k, k)$, then the second summand in (11.24) definitely has a lower degree than the first summand. This yields (1).

If, however, $\deg(A_k - B_k, k) > \deg(A_k + B_k, k)$, then the situation is slightly more complicated. Let $\deg(g_k, k) = m \in \mathbb{N}_{\geq 0}$, with $g_k = c k^m + \ldots$. Then it follows for the highest coefficient in (11.24) that

```
In[1]:= eq =
```
$$C == \mathtt{AplusB}\,\frac{g[k+1] + g[k]}{2} + \mathtt{AminusB}\,\frac{g[k+1] - g[k]}{2}\, /.$$
$$\{\mathtt{g[k\_]} \to ck^m, \mathtt{AplusB} \to bk^{n-1}, \mathtt{AminusB} \to ak^n\}$$
$$Out[1] = C == \frac{1}{2}\,a\,k^n\left(c\,(k+1)^m - c\,k^m\right) + \frac{1}{2}\,b\,k^{n-1}\left(c\,k^m + c\,(k+1)^m\right)$$

and, using the binomial formula, we get

```
In[2]:= eq = eq/. (k + 1)^m → k^m + m k^(m-1) //ExpandAll
```
$$Out[2]= \quad C == \frac{1}{2} a c m k^{m+n-1} + \frac{1}{2} b c m k^{m+n-2} + b c k^{m+n-1}$$

hence

$$C_k = \left(b + a\frac{m}{2}\right) c k^{m+n-1} + \dots .$$

This results in (2). □

Actually the algorithm of Theorem 11.9 does not only give a degree bound, but in most cases it provides *the exact degree* of the resulting polynomial. It is only in case (2b) that there are two possible options.

To compute the polynomial solution g_k, we must finally—as in Session 11.8—substitute a generic polynomial of the desired degree in (11.17) and compute its coefficients by linear algebra. If the degree bound is not a nonnegative integer or if the linear system has no solution, then we have proved that the recurrence equation (11.17) does not have polynomial solutions. For our summation problem, the latter means that a_k does not possess an antidifference s_k which is a hypergeometric term.

Session 11.10. *The algorithm presented in Theorem 11.9 can easily be implemented:*

```
In[1]:= Clear[DegreeBound]
        DegreeBound[A_, B_, C_, k_] :=
          Module[{pol1, pol2, deg1, deg2, a, b},
            pol1 = Collect[A - B, k];
            pol2 = Collect[A + B, k];
            If[pol1 === 0, deg1 = -1,
              deg1 = Exponent[pol1, k]];
            If[pol2 === 0, deg2 = -1, deg2 = Exponent[pol2, k]];
            If[deg1 ≤ deg2, Print["Part 1"];
              Return[Exponent[C, k] - deg2]];
            a = Coefficient[pol1, k, deg1];
            If[deg2 < deg1 - 1, b = 0,
              b = Coefficient[pol2, k, deg2]];
            If[Not[IntegerQ[-2 * b/a] && -2 * b/a ≥ 0],
              Print["Part 2a"];
              Return[Exponent[C, k] - deg1 + 1],
              Print["Part 2b"];
              Return[Max[-2 * b/a, Exponent[C, k] - deg1 + 1]]]
          ]
```

Note that to check which part of Theorem 11.9 is selected, we have—for the moment— incorporated suitable messages using the `Print` *command.*

As a first application, we check the function `DegreeBound` *with the examples of Session 11.8. For* $a_k = (-1)^k \binom{n}{k}$ *we derived the recurrence equation* $(k - n) g_{k+1} - (k + 1) g_k = k + 1$. *For this input we obtain*

```
In[2]:= DegreeBound[k - n, - (k + 1), k + 1, k]
```
Part 2a
$$Out[2]= 1$$

Hence, this example led to part (2a) of the algorithm with the result $\deg(g_k, k) = 1$, *in accordance with our choice in Session 11.8. Of course the existence of a polynomial solution is not guaranteed after the computation of the degree bound.*

For $a_k = k\,k!$ *we received the recurrence equation*

$$k(k+1)^2 g_{k+1} - k(k+1)g_k = k^2(k+1)$$

with the degree bound

```
In[3]:= DegreeBound[k(k+1)^2,-k(k+1),k^2(k+1),k]
Part 1
Out[3]= 0
```

Finally we consider again $a_k = k!$ *with the recurrence equation* $(k+1)g_{k+1} - g_k = 1$ *and the degree bound*

```
In[4]:= DegreeBound[k+1,-1,1,k]
Part 1
Out[4]= -1
```

This result shows again—but this time completely automatically—that $k!$ *is not Gosper summable.*

The following procedure REtoPol *is an implementation of Theorem 11.9 and therefore yields a polynomial solution of the recurrence (11.23):*

```
In[5]:= Clear[REtoPol]
        REtoPol[A_,B_,C_,k_] :=
          Module[{deg,g,a,j,rec,sol},
            deg = DegreeBound[A,B,C,k];
            If[deg < 0,Return["no polynomial solution"]];
            g = Sum[a[j]*k^j,{j,0,deg}];
            rec = Collect[A*(g/.k->k+1)+B*g-C,k];
            sol = Solve[CoefficientList[rec,k] == 0,
                Table[a[j],{j,0,deg}]];
            If[sol === {},Return["no polynomial solution"]];
            g/.sol[[1]]/.a[_] -> 0
          ]
```

In the last step we set all coefficients of the computed polynomial which are not uniquely determined, to zero, since we need only one solution. This applies to case (2b) of Theorem 11.9 since in this case the degree of g_k *is not exactly known.*

Next, we solve the three examples in one step each:

```
In[6]:= REtoPol[k-n,-(k+1),k+1,k]
Part 2a
```
$$Out[6]= -\frac{k}{n}$$

```
In[7]:= REtoPol[k(k+1)^2,-k(k+1),k^2(k+1),k]
Part 1
Out[7]= 1
```

```
In[8]:= REtoPol[k+1,-1,1,k]
Part 1
Out[8]= "no polynomial solution"
```

As a further example, we would like to ask the question whether the harmonic numbers H_k can be represented by a hypergeometric term.[14] *Since $H_k = \sum_k \frac{1}{k+1}$, we set $a_k = \frac{1}{k+1}$:*

$$In[9] := \quad a = \frac{1}{k+1}$$

$$Out[9] = \quad \frac{1}{k+1}$$

We get

$$In[10] := \quad rat = FunctionExpand\left[\frac{(a/.k \to k+1)}{a}\right]$$

$$Out[10] = \quad \frac{k+1}{k+2}$$

$$In[11] := \quad u = Numerator[rat]; \ v = Denominator[rat];$$

hence $u_{k-1} = k$ and $v_k = k+2$. It easily follows that $h_k = 1$:

$$In[12] := \quad h = 1;$$

This yields the recurrence equation for g_k

$$In[13] := \quad RE = h*u*g[k+1] - (h/.k \to k+1)*v*g[k] == h*(h/.k \to k+1)*v$$

$$Out[13] = \quad (k+1) g(k+1) - (k+2) g(k) == k+2$$

with the degree bound

$$In[14] := \quad DegreeBound[Coefficient[RE[[1]],g[k+1]],$$
$$\qquad\qquad Coefficient[RE[[1]],g[k]],RE[[2]],k]$$

Part 2b

$$Out[14] = \quad 1$$

which leaves open the possibility that a polynomial g_k exists. Nevertheless, we obtain

$$In[15] := \quad REtoPol[Coefficient[RE[[1]],g[k+1]],$$
$$\qquad\qquad Coefficient[RE[[1]],g[k]],RE[[2]],k]$$

Part 2b

$$Out[15] = \quad \text{"no polynomial solution"}$$

which proves that $\frac{1}{k+1}$ is not Gosper summable. This is clearly equivalent to the fact that H_k is no hypergeometric term.

To complete the algorithm for the identification of hypergeometric antidifferences, an algorithm for the computation of the dispersion N according to (11.20) is still missing. The following algorithm fills this gap and determines the *dispersion set*

$$\{j \in \mathbb{N}_{\geq 0} \mid \gcd(u_{k-1}, v_{k+j}) \neq 1\}.$$

The computation of its maximum to get N is then trivial.

Theorem 11.11. (Dispersion Set) The following algorithm determines the dispersion set

$$J := \{j \in \mathbb{N}_{\geq 0} \mid \gcd(q_k, r_{k+j}) \neq 1\}$$

of two polynomials $q_k, r_k \in \mathbb{K}[k]$ under the condition that a factorization algorithm in $\mathbb{K}[x]$ is available. This algorithm is applicable in particular if $q_k, r_k \in \mathbb{Q}(a_1, a_2, ..., a_p)[k]$.

---

[14] In the continuous case this corresponds to the question whether the logarithmic function $\ln x$ is a hyperexponential term. This question can be clearly denied by a similar approach, compare [Koe2014, Chapter 11].

1. Input: two polynomials $q_k, r_k \in \mathbb{K}[k]$.
2. Factorize q_k and r_k over $\mathbb{K}$.
3. Set $J := \{\}$. For every pair of irreducible factors s_k of q_k and t_k of r_k compute the set $D := \texttt{PrimeDispersion[s,t,k]}$ in the following steps:
   a. If the degrees $m := \deg(s_k, k)$ and $n := \deg(t_k, k)$ differ from each other, then return $D := \{\}$.
   b. Compute the coefficients $a := \text{coeff}(s_k, k, n)$, $b := \text{coeff}(s_k, k, n-1)$, $c := \text{coeff}(t_k, k, n)$ and $d := \text{coeff}(t_k, k, n-1)$.
   c. If $j := \frac{bc-ad}{acn} \notin \mathbb{N}_{\geq 0}$, then return $D := \{\}$.
   d. Check whether $cs_k - at_{k+j} \equiv 0$. If this is the case, then return $D := \{j\}$, otherwise return $D := \{\}$.
   e. Set $J := J \cup D$.
4. Output: J.

Proof. First we would like to show that the dispersion of two *irreducible* polynomials q_k and r_k can be computed by the subroutine $\texttt{PrimeDispersion[q,r,k]}$.

Let us assume that q_k and r_k have the dispersion $j \geq 0$. In this case q_k and r_k must have the same degree and are multiples of each other. In order to see this, note that the relation

$$\gcd(q_k, r_{k+j}) = g_k \not\equiv 1$$

implies that q_k has the factor g_k and r_k has the factor g_{k-j}. Since by assumption q_k and r_k are irreducible, the degree of g_k must be equal to the common degree n of q_k and r_k, which proves our claim.

Therefore the two polynomials

$$q_k = ak^n + bk^{n-1} + \dots$$

and

$$r_k = ck^n + dk^{n-1} + \dots$$

have the dispersion $j \in \mathbb{N}_{\geq 0}$ if and only if

$$\frac{c}{a} q_k \equiv r_{k+j} = c(k+j)^n + d(k+j)^{n-1} + \dots = ck^n + (cnj+d)k^{n-1} + \dots, \qquad (11.26)$$

where the binomial formula was used again.

The resulting identity (11.26) can only be valid if the coefficients of k^{n-1} on both sides agree. Hence we get

$$\frac{bc}{a} = cnj + d$$

and therefore

$$j = \frac{bc - ad}{acn}. \qquad (11.27)$$

This implies that j, given by (11.27), must be a nonnegative integer. If this is the case, then we can compute r_{k+j} and check whether (11.26) is valid.

To compute the complete dispersion set of q_k and r_k, the algorithm uses the subroutine `PrimeDispersion` for every pair of irreducible factors and collects the computed values in the set J. □

Session 11.12. *We can implement this algorithm in Mathematica as follows:*

```
In[1]:= Clear[PrimeDispersion]
        PrimeDispersion[q_, r_, k_] :=
          Module[{s, t, n, a, b, c, d, j},
            s = Collect[q, k];
            t = Collect[r, k];
            n = Exponent[s, k];
            If[n === 0 || Not[n == Exponent[t, k]],
              Return[{}]];
            a = Coefficient[s, k, n];
            b = Coefficient[s, k, n-1];
            c = Coefficient[t, k, n];
            d = Coefficient[t, k, n-1];
            j = Together[(b*c - a*d)/(a*c*n)];
            If[Not[IntegerQ[j] && j >= 0], Return[{}]];
            If[Together[c*s - a*(t/.k -> k+j)] === 0,
              Return[{j}], Return[{}]]
          ]
```

As an example application, we first compute the dispersion of two linear polynomials:

```
In[2]:= PrimeDispersion[k, k - 1234, k]
Out[2]= {1234}
```

In this case, of course, the dispersion is visible to the naked eye. For the next example of two irreducible polynomials of second degree over $\mathbb{Q}(m)$, this is clearly not the case:

```
In[3]:= q = Expand[3k² + m k - 5m + 1/.k -> k + 4321]
Out[3]= 3 k² + m k + 25926 k + 4316 m + 56013124
```

```
In[4]:= r = Expand[n (3k² + m k - 5m + 1)]
Out[4]= 3 k² + m k - 5 m + 1
```

```
In[5]:= PrimeDispersion[q, r, k]
Out[5]= {4321}
```

Now we can finally compute the dispersion set of two arbitrary polynomials:

```
In[6]:= Clear[DispersionSet]
        DispersionSet[q_,r_,k_] :=
          Module[{f,g,m,n,i,j,result,tmp,op1,op2},
            f = Factor[q];
            g = Factor[r];
            If[Not[Head[f] === Times],m = 1,m = Length[f]];
            If[Not[Head[g] === Times],n = 1,n = Length[g]];
            result = {};
            Do[
              If[Head[f] === Times,op1 = f[[i]],op1 = f];
              If[Head[op1] === Power,op1 = op1[[1]]];
              Do[
                If[Head[g] === Times,op2 = g[[j]],op2 = g];
                If[Head[op2] === Power,op2 = op2[[1]]];
                tmp = PrimeDispersion[op1,op2,k];
                If[Not[tmp === {}],AppendTo[result,tmp[[1]]]],
                {j,1,n}],
              {i,1,m}];
            Union[result]
          ]
```

We compute the dispersion set of two polynomials with linear factors:

```
In[7]:= DispersionSet[k²(k+4),k(k-3)³,k]
Out[7]= {0,3,4,7}
```

Finally, let us compute the dispersion set of two complicated polynomials over $\mathbb{Q}(a)$:

```
In[8]:= q = Expand[k(3k²+a)(k-22536)(k³+a)]
```
$Out[8]= a^2 k^2 - 22536\, a^2\, k + a\, k^5 - 22533\, a\, k^4 - 67608\, a\, k^3 + 3\, k^7 - 67608\, k^6$

```
In[9]:= r = Expand[k(3k²+a)(k-22536)(k³+a)/.{k→k-345}]
```
$Out[9]= 3\, k^7 - 74853\, k^6 + a\, k^5 + 147447135\, k^5 - 24258\, a\, k^4 -$
$\qquad 125017313625\, k^4 + 32218182\, a\, k^3 + 57012120995625\, k^3 +$
$\qquad a^2 k^2 - 16432603920\, a\, k^2 - 14674906639659375\, k^2 -$
$\qquad 23226\, a^2\, k + 3747840275025\, a\, k + 2018054986820803125\, k +$
$\qquad 7893945\, a^2 - 321335266839750\, a - 115747288568266921875$

```
In[10]:= DispersionSet[q,r,k]
Out[10]= {345,22881}
```

This completes the last step in the algorithm for computing antidifferences of hypergeometric terms.

In Exercise 11.4 you are asked to combine all steps and implement your version of the procedure `Antidifference` which executes the presented variant of Gosper's algorithm.

11.5 Definite Summation of Hypergeometric Terms

In this section we would like to consider the definite sum of hypergeometric terms once again. In Section 11.1 we have developed Fasenmyer's method for this purpose. This algorithm was suitable for the computation of a holonomic sequence for

$$s_n = \sum_{k=-\infty}^{\infty} F(n,k),$$

if $F(n,k)$ is a hypergeometric term with respect to both variables n and k, hence

$$\frac{F(n+1,k)}{F(n,k)} \in \mathbb{K}(n,k) \quad \text{and} \quad \frac{F(n,k+1)}{F(n,k)} \in \mathbb{K}(n,k).$$

If the resulting recurrence equation is of first order, then this enables us to represent s_n as a hypergeometric term.

But due to its high complexity, Fasenmyer's method was not capable to solve really difficult questions.

In this section we will consider a much more efficient method for the same purpose, which was developed by Doron Zeilberger [Zei1990a] and is based on the algorithm of the last section.

Let us start, however, with the message that a *direct* application of indefinite summation is not successful. In this section we generally assume, again, that $F(n,k)$ has *finite support*, i.e., for every fixed $n \in \mathbb{N}_{\geq 0}$ the summand $F(n,k)$ deviates from zero for only a finite set of arguments $k \in \mathbb{Z}$.

Theorem 11.13. Let $F(n,k)$ be a hypergeometric term with respect to n and k which is Gosper summable with respect to k with an antidifference $s_k = G(n,k)$ which is finite for all $k \in \mathbb{Z}$. Let further $F(n,k)$ be well defined for all $n \in \mathbb{N}_{\geq 0}$ and have finite support. Then

$$\sum_{k=-\infty}^{\infty} F(n,k) = 0 \tag{11.28}$$

for all except finitely many $n \in \mathbb{N}_{\geq 0}$. In detail: If $G(n,k) = R(n,k)F(n,k)$ is a hypergeometric antidifference, then (11.28) is valid for all $n \in \mathbb{N}_{\geq 0}$ for which the denominator of the rational function $R(n,k) \in \mathbb{K}(n,k)$ is not identically zero.

Proof. By assumption $F(n,k)$ is Gosper summable with respect to k, hence there is a hypergeometric antidifference $G(n,k)$:

$$F(n,k) = G(n,k+1) - G(n,k).$$

We sum this equation over all $k \in \mathbb{Z}$. This yields

$$\sum_{k=-\infty}^{\infty} F(n,k) = \sum_{k=-\infty}^{\infty} \left(G(n,k+1) - G(n,k)\right) = 0,$$

because the right-hand side is a telescoping sum. Since $F(n,k)$ has finite support, the considered sum is finite. However, it can happen that $G(n,k)$ has singularities at certain values of $n \in \mathbb{N}_{\geq 0}$. Since $G(n,k) = R(n,k)F(n,k)$ is a rational multiple of $F(n,k)$, the singularities of $G(n,k)$ are the poles of $R(n,k)$. $\qquad\square$

Example 11.14. In Session 11.8 we proved that $F(n,k) = (-1)^k \binom{n}{k}$ is Gosper summable with

$$G(n,k) = -\frac{k}{n} F(n,k).$$

Using Theorem 11.13 this implies immediately that for $n \in \mathbb{N}$ (but not for $n = 0$)

$$\sum_{k=-\infty}^{\infty} (-1)^k \binom{n}{k} = \sum_{k=0}^{n} (-1)^k \binom{n}{k} = 0 \,.$$

Note that in Session 11.8 we got this result by substituting the natural summation bounds. △

By contraposition, Theorem 11.13 yields the following result.

Corollary 11.15. Let $F(n,k)$ be a hypergeometric term with respect to n and k which is well-defined for all $n \in \mathbb{N}_{\geq 0}$ and has finite support. If $\sum_{k=-\infty}^{\infty} F(n,k)$ can be represented by a hypergeometric term (which is obviously different from zero), then $F(n,k)$ is not Gosper summable with respect to k. $\qquad\square$

Example 11.16. Because of

$$\sum_{k=0}^{n} \binom{n}{k} = 2^n \,,$$

$$\sum_{k=0}^{n} k \binom{n}{k} = n 2^{n-1}$$

and

$$\sum_{k=0}^{n} \binom{n}{k}^2 = \binom{2n}{n}$$

the summands $\binom{n}{k}$, $k \binom{n}{k}$ as well as $\binom{n}{k}^2$ cannot be Gosper summable with respect to k. △

Zeilberger's idea [Zei1990b] is that we do not apply the algorithm of the previous section for computing a hypergeometric antidifference to $F(n,k)$, but for a suitable $J \in \mathbb{N}$ to the expression

$$a_k := F(n,k) + \sum_{j=1}^{J} \sigma_j(n) F(n+j,k) \tag{11.29}$$

with not yet determined variables σ_j $(j = 1, \ldots, J)$ depending on n, but not on k. Because of

$$\frac{a_{k+1}}{a_k} = \frac{F(n,k+1) + \sum\limits_{j=1}^{J} \sigma_j(n) F(n+j,k+1)}{F(n,k) + \sum\limits_{j=1}^{J} \sigma_j(n) F(n+j,k)}$$

$$= \frac{F(n,k+1)}{F(n,k)} \cdot \frac{1 + \sum\limits_{j=1}^{J} \sigma_j(n) \frac{F(n+j,k+1)}{F(n,k+1)}}{1 + \sum\limits_{j=1}^{J} \sigma_j(n) \frac{F(n+j,k)}{F(n,k)}} = \frac{u_k}{v_k} \in \mathbb{K}(n,k) \,, \tag{11.30}$$

a_k is a hypergeometric term, and therefore the algorithm of the last section is applicable.

By (11.30) v_k contains a divisor $w_k(\sigma_j)$ which linearly includes the variables σ_j $(j = 1, \ldots, J)$, and u_k contains the divisor $w_{k+1}(\sigma_j)$. This implies that u_{k-1} and v_k have the common divisor $w_k(\sigma_j) = 1 + \sum\limits_{j=1}^{J} \sigma_j(n) \frac{F(n+j,k)}{F(n,k)}$. Therefore in the current situation the dispersion set of u_{k-1} and v_k always contains the number 0, hence $N \geq 0$.

We again choose h_k according to (11.21), hence

$$h_k := \gcd\left(\prod_{j=0}^{N} u_{k-1-j}, \prod_{j=0}^{N} v_{k+j}\right) \in \mathbb{K}(n)[k].$$

It remains to find a polynomial solution $g_k \in \mathbb{K}(n)[k]$ of the inhomogeneous recurrence equation (11.17),

$$h_k u_k g_{k+1} - h_{k+1} v_k g_k = h_k h_{k+1} v_k. \tag{11.31}$$

If $N = 0$,[15] then $h_k | v_k$ and $h_k | u_{k-1}$, so that the recurrence equation (11.31) for g_k can be divided by $h_k \cdot h_{k+1}$, which results in the much simpler form

$$\frac{u_k}{h_{k+1}} g_{k+1} - \frac{v_k}{h_k} g_k = v_k. \tag{11.32}$$

Since $\frac{u_k}{h_{k+1}}, \frac{v_k}{h_k} \in \mathbb{K}(n)[k]$ this equation has polynomial coefficients.

If $N > 0$, then the common divisor $\gcd(h_k u_k, h_{k+1} v_k g_k, h_k h_{k+1} v_k)$ of the coefficients of the recurrence equation (11.31) should be canceled to decrease the complexity. Furthermore, note that the variables σ_j ($j = 1, \ldots, J$) always occur linearly on the right-hand side of (11.31) or the reduced equation (11.32), respectively.

In the usual way we find a degree bound m for g_k, and Zeilberger's essential observation is that the linear system for computing the coefficients of the generic polynomial

$$g_k = \alpha_0 + \alpha_1 k + \alpha_2 k^2 + \cdots + \alpha_m k^m$$

contains the coefficients α_l ($l = 0, \ldots, m$) of g_k as well as the *unknowns* σ_j ($j = 1, \ldots, J$) linearly. Therefore we can solve this linear system for all these variables and obtain $g_k \in \mathbb{K}(n)[k]$ as well as suitable rational functions $\sigma_j \in \mathbb{K}(n)$ ($j = 1, \ldots, J$).

If this procedure is successful, then it results in a hypergeometric term $G(n, k) = \frac{g_k}{h_k} a_k$ and rational functions $\sigma_j \in \mathbb{K}(n)$ ($j = 1, \ldots, J$) which satisfy the equation

$$G(n, k+1) - G(n, k) = a_k = F(n, k) + \sum_{j=1}^{J} \sigma_j(n) F(n+j, k). \tag{11.33}$$

By summation, we therefore get

$$\sum_{k=-\infty}^{\infty} a_k = \sum_{k=-\infty}^{\infty} \left(F(n, k) + \sum_{j=1}^{J} \sigma_j(n) F(n+j, k) \right)$$

$$= s_n + \sum_{j=1}^{J} \sigma_j(n) s_{n+j} = \sum_{k=-\infty}^{\infty} \left(G(n, k+1) - G(n, k) \right) = 0,$$

since the right-hand side is again a telescoping sum. After multiplication by the common denominator we finally get the holonomic recurrence equation of order J for s_n sought.

Since the algorithm of the last section is a decision procedure, its application is always successful if an equation of the form (11.33) for $F(n, k)$ is valid. Luckily this is almost always the

---

[15] Note that this is mostly the case.

case. However, it is not guaranteed that this method will always find the holonomic recurrence equation of lowest order for s_n. But one can prove that under certain slight constraints on the allowed input, Zeilberger's method terminates (see e.g. [Koe2014, Theorem 7.10]). Therefore we always start with $J = 1$, and in case of failure we increase J by 1 until the algorithm stops.

Session 11.17. *We load the package* SpecialFunctions *which contains the procedures* PrimeDispersion, DispersionSet, DegreeBound *and* REtoPol *that were introduced in the last section. The described Zeilberger algorithm can now be implemented as follows:* [16]

```
In[1]:= Clear[SumRecursion]
        SumRecursion[F_, k_, S_[n_]] :=
        Module[{a, σ, ratk, ratn, rat, denominator, u, v, M,
            dis, h, j, deg, rec, sol, A, B, CC, gcd, α, g, RE}, RE = {};
        Do[
        ratk = FunctionExpand[(F/.k → k+1)/F];
        ratn = FunctionExpand[(F/.n → n+1)/F];
        denominator = 1+
          Sum[σ[j] * Product[ratn/.n → n+i, {i, 0, j-1}], {j, J}];
        rat = Together[ratk * (denominator/.k → k+1)/denominator];
        u = Numerator[rat]; v = Denominator[rat];
        If[Not[PolynomialQ[u, k] && PolynomialQ[v, k]], Return[]];
        M = DispersionSet[u/.k → k-1, v, k];
        dis = Max[M];
        h = PolynomialGCD[Product[u/.k → k-1-j, {j, 0, dis}],
            Product[v/.k → k+j, {j, 0, dis}]];
        If[dis == 0,
          A = Together[u/(h/.k → k+1)]; B = -Together[v/h]; CC = v,
          A = h*u; B = -(h/.k → k+1) *v; CC = h*(h/.k → k+1) *v];
        gcd = PolynomialGCD[A, B, CC]; A = Together[A/gcd];
        B = Together[B/gcd]; CC = Together[CC/gcd];
        deg = DegreeBound[A, B, CC, k];
        If[deg >= 0,
          g = Sum[α[j] * k^j, {j, 0, deg}];
          rec = Collect[A * (g/.k → k+1) +B*g-CC, k];
          sol = Solve[CoefficientList[rec, k] == 0,
              Union[Table[α[j], {j, 0, deg}],
                  Table[σ[j], {j, 1, J}]]];
          If[Not[sol === {}],
            RE = S[n] + Sum[σ[j] * S[n+j]/.sol[[1]], {j, 1, J}];
            RE = Numerator[Together[RE]]; RE = Collect[RE, S[___]];
            RE = Map[Factor, RE]; Return[]
          ]
        ], {J, 1, 5}];
        If[Not[PolynomialQ[u, k] && PolynomialQ[v, k]],
          Return["input is no hypergeometric term"],
          If[RE === {},
            "There is no recurrence of order 5", RE == 0]]]
```

---

[16] The command SumRecursion is also contained in the SpecialFunctions package. Instead of FunctionExpand the built-in function uses SimplifyCombinatorial which is a more reliable normal function, see [Koe2014, Algorithm 2.2].

Let us test the algorithm by again computing the recurrence equation valid for $s_n = \sum_{k=0}^{n} \binom{n}{k}$:

$In[2]:=$ **SumRecursion[Binomial[n,k],k,S[n]]**
$Out[2]=$ $2\,S(n) - S(n+1) == 0$

Let us also check some further recurrence equations for sums for which Fasenmyer's algorithm was already successful:

$In[3]:=$ **SumRecursion[Binomial[n,k]^2,k,S[n]]**
$Out[3]=$ $2\,(2\,n+1)\,S(n) - (n+1)\,S(n+1) == 0$

$In[4]:=$ **SumRecursion[(-1)^{k}Binomial[n,k]^2,k,S[n]]**
$Out[4]=$ $4\,(n+1)\,S(n) + (n+2)\,S(n+2) == 0$

But now we can solve much more difficult problems! The following computations find recurrence equations for the sums

$$\sum_{k=0}^{n}\binom{n}{k}^3, \qquad \sum_{k=0}^{n}\binom{n}{k}^4 \quad and \quad \sum_{k=0}^{n}\binom{n}{k}^5,$$

which were not accessible via Fasenmyer's algorithm:

$In[5]:=$ **SumRecursion[Binomial[n,k]^3,k,S[n]]**
$Out[5]=$ $\left(7\,n^2 + 21\,n + 16\right)S(n+1) + 8\,S(n)\,(n+1)^2 - (n+2)^2\,S(n+2) == 0$

$In[6]:=$ **SumRecursion[Binomial[n,k]^4,k,S[n]]**
$Out[6]=$ $2\,(2n+3)\,(3n^2+9n+7)\,S(n+1) + (n+2)^3\,(-S(n+2)) + 4\,(n+1)\,(4n+3)\,(4n+5)\,S(n) == 0$

$In[7]:=$ **SumRecursion[Binomial[n,k]^5,k,S[n]]**
$Out[7]=$ $32\,(55\,n^2 + 253\,n + 292)\,S(n)\,(n+1)^4 +$
$\qquad (-19415\,n^6 - 205799\,n^5 - 900543\,n^4 - 2082073\,n^3 -$
$\qquad\qquad 2682770\,n^2 - 1827064\,n - 514048)\,S(n+1) +$
$\qquad (-1155\,n^6 - 14553\,n^5 - 75498\,n^4 - 205949\,n^3 -$
$\qquad\qquad 310827\,n^2 - 245586\,n - 79320)\,S(n+2) +$
$\qquad (n+3)^4\,(55\,n^2 + 143\,n + 94)\,S(n+3) == 0$

We finally compute the recurrence equation of the Legendre polynomials $P_n(x)$ using our procedure SumRecursion, *see (11.5):*

$In[8]:=$ **SumRecursion$\left[\right.$Binomial[n,k]Binomial[-n-1,k]**
$\qquad \left(\dfrac{1-x}{2}\right)^k$**,k,P[n]$\left.\right]$**
$Out[8]=$ $-(2\,n+3)\,x\,P(n+1) + (n+1)\,P(n) + (n+2)\,P(n+2) == 0$

For the Laguerre polynomials $L_n^{(\alpha)}(x)$, see Exercises 10.14 and 11.2, we find

$In[9]:=$ **SumRecursion$\left[\dfrac{(-1)^k}{k!}\right.$Binomial[n+$\alpha$,n-k]x^k,k,L[n]$\left.\right]$**
$Out[9]=$ $L(n)\,(\alpha+n+1) - L(n+1)\,(\alpha+2n-x+3) + (n+2)\,L(n+2) == 0$

Further computations will be executed in the exercises.

In the last part of this chapter we will consider the question whether and to what extent we can trust the results given by Zeilberger's algorithm. Of course we have proved that the algorithm is doing what we want. However, it is not easy to prove that our implementation is error-free.

If you have ever implemented a somehow complex algorithm you will realize this.[17] Therefore it would be good if we could check every result of our algorithm *independently*. Indeed,
Zeilberger's algorithm provides such a verification method!

For the sake of simplicity we restrict ourselves to the most interesting case that Zeilberger's algorithm returns a recurrence equation of *first order*. Note, however, that the verification method
can be generalized to higher order. In our specific case, s_n is therefore a hypergeometric term
which we can compute. Hence we find a formula of the form

$$\sum_{k=-\infty}^{\infty} F(n,k) = t_n,$$

where t_n is a hypergeometric term. Division of this equation by t_n creates an equation of the
type

$$\sum_{k=-\infty}^{\infty} \tilde{F}(n,k) = 1,$$

where $\tilde{F}(n,k) = F(n,k)/t_n$ is again a hypergeometric term with respect to both n and k. For
simplicity reasons we write again $F(n,k)$ for $\tilde{F}(n,k)$.

Next we will show that a summation formula

$$s_n = \sum_{k=-\infty}^{\infty} F(n,k) = 1 \tag{11.34}$$

with hypergeometric term $F(n,k)$ can be proved *by simple rational arithmetic*. This method
goes back to Wilf and Zeilberger [WZ1992] and is therefore called the *WZ method*.

For $s_n = \sum_{k=-\infty}^{\infty} F(n,k)$ according to (11.34), the recurrence equation $s_{n+1} - s_n = 0$ is valid. If successful, the Zeilberger-type application of Gosper's algorithm to $a_k = F(n+1,k) + \sigma_1(n)F(n,k)$
therefore yields $\sigma_1(n) = -1$ and hence the symmetric equation

$$F(n+1,k) - F(n,k) = G(n,k+1) - G(n,k), \tag{11.35}$$

where $G(n,k)$ is a hypergeometric antidifference of $F(n+1,k) - F(n,k)$ with respect to the summation variable k. Therefore $G(n,k)$ is a rational multiple of the input function

$$r(n,k) = \frac{G(n,k)}{F(n+1,k) - F(n,k)} \in \mathbb{K}(n,k).$$

This also implies that $R(n,k) = \frac{G(n,k)}{F(n,k)} \in \mathbb{K}(n,k)$, since

$$R(n,k) = \frac{G(n,k)}{F(n,k)} = r(n,k) \cdot \frac{F(n+1,k) - F(n,k)}{F(n,k)} = r(n,k) \cdot \left(\frac{F(n+1,k)}{F(n,k)} - 1 \right).$$

We call $R(n,k)$ the *rational certificate* of the identity (11.34).

If we write

$$r_k(n,k) := \frac{F(n,k+1)}{F(n,k)} \qquad \text{and} \qquad r_n(n,k) := \frac{F(n+1,k)}{F(n,k)}$$

---

[17] This problem is abundant in every kind of software development.

for the rational input quotients of $F(n,k)$, then (11.35) yields, after division by $F(n,k)$, the purely rational identity

$$r_n(n,k) - 1 = R(n,k+1) \cdot r_k(n,k) - R(n,k),$$ (11.36)

whose validity can be checked easily.

Obviously (11.36) is equivalent to (11.35), hence for the validity of (11.35) it is sufficient to prove (11.36). Then from (11.35) it follows that

$$s_{n+1} - s_n = \sum_{k=-\infty}^{\infty} \Big(F(n+1,k) - F(n,k) \Big) = \sum_{k=-\infty}^{\infty} \Big(G(n,k+1) - G(n,k) \Big) = 0.$$

Using the initial value $s_0 = 1$, we therefore get (11.34). Summing up, if $R(n,k)$ is known, then (11.34) can be proved simply by checking the rational identity (11.36) and the initial value $s_0 = 1$.

Let us consider some examples for how this works.

Session 11.18. *We would like to give a WZ proof of the binomial formula*

$$(x+y)^n = \sum_{k=0}^{n} \binom{n}{k} x^k y^{n-k}.$$ (11.37)

For this purpose we write

$$F(n,k) := \binom{n}{k} \frac{x^k y^{n-k}}{(x+y)^n},$$

and we have to prove that

$$s_n = \sum_{k=0}^{n} F(n,k) = \sum_{k=-\infty}^{\infty} F(n,k) = 1$$

is valid. We therefore set

$$\mathit{In[1]} := \mathbf{F} = \frac{\mathbf{Binomial[n,k]x^k y^{n-k}}}{\mathbf{(x+y)^n}}$$

$$\mathit{Out[1]} = x^k \binom{n}{k} y^{n-k} (x+y)^{-n}$$

and use the function

$$\mathit{In[2]} := \mathbf{R} = \frac{\mathbf{k\,y}}{\mathbf{(k-n-1)\,(x+y)}}$$

$$\mathit{Out[2]} = \frac{k\,y}{(k-n-1)\,(x+y)}$$

as a rational certificate which comes out of the blue. However, as far as the effect of the verification mechanism is concerned, it is completely irrelevant where this information comes from. Later we will show how to compute the certificate from $F(n,k)$ as given by (11.34).

To show (11.36), we first compute $r_k(n,k)$ and $r_n(n,k)$:

$$\mathit{In[3]} := \mathbf{ratk = FunctionExpand}\Big[\frac{\mathbf{F/.k \to k+1}}{\mathbf{F}}\Big]$$

$$\mathit{Out[3]} = \frac{x\,(n-k)}{(k+1)\,y}$$

$In[4] :=$ **ratn = FunctionExpand** $\left[\frac{F/.n \to n+1}{F}\right]$

$Out[4]=$ $\dfrac{(n+1)\,y}{(-k+n+1)\,(x+y)}$

Then (11.36) can be checked by the computation[18]

$In[5] :=$ **Together[ratn - 1 - ((R/.k → k + 1) ratk - R)]**

$Out[5]=$ 0

Therefore (11.35) follows, and by summation over $k = -\infty, \ldots, \infty$ we obtain the recurrence equation $s_{n+1} - s_n = 0$. Since

$In[6] :=$ **Sum[F/.n → 0, {k, 0, 0}]**

$Out[6]=$ 1

we have $s_0 = 1$. This completes the proof of (11.37) for $n \in \mathbb{N}_{\geq 0}$.

Similarly the computations

$In[7] :=$ **F** $= \dfrac{k\,\mathbf{Binomial[n, k]}}{n\,2^{n-1}}$

$Out[7]=$ $\dfrac{k\,2^{1-n}\binom{n}{k}}{n}$

$In[8] :=$ **R** $= \dfrac{k - 1}{2\,(k - n - 1)}$

$Out[8]=$ $\dfrac{k-1}{2\,(k-n-1)}$

$In[9] :=$ **ratk = FunctionExpand** $\left[\dfrac{F/.k \to k+1}{F}\right]$

$Out[9]=$ $\dfrac{n-k}{k}$

$In[10] :=$ **ratn = FunctionExpand** $\left[\dfrac{F/.n \to n+1}{F}\right]$

$Out[10]=$ $\dfrac{n}{2\,(-k+n+1)}$

$In[11] :=$ **Together[ratn - 1 - ((R/.k → k + 1) ratk - R)]**

$Out[11]=$ 0

$In[12] :=$ **Sum[F/.n → 1, {k, 0, 1}]**

$Out[12]=$ 1

prove the identity

$$\sum_{k=0}^{n} k\binom{n}{k} = n\,2^{n-1}$$

for $n \in \mathbb{N}$, and the computations

$In[13] :=$ **F** $= \dfrac{\mathbf{Binomial[n, k]}^2}{\mathbf{Binomial[2n, n]}}$

$Out[13]=$ $\dfrac{\binom{n}{k}^2}{\binom{2n}{n}}$

$In[14] :=$ **R** $= \dfrac{k^2\,(2k - 3n - 3)}{2\,(k - n - 1)^2\,(2n + 1)}$

---

[18] This can even be done by hand easily!

$$\text{Out[14]} = \frac{k^2 (2k - 3n - 3)}{2 (k - n - 1)^2 (2n + 1)}$$

$\text{In[15]} := \textbf{ratk = FunctionExpand}\left[\frac{\textbf{F/.k} \to \textbf{k+1}}{\textbf{F}}\right]$

$$\text{Out[15]} = \frac{(n - k)^2}{(k + 1)^2}$$

$\text{In[16]} := \textbf{ratn = FunctionExpand}\left[\frac{\textbf{F/.n} \to \textbf{n+1}}{\textbf{F}}\right]$

$$\text{Out[16]} = \frac{(n + 1)^3}{2 (2n + 1) (-k + n + 1)^2}$$

$\text{In[17]} := \textbf{Together[ratn - 1 - ((R/.k} \to \textbf{k+1) ratk - R)]}$

$\text{Out[17]} = 0$

$\text{In[18]} := \textbf{Sum[F/.n} \to \textbf{0, \{k, 0, 0\}]}$

$\text{Out[18]} = 1$

prove the identity

$$\sum_{k=0}^{n} \binom{n}{k}^2 = \binom{2n}{n}$$

for $n \in \mathbb{N}_{\geqq 0}$. Finally we get the Chu-Vandermonde identity

$$_2F_1\left(\begin{matrix} a, -n \\ b \end{matrix} \middle| 1\right) = \frac{(b - a)_n}{(b)_n}$$

by

$\text{In[19]} := \textbf{F} = \frac{\textbf{HyperTerm[\{-n, a\}, \{b\}, 1, k]}}{\frac{\textbf{Pochhammer[b-a, n]}}{\textbf{Pochhammer[b, n]}}}$

$$\text{Out[19]} = \frac{(a)_k (b)_n (-n)_k}{k! (b)_k (b - a)_n}$$

$\text{In[20]} := \textbf{R} = \frac{\textbf{k (b + k - 1)}}{\textbf{(k - n - 1) (-a + b + n)}}$

$$\text{Out[20]} = \frac{k (b + k - 1)}{(k - n - 1) (-a + b + n)}$$

$\text{In[21]} := \textbf{ratk = FunctionExpand}\left[\frac{\textbf{F/.k} \to \textbf{k+1}}{\textbf{F}}\right]$

$$\text{Out[21]} = \frac{(-n - 1) (b + n)}{(k - n - 1) (-a + b + n)}$$

$\text{In[22]} := \textbf{ratn = FunctionExpand}\left[\frac{\textbf{F/.n} \to \textbf{n+1}}{\textbf{F}}\right]$

$$\text{Out[22]} = \frac{(n + 1) (b + n)}{(-a + b + n) (-k + n + 1)}$$

$\text{In[23]} := \textbf{Together[ratn - 1 - ((R/.k} \to \textbf{k+1) ratk - R)]}$

$\text{Out[23]} = 0$

$\text{In[24]} := \textbf{Sum[F/.n} \to \textbf{0, \{k, 0, 0\}]}$

$\text{Out[24]} = 1$

In all the above cases the rational certificates were just given by a black box. Now we will compute them. Because of $R(n, k) = \frac{G(n,k)}{F(n,k)}$ this can be done by a simple application of Gosper's algorithm to the function $F(n + 1, k) - F(n, k)$ which yields $G(n, k)$. This can be realized by the function `WZCertificate`. *Using Gosper's algorithm via* `AntiDifference` *from Exercise 11.4, this function is implemented by the following short program:*

```
In[25]:= Clear[WZCertificate]
         WZCertificate[F_,k_,n_] :=
           Module[{a,G},
             a = (F/.n→n+1) - F;
             G = AntiDifference[a,k];
             Factor[FunctionExpand[G/F]]
           ]
```

Note that the SpecialFunctions *package contains a more efficient version of* WZCertificate, *based on the computation of* $\frac{a_{k+1}}{a_k}$ *from* $r_k(n,k)$ *and* $r_n(n,k)$ *by the formula*

$$\frac{a_{k+1}}{a_k} = \frac{F(n+1,k+1)-F(n,k+1)}{F(n+1,k)-F(n,k)} = \frac{F(n,k+1)}{F(n,k)} \cdot \frac{\frac{F(n+1,k+1)}{F(n,k+1)}-1}{\frac{F(n+1,k)}{F(n,k)}-1} .$$

Using WZCertificate *we can compute the rational certificates that we had used before:*

```
In[26]:= WZCertificate[ Binomial[n,k]x^k y^(n-k) / (x+y)^n , k, n]
```

$$Out[26] = \frac{k\,y}{(k-n-1)\,(x+y)}$$

```
In[27]:= WZCertificate[ k Binomial[n,k] / (n 2^(n-1)) , k, n]
```

$$Out[27] = \frac{k-1}{2\,(k-n-1)}$$

```
In[28]:= WZCertificate[ Binomial[n,k]^2 / Binomial[2n,n] , k, n]
```

$$Out[28] = \frac{k^2\,(2\,k-3\,n-3)}{2\,(k-n-1)^2\,(2\,n+1)}$$

```
In[29]:= WZCertificate[ HyperTerm[{-n,a},{b},1,k] / (Pochhammer[b-a,n] / Pochhammer[b,n]) , k, n]
```

$$Out[29] = \frac{k\,(b+k-1)}{(k-n-1)\,(-a+b+n)}$$

Please keep in mind that the WZ method works if and only if Zeilberger's algorithm finds a recurrence equation of first order.

11.6 Additional Remarks

Fasenmyer's method [Fas1945] for definite summation was already published in 1945 and then ignored for a long time. It was only after the development of the first computer algebra systems that such algorithmic methods were getting relevant again. First Gosper [Gos1978] with his algorithm for indefinite summation of hypergeometric terms and finally Zeilberger [Zei1990a] have reanimated this topic. Since the 1990s this is a rather active research topic. The simplified variant of Gosper's algorithm treated in Section 11.3 was developed in the diploma thesis of Harald Böing [Böi1998].

In the meantime there are exact a-priori-criteria for the success of Zeilberger's algorithm [Abr2003]. Also the order of Zeilberger's recurrence equation can be computed in advance [MM2005].

Furthermore, there are similar algorithms for the integration of hyperexponential functions [AZ1991]. This topic is covered in detail in [Koe2014]. The underlying normal form consists again of a holonomic differential equation. Also for multiple sums and integrals there are algorithms ([WZ1992], [Weg1997], [Spr2004], [Tef1999], [Tef2002]) that generalize Fasenmyer's method in suitable ways.

There is also a summation theory in extensions of difference fields which therefore forms a discrete analog of Risch's integration algorithm (compare next chapter). This theory is based on work by Karr [Karr1981].

11.7 Exercises

Exercise 11.1. (American Mathematical Monthly) *The journal American Mathematical Monthly has a Problems Section. As Problem # 10473 the following question was posed:*

Prove that there are infinitely many positive integers m such that

$$s_m := \frac{1}{5 \cdot 2^m} \sum_{k=0}^{m} \binom{2m+1}{2k} 3^k$$

is an odd integer.

Prove the following assertions using Fasenmyer's algorithm and therefore solve this problem:

(a) *The sum s_m satisfies the following recurrence equation of second order:*

$$s_{m+2} - 4 s_{m+1} + s_m = 0.$$

(b) *The sum s_m also satisfies the following recurrence equation of third order:*

$$s_{m+3} - 5 s_{m+2} + 5 s_{m+1} - s_m = 0.$$

(c) *For all $n \in \mathbb{N}_{\geq 0}$ the numbers $5s_{3n}, s_{3n+1}, 5s_{3n+2}$ are odd integers. In particular s_{3n+1} is odd. Hint: Induction.*

Exercise 11.2. (Holonomic Differential Equations) *Adapt the procedures* kfreeRE *and* FasenmyerRE *for the computation of a holonomic differential equation for the sum over a term $F(x, k)$ which is hypergeometric with respect to the discrete variable k and hyperexponential with respect to the continuous variable x, i.e.*

$$\frac{F(x, k+1)}{F(x, k)} \in \mathbb{K}(x, k) \qquad \text{and} \qquad \frac{F'(x, k)}{F(x, k)} \in \mathbb{K}(x, k),$$

and implement the corresponding functions kfreeDE *and* FasenmyerDE. *Based on your implementation, compute holonomic differential equations for the following sums:*

(a) $e^x = \displaystyle\sum_{k=0}^{\infty} \frac{1}{k!} x^k$;

(b) $e^{\frac{1+x}{1-x}} = \displaystyle\sum_{k=0}^{\infty} \frac{1}{k!} \left(\frac{1+x}{1-x}\right)^k$;

(c) $\sin x = \displaystyle\sum_{k=0}^{\infty} \frac{(-1)^k}{(2k+1)!} x^{2k+1};$

(d) **(Bessel functions)** $\displaystyle\sum_{k=0}^{\infty} \frac{1}{k!^2} x^k;$

(e) $\displaystyle\sum_{k=0}^{\infty} \frac{1}{k!^3} x^k;$

(f) **(Legendre polynomials)** $P_n(x) = \dfrac{1}{2^n} \displaystyle\sum_{k=0}^{n} (-1)^k \binom{n}{k} \binom{2n-2k}{n} x^{n-2k};$

(g) **(Laguerre polynomials)** $L_n^{(\alpha)}(x) = \displaystyle\sum_{k=0}^{n} \frac{(-1)^k}{k!} \binom{n+\alpha}{n-k} x^k.$

Solve the resulting differential equations with `DSolve` *and compare your results with* `FPS`.

Exercise 11.3. (Identification of the Legendre Polynomials) *Use the functions* `FasenmyerRE`, `FasenmyerDE` *and* `SumToHypergeometric` *for the identification of the Legendre polynomials by proving that the following series all constitute the same polynomial family. Compute all necessary initial values.*

$$
\begin{aligned}
P_n(x) &= \sum_{k=0}^{n} \binom{n}{k} \binom{-n-1}{k} \left(\frac{1-x}{2}\right)^k \\[2mm]
&= {}_2F_1 \left(\begin{array}{c} -n, n+1 \\ 1 \end{array} \middle| \frac{1-x}{2} \right) \\[2mm]
&= \frac{1}{2^n} \sum_{k=0}^{n} \binom{n}{k}^2 (x-1)^{n-k} (x+1)^k \\[2mm]
&= \left(\frac{1-x}{2}\right)^n {}_2F_1 \left(\begin{array}{c} -n, -n \\ 1 \end{array} \middle| \frac{1+x}{1-x} \right) \\[2mm]
&= \frac{1}{2^n} \sum_{k=0}^{\lfloor n/2 \rfloor} (-1)^k \binom{n}{k} \binom{2n-2k}{n} x^{n-2k} \\[2mm]
&= \binom{2n}{n} \left(\frac{x}{2}\right)^n {}_2F_1 \left(\begin{array}{c} -n/2, -n/2+1/2 \\ -n+1/2 \end{array} \middle| \frac{1}{x^2} \right)
\end{aligned}
$$

Exercise 11.4. (Antidifference) *Implement a function* `Antidifference[a_k, k]` *which computes an antidifference* s_k *of* a_k *which is a hypergeometric term, whenever such an antidifference exists. For this purpose use the subroutines* `REtoPol` *and* `DispersionSet`. *Check your function* `AntiDifference` *with the following examples:*

(a) $a_k = (-1)^k \binom{n}{k};$

(b) $a_k = k\,k!;$

(c) $a_k = k!;$

(d) $a_k = \frac{1}{k+1};$

(e) $a_k = \binom{n}{k};$

(f) $a_k = \binom{k}{n};$

(g) $a_k = (a)_n;$

(i) $a_k = \frac{1}{k(k+10)};$

(j) $a_k = (k+10)! - k!.$

Hint: Since the Mathematica function `FunctionExpand` *does not work properly in complicated situations, you can use the function* `SimplifyCombinatorial` *from the package* `SpecialFunctions`.

Exercise 11.5. (SIAM Review) *In SIAM Review 36, 1994, Problem 94-2, the following question was posed:*

Determine the infinite sum

$$S = \sum_{n=1}^{\infty} \frac{(-1)^{n+1}(4n+1)(2n-1)!!}{2^n(2n-1)(n+1)!},$$

where $(2n-1)!! = 1 \cdot 3 \cdots (2n-1)$.

Solve the above problem using Gosper's algorithm. Hint: You can use Stirling's formula

$$\lim_{n \to \infty} \frac{n!}{\sqrt{n}\,(n/e)^n} = \sqrt{2\pi}.$$

Exercise 11.6. (Summation of Polynomials) *Show that Gosper's algorithm, when applied to a polynomial, always reaches step (2b) of Theorem 11.25. Compare the computation times and the results of the algorithm of Session 11.6 with those of Gosper's algorithm on the basis of a dense polynomial with a high degree.*

Exercise 11.7. (Clausen's Formula) *Use Zeilberger's algorithm to prove Clausen's Formula*

$$_2F_1\left(\begin{array}{c} a, b \\ a+b+1/2 \end{array}\middle|\, x\right)^2 = {}_3F_2\left(\begin{array}{c} 2a, 2b, a+b \\ 2a+2b, a+b+1/2 \end{array}\middle|\, x\right),$$

which reveals under which conditions the square of a Gaussian hypergeometric function $_2F_1$ equals a Clausen series $_3F_2$. Hint: Represent the left-hand side as Cauchy product and compute the recurrence equation of the inner sum. This automatically yields the right-hand side. Use Fasenmyer's algorithm for the same purpose and compare the computation times.

Exercise 11.8. (Zeilberger Algorithm) *Compute, whenever possible, representations by hypergeometric terms for the following sums (we always assume $n \in \mathbb{N}_{\geq 0}$):*

(a) $\displaystyle\sum_{k=0}^{n} \frac{1}{(k+1)(k+2)\cdots(k+10)}$;

(b) $\displaystyle\sum_{k=1}^{n} \frac{2^k\left(k^3 - 3k^2 - 3k - 1\right)}{k^3(k+1)^3}$;

(c) $\displaystyle\sum_{k=0}^{n} \frac{(-1)^k}{\binom{m}{k}}$;

(d) $\displaystyle\sum_{k=0}^{n} \frac{\binom{j}{k}}{\binom{m}{k}}$;

(e) $\displaystyle\sum_{k=0}^{n} \binom{n}{k}^2 \binom{3n+k}{2n}$;

(f) **(Chu-Vandermonde)** $\displaystyle{}_2F_1\left(\begin{array}{c} a, -n \\ b \end{array}\middle|\, 1\right) = \sum_{k=0}^{n} \frac{(a)_k(-n)_k}{(b)_k k!}$;

(g) **(Kummer)** $_2F_1\left(\begin{array}{c} a,-n \\ 1+a+n \end{array}\middle| -1\right);$

(h) **(Pfaff-Saalschütz)** $_3F_2\left(\begin{array}{c} a,b,-n \\ c,1+a+b-c-n \end{array}\middle| 1\right);$

(i) **(Dixon)** $_3F_2\left(\begin{array}{c} a,b,-n \\ 1+a-b,1+a-c \end{array}\middle| 1\right);$

(j) **(Dougall)** $_7F_6\left(\begin{array}{c} a,1+a/2,b,c,d,1+2a-b-c-d+n,-n \\ a/2,1+a-b,1+a-c,1+a-d,b+c+d-a-n,1+a+n \end{array}\middle| 1\right).$

Exercise 11.9. (WZ Method) *Show the following identities using the WZ method by computing the rational certificate $R(n,k)$ and proving the corresponding rational identity:*

(a) $\displaystyle\sum_{k=0}^{m}\binom{n}{m+k}\binom{m+k}{2k}4^k = \frac{4^m\binom{n}{m}\binom{n-1/2}{m}}{\binom{2m}{m}};$

(b) $\displaystyle\sum_{k=0}^{n}\binom{m-r+s}{k}\binom{n+r-s}{n-k}\binom{r+k}{n+m} = \binom{r}{m}\binom{s}{n};$

(c) $\displaystyle\sum_{k=-n}^{n}(-1)^k\binom{n+a}{n+k}\binom{a+n}{a+k} = \binom{n+a}{n};$

(d) $\displaystyle\sum_{k=1}^{n}k\binom{n}{k}\binom{s}{k} = s\binom{n+s-1}{n-1};$

(e) Chu-Vandermonde identity (e) from Exercise 11.8;

(f) Kummer identity (f) from Exercise 11.8;

(g) Pfaff-Saalschütz identity (g) from Exercise 11.8;

(h) Dixon identity (h) from Exercise 11.8;

(i) Dougall identity (i) from Exercise 11.8.

Exercise 11.10. (Chu-Vandermonde Identity) *Check which of the following identities are special cases of the Chu-Vandermonde identity*

$$_2F_1\left(\begin{array}{c} A,-n \\ B \end{array}\middle| 1\right) = \frac{(B-A)_n}{(B)_n}.$$

(a) $\displaystyle\sum_{k=0}^{n}\binom{a}{k}\binom{b}{n-k} = \binom{a+b}{n};$

(b) $\displaystyle\sum_{k=0}^{n}\binom{n}{k}\binom{s}{t-k} = \binom{n+s}{t};$

(c) $\displaystyle\sum_{k=0}^{n}(-1)^k\binom{n}{k}\binom{2n-k}{m-k} = \binom{n}{m};$

(d) $\displaystyle\sum_{k=0}^{n}\binom{n}{k} = 2^n;$

(e) $\displaystyle\sum_{k=0}^{n}k\binom{n}{k} = n\,2^{n-1};$

(f) $\displaystyle\sum_{k=0}^{n}\binom{n}{k}\binom{s}{t+k} = \binom{n+s}{n+t}.$

Chapter 12
Algorithmic Integration

12.1 The Bernoulli Algorithm for Rational Functions

In the previous chapter, we studied, among other things, the algorithmic computation of anti-differences in the special case where these antidifferences are hypergeometric terms.

In this chapter we would like to consider algorithmic integration. A completely analogous situation to Gosper's algorithm as described in the previous chapter is given when we want to compute antiderivatives of *hyperexponential functions* $f(x)$ for which $f'(x)/f(x) \in \mathbb{K}(x)$, whose antiderivative is again hyperexponential. This problem was considered by Almkvist and Zeilberger [AZ1991], see also [Koe2014, Chapter 11].

In a more general context Risch [Ris1969]–[Ris1970] gave an algorithm for the computation of an antiderivative of *exp-log-functions*. These are functions that can be defined by rational functions with the aid of finitely many applications of the exponential and the logarithm function so that they are elements of a suitable finite field extension of the rational functions. *Risch's algorithm* also decides whether or not such an antiderivative exists. It was completed and generalized by Bronstein, Davenport, Rothstein, Trager and others, see e.g. [Bro1997].

This general algorithm, however, is rather comprehensive and algebraically much more elaborate as in the discrete case, although the basic ideas are rather similar. Due to a representation theorem which is valid for every antiderivative and was given already in the early 19th century by Liouville [Lio1835], it is possible—similarly as in Gosper's algorithm—to write down an ansatz whose coefficients can be found by solving a linear system.[1]

The integration of exp-log-functions (without considering algebraic integrands) is the topic of the book [Bro1997] which contains many further references. The essential ideas of the algebraic case can be found in [GCL1992].

In this chapter we only consider the integration of rational functions for presenting the essential ideas of Risch's algorithm.

Note that Gosper's algorithm is able to sum certain rational functions, namely if the antidifference is rational or a hypergeometric term. However, not every rational function can be summed in such a way, as the example of harmonic numbers $H_k = \sum_k \frac{1}{k+1}$ showed. This situation was not considered in detail in the previous chapter, since suitable field extensions would have been

---

[1] This is valid for purely logarithmic integrands. The cases of exponential and algebraic extensions are more complicated.

© Springer Nature Switzerland AG 2021

W. Koepf, *Computer Algebra*, Springer Undergraduate Texts in Mathematics and Technology, https://doi.org/10.1007/978-3-030-78017-3_12

necessary for extending the summation to general rational functions. In this chapter, we will demonstrate how this can be accomplished in detail for the case of integration of rational functions.

In many textbooks on calculus the following method for the integration of rational functions is presented, which essentially goes back to Leibniz and was proved by Johann Bernoulli.

Theorem 12.1. Let the rational function $r(x) = \frac{p(x)}{q(x)} \in \mathbb{R}(x)$, $p(x), q(x) \in \mathbb{R}[x]$ with $\gcd(p(x), q(x)) = 1$ be given. Then there exists an irreducible factorization

$$q(x) = c \cdot \prod_{i=1}^{m} (x - a_i)^{e_i} \cdot \prod_{j=1}^{n} (x^2 + b_j x + c_j)^{f_j}$$

of the denominator polynomial over $\mathbb{R}$, where $c, a_i, b_j, c_j \in \mathbb{R}$ and $e_i, f_j \in \mathbb{N}$. This *real factorization* generates a *real partial fraction decomposition* for $r(x)$ of the form

$$r(x) = P(x) + \sum_{i=1}^{m} \sum_{k=1}^{e_i} \frac{A_{ik}}{(x - a_i)^k} + \sum_{j=1}^{n} \sum_{k=1}^{f_j} \frac{B_{jk}x + C_{jk}}{(x^2 + b_j x + c_j)^k} \tag{12.1}$$

with $P(x) \in \mathbb{R}[x]$, $A_{ik}, B_{jk}, C_{jk} \in \mathbb{R}$, which can be be integrated termwise (see e.g. [Bro1997, Section 2.1]). $\qquad \square$

However, the problem of this theorem is—and similar statements are valid for many existence theorems in mathematics—that although the existence of a real factorization is secured, an algorithm for the computation of the factors is missing. In Galois theory one shows that the zeros of polynomials of degree 5 and higher can generally not be written in terms of radicals. Therefore, in general, the linear factors a_i ($i = 1, \ldots, m$) cannot be computed algorithmically.

To consider the zeros of the denominator as algebraic numbers is not feasible either since this increases the complexity considerably and might present the result in an unnecessarily complicated form, even if no algebraic numbers are necessary for this purpose, as in

$$\int \frac{x^3}{x^4 - 1} \, dx = \frac{1}{4} \ln(x^4 - 1). \tag{12.2}$$

Since a given integrand cannot contain *all* real numbers, but only finitely many, the integrand might even be a member of $\mathbb{Q}(x)$. This raises the question whether there is an algorithm for this situation which does not need the full factorization of the denominator polynomial over $\mathbb{R}$.

We will see that for this purpose we will not even need a full factorization of the denominator polynomial over $\mathbb{Q}$. Much rather, a squarefree factorization will turn out to be sufficient. In the next section we will introduce some algebraic prerequisites that form the algebraic background for the logarithmic terms arising from the termwise integration of (12.1).

12.2 Algebraic Prerequisites

As for differentiation, which we had implemented in Section 2.7, and for squarefree factorization, which was considered in Section 6.8, we will define integration as a purely algebraic routine without using the limit concept. Of course this is accomplished by considering integration as the inverse of differentiation. This time, however, we will not only consider polynomials, but rational and certain transcendental functions. Therefore we will extend the notions given in Section 6.8 towards a more general framework.

Definition 12.2. (Differential Field) Let $\mathbb{K}$ be a field and let $D : \mathbb{K} \to \mathbb{K}$ be a mapping with the following properties:

(D_1) **(Sum Rule)** $D(f + g) = D(f) + D(g)$;
(D_2) **(Product Rule)** $D(f \cdot g) = D(f) \cdot g + f \cdot D(g)$.

Then D is called a *differential operator* or a *derivative*, and $(\mathbb{K}, D)$ is called a *differential field*.

The set $\mathbb{K}_0 := \{c \in \mathbb{K} \mid D(c) - 0\}$ is called *constant field* of $(\mathbb{K}, D)$.[2]

If the equation $D(f) = g$ is valid for $f, g \in \mathbb{K}$, then we call f an *antiderivative* or an *integral* of g, and we write $f = \int g$.[3] △

Note that (as in calculus) the notation $\int g$ is not unique since with f every function of the form $f + c$ $(c \in \mathbb{K}_0)$ is an antiderivative, too.

Theorem 12.3. Every derivative has the following well-known properties:

(D_3) **(Constant Rule)** $D(0) = 0$ and $D(1) = 0$;
(D_4) **($\mathbb{K}_0$-linearity)** $D(af + bg) = aD(f) + bD(g)$ for all $a, b \in \mathbb{K}_0$;
(D_5) **(Quotient Rule)** $D\left(\frac{f}{g}\right) = \frac{D(f)g - fD(g)}{g^2}$ for $g \neq 0$;
(D_6) **(Power Rule)** $D(f^n) = n f^{n-1} \cdot D(f)$ for $n \in \mathbb{Z}$;
(D_7) **(Constant Field)** The constants $\mathbb{K}_0$ form a field;
(D_8) **(Partial Integration)** $\int f D(g) = f \cdot g - \int D(f)g$.

Proof. (D_3): The sum rule implies $D(f) = D(f + 0) = D(f) + D(0)$, hence $D(0) = 0$.
Let $f \neq 0$. Then the product rule implies $D(f) = D(f \cdot 1) = D(f) \cdot 1 + f \cdot D(1)$, hence $D(1) = 0$.
(D_4): Let $a, b \in \mathbb{K}_0$, i.e., $D(a) = D(b) = 0$. Then it follows from the product rule that

$$
\begin{aligned}
D(af + bg) &= D(af) + D(bg) \\
&= D(a) \cdot f + a \cdot D(f) + D(b) \cdot g + b \cdot D(g) \\
&= a \cdot D(f) + b \cdot D(g).
\end{aligned}
$$

(D_5): Let $g \neq 0$ be given. Since $0 = D(1) = D\left(g \cdot \frac{1}{g}\right) = D(g) \cdot \frac{1}{g} + g \cdot D\left(\frac{1}{g}\right)$, we get $D(\frac{1}{g}) = -\frac{D(g)}{g^2}$.
Hence it follows

$$
D\left(\frac{f}{g}\right) = D(f) \cdot \frac{1}{g} + f \cdot D\left(\frac{1}{g}\right) = D(f) \cdot \frac{1}{g} - f \cdot \frac{D(g)}{g^2} = \frac{D(f)g - fD(g)}{g^2}.
$$

(D_6): This statement is proved by mathematical induction. For this purpose, we first prove the statement for all $n \in \mathbb{N}_{\geq 0}$. The base case for $n = 0$ is clear. Now assume that the statement is valid for some $n \in \mathbb{N}$. Then the product rule implies

$$
D(f^{n+1}) = D(f \cdot f^n) = D(f) \cdot f^n + f \cdot D(f^n) = D(f) \cdot f^n + f \cdot n f^{n-1} \cdot D(f) = (n + 1) f^n D(f),
$$

---

[2] In the following theorem we will prove that $\mathbb{K}_0$ is a field indeed.
[3] We do not use the dx notation in this chapter. The integration variable is defined by the context.

which proves the assertion. For $n = -m < 0$ the statement follows using the quotient rule:

$$D\left(\frac{1}{f^m}\right) = -\frac{D(f^m)}{f^{2m}} = -\frac{m f^{m-1} D(f)}{f^{2m}} = -m f^{-m-1} D(f).$$

(D_7): (D_3) implies that $0 \in \mathbb{K}_0$ and $1 \in \mathbb{K}_0$. From (D_1) and (D_2) it follows further that with $a, b \in \mathbb{K}_0$ the elements $a + b \in \mathbb{K}_0$ and $a \cdot b \in \mathbb{K}_0$. Finally $D(a) = 0$ implies $D(-a) = 0$, since $0 = D(0) = D(a + (-a)) = D(a) + D(-a)$, and therefore $a \in \mathbb{K}_0$ implies $-a \in \mathbb{K}_0$. The quotient rule shows similarly that $\frac{1}{a} \in \mathbb{K}_0$ if $a \neq 0$. Hence $\mathbb{K}_0$ is a subfield of $\mathbb{K}$.

(D_8): This follows (as in calculus) from the product rule.   □

We would like to remark that the algebraic concept of derivatives and antiderivatives can also be defined in integral domains R with unit 1. In this setting, for the quotient rule one must assume that g is a unit. The resulting structure (R, D) is called a *differential algebra*.

We further remark that the proved differentiation rules are now complete besides the *chain rule*. Of course, a general chain rule cannot be valid in our algebraic framework, since very complicated field extensions might be necessary in order to be able to consider general compositions. In our context, the most exhaustive substitute for the chain rule is the power rule (D_6).

Next we show that in the field $\mathbb{Q}(x)$ of rational functions the derivative—which is uniquely determined by the additional property $D(x) = 1$—has the usual properties and is therefore the derivative that we know from calculus. Note that for the sake of simplicity we often omit the argument of polynomials in this chapter.

Theorem 12.4. Let D be a derivative in the field $\mathbb{Q}(x)$ with the property $D(x) = 1$.

(a) For $r \in \mathbb{Q}$ we have $D(r) = 0$.

(b) For every polynomial $p = \sum\limits_{k=0}^{n} a_k x^k \in \mathbb{Q}[x]$ we obtain

$$D(p) = \sum_{k=1}^{n} k a_k x^{k-1}.$$

In particular, for non-constant $p \in \mathbb{Q}[x]$ the degree condition $\deg(D(p), x) = \deg(p, x) - 1$ is valid.

(c) The constant field is $\mathbb{Q}(x)_0 = \mathbb{Q}$.

(d) In $\mathbb{Q}(x)$ the mapping D is the usual differential operator.

(e) For every polynomial $p = \sum\limits_{k=0}^{n} a_k x^k \in \mathbb{Q}[x]$ we furthermore obtain

$$\int p = \sum_{k=0}^{n} \frac{a_k}{k+1} x^{k+1} \in \mathbb{Q}[x].$$

Proof. (a) From $D(0) = 0$ and the sum rule we get by double induction $D(n) = 0$ for $n \in \mathbb{Z}$. The quotient rule then implies for $r = \frac{m}{n} \in \mathbb{Q}$ with $m, n \in \mathbb{Z}$, $n \neq 0$, that

$$D\left(\frac{m}{n}\right) = \frac{D(m)n - m D(n)}{n^2} = 0.$$

(b) First $D(x) = 1$ and the power rule imply the relation $D(x^k) = kx^{k-1}$. Then $D(a_0) = 0$ together with the sum and product rules imply for $p = \sum_{k=0}^{n} a_k x^k \in \mathbb{Q}[x]$:[4]

$$D(p) = D(a_0 + \sum_{k=1}^{n} a_k x^k) = \sum_{k=1}^{n} D(a_k x^k) = \sum_{k=1}^{n} \left(D(a_k) x^k + a_k D(x^k) \right) = \sum_{k=1}^{n} k a_k x^{k-1} .$$

(c) Because of (a) it remains to prove that $D(r) = 0 \Rightarrow r \in \mathbb{Q}$. Therefore let $r \in \mathbb{Q}(x)$ with $r = p/q$ be given, where $p, q \in \mathbb{Q}[x]$ with $q \neq 0$ and $\gcd(p, q) = 1$. The quotient rule implies

$$D(r) = \frac{D(p)q - pD(q)}{q^2} = 0 ,$$

hence $D(p)q - pD(q) = 0$ or

$$D(p) = \frac{pD(q)}{q} ,$$

respectively. Since by assumption p and q have no common divisor and since $D(p) \in \mathbb{Q}[x]$, it follows that q must be a divisor of $D(q)$:

$$\frac{D(q)}{q} = s \in \mathbb{Q}[x] ,$$

hence $D(q) = s \cdot q$ with $\deg(s, x) \geq 0$. If $\deg(q, x) > 0$, then it follows from the degree formula given in (b) that

$$\deg(q, x) - 1 = \deg(s, x) + \deg(q, x) .$$

Since this is impossible, either $\deg(q, x) = 0$ or $q \in \mathbb{Q}$ must be valid. Therefore $D(r) = 0$ implies that $D(p) = 0$. With (b) we finally get that $p \in \mathbb{Q}$, as announced.
(d) Since by (b) polynomials have the usual derivative, this follows directly from the quotient rule.
Lastly, (e) follows directly from the definition of the integral, using (b).   □

Note that the antiderivatives of polynomials are again polynomials. Therefore the question arises—and we know the answer already from calculus—whether every $r \in \mathbb{Q}(x)$ possesses an antiderivative $\int r \in \mathbb{Q}(x)$. We will give an algebraic proof that this is not the case.

Theorem 12.5. There is no $r \in \mathbb{Q}(x)$ with $D(r) = \frac{1}{x}$. In other words, the function $\frac{1}{x}$ has no rational antiderivative.

Proof. Let us assume that there is $r = p/q \in \mathbb{Q}(x)$ with $p, q \in \mathbb{Q}[x]$, $\gcd(p, q) = 1$ such that $D(\frac{p}{q}) = \frac{1}{x}$. Then

$$\frac{D(p)q - pD(q)}{q^2} = \frac{1}{x}$$

or

$$x \left(D(p)q - pD(q) \right) = q^2 . \tag{12.3}$$

Therefore x must be a divisor of q^2 and hence also of q. This implies that there are $n \in \mathbb{N}$ and $s \in \mathbb{Q}[x]$ with $q = x^n s$ such that $\gcd(s, x) = 1$.[5] If we substitute this in (12.3), we get

---

[4] The claim (b) therefore essentially follows from the fact that 1 and x generate the polynomial ring. To define an arbitrary derivative, it is therefore sufficient to set $D(x)$ arbitrarily, see Exercise 12.1.

[5] The number n is the order of the zero of q.

$$x^{n+1} D(p)s - xp\Big(nx^{n-1} s + x^n D(s)\Big) = x^{2n} s^2 .$$

After dividing by x^n and sorting we have

$$nps = xD(p)s - xpD(s) - x^n s^2 .$$

This means, however, that the polynomial p has the divisor x since $\gcd(s, x) = 1$, which contradicts our basic assumption that p and q are relatively prime. Therefore we have shown that such p and q cannot exist.   □

We would like to remark that this proof has similarities to the common proof of the irrationality of $\sqrt{2}$. The theorem therefore implies that for the integration of rational functions a field extension is necessary: we must introduce logarithms. The algebraic concept of logarithms can be defined as follows.

Definition 12.6. Let $(\mathbb{K}, D)$ be a differential field and let $(\mathbb{E}, D)$ with $\mathbb{K} \neq \mathbb{E}$ be a *differential extension field* of $(\mathbb{K}, D)$, hence an extension field of $\mathbb{K}$ which extends D into $\mathbb{E}$. If for given $\vartheta \in \mathbb{E} \setminus \mathbb{K}$ there is an element $u \in \mathbb{K}$ such that

$$D(\vartheta) = \frac{D(u)}{u}$$

is valid, then ϑ is called *logarithmic over* $\mathbb{K}$ and we write $\vartheta = \ln u$. The field $(\mathbb{E}, D)$ is then called a *logarithmic field extension* of $(\mathbb{K}, D)$. △

Note that the definition of logarithmic terms just models the chain rule for logarithmic functions.

Example 12.7. We consider several integrations which we cannot derive (algebraically) yet, but which can be easily checked by differentiation. Our goal is to analyze the necessary field extensions.
(a) Obviously

$$\int \frac{1}{x} = \ln x \in \mathbb{Q}(x, \ln x) .$$

The extension field of $\mathbb{Q}(x)$ which is necessary for the representation of the antiderivative is therefore the field of rational functions in x and in the logarithmic element $\ln x$. It is therefore sufficient to extend $\mathbb{Q}(x)$ by one single logarithmic element.
(b)

$$\int \frac{1}{x^2 + x} = \int \left(\frac{1}{x} - \frac{1}{1+x} \right) = \ln x - \ln(1+x) \in \mathbb{Q}(x, \ln x, \ln(1+x)) .$$

For the representation of this antiderivative two logarithmic elements are needed.
(c)

$$\int \frac{1}{x - x^3} = \int \left(\frac{1}{x} + \frac{x}{1-x^2} \right) = \ln x - \frac{1}{2} \ln(1 - x^2) \in \mathbb{Q}(x, \ln x, \ln(1 - x^2)) .$$

However, we can also write

$$\int \frac{1}{x - x^3} = \ln x - \frac{1}{2} \ln(1+x) - \frac{1}{2} \ln(1-x) \in \mathbb{Q}(x, \ln x, \ln(1+x), \ln(1-x)) .$$

In this case two or three logarithmic elements are needed for the antiderivative, depending on the chosen representation.
(d) In the example

$$\int \frac{1}{x^2-3} = \frac{1}{2\sqrt{3}} \ln(x-\sqrt{3}) - \frac{1}{2\sqrt{3}} \ln(x+\sqrt{3}) \in \mathbb{Q}(x)(\sqrt{3})(\ln(x-\sqrt{3}), \ln(x+\sqrt{3})) \,,$$

the representation of the antiderivative needs an algebraic extension in addition to the two logarithmic terms.

(e) Finally we consider the example

$$I = \int \frac{x^3}{x^4-1} \,.$$

In (12.2) we had already seen that $I = \frac{1}{4} \ln(x^4 - 1)$. This representation of the integral is an element of $\mathbb{Q}(x, \ln(x^4 - 1))$. The resulting representation does, however, depend on the degree of factorization of the denominator. The full factorization $x^4 - 1 = (x+1)(x-1)(x+i)(x-i)$ over $\mathbb{Q}(x)(i)$ yields the partial fraction decomposition

$$\frac{x^3}{x^4-1} = \frac{1}{4}\frac{1}{x+1} + \frac{1}{4}\frac{1}{x-1} + \frac{1}{4}\frac{1}{x+i} + \frac{1}{4}\frac{1}{x-i}$$

and therefore the representation

$$I = \frac{1}{4}\ln(x+1) + \frac{1}{4}\ln(x-1) + \frac{1}{4}\ln(x+i) + \frac{1}{4}\ln(x-i) \,.$$

in $\mathbb{Q}(i)(x, \ln(x+1), \ln(x-1), \ln(x+i), \ln(x-i))$. However, if we combine the last two integrals, then we get

$$I = \frac{1}{4}\ln(x+1) + \frac{1}{4}\ln(x-1) + \frac{1}{4}\ln(x^2+1) \in \mathbb{Q}(x, \ln(x+1), \ln(x-1), \ln(x^2+1)) \,.$$

This example vividly shows that the required extension field is by no means uniquely determined. $\triangle$

We would like to mention that the resulting logarithmic terms do not contain absolute values, which is often the case in textbooks on calculus. This, however, is not really meaningful since even in calculus it is not possible to integrate a function in an interval containing a singularity. In our algebraic context logarithms have no branches. Of course, if we wanted to evaluate antiderivatives for $x \in \mathbb{Q}$ or $x \in \mathbb{R}$, then analytical concepts would be needed.

The integration of the partial fraction decomposition (12.1) obviously yields a rational and a logarithmic part. In the next section we will show how the rational part can be computed over $\mathbb{Q}(x)$, whereas in the subsequent section we will take care of the computation of the logarithmic part. In this latter context, we will also deal with the question which field extensions are unavoidably necessary for the representation of the antiderivative.

12.3 Rational Part

For simplicity reasons let $\mathbb{K}$ denote a subfield of $\mathbb{C}$ in the sequel. If we want to integrate $r(x) = \frac{p(x)}{q(x)} \in \mathbb{K}(x)$, then the first step is a polynomial division which separates a polynomial part $P(x)$. The remaining integrand is then still of the form $\frac{p(x)}{q(x)}$, but now with $\deg(p(x), x) < \deg(q(x), x)$. Therefore we will assume this condition without loss of generality in the sequel.

Since we want (and have) to accept algebraic extensions, it makes sense to consider the Bernoulli algorithm again under this viewpoint. In a suitable algebraic extension field, the real partial fraction decomposition (12.1) can be split further by factorizing the quadratic denominator terms. This yields the form

$$r(x) = \sum_{i=1}^{M} \sum_{k=1}^{e_i} \frac{A_{ik}}{(x - a_i)^k} \tag{12.4}$$

with $A_{ik}, a_i \in \mathbb{C}$, where a_i, $(i = 1, \dots, M)$ denote the complex zeros of the denominator polynomial and e_i are the corresponding orders of these zeros. Integration of the summands with $k > 1$ yields rational functions, whereas the summands with $k = 1$ yield the logarithmic terms. Hence we get a decomposition of the form

$$\int r(x) = \sum_{i=1}^{M} \frac{p_i(x)}{(x - a_i)^{e_i - 1}} + \int \sum_{i=1}^{M} \frac{A_{i1}}{x - a_i} \tag{12.5}$$

with $\deg(p_i(x), x) < e_i - 1$.

By the results of Section 6.8 about the squarefree factorization we can identify the above denominators and summarize

$$\int r(x) = \frac{a(x)}{b(x)} + \int \frac{c(x)}{d(x)} \,, \tag{12.6}$$

with $\deg(a(x), x) < \deg(b(x), x)$ and $\deg(c(x), x) < \deg(d(x), x)$, as well as

$$b(x) = \gcd(q(x), q'(x)) \in \mathbb{K}[x] \,, \tag{12.7}$$

where

$$d(x) = \frac{q(x)}{b(x)} \in \mathbb{K}[x] \tag{12.8}$$

is the *squarefree part* of the denominator $q(x)$.

In Session 12.8 and Theorem 12.9 we will show that the polynomials $a(x), c(x) \in \mathbb{K}[x]$ have coefficients in $\mathbb{K}$, too. Equation (12.6) therefore provides a decomposition of the integral $\int r(x)$ into a *rational part* $\frac{a(x)}{b(x)}$ and a *trancendental* or *logarithmic part* $\int \frac{c(x)}{d(x)}$ with *squarefree* denominator $d(x)$. As already mentioned, the computation of the transcendental part will be considered in the next section.

Next we would like to discuss how the collected knowledge can be combined into an algorithm for the computation of representation (12.6) in an example case.

Session 12.8. *Let us consider the following function $r(x)$:*

```
In[1]:= r =
            (3 x¹⁶ – 19 x¹⁵ + 43 x¹⁴ – 20 x¹³ – 91 x¹² + 183 x¹¹ – 81 x¹⁰ –
                166 x⁹ + 271 x⁸ – 101 x⁷ – 127 x⁶ + 168 x⁵ – 53 x⁴ – 31 x³ +
                41 x² – 2 x – 2) /
            (4 x¹⁴ – 20 x¹³ + 28 x¹² + 24 x¹¹ – 108 x¹⁰ + 84 x⁹ + 76 x⁸ –
                176 x⁷ + 76 x⁶ + 84 x⁵ – 108 x⁴ + 24 x³ + 28 x² – 20 x + 4)
```

$$Out[1]= \Big(3 x^{16} - 19 x^{15} + 43 x^{14} - 20 x^{13} - 91 x^{12} + 183 x^{11} - 81 x^{10} - 166 x^9 +$$
$$271 x^8 - 101 x^7 - 127 x^6 + 168 x^5 - 53 x^4 - 31 x^3 + 41 x^2 - 2 x - 2\Big)\Big/$$
$$\Big(4 x^{14} - 20 x^{13} + 28 x^{12} + 24 x^{11} - 108 x^{10} + 84 x^9 + 76 x^8 - 176 x^7 +$$
$$76 x^6 + 84 x^5 - 108 x^4 + 24 x^3 + 28 x^2 - 20 x + 4\Big)$$

Of course Mathematica is able to integrate r(x):

$In[2]:=$ **int** $= \int$ **r dx**

$Out[2]= \dfrac{1}{4}\left(x^3 - 2x^2 + 2x - \dfrac{4\tan^{-1}\left(\frac{2x-1}{\sqrt{3}}\right)}{9\sqrt{3}} - \log(x-1) - \right.$

$\dfrac{1}{3}\log(x+1) + \dfrac{2}{3}\log((x-1)^2 + x) - \dfrac{13}{32(x-1)} + \dfrac{53}{288(x+1)} - $

$\dfrac{71}{128(x-1)^2} + \dfrac{13}{384(x+1)^2} + \dfrac{35}{48(x-1)^3} + \dfrac{1}{288(x+1)^3} - $

$\left. \dfrac{15}{32(x-1)^4} + \dfrac{1}{10(x-1)^5} + \dfrac{1}{8(x-1)^6} - \dfrac{1}{7(x-1)^7} \right)$

The rational part of the integral is therefore given by

$In[3]:=$ **rat** = **int** /. {**ArcTan**[_] → 0, **Log**[_] → 0}

$Out[3]= \dfrac{1}{4}\left(x^3 - 2x^2 + 2x - \dfrac{13}{32(x-1)} + \dfrac{53}{288(x+1)} - \right.$

$\dfrac{71}{128(x-1)^2} + \dfrac{13}{384(x+1)^2} + \dfrac{35}{48(x-1)^3} + \dfrac{1}{288(x+1)^3} - $

$\left. \dfrac{15}{32(x-1)^4} + \dfrac{1}{10(x-1)^5} + \dfrac{1}{8(x-1)^6} - \dfrac{1}{7(x-1)^7} \right)$

Next we compute this representation ourselves. The input function r(x) has the numerator

$In[4]:=$ **p** = **Numerator**[**r**]

$Out[4]= 3x^{16} - 19x^{15} + 43x^{14} - 20x^{13} - 91x^{12} + 183x^{11} - 81x^{10} - $
$\qquad 166x^9 + 271x^8 - 101x^7 - 127x^6 + 168x^5 - 53x^4 - 31x^3 + 41x^2 - 2x - 2$

and the denominator

$In[5]:=$ **q** = **Denominator**[**r**]

$Out[5]= 4x^{14} - 20x^{13} + 28x^{12} + 24x^{11} - 108x^{10} + 84x^9 + $
$\qquad 76x^8 - 176x^7 + 76x^6 + 84x^5 - 108x^4 + 24x^3 + 28x^2 - 20x + 4$

The polynomial part P(x) is therefore given by

$In[6]:=$ **Pol** = **PolynomialQuotient**[**p**, **q**, **x**]

$Out[6]= \dfrac{3x^2}{4} - x + \dfrac{1}{2}$

and we get the reduced numerator

$In[7]:=$ **p** = **PolynomialRemainder**[**p**, **q**, **x**]

$Out[7]= 4x^4 + 4x^2 + 12x - 4$

With the help of (12.7) and (12.8) we can compute the denominators b(x) and d(x) of representation (12.6):

$In[8]:=$ **b** = **PolynomialGCD**[**q**, **D**[**q**, **x**]]

$Out[8]= 4x^{10} - 16x^9 + 12x^8 + 32x^7 - 56x^6 + 56x^4 - 32x^3 - 12x^2 + 16x - 4$

$In[9]:=$ **d** = **Together**$\left[\dfrac{q}{b}\right]$

$Out[9]= x^4 - x^3 + x - 1$

Note that these denominators are not in factorized form. This is not necessary, however, and it is desirable not to compute unnecessary factorizations.

Next we write

$$In[10]:= \mathbf{a} = \sum_{k=0}^{\text{Exponent[b,x]}-1} \alpha_k \mathbf{x}^k$$

$$Out[10]= \alpha_9\, x^9 + \alpha_8\, x^8 + \alpha_7\, x^7 + \alpha_6\, x^6 + \alpha_5\, x^5 + \alpha_4\, x^4 + \alpha_3\, x^3 + \alpha_2\, x^2 + \alpha_1\, x + \alpha_0$$

$$In[11]:= \mathbf{c} = \sum_{k=0}^{\text{Exponent[d,x]}-1} \beta_k \mathbf{x}^k$$

$$Out[11]= \beta_3\, x^3 + \beta_2\, x^2 + \beta_1\, x + \beta_0$$

with undetermined coefficients α_k and β_k for the numerator polynomials $a(x)$ and $c(x)$, since we know a degree bound for each.

Equation (12.6) is equivalent to the identity

$$p(x) - q(x)\left(\left(\frac{a(x)}{b(x)}\right)' + \frac{c(x)}{d(x)}\right) = 0\,.$$

Note that by (12.7) and (12.8) (after reduction) this is a polynomial identity (see Theorem 12.9) which in the given case has 91 summands:

$$In[12]:= \mathbf{s} = \text{Together}\left[\mathbf{p} - \mathbf{q}\left(\mathbf{D}\left[\frac{\mathbf{a}}{\mathbf{b}}, \mathbf{x}\right] + \frac{\mathbf{c}}{\mathbf{d}}\right)\right];$$

$$In[13]:= \text{Length[s]}$$
$$Out[13]= 91$$

Now we equate the coefficients and solve for the unknowns α_k and β_k:

$$In[14]:= \mathbf{sol} = \text{Solve[CoefficientList[s,x]} == 0,$$
$$\text{Join[Table}[\alpha_k, \{k, 0, \text{Exponent[b,x]} - 1\}],$$
$$\text{Table}[\beta_k, \{k, 0, \text{Exponent[d,x]} - 1\}]]]$$

$$Out[14]= \left\{\left\{\alpha_0 \to \frac{67}{70}, \alpha_1 \to -\frac{16}{315}, \alpha_2 \to -\frac{4631}{630}, \alpha_3 \to \frac{872}{315},\right.\right.$$
$$\alpha_4 \to \frac{392}{45}, \alpha_5 \to -\frac{268}{45}, \alpha_6 \to -\frac{26}{9}, \alpha_7 \to \frac{28}{9}, \alpha_8 \to -\frac{2}{9},$$
$$\left.\left.\alpha_9 \to -\frac{2}{9}, \beta_0 \to \frac{1}{18}, \beta_1 \to -\frac{1}{2}, \beta_2 \to -\frac{1}{18}, \beta_3 \to 0\right\}\right\}$$

Therefore this computation yields the rational part $\frac{a(x)}{b(x)}$ as

$$In[15]:= \text{Together}\left[\frac{\mathbf{a}}{\mathbf{b}} \,/.\, \text{sol[[1]]}\right]$$
$$Out[15]= (-140\, x^9 - 140\, x^8 + 1960\, x^7 - 1820\, x^6 - 3752\, x^5 + 5488\, x^4 +$$
$$1744\, x^3 - 4631\, x^2 - 32\, x + 603)/(2520\,(x-1)^7\,(x+1)^3)$$

which clearly equals Mathematica's computation:

$$In[16]:= \text{Together}\left[\mathbf{rat} - \int \mathbf{Pol}\ \mathbf{dx}\right]$$
$$Out[16]= (-140\, x^9 - 140\, x^8 + 1960\, x^7 - 1820\, x^6 - 3752\, x^5 + 5488\, x^4 +$$
$$1744\, x^3 - 4631\, x^2 - 32\, x + 603)/(2520\,(x-1)^7\,(x+1)^3)$$

For the transcendental part $\frac{c(x)}{d(x)}$, we find

$$In[17]:= \mathbf{trans} = \text{Together}\left[\frac{\mathbf{c}}{\mathbf{d}} \,/.\, \text{sol[[1]]}\right]$$
$$Out[17]= \frac{-x^2 - 9\,x + 1}{18\,(x^4 - x^3 + x - 1)}$$

which (after integration) yields the purely logarithmic part:

$$In[18]:= \int \mathbf{trans}\ \mathbf{dx}$$

$$Out[18] = \frac{1}{18} \left(-\frac{2 \tan^{-1}\left(\frac{2x-1}{\sqrt{3}}\right)}{\sqrt{3}} - \frac{9}{2} \log(x-1) - \frac{3}{2} \log(x+1) + 3 \log\left(x^2 - x + 1\right) \right)$$

In Mathematica's output this part contains an inverse tangent term which—in a suitable algebraic extension field—can be represented by certain complex logarithms.

We would like to point out that the given algorithm shows in particular that the two sought polynomials $a(x)$ and $c(x)$ have coefficients in the ground field $\mathbb{K}$, i.e. $a(x), c(x) \in \mathbb{K}[x]$, since the linear system to be solved obviously generates coefficients in this field. In our specific case the input polynomial was an element of $\mathbb{Q}(x)$, so that $a(x), b(x), c(x), d(x) \in \mathbb{Q}[x]$.

We can implement the complete algorithm by collecting the above steps:

```
In[19]:= Clear[RationalDecomposition]
         RationalDecomposition[r_, x_] :=
           Module[{p, q, Pol, a, b, c, d, s, sol, k, α, β},
             p = Numerator[r];
             q = Denominator[r];
             Pol = PolynomialQuotient[p, q, x];
             p = PolynomialRemainder[p, q, x];
             b = PolynomialGCD[q, D[q, x]];
             d = Together[q/b];
             a = Sum[α[k] x^k, {k, 0, Exponent[b, x] - 1}];
             c = Sum[β[k] x^k, {k, 0, Exponent[d, x] - 1}];
             s = Together[p - q * (D[a/b, x] + c/d)];
             sol = Solve[CoefficientList[s, x] == 0,
                 Join[Table[α[k], {k, 0, Exponent[b, x] - 1}],
                   Table[β[k], {k, 0, Exponent[d, x] - 1}]]];
             a = a/.sol[[1]];
             c = c/.sol[[1]];
             {Integrate[Pol, x] + Together[a/b], Together[c/d]}
           ]
```

The function `RationalDecomposition` *computes the rational decomposition for arbitrary $r(x) \in \mathbb{Q}(x)$ and returns a list whose first element is the rational part $\int P(x) + \frac{a(x)}{b(x)}$ and whose second element is the integrand $\frac{c(x)}{d(x)}$ of the logarithmic part. For our above example we again obtain*

```
In[20]:= RationalDecomposition[r, x]
```
$$Out[20] = \left\{ \frac{x^3}{4} - \frac{x^2}{2} + \frac{x}{2} + \right.$$
$$(-140 x^9 - 140 x^8 + 1960 x^7 - 1820 x^6 - 3752 x^5 + 5488 x^4 +$$
$$1744 x^3 - 4631 x^2 - 32 x + 603)/$$
$$\left. (2520 (x-1)^7 (x+1)^3), \frac{-x^2 - 9x + 1}{18 (x^4 - x^3 + x - 1)} \right\}$$

We give two more examples:

```
In[21]:= RationalDecomposition[1/x, x]
```
$$Out[21] = \left\{ 0, \frac{1}{x} \right\}$$

```
In[22]:= RationalDecomposition[x^2/(1+x^2)^2, x]
```
$$Out[22] = \left\{ -\frac{x}{2(x^2+1)}, \frac{1}{2(x^2+1)} \right\}$$

In the first case the rational part equals 0.

Now we would like to show that this algorithm, which was published by Ostrogradsky [Ost1845] and therefore is already 175 years old, is always successful.

Theorem 12.9. (Rational Decomposition) Let $r(x) = \frac{p(x)}{q(x)} \in \mathbb{K}(x)$ be a rational function with $p(x), q(x) \in \mathbb{K}[x]$ and $\gcd(p(x), q(x)) = 1$. Let further $b(x)$ and $d(x)$ be defined according to (12.7) and (12.8), hence by

$$b(x) = \gcd(q(x), q'(x)) \qquad \text{and} \qquad d(x) = \frac{q(x)}{b(x)} .$$

Then there are uniquely determined polynomials $P(x), a(x), c(x) \in \mathbb{K}[x]$ of degree $\deg(a(x), x) < \deg(b(x), x)$ and $\deg(c(x), x) < \deg(d(x), x)$ with

$$\int r(x) = \frac{a(x)}{b(x)} + \int P(x) + \int \frac{c(x)}{d(x)} .$$

The polynomial $P(x)$ is given by polynomial division. For $\deg(p(x), x) < \deg(q(x), x)$, i.e. if $P(x) = 0$, the above equation is equivalent to the polynomial identity

$$p(x) - q(x)\left(\left(\frac{a(x)}{b(x)}\right)' + \frac{c(x)}{d(x)}\right) = 0 . \tag{12.9}$$

The coefficients of the polynomials $a(x)$ and $c(x)$ can be computed by equating the coefficients in (12.9) using an ansatz with undetermined coefficients.

Proof. In a first step the polynomial part $P(x)$ is separated by polynomial division. The remaining numerator polynomial has degree smaller than $\deg(q(x), x)$. Therefore in the sequel we can assume that $r(x) = \frac{p(x)}{q(x)} \in \mathbb{K}(x)$ with $\deg(p(x), x) < \deg(q(x), x)$.

Now let $b(x) = \gcd(q(x), q'(x))$ and $d(x) = \frac{q(x)}{b(x)}$. Obviously these polynomials are members of $\mathbb{K}[x]$. Next we must show that (12.9) is a polynomial equation. Since $q(x) = b(x) d(x)$, we get

$$q(x)\frac{c(x)}{d(x)} = b(x) \cdot c(x) \in \mathbb{K}[x] .$$

It is a little more complicated to show that $q(x)\left(\frac{a(x)}{b(x)}\right)' \in \mathbb{K}[x]$. Partial integration yields

$$\int q(x)\left(\frac{a(x)}{b(x)}\right)' = q(x)\frac{a(x)}{b(x)} - \int q'(x)\frac{a(x)}{b(x)} . \tag{12.10}$$

The relation $b(x) = \gcd(q(x), q'(x))$ implies that both $q(x)\frac{a(x)}{b(x)} \in \mathbb{K}[x]$ and $q'(x)\frac{a(x)}{b(x)} \in \mathbb{K}[x]$. Therefore by (12.10), every antiderivative of $q(x)\left(\frac{a(x)}{b(x)}\right)'$ is a polynomial. Differentiation implies that $q(x)\left(\frac{a(x)}{b(x)}\right)' \in \mathbb{K}[x]$, as claimed.

Since the complex variant (12.5) of Bernoulli's Theorem 12.1 guarantees that there is a unique representation of the form (12.6), this representation will be found by solving the resulting linear system. $\square$

12.4 Logarithmic Case

To compute the logarithmic part, we can now assume that $r(x) = \frac{c(x)}{d(x)}$ with $c(x), d(x) \in \mathbb{K}[x]$, where $d(x)$ is squarefree and $\deg(c(x), x) < \deg(d(x), x)$. Furthermore we assume without loss of generality that $d(x)$ is monic. Of course we know how to generate the result for $\mathbb{K} = \mathbb{C}$ or in the smallest splitting field $\mathbb{K}_{d(x)}$ of $d(x)$ over $\mathbb{K}$, respectively: in this case, $d(x)$ has a complete factorization into linear factors $d(x) = \prod_{k=1}^{m} (x - a_k)$ which generates a partial fraction decomposition of the form

$$\frac{c(x)}{d(x)} = \sum_{k=1}^{m} \frac{\gamma_k}{x - a_k} \qquad (\gamma_k \in \mathbb{K}_{d(x)}) \, . \tag{12.11}$$

Since $d(x)$ is squarefree, the zeros a_k $(k = 1, \dots, m)$ are pairwise different and we therefore get the representation of the integral

$$\int \frac{c(x)}{d(x)} = \sum_{k=1}^{m} \gamma_k \ln(x - a_k) \, . \tag{12.12}$$

This representation is a member of the extension field $\mathbb{K}_{d(x)}(x, \ln(x - a_1), \dots, \ln(x - a_m))$ of $\mathbb{K}(x)$. However, for the example

$$\int \frac{1}{x^3 + x} = \ln x - \frac{1}{2} \ln(x + i) - \frac{1}{2} \ln(x - i)$$

the extension of $\mathbb{Q}$ by the algebraic number i is not necessary: since the coefficients of the two logarithmic terms $\ln(x \pm i)$ agree, we can simplify the result using the laws of logarithms[6]

$$-\frac{1}{2} \ln(x + i) - \frac{1}{2} \ln(x - i) = -\frac{1}{2} \ln\left((x + i)(x - i)\right) = -\frac{1}{2} \ln(x^2 + 1) \, ,$$

so that, finally,

$$\int \frac{1}{x^3 + x} = \ln x - \frac{1}{2} \ln(x^2 + 1) \, .$$

In the given form, the algebraic field extension by i is no longer necessary. Obviously the arising logarithms can always be merged if they have *identical* coefficients. Hence—to avoid unnecessary field extensions—we must search for the *different* solutions γ_k of equation (12.11). How can these coefficients be determined? We would like to show that for γ_k in (12.11) the formula $\gamma_k = \frac{c(a_k)}{d'(a_k)}$ is valid. For this purpose we use arguments from complex function theory.

Let a be one of the zeros of $d(x)$. Since $d(x)$ is squarefree, all zeros are simple and $\frac{c(x)}{d(x)}$ has a Laurent series representation of the form

$$\frac{c(x)}{d(x)} = \frac{\gamma}{x - a} + p(x) \, , \tag{12.13}$$

where $p(x)$ denotes a power series.[7]

---

[6] We use these rules in a formal manner although their validity, in particular in the complex domain, is not secured. In our concrete example, however, the correctness of our computation can be verified by differentiation. Anyway, we will not use the current preconsiderations in the proof of the subsequent theorem.

[7] Integration of (12.13) shows that the values γ_k are the *residues* of $\frac{c(x)}{d(x)}$ at the zeros a_k of $d(x)$.

Differentiation of the identity

$$(x-a)c(x) = \gamma d(x) + (x-a)d(x)p(x)$$

generates[8]

$$c(x) + (x-a)c'(x) = \gamma d'(x) + d(x)p(x) + (x-a)\Big(d'(x)p(x) + d(x)p'(x)\Big),$$

and since $d(a) = 0$, the substitution of $x = a$ yields the following formula for the computation of γ:

$$\gamma = \frac{c(a)}{d'(a)}.$$

In particular, equation (12.12) can be written in the form

$$\int \frac{c(x)}{d(x)} = \sum_{\{a\,|\,d(a)=0\}} \frac{c(a)}{d'(a)} \ln(x-a). \tag{12.14}$$

Session 12.10. *In complicated cases, Mathematica returns antiderivatives in exactly this form:*

$In[1]:= \displaystyle\int \frac{x}{1+x+x^7}\, dx$

$Out[1]= \text{RootSum}\left[\#1^7 + \#1 + 1\&,\ \dfrac{\log(x-\#1)\,\#1}{7\,\#1^6 + 1}\&\right]$

For the representation of those sums, the Mathematica function RootSum *is used.*

Sometimes Mathematica is able to partition such sums:

$In[2]:= \displaystyle\int \frac{x}{1+x+x^5}\, dx$

$Out[2]= -\dfrac{1}{7}\,\text{RootSum}\left[\#1^3 - \#1^2 + 1\&,\ \dfrac{3\,\log(x-\#1)\,\#1^2 - 5\,\log(x-\#1)\,\#1 + \log(x-\#1)}{3\,\#1^2 - 2\,\#1}\&\right]$

$\qquad + \dfrac{3}{14}\,\log(x^2 + x + 1) - \dfrac{\tan^{-1}\left(\frac{2x+1}{\sqrt{3}}\right)}{7\sqrt{3}}$

The above decomposition uses the factorization

$In[3]:= \textbf{Factor[1+x+x}^5\textbf{]}$

$Out[3]= \left(x^2 + x + 1\right)\left(x^3 - x^2 + 1\right)$

of the denominator or the resulting partial fraction decomposition

$In[4]:= \textbf{term = Apart}\left[\dfrac{x}{1+x+x^5}, x\right]$

$Out[4]= \dfrac{3x+1}{7\,(x^2 + x + 1)} + \dfrac{-3x^2 + 5x - 1}{7\,(x^3 - x^2 + 1)}$

respectively, whose first summand creates the two individual summands of the integral

$In[5]:= \displaystyle\int \textbf{term[[1]] } dx$

$Out[5]= \dfrac{1}{7}\left(\dfrac{3}{2}\,\log\left(x^2 + x + 1\right) - \dfrac{\tan^{-1}\left(\frac{2x+1}{\sqrt{3}}\right)}{\sqrt{3}}\right)$

---

[8] We use the fact from function theory that every power series $p(x)$ is differentiable. However, one could—similarly as for polynomials—define derivatives for power series algebraically.

whereas the second summand has an antiderivative which can be represented by RootSum *again:*

$In[6] :=$ $\int$ **term[[2]] dx**

$Out[6] =$ $-\dfrac{1}{7}$ $\text{RootSum}\Big[\#1^3 - \#1^2 + 1 \&,$

$$\dfrac{3 \log(x-\#1) \, \#1^2 - 5 \log(x-\#1) \, \#1 + \log(x-\#1)}{3 \, \#1^2 - 2 \, \#1} \&\Big]$$

The method discussed above requires the computation of all zeros a_k of the denominator $d(x)$. This, however, should be avoided as much as possible. Therefore it is our goal to compute the *different* solutions γ of the equation

$$0 = c(a) - \gamma d'(a),$$

where a runs through the zeros of $d(x)$, if possible without explicitly computing the set $\{a \mid d(a) = 0\}$. In Theorem 7.38 we had seen that these are exactly the different zeros of the polynomial

$$R(z) := \mathrm{res}(c(x) - z d'(x), d(x)) \in \mathbb{K}[z],$$

namely the *common* zeros of $c(x) - z d'(x)$ and $d(x)$. On the other hand, every *multiple* zero of $R(z)$ results in logarithmic terms that can be combined. Therefore we finally get

$$\int \frac{c(x)}{d(x)} = \sum_{k=1}^{n} z_k \ln v_k(x),$$

where z_k are the different zeros of $R(z)$ and the polynomials $v_k(x)$ are monic, squarefree and pairwise coprime.

Session 12.11. *We use this method to compute* $\int \frac{1}{x^3+x}$ *again:*

$In[1] :=$ **c = 1; d = x^3 + x;**

$In[2] :=$ **res = Resultant [c - z D[d, x], d, x]**
$Out[2] =$ $(1-z)(2z+1)^2$

This resultant provides the coefficients of the partial fraction decomposition. The given factorization shows that these coefficients are $\gamma_1 = 1$ *and* $\gamma_2 = -\frac{1}{2}$, *where* γ_2 *is a double zero which therefore collects two logarithmic terms.*

The next theorem will show how the associated functions $v_k(x)$ *can be computed: the calculations*

$In[3] :=$ **PolynomialGCD [c - z D[d, x] /. z → 1, d]**
$Out[3] =$ x

$In[4] :=$ **PolynomialGCD $\Big[$c - z D[d, x] /. z → $-\dfrac{1}{2}$, d$\Big]$**
$Out[4] =$ $\dfrac{x^2}{2} + \dfrac{1}{2}$

yield $v_1(x) = x$ *and* $v_2(x) = x^2 + 1$, *if* $v_k(x)$ *are chosen monic.*

The following theorem due to Rothstein [Rot1976] and Trager [Tra1976] shows that the above method always yields a representation of the sought antiderivative and reasons why all the different zeros of $R(z)$ are in fact needed. Furthermore it provides an algorithm to compute the polynomials $v_k(x)$.

Theorem 12.12. Let $\mathbb{E}(x)$ be a differential field over a constant field $\mathbb{E}$ and let $c(x), d(x) \in \mathbb{E}[x]$ be given with $\gcd(c(x), d(x)) = 1$. Let further $d(x)$ be monic and squarefree and let $\deg(c(x), x) < \deg(d(x), x)$. Assume that

$$\int \frac{c(x)}{d(x)} = \sum_{k=1}^{n} z_k \ln v_k(x),\tag{12.15}$$

where $z_k \in \mathbb{E} \setminus \{0\}$ $(k = 1, \ldots, n)$ are pairwise different constants and $v_k(x) \in \mathbb{E}[x]$ $(k = 1, \ldots, n)$ are monic, squarefree and pairwise coprime polynomials of positive degree.

Then the numbers z_k are the different solutions of the polynomial equation

$$R(z) = \mathrm{res}(c(x) - z\,d'(x), d(x)) = 0$$

and $v_k(x)$ are the polynomials

$$v_k(x) = \gcd(c(x) - z_k\,d'(x), d(x)) \in \mathbb{E}[x].$$

Proof. We differentiate equation (12.15) and get

$$\frac{c(x)}{d(x)} = \sum_{k=1}^{n} z_k \frac{v_k'(x)}{v_k(x)}.$$

If we set

$$u_k(x) := \frac{\prod\limits_{j=1}^{n} v_j(x)}{v_k(x)} = \prod_{\substack{j=1 \\ j \neq k}}^{n} v_j(x) \in \mathbb{E}(x),\tag{12.16}$$

then, after multiplication by $d(x) \cdot \prod\limits_{j=1}^{n} v_j(x)$, this yields

$$c(x) \prod_{j=1}^{n} v_j(x) = d(x) \sum_{k=1}^{n} z_k v_k'(x) u_k(x).\tag{12.17}$$

We will now show that $d(x) = \prod\limits_{j=1}^{n} v_j(x)$. Since $\gcd(c(x), d(x)) = 1$, it follows from (12.17) that $d(x) \mid \prod\limits_{j=1}^{n} v_j(x)$. To prove the converse divisibility assertion, we first state that for every $j = 1, \ldots, n$, (12.17) implies the relation

$$v_j(x) \mid d(x) \sum_{k=1}^{n} z_k v_k'(x) u_k(x).$$

Since $v_j(x)$ is a factor of $u_k(x)$ for every $k \neq j$, it follows for the summand with $k = j$ that

$$v_j(x) \mid d(x) v_j'(x) u_j(x).$$

Furthermore, we have $\gcd(v_j(x), v_j'(x)) = 1$ since $v_j(x)$ are squarefree, and also $\gcd(v_j(x), u_j(x)) = 1$, since the polynomials $v_j(x)$ are pairwise coprime. Therefore we obtain for all $j = 1, \ldots, n$ the conclusion $v_j(x) \mid d(x)$, and using the coprimality again, we finally get $\prod\limits_{j=1}^{n} v_j(x) \mid d(x)$. Since all involved polynomials are monic we have shown that

$$d(x) = \prod_{k=1}^{n} v_k(x) \tag{12.18}$$

is valid. Therefore the relation

$$c(x) = \sum_{k=1}^{n} z_k v_k'(x) u_k(x)$$

follows from (12.17). Differentiation of (12.18) yields

$$d'(x) = \sum_{k=1}^{n} v_k'(x) u_k(x),$$

hence we obtain

$$c(x) - z_j d'(x) = \sum_{k=1}^{n} z_k v_k'(x) u_k(x) - z_j \sum_{k=1}^{n} v_k'(x) u_k(x) = \sum_{k=1}^{n} (z_k - z_j) v_k'(x) u_k(x). \tag{12.19}$$

Because of (12.16), we have $v_j(x) \mid u_k(x)$ for $j \neq k$, which implies that each summand in the above sum is divisible by $v_j(x)$ unless $j = k$. Since for $j = k$ the summand vanishes due to the first factor $z_k - z_j$, we therefore get for all $j = 1, \ldots, n$

$$v_j(x) \mid c(x) - z_j d'(x). \tag{12.20}$$

From (12.18) and (12.20) it now follows that $v_j(x)$ is a *common* divisor of $d(x)$ and of $c(x) - z_j d'(x)$. Finally, we want to show that $v_j(x)$ is the *greatest* common divisor of these two polynomials. For this purpose, by (12.18) it is sufficient to show that for $j \neq i$ the relation $\gcd(c(x) - z_j d'(x), v_i(x)) = 1$ is valid.

We get

$$\gcd(c(x) - z_j d'(x), v_i(x)) = \gcd\left(\sum_{k=1}^{n} (z_k - z_j) v_k'(x) u_k(x), v_i(x) \right)$$

$$= \gcd((z_i - z_j) v_i'(x) u_i(x), v_i(x)),$$

where the last equality follows since $v_i(x)$ is a factor of $u_j(x)$ for every $j \neq i$. The last term equals 1 for $j \neq i$ since $z_i \neq z_j$ (z_k are pairwise different), $\gcd(v_i'(x), v_i(x)) = 1$ (v_k are squarefree) and $\gcd(u_i(x), v_i(x)) = 1$ (v_k are pairwise different). Hence we have proved that

$$v_j(x) = \gcd(c(x) - z_j d'(x), d(x)) \qquad \text{for all } j = 1, \ldots, n.$$

Therefore $c(x) - z_j d'(x)$ and $d(x)$ have a nontrivial common factor, hence

$$\mathrm{res}(c(x) - z_j d'(x), d(x)) = 0.$$

Now we have finally shown all equations of the theorem. It remains to prove that indeed all zeros of $R(z)$ in (12.15) are needed.

To this end, assume that z is an arbitrary zero of $R(z)$.[9] Then $\mathrm{res}(c(x) - z d'(x), d(x)) = 0$ and hence

$$\gcd(c(x) - z d'(x), d(x)) = g(x)$$

---

[9] In general, $z \in \mathbb{E}_R$ could be a member of the splitting field of $R(z)$. We will see, however, that $z = z_k$ must be valid, and since we assumed that $z_k \in \mathbb{E}$, all computations take place in $\mathbb{E}$.

with $\deg(g(x), x) > 0$. In particular $z \neq 0$ since $c(x)$ and $d(x)$ are relatively prime.

Now if $h(x)$ is an irreducible factor of $g(x)$, then $h(x) \mid d(x)$, and thanks to (12.18) there is exactly one $j \in \{1, \ldots, n\}$ with $h(x) \mid v_j(x)$. Furthermore $h(x) \mid c(x) - z d'(x)$, and due to (12.19) we get

$$h(x) \mid \sum_{k=1}^{n} (z_k - z) v'_k(x) u_k(x).$$

Because $h(x) \mid v_j(x) \mid u_k(x)$ for all $j \neq k$, it follows that

$$h(x) \mid (z_j - z) v'_j(x) u_j(x).$$

Since $h(x)$ as a divisor of $v_j(x)$ cannot be a divisor of $v'_j(x)$ nor of $u_j(x)$, we see that $z = z_j$. Hence we have proved that the arbitrarily chosen zero z of $R(z)$ must be one of the given numbers $z_j \in \mathbb{E}$. This also shows that the given numbers z_j ($j = 1, \ldots, n$) must be exactly the different zeros of $R(z)$. $\qquad\square$

The obvious disadvantage of Theorem 12.12 is the fact that we a priori demand an integral representation of the desired form. This representation, however, might only be available in a suitable algebraic extension field. This is the reason why we denoted the differential field by $\mathbb{E}$ in Theorem 12.12. We would therefore like to show that the Rothstein-Trager algorithm generates the *smallest possible* algebraic extension. Before we will do so, let us consider some more examples with *Mathematica*.

Session 12.13. *We again consider the example $\int \frac{1}{x^3+x}$ and complete the computations given in Session 12.11. After entering the numerator and denominator as*

```
In[1]:= c = 1; d = x^3 + x;
```

we compute the resultant $R(z)$,

```
In[2]:= res = Resultant[c - z D[d, x], d, x]
Out[2]= (1 - z)(2z + 1)^2
```

and search for its zeros using `Solve`*:*

```
In[3]:= sol = Solve[res == 0, z]
Out[3]= {{z → -1/2}, {z → -1/2}, {z → 1}}
```

One of the zeros is multiple, but we are only interested in the different zeros. For this reason, we remove multiple zeros using `Union`*:*

```
In[4]:= sol = Union[Solve[res == 0, z]]
Out[4]= {{z → -1/2}, {z → 1}}
```

This yields the list of values z_k:

```
In[5]:= zlist = z/.sol
Out[5]= {-1/2, 1}
```

The list of polynomials $v_k(x)$ is given by[10]

---

[10] In the following computations we make use of the fact that many *Mathematica* functions have the attribute `Listable`—consider for example `??PolynomialGCD`—and can therefore be applied (without the use of `Map`) to lists.

```
In[6]:= vlist = PolynomialGCD[c - z D[d, x] /. sol, d]
```
$$Out[6]= \left\{\frac{x^2}{2} + \frac{1}{2}, x\right\}$$

Note that the polynomials returned by Mathematica are not monic. For this purpose we use the auxiliary function makemonic:[11]

```
In[7]:= makemonic[p_, x_] := Module[{n},
            n = Exponent[p, x];
            Expand[p/Coefficient[p, x, n]]
        ]
```

```
In[8]:= makemonic[vlist, x]
```
$$Out[8]= \left\{x^2 + 1, x\right\}$$

The integral to be computed can be written as the scalar product $\sum_{k=1}^{n} z_k \ln v_k(x) = z \cdot \ln v(x)$:

```
In[9]:= zlist . Log[makemonic[vlist, x]]
```
$$Out[9]= \log(x) - \frac{1}{2} \log(x^2 + 1)$$

Next we consider a more complicated example which we construct from a desired integral representation. Let the integral be given by

```
In[10]:= int = 1/2 Log[x^3 + x - 1] - 2/3 Log[x^5 + 2x^2 - 3]
```
$$Out[10]= \frac{1}{2} \log(x^3 + x - 1) - \frac{2}{3} \log(x^5 + 2x^2 - 3)$$

This yields the input function

```
In[11]:= r = Together[D[int, x]]
```
$$Out[11]= \frac{-11x^7 - 17x^5 + 22x^4 - 37x^2 + 16x - 9}{6(x^3 + x - 1)(x^5 + 2x^2 - 3)}$$

We compute the numerator and the denominator as

```
In[12]:= c = Numerator[r]/6; d = Denominator[r]/6;
```

respecting $d(x)$ *to be monic. The resultant is then given by*

```
In[13]:= res = Resultant[c - z D[d, x], d, x]
```
$$Out[13]= \frac{\left(-850392 z^3 + 1275588 z^2 - 637794 z + 106299\right)}{1679616} \cdot \left(239735160864 z^5 + 799117202880 z^4 + \right.$$
$$\left. 1065489603840 z^3 + 710326402560 z^2 + 236775467520 z + 31570062336\right)$$

and has the different zeros

```
In[14]:= sol = Union[Solve[res == 0, z]]
```
$$Out[14]= \left\{\left\{z \to -\frac{2}{3}\right\}, \left\{z \to \frac{1}{2}\right\}\right\}$$

Therefore we get the coefficients

```
In[15]:= zlist = z /. sol
```
$$Out[15]= \left\{-\frac{2}{3}, \frac{1}{2}\right\}$$

and the arguments of the logarithmic terms are

```
In[16]:= vlist = PolynomialGCD[c - z D[d, x] /. sol, d]
```

---

[11] This is not absolutely necessary, and without this step we just generate another integration constant. However, we would like to standardize our output.

$Out[16]= \left\{ \frac{1}{6}\left(x^5+2\,x^2-3\right), \frac{x^3}{6}+\frac{x}{6}-\frac{1}{6} \right\}$

$In[17]:=$ **makemonic[vlist,x]**

$Out[17]= \left\{x^5+2\,x^2-3, x^3+x-1\right\}$

Hence the solution integral reads

$In[18]:=$ **zlist . Log[makemonic[vlist,x]]**

$Out[18]= \dfrac{1}{2}\log\left(x^3+x-1\right)-\dfrac{2}{3}\log\left(x^5+2\,x^2-3\right)$

Mathematica's Integrate *command yields the same solution with a slightly different representation:*

$In[19]:= \displaystyle\int$ **r dx**

$Out[19]= \dfrac{1}{6}\left(3\,\log\left(x^3+x-1\right)-4\,\log\left(x^5+2\,x^2-3\right)\right)$

Of course this method is only better than the method given in Session 12.10 if the polynomial $R(z)$ has a reasonable factorization. For the example given there,

$In[20]:=$ **c = x; d = 1 + x + x^7;**

we get the resultant

$In[21]:=$ **res = Resultant[c - z D[d, x], d, x]**

$Out[21]= -870199\,z^7+37044\,z^5-9604\,z^4-z-1$

which has no factorization over $\mathbb{Q}$:

$In[22]:=$ **Factor[res]**

$Out[22]= -870199\,z^7+37044\,z^5-9604\,z^4-z-1$

Therefore in such a case the representation (12.14) is preferable over the Rothstein-Trager algorithm, since the latter also generates a sum over the (unknown) zeros of a polynomial.

We can collect the computational steps of the Rothstein-Trager algorithm in the following Mathematica function:

```
In[23]:= Clear[RothsteinTrager]
        RothsteinTrager[r_, x_] :=
          Module[{c, d, dmonic, z, res, sol, zlist, vlist},
            c = Numerator[r];
            d = Denominator[r];
            dmonic = makemonic[d, x];
            c = Together[c * dmonic/d];
            d = dmonic;
            res = Resultant[c - z * D[d, x], d, x];
            sol = Union[Solve[res == 0, z]];
            zlist = z/.sol;
            vlist = PolynomialGCD[c - z * D[d, x]/.sol,
                d, Extension → Automatic];
            zlist . Log[makemonic[vlist, x]]
          ]
```

The examples just given can now be computed in a single step:

$In[24]:=$ **RothsteinTrager** $\left[\dfrac{1}{x^3+x},x\right]$

$Out[24]=$ $\log(x)-\dfrac{1}{2}\log\left(x^2+1\right)$

$In[25]:=$ **RothsteinTrager** $[r,x]$

$Out[25]=$ $\dfrac{1}{2}\log\left(x^3+x-1\right)-\dfrac{2}{3}\log\left(x^5+2x^2-3\right)$

In the example

$In[26]:=$ **RothsteinTrager** $\left[\dfrac{1}{x^2-2},x\right]$

$Out[26]=$ $\dfrac{\log\left(x-\sqrt{2}\right)}{2\sqrt{2}}-\dfrac{\log\left(x+\sqrt{2}\right)}{2\sqrt{2}}$

$R(z)$ has no factorization over $\mathbb{Q}$. Therefore our computations have automatically detected the necessary algebraic field extension. Note that in our implementation the correct computation of the greatest common divisor in $\mathbb{Q}(\sqrt{2})[x]$ is guaranteed by using the option Extension→Automatic.

Let us continue with the example from Session 12.8 on page 336. The logarithmic part of the rational function considered there was computed as

$$\frac{c(x)}{d(x)}=\frac{-x^2-9x+1}{18(x^4-x^3+x-1)}\ .$$

Now, using the Rothstein-Trager algorithm, we obtain

$In[27]:=$ **RT = RothsteinTrager** $\left[\dfrac{-x^2-9x+1}{18(x^4-x^3+x-1)},x\right]$

$Out[27]=$ $-\dfrac{1}{4}\log(x-1)-\dfrac{1}{12}\log(x+1)+$

$\qquad\qquad \dfrac{1}{54}\left(9+i\sqrt{3}\right)\log\left(x-\dfrac{i\sqrt{3}}{2}-\dfrac{1}{2}\right)+\dfrac{1}{54}\left(9-i\sqrt{3}\right)\log\left(x+\dfrac{i\sqrt{3}}{2}-\dfrac{1}{2}\right)$

i.e., a representation in $\mathbb{Q}(x)(\sqrt{3},i)(\ln(x-1),\ln(x+1),\ln(x-\frac{i\sqrt{3}}{2}-\frac{1}{2}),\ln(x+\frac{i\sqrt{3}}{2}-\frac{1}{2}))$ which is complex, although there is also a real representation[12]

$In[28]:=$ **int** $=\displaystyle\int\dfrac{-x^2-9x+1}{18(x^4-x^3+x-1)}\,dx$

$Out[28]=$ $\dfrac{1}{18}\left(3\log\left(x^2-x+1\right)-\dfrac{9}{2}\log(1-x)-\dfrac{3}{2}\log(x+1)-\dfrac{2\tan^{-1}\left(\frac{2x-1}{\sqrt{3}}\right)}{\sqrt{3}}\right)$

Later on, we will come back to the question how a real antiderivative can be computed.

What is the difference of the two outputs in our concrete example? Of course this difference must be constant, namely an integration constant. Let us first represent the two results graphically. We start with the output of the Integrate *command:*

$In[29]:=$ **Plot[int, {x, -5, 5}]**

---

[12] Please note that *Mathematica*'s integration output depends on the version number.

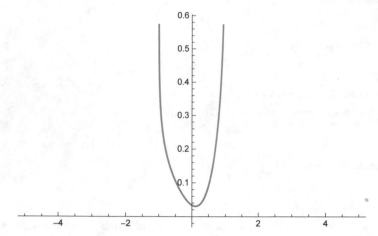

Next, we plot the Rothstein-Trager solution:

$In[30]:=$ `Plot[RT, {x, -5, 5}]`

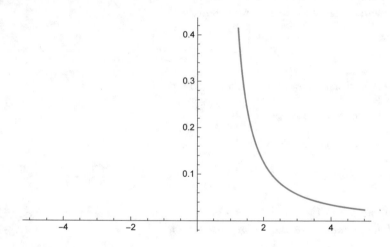

Since the logarithmic output functions do not have absolute values as arguments, they are real in certain intervals and complex elsewhere. The two chosen branches must therefore differ by a complex constant since they are real in different intervals. Let us compute their difference. In a first step, we check that the difference of the two integrals is indeed constant:

$In[31]:=$ `D[RT - int, x] //Together`
$Out[31]=$ 0

Their difference can therefore be computed by substituting $x = 0$:[13]

$In[32]:=$ `const = RT - int /.x → 0`
$Out[32]= -\dfrac{i\pi}{4} - \dfrac{\pi}{54\sqrt{3}} + \dfrac{1}{81}i\left(9 - i\sqrt{3}\right)\pi - \dfrac{1}{81}i\left(9 + i\sqrt{3}\right)\pi$

For simplification we can use the `ComplexExpand` *command:*

$In[33]:=$ `const = ComplexExpand[const]`

---

[13] It turns out that in the two relevant intervals the complex part of the constant difference changes. Check this by computing the value at $x = 2$, too!

$$Out[33] = \frac{\pi}{18\sqrt{3}} - \frac{i\pi}{4}$$

$In[34] := $ **N[const]**

$Out[34] = 0.100767 - 0.785398i$

Since Mathematica cannot perfectly simplify complex functions, one needs some experience to find the best simplification. For example, we did not succeed in simplifying the difference RT-int *directly, even after having introduced some assumptions.*

Now we show that the Rothstein-Trager generates the smallest possible algebraic extension.

Theorem 12.14. Let $\mathbb{K}(x)$ be a differential field over the constant field $\mathbb{K}$. Let further $c(x), d(x) \in \mathbb{K}[x]$ with $\gcd(c(x), d(x)) = 1$, $d(x)$ monic and squarefree with $\deg(c(x), x) < \deg(d(x), x)$. Let $\mathbb{E}$ finally be the smallest algebraic extension field of $\mathbb{K}$ with the property that the integral can be represented in the form

$$\int \frac{c(x)}{d(x)} = \sum_{k=1}^{N} Z_k \ln V_k(x), \tag{12.21}$$

where $Z_k \in \mathbb{E}$ and $V_k(x) \in \mathbb{E}[x]$. Then

$$\mathbb{E} = \mathbb{K}(z_1, \dots, z_n),$$

such that $z_1, \dots, z_n$ are the different zeros of the polynomial

$$R(z) = \operatorname{res}(c(x) - z\,d'(x), d(x)) \in \mathbb{K}[z],$$

i.e., $\mathbb{E}$ is the smallest splitting field of $\mathbb{K}$ with respect to $R(z)$. The following relations are valid again:

$$\int \frac{c(x)}{d(x)} = \sum_{k=1}^{n} z_k \ln v_k(x) \tag{12.22}$$

and

$$v_k(x) = \gcd(c(x) - z_k\,d'(x), d(x)) \in \mathbb{E}[x].$$

Proof. If the polynomials $V_k(x)$ do not satisfy the conditions of Theorem 12.12 (monic, square-free, pairwise different, coprime, positive degree), then these can be reached by rearrangement using the logarithm laws (see [GCL1992, Theorem 11.8]). After this rearrangement we have a representation of the same form satisfying these conditions, which is given by (12.21) (or by (12.22)).

Then Theorem 12.12 can be applied and states that the values z_k $(k = 1, \dots, n)$ are the different zeros of the resultant polynomial $R(z)$. Since by definition $R(z)$ lies in $\mathbb{K}[z]$, $\mathbb{E}$ must be a spitting field of $R(z)$. Since in the proof of Theorem 12.12 it was shown that all zeros of $R(z)$ are needed in the representation (12.20), $\mathbb{E}$ is the smallest possible algebraic extension field. $\qquad\square$

To complete our discussions we will show how—in the case of real integrands—complex representations can be avoided. For this purpose the occurrent logarithms are replaced by inverse tangent functions. First we define in an obvious way how inverse tangent terms are algebraically given.

Definition 12.15. Let $(\mathbb{K}, D)$ be a differential field and let $(\mathbb{E}, D)$ be a differential extension field of $(\mathbb{K}, D)$. If for a given $\varphi \in \mathbb{E} \setminus \mathbb{K}$ there is an element $u \in \mathbb{K}$ such that

$$D(\varphi) = \frac{D(u)}{1 + u^2}$$

is valid, then φ is called an *inverse tangent* over $\mathbb{K}$ and we write $\varphi = \arctan u$. △

Thus we have again defined the chain rule, this time for the inverse tangent function.

If the denominator polynomial $d(x) \in \mathbb{R}[x]$ has complex zeros, then those zeros always occur as a pair of complex conjugate numbers, see Exercise 12.2. Therefore when integrating $\frac{c(x)}{d(x)}$ pairs of logarithms with complex arguments occur, and it is easy to show that the associated coefficients are also complex conjugates. Note that the complex logarithms with real coefficients can be represented by real logarithms due to

$$c \ln(a + ib) + c \ln(a - ib) = c \ln\big((a + ib)(a - ib)\big) = c \ln(a^2 + b^2).$$

Now we study how we can collect logarithms with purely complex coefficients in terms of real inverse tangent functions. To this end, let $u(x) \in \mathbb{K}(x)$ with $u(x)^2 \not\equiv -1$. Then

$$i D\big(\ln(u(x) + i) - \ln(u(x) - i)\big) = 2 D \arctan u(x), \tag{12.23}$$

where D denotes the derivative with respect to the variable x. This follows from the computation

$$
\begin{aligned}
i D\big(\ln(u(x) + i) - \ln(u(x) - i)\big) &= i \left(\frac{u'(x)}{u(x) + i} - \frac{u'(x)}{u(x) - i} \right) \\
&= 2 \frac{u'(x)}{u(x)^2 + 1} = 2 D \arctan u(x).
\end{aligned}
$$

Note that by (12.23) the two functions $i\big(\ln(u(x) + i) - \ln(u(x) - i)\big)$ and $2 \arctan u(x)$ have the same derivative and therefore differ by a constant.

Session 12.16. *If $u \in \mathbb{R}$, then this difference equals π for Mathematica's choices of the branches of the logarithm and the inverse tangent function. For $u(x) = x$, for example, we get*

```
In[1]:= Plot[{i(Log[x+i] - Log[x-i]), 2ArcTan[x]}, {x, -2, 2}]
```

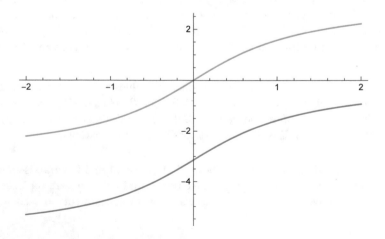

If $u(x)$ has real poles, then $\arctan u(x)$ has jump discontinuities:

$In[2]:=$ **Plot** $\Big[$ **Evaluate** $\Big[$ $\{$ $\mathbb{i}$ **(Log[u + i̇] - Log[u - i̇])**, **2 ArcTan[u]** $\}$ **/.**

$$u \rightarrow \frac{x^3 - 2x}{x^2 - 1}\Big], \{x, -2, 2\}\Big]$$

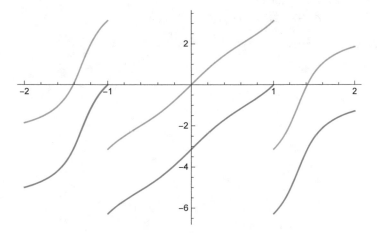

Next, we compute the difference $2 \arctan z - i\Big(\ln(z+i) - \ln(z-i)\Big)$:

$In[3]:=$ **simp = ComplexExpand[2 ArcTan[z] - i̇ (Log[z + i̇] - Log[z - i̇])]**
$Out[3]=$ $-\arg(1 - iz) + \arg(1 + iz) - \arg(z - i) + \arg(z + i)$

$In[4]:=$ **FullSimplify[Table[simp, {z, -3, 3}]]**
$Out[4]=$ $\{\pi, \pi, \pi, \pi, \pi, \pi, \pi\}$

Unfortunately Mathematica *does not find this constant symbolically.*

Of course the choice of an integration constant is not very important. However, we must expect that in different intervals different integration constants apply. This will be considered in Session 12.17.

Using the conversion (12.23) in integrated form, all logarithms which have purely imaginary coefficients and occur as pairwise conjugate terms can be transformed:

$$
\begin{aligned}
ci\ln(a+ib) - ci\ln(a-ib) &= ci\ln\left(\frac{a+ib}{a-ib}\right) \\
&= ci\ln\left(\frac{\frac{a}{b}+i}{\frac{a}{b}-i}\right) = ci\ln(\frac{a}{b}+i) - ci\ln(\frac{a}{b}-i) \\
&\equiv 2c\arctan\frac{a}{b}, \qquad\qquad (12.24)
\end{aligned}
$$

Here by $\equiv$ we denote equality modulo a suitable integration constant. Note that the application of the logarithm rule $\ln x - \ln y = \ln\frac{x}{y}$ may again result in a change of the branch of the logarithm.

Session 12.17. *As an example, we compute*

$$\int \frac{x^4 - 3x^2 + 6}{x^6 - 5x^4 + 5x^2 + 4}.$$

For this purpose we again load the function RothsteinTrager *from Session 12.13 and compute*

$In[1]:=$ $\mathbf{r} = \dfrac{x^4 - 3x^2 + 6}{x^6 - 5x^4 + 5x^2 + 4}$

$Out[1]=$ $\dfrac{x^4 - 3x^2 + 6}{x^6 - 5x^4 + 5x^2 + 4}$

$In[2]:=$ $\mathbf{RT} = \mathbf{RothsteinTrager[r, x]}$

$Out[2]=$ $\dfrac{1}{2}\, i\, \log(x^3 + i\, x^2 - 3\, x - 2\, i) - \dfrac{1}{2}\, i\, \log(x^3 - i\, x^2 - 3\, x + 2\, i)$

We plot the resulting integral:

$In[3]:=$ $\mathbf{Plot[RT, \{x, -3, 3\}]}$

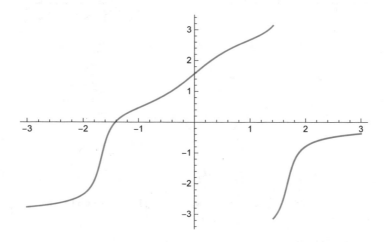

The conversion according to (12.24),

$In[4]:=$ $\mathbf{int = ArcTan}\left[\dfrac{\mathbf{ComplexExpand[Re[x^3 + i\, x^2 - 3\, x - 2\, i]]}}{\mathbf{ComplexExpand[Im[x^3 + i\, x^2 - 3\, x - 2\, i]]}}\right]$

$Out[4]=$ $\tan^{-1}\!\left(\dfrac{x^3 - 3\, x}{x^2 - 2}\right)$

yields another graph. We plot the difference of the two functions:

$In[5]:=$ $\mathbf{Plot[int - RT, \{x, -3, 3\}]}$

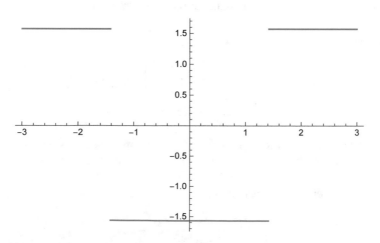

Both graphs agree modulo π, but only in suitable intervals which are defined by the two zeros $x = \pm\sqrt{2}$ of the denominator $x^2 - 2$ of the inverse tangent solution. Both integration functions

have a rather uncomfortable property: they are discontinuous, although there is also a continuous antiderivative. So if the fundamental theorem of calculus is applied to compute the definite integral

$$\int_1^2 \frac{x^4 - 3x^2 + 6}{x^6 - 5x^4 + 5x^2 + 4}$$

which is positive due to the positivity of the integrand, see the following plot,

`In[6]:= Plot[r, {x, 1, 2}]`

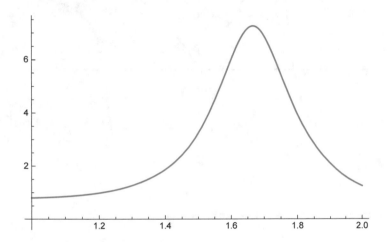

then we get for the Rothstein-Trager solution:

`In[7]:= definite = (RT/.x → 2) - (RT/.x → 1)`

$Out[7] = -\frac{1}{2} i \log(-2 - i) + \frac{1}{2} i \log(-2 + i) - \frac{1}{2} i \log(2 - 2i) + \frac{1}{2} i \log(2 + 2i)$

`In[8]:= N[definite]`

$Out[8] = -3.46334 + 0.\, i$

For the inverse tangent solution, we find

`In[9]:= definite = (int/.x → 2) - (int/.x → 1)`

$Out[9] = \frac{\pi}{4} - \tan^{-1}(2)$

`In[10]:= N[definite]`

$Out[10] = -0.321751$

Hence in both cases we have negative values, which of course agree with the correct integral value modulo a multiple of π. Since the antiderivative is discontinuous, an application of the fundamental theorem of calculus is not admissible!

Mathematica's integration procedure

`In[11]:= definite = ` $\int_1^2 r \, dx$

$Out[11] = \frac{5\pi}{4} - \tan^{-1}(2)$

`In[12]:= N[definite]`

$Out[12] = 2.81984$

obviously computes the correct value, although the underlying antiderivative is also discontinuous and not suitable for this computation:

$In[13]:=$ **int** $= \int \mathbf{r} \, \mathbf{dx}$

$Out[13]= \dfrac{1}{2} \tan^{-1}\left(\dfrac{x\,(x^2-3)}{x^2-2}\right) - \dfrac{1}{2} \tan^{-1}\left(\dfrac{x\,(x^2-3)}{2-x^2}\right)$

$In[14]:=$ **Plot[int, {x, -3, 3}]**

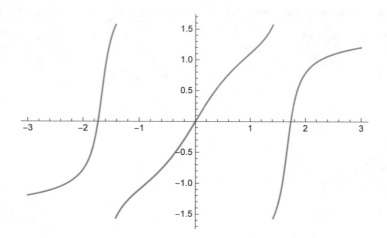

$In[15]:=$ **definite = (int/.x → 2) - (int/.x → 1)**

$Out[15]= \dfrac{\pi}{4} - \tan^{-1}(2)$

$In[16]:=$ **N[definite]**

$Out[16]= -0.321751$

For the computation of definite integrals using their antiderivatives, it is therefore helpful to find *continuous* antiderivatives so that this application is admissible. Clearly the discontinuity in the previous examples was caused by the rationality of the arguments of the occurring inverse tangent functions. Every pole of such a rational function might cause a change of branch. This can be avoided by writing the resulting antiderivative as the sum of inverse tangent functions whose arguments are polynomials. Such a sum is automatically continuous. This can be reached by the following algorithm due to Rioboo [Rio1991].

Theorem 12.18. Let $\mathbb{K}$ be a field with $i \notin \mathbb{K}$. Let further $a(x), b(x) \in \mathbb{K}[x]$ be non-zero polynomials. Finally, let $c(x), d(x) \in \mathbb{K}[x]$, $c(x) \neq 0$, such that $b(x)d(x) - a(x)c(x) = g(x) = \gcd(a(x), b(x))$, i.e., $g(x), c(x)$ and $d(x)$ can be computed by the extended Euclidean algorithm.

Then the following equations are valid:

$$
\begin{aligned}
Di\ln \frac{a(x)+ib(x)}{a(x)-ib(x)} \;&=\; Di\ln \frac{-b(x)+ia(x)}{-b(x)-ia(x)} \\[2mm]
&=\; 2D\arctan \frac{a(x)d(x)+b(x)c(x)}{g(x)} + Di\ln \frac{d(x)+ic(x)}{d(x)-ic(x)} .
\end{aligned}
\tag{12.25}
$$

These equations yield a recursive algorithm for the representation of

$$
i\ln(a(x)+ib(x)) - i\ln(a(x)-ib(x)) = i\ln \frac{a(x)+ib(x)}{a(x)-ib(x)}
$$

as a sum of inverse tangent terms with polynomial arguments.

Proof. The first equation follows from the computation

```
In[1]:= Together[
          D[Log[(a[x]+I b[x])/(a[x]-I b[x])]-
            Log[(-b[x]+I a[x])/(-b[x]-I a[x])],
          x]]
Out[1]= 0
```

Therefore let $g(x) := \gcd(a(x), b(x)) \in \mathbb{K}[x]$. Then the extended Euclidean algorithm computes $c(x), d(x) \in \mathbb{K}[x]$ with $b(x)d(x) - a(x)c(x) = g(x)$.

In the sequel we write $p(x) := \frac{a(x)d(x)+b(x)c(x)}{g(x)}$. Because $g(x)$ is a divisor of both $a(x)$ and $b(x)$, obviously $p(x) \in \mathbb{K}[x]$. Since

$$
\begin{aligned}
\frac{a(x)+lb(x)}{a(x)-ib(x)} &= \left(\frac{d(x)-ic(x)}{d(x)+ic(x)} \cdot \frac{a(x)+ib(x)}{a(x)-ib(x)}\right) \cdot \frac{d(x)+ic(x)}{d(x)-ic(x)} \\
&= \left(\frac{a(x)d(x)+b(x)c(x)+i(b(x)d(x)-a(x)c(x))}{a(x)d(x)+b(x)c(x)-i(b(x)d(x)-a(x)c(x))}\right) \cdot \frac{d(x)+ic(x)}{d(x)-ic(x)} \\
&= \left(\frac{a(x)d(x)+b(x)c(x)+ig(x)}{a(x)d(x)+b(x)c(x)-ig(x)}\right) \cdot \frac{d(x)+ic(x)}{d(x)-ic(x)} \\
&= \frac{p(x)+i}{p(x)-i} \cdot \frac{d(x)+ic(x)}{d(x)-ic(x)},
\end{aligned}
$$

using (12.23) we get

$$
Di\ln\frac{a(x)+ib(x)}{a(x)-ib(x)} = 2D\arctan p(x) + Di\ln\frac{d(x)+ic(x)}{d(x)-ic(x)},
$$

i.e. the second equation of (12.25).

The recursive algorithm for the computation of a continuous antiderivative is invoked as follows: if $\deg(a(x), x) < \deg(b(x), x)$, then the role of $a(x)$ and $b(x)$ can be swapped by using the first equation in (12.25). Therefore we can assume, without loss of generality, that $\deg(b(x), x) \leq \deg(a(x), x)$.

If now $b(x) \mid a(x)$, then we use the transformation (12.24) to get

$$
Di\ln\frac{a(x)+ib(x)}{a(x)-ib(x)} = 2D\arctan\frac{a(x)}{b(x)}. \tag{12.26}
$$

This is already a representation by an inverse tangent term with polynomial argument, and we are done.

In the sequel we assume that $b(x)$ is no divisor of $a(x)$. Using the extended Euclidean algorithm we compute the greatest common divisor $g(x) = \gcd(a(x), b(x))$ as well as $c(x), d(x) \in \mathbb{K}[x]$ such that $b(x)d(x) - a(x)c(x) = g(x)$ is valid. It turns out that this implies $\deg(d(x), x) < \deg(a(x), x)$, see Exercise 8.9. Then the second equation in (12.25) yields the recursive call

$$
Di\ln\frac{a(x)+ib(x)}{a(x)-ib(x)} = 2D\arctan p(x) + Di\frac{d(x)+ic(x)}{d(x)-ic(x)}
$$

with a (further) inverse tangent summand with a polynomial argument. The relation $\deg(d(x), x) < \deg(a(x), x)$ implies that this algorithm terminates. $\qquad\square$

Session 12.19. *We implement the Rioboo algorithm as follows:*

```
In[1]:= Clear[LogToArcTan]

      LogToArcTan[a_,b_,x_] :=
        LogToArcTan[-b,a,x]/;Exponent[a,x] < Exponent[b,x]

      LogToArcTan[a_,b_,x_] :=
        2 ArcTan[a/b]/;PolynomialQ[Together[a/b],x]

      LogToArcTan[a_,b_,x_] := Module[{g,c,d},
        {g,{d,c}} = PolynomialExtendedGCD[b,-a,x];
        2 ArcTan[Together[(a*d+b*c)/g]] + LogToArcTan[d,c,x]
      ]
```

For our example function from Session 12.17 we get

$In[2]:= $ **rioboo = LogToArcTan[x^3 - 3x, x^2 - 2, x]**

$Out[2]= 2\tan^{-1}(x^3) + 2\tan^{-1}\left(\frac{1}{2}\left(x^5 - 3x^3 + x\right)\right) + 2\tan^{-1}(x)$

which is a continuous representation as a sum of inverse tangent functions with polynomial arguments.

$In[3]:= $ **Plot$\left[\dfrac{\text{rioboo}}{2}, \{x, -3, 3\}\right]$**

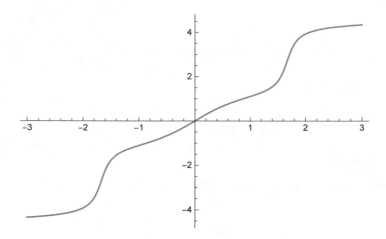

In the plot, we see that the algorithm has automatically produced yet another desirable property of the antiderivative: The integrated function $\frac{x^4-3x^2+6}{x^6-5x^4+5x^2+4}$ is an even function, therefore there is an odd antiderivative. This is what Rioboo's algorithm yields indeed.

12.5 Additional Remarks

The combination of the algorithms considered in this chapter leads to a complete algorithm for rational integration. Up to some efficiency-raising adaptions this is the current state of the art, and we have seen that this algorithm is not completely available in *Mathematica*.

In Risch's algorithm for the integration of elementary functions the algebraic, logarithmic and exponential differential field extensions are allowed in both the input and the output. As in the rational case, a structure theorem assures what a potential antiderivative looks like: for the generation of such an antiderivative only logarithmic field extensions are necessary. However, more complicated towers of field extensions are possible and the theory is algebraically more demanding. Eventually the antiderivative can again be found by a sort of equating coefficients. An essential aid for this purpose is a recursive scheme, the Hermite reduction, see Exercise 12.3. Recommended literature on this topic are the textbooks [GCL1992] and [Bro1997].

12.6 Exercises

Exercise 12.1. *Show that all possible differential operators in $\mathbb{Q}(x)$ are given by the two conditions $D(1) = 0$ and $D(x) = F(x)$, where $F(x) \in \mathbb{Q}(x)$ is an arbitrary rational function.*

Exercise 12.2. *Show that if a real polynomial $a(x) \in \mathbb{R}[x]$ has a complex root $z_0 \in \mathbb{C} \setminus \mathbb{R}$, then the conjugate complex number $\overline{z_0}$ is also a root of $a(x)$.*

Exercise 12.3. (Hermite Reduction)   *Let $r(x) = \frac{p(x)}{q(x)} \in \mathbb{K}(x)$ be given with $\deg(p(x), x) < \deg(q(x), x)$. An alternative algorithm for finding the rational decomposition of $r(x)$ is the so-called Hermite reduction which works as follows:*

(a)   *First one computes a squarefree factorization of $q(x) = \prod_{k=1}^{m} q_k(x)^k$, where $q_k(x)$ are pairwise coprime, monic and squarefree. This can be accomplished in Mathematica by* `FactorSquareFree` *or* `FactorSquareFreeList`.

(b)   *Next one computes a partial fraction decomposition of $r(x)$ with respect to the squarefree factorization*

$$r(x) = \sum_{k=1}^{m} \sum_{j=1}^{k} \frac{r_{jk}}{q_k(x)^j} .$$

For this purpose one cannot use `Apart` *since this function automatically factorizes the denominator. However, the partial fraction decomposition can be computed by solving a linear system.*

(c)   *The resulting representation is a sum which can be simplified termwise. The summands have the form $\frac{r_{jk}}{q_k(x)^j}$ with $j \geq 1$. If $j = 1$, then this yields a term for the logarithmic part, and we are done. Otherwise we continue with (d).*

(d)   *Now we reduce the exponent j recursively. After an application of the extended Euclidean algorithm (for example using* `PolynomialExtendedGCD`), $\gcd(q_k, q_k') = 1$ *implies a representation of the form*

$$s \cdot q_k + t \cdot q_k' = r_{jk}$$

with $s, t \in \mathbb{K}[x]$.

(e)   *With the aid of the identity[14]*

$$\int \frac{r_{jk}}{q_k(x)^j} = \int \frac{s(x)}{q_k(x)^{j-1}} + \int \frac{t(x) q_k'(x)}{q_k(x)^j} = -\frac{t(x)}{(j-1) q_k(x)^{j-1}} + \int \frac{s(x) + \frac{t'(x)}{j-1}}{q_k(x)^{j-1}}$$

j can finally be reduced by at least 1. This either stops with $j = 0$ and yields a rational part, or it stops with $j = 1$ and yields a logarithmic part.

---

[14] Prove this identity!

Apply Hermite reduction to the example function

$$r(x) = \frac{36x^6 + 126x^5 + 183x^4 + \frac{13807}{6}x^3 - 407x^2 - \frac{3242}{5}x + \frac{3044}{15}}{x^7 + \frac{17x^6}{15} - \frac{263x^5}{900} - \frac{1349x^4}{2250} + \frac{2x^3}{1125} + \frac{124x^2}{1125} + \frac{4x}{1125} - \frac{8}{1125}}$$

and compare your result with the one of `RationalDecomposition`.

Exercise 12.4. Compare the results for $\int \frac{1}{x^4+4}$

(a) of Mathematica;
(b) of the Rothstein-Trager algorithm;
(c) by applying a rational partial fraction decomposition of the integrand.

Exercise 12.5. Give two more—relatively simple—examples for the Rioboo algorithm.

Exercise 12.6. In calculus courses sometimes the $\tan \frac{x}{2}$ substitution is used with the help of which every trigonometric rational function $R(x) \in \mathbb{R}(\sin x, \cos x)$ can be integrated since the substitution creates a rational integrand. Give an example which shows that this substitution generally leads to discontinuous antiderivatives.

References

Abr2003. Abramov, S. A.: When does Zeilberger's algorithm succeed? *Advances in Applied Mathematics* **30**, 2003, 424–441. 324

AGP1994. Alford, W. R., Granville, A., Pomerance, C.: There are infinitely many Carmichael numbers. *J. Ann. Math. II. Ser.* **139**, 1994, 703–722. 82

AKS2004. Agarwal, M., Kayal, N., Saxena, N.: Primes is in *P. Annals of Mathematics* **160**, 2004, 781–793. 87

AZ1991. Almkvist, G., Zeilberger, D.: The method of differentiating under the integral sign. *J. Symbolic Computation* **10**, 1990, 571–591. 325, 329

Bai1935. Bailey, W. N.: *Generalized Hypergeometric Series.* Cambridge University Press, England, 1935. Reprinted 1964 by Stechert-Hafner Service Agency, New York/London. 262

BCS1997. Bürgisser, P., Clausen, M., Shokrollahi, M. A.: *Algebraic Complexity Theory.* Grundlehren der mathematischen Wissenschaften 315. Springer, Berlin/Heidelberg/New York, 1997. 87, 110

BDG1988. Balcázar, J. L., Díaz, J., Gabarró, J.: *Structural Complexity.* Springer, Heidelberg, 1988. 87, 110

Ber1967. Berlekamp, E. R.: Factoring polynomials over finite fields. *Bell System Technical Journal* **46**, 1967, 1853–1859. 197

BHKS2005. Belabas, K., van Hoeij, M., Klüners, J., Steel, A.: Factoring polynomials over global fields. *Journal de Théorie des Nombres de Bordeaux* **21**, 2009, 15–39. 217

Böi1998. Böing, H.: *Theorie und Anwendungen zur q-hypergeometrischen Summation.* Diploma thesis, 1998, Freie Universität Berlin. www.mathematik.uni-kassel.de/~koepf/Diplome. 324

Bro1997. Bronstein, M.: *Symbolic Integration I.* Springer, Berlin, 1997. 329, 330, 359

BS2011. Biham, E., Shamir, A.: *Differential Cryptanalysis of the Data Encryption Standard.* Springer, Berlin, 2011. 105

Buc1999. Buchmann, J.: *Einführung in die Kryptographie.* Springer, Berlin/Heidelberg, 1999. 87

CC1986. Chudnovski, D. V., Chudnovski, G. V.: On expansion of algebraic functions in power and Puiseux series I. *Journal of Complexity* **2**, 1986, 271–294. 281

CC1987. Chudnovski, D. V., Chudnovski, G. V.: On expansion of algebraic functions in power and Puiseux series II. *Journal of Complexity* **3**, 1987, 1–25. 281

Chi2000. Childs, L.: *A Concrete Introduction to Higher Algebra.* Springer, New York, 2nd edition, 2000. First edition 1979. VII, 88, 153, 173

CLO1997. Cox, D., Little, J., O'Shea, D.: *Ideals, Varieties and Algorithms.* Springer, New York, 1997. VI, 186

DeB1984. De Branges, L.: A proof of the Bieberbach conjecture. *Acta Math.* **154**, 1985, 137–152. 19

DH1976. Diffie, W., Hellman, M. E.: New directions in Cryptography. *IEEE Transactions on Information Theory* **6**, 1976, 644–654. 106

Eng1995. Engel, A.: Kann man mit DERIVE geometrische Sätze beweisen? In: Koepf, W. (Ed.): *Der Mathematik-Unterricht* 41, Heft 4, Juli 1995: *DERIVE-Projekte im Mathematikunterricht,* 38–47. 233

Erm1982. Ermolaev, H.: *Mikhail Sholokhov and his Art.* Princeton University Press, 1982. 95

Fas1945. Fasenmyer, M. C.: *Some Generalized Hypergeometric Polynomials.* Ph. D. Dissertation, University of Michigan, 1945. 290, 324

For1996. Forster, O.: *Algorithmische Zahlentheorie.* Vieweg, Braunschweig/Wiesbaden, 1996. 87

FR1998. Fowler, D., Robson, E.: Square Root Approximations in Old Babylonian Mathematics: YBC 7289 in Context. *Historia Mathematica* **25**, 1998, 366–378. 61

GCL1992. Geddes, K. O., Czapor, S. R., Labahn, G.: *Algorithms for Computer Algebra.* Kluwer, Boston/Dordrecht/London, 1992. VII, 182, 186, 217, 226, 234, 329, 351, 359

GG1999. von zur Gathen, J., Gerhard, J.: *Modern Computer Algebra.* Cambridge University Press, Cambridge, 1999. VII, 56, 62, 87, 137, 148, 153, 186, 278

© Springer Nature Switzerland AG 2021

W. Koepf, *Computer Algebra*, Springer Undergraduate Texts in Mathematics and Technology, https://doi.org/10.1007/978-3-030-78017-3

GKP1994.   Graham, R. L., Knuth, D. E., Patashnik, O.: *Concrete Mathematics. A Foundation for Computer Science*. Addison-Wesley, Reading, Massachussets, Second edition, 1994. First edition 1988. 34

Gos1978.   Gosper Jr., R. W.: Decision procedure for indefinite hypergeometric summation. *Proc. Natl. Acad. Sci. USA* **75**, 1978, 40–42. 302, 304, 324

Henn2003.   Henn, H.-W.: *Elementare Geometrie und Algebra*. Vieweg, Braunschweig/Wiesbaden, 2003. 187

Her1975.   Herstein, I. N.: *Topics in Algebra*. Wiley, New York, 1975. 159

Hoe2002.   van Hoeij, M.: Factoring polynomials and the knapsack problem. *Journal of Number Theory* **95**, 2002, 167–189. 217

JS1992.   Jenks, R. D., Sutor, R. S.: *AXIOM – The Scientific Computation System*. Springer, New York and NAG, Ltd., Oxford, 1992. 281

Kar1962.   Karatsuba, A.: Multiplication of multidigit numbers on automata. *Soviet Physics Doklady* **7**, English edition 1967, 595–596. 44, 62

Karr1981.   Karr, M.: Summation in finite terms. *Journal of the ACM* **28**, 1981, 305–350. 325

KBM1995.   Koepf, W., Bernig, A., Melenk, H.: Trigsimp, a REDUCE Package for the Simplification and Factorization of Trigonometric and Hyperbolic Functions. In: Koepf, W. (Ed.): *REDUCE Packages on Power Series, Z-Transformation, Residues and Trigonometric Simplification*. Konrad-Zuse-Zentrum Berlin (ZIB), Technical Report TR 95-3, 1995. 231

Koe1992.   Koepf, W.: Power series in computer algebra. *J. Symbolic Computation* **13**, 1992, 581–603. 268, 281

Koe1993a.   Koepf, W.: *Mathematik mit DERIVE*. Vieweg, Braunschweig/Wiesbaden, 1993. 242, 277

Koe1993b.   Koepf, W.: Eine Vorstellung von Mathematica und Bemerkungen zur Technik des Differenzierens. *Didaktik der Mathematik* **21**, 1993, 125–139. 34

Koe1994.   Koepf, W.: *Höhere Analysis mit DERIVE*. Vieweg, Braunschweig/Wiesbaden, 1994. 15, 274

Koe1996.   Koepf, W.: Closed form Laurent-Puiseux series of algebraic functions. *Applicable Algebra in Engineering, Communication and Computing* **7**, 1996, 21–26. 281

Koe2010.   Koepf, W.: Efficient computation of truncated power series: Direct approach versus Newton's method. Utilitas Mathematica **83**, 2010, 37–55. 277, 280

Koe2014.   Koepf, W.: *Hypergeometric Summation*. Springer Universitext. Springer, London, 2014, 2nd edition. First edition Vieweg, Braunschweig/Wiesbaden, 1998. 282, 311, 318, 325, 329

Kro1882.   Kronecker, L.: Grundzüge einer arithmetischen Theorie der algebraischen Grössen. *J. Reine Angew. Math.* **92**, 1882, 1–122. 140

Kro1883.   Kronecker, L.: Die Zerlegung der ganzen Grössen eines natürlichen Rationalitäts-Bereichs in ihre irreduciblen Factoren. *J. Reine Angew. Math.* **94**, 1883, 344–348. 140

Lam1844.   Lamé, G.: Note sur la limite du nombre des divisions dans la recherche du plus grand commun diviseur entre deux nobres entiers. *Comptes Rendues Acad. Sci. Paris* **19**, 1844, 867–870. 62

Lew2000.   Lewand, R. E.: *Cryptological Mathematics*. The Mathematical Association of America, 2000. 95, 110

Lint2000.   van Lint, J. H.: Die Mathematik der Compact Disc. In: Aigner, M., Behrends, E. (Eds.): *Alles Mathematik*, Vieweg, Braunschweig/Wiesbaden, 2000, 11–19. 110

Lio1835.   Liouville, J.: Mémoire sur l'intégration d'une classe de fonctions transcendantes. J. Reine Angew. Math. **13**, 1835, 93–118. 329

LLL1982.   Lenstra, A. K., Lenstra, H. W., Jr., Lovász, L.: Factoring polynomials with rational coefficients, *Math. Ann.* **261**, 1982, 515–534. VI, 217

Mae1997.   Maeder, R. E.: *Programming in Mathematica*. Addison Wesley, Reading, Massachussets, 3rd edition, 1997. First edition 1990. 34

Mig1992.   Mignotte, M.: *Mathematics for Computer Algebra*. Springer, New York, 1992. 62, 120

MM2001.   Mulholland, J., Monagan, M.: Algorithms for trigonometric polynomials. *Proc. of ISSAC 2001*, ACM Press, New York, 2001, 245–252. 231

MM2005.   Mohammed, M., Zeilberger, D.: Sharp upper bounds for the orders of the recurrences outputted by the Zeilberger and q-Zeilberger algorithms. *J. Symbolic Computation* **39**, 2005, 201–207. 324

Nor1975.   A. C. Norman: Computing with formal power series. *Transactions on Mathematical Software* **1**, ACM Press, New York, 1975, 346–356. 281

Ost1845.   Ostrogradsky, M. W.: De l'intégration des fractions rationelles. Bulletin de la Classe Physico-Mathématiques de l'Académie Impériale des Sciences de St. Pétersbourg **IV**, 1845, 145–167, 286–300. 340

PGP.   PGP Homepage: www.pgpi.org. 103

Ric1968.   Richardson, D.: Some unsolvable problems involving elementary functions of a real variable. *J. Symbolic Logic* **33**, 1968, 511–520. 227

Rie1994.   Riesel, Hans: *Prime Numbers and Computer Methods for Factorization*. Progress in Mathematics, Birkhäuser, Boston, Second edition, 1994. First edition 1985. 85, 87

Rio1991.   Rioboo, R.: *Quelques aspects du calcul exact avec des nombres réels*. Thèse de Doctorat de l'Université de Paris 6, Informatiques, 1991. 356

Ris1969. Risch, R.: The problem of integration in finite terms. *Trans. Amer. Math. Soc.* **139**, 1969, 167–189. 329

Ris1970. Risch, R.: The solution of the problem of integration in finite terms. *Bull. Amer. Math. Soc.* **76**, 1970, 605–608. 329

Rob1936. Robertson, M. S.: A remark on the odd schlicht functions. *Bull. Amer. Math. Soc.* **42**, 1936, 366–370. 19

Rob1978. Robertson, M. S.: Complex powers of *p*-valent functions and subordination. Complex Anal., Proc. S.U.N.Y. Conf., Brockport 1976, Lect. Notes Math. **36**, 1978, 1–33. 20

Rob1989. Robertson, M. S.: A subordination problem and a related class of polynomials. In: Srivastava, H. M., Owa, Sh. (Eds.): *Univalent Functions, Fractional Calculus, and Their Applications.* Ellis Horwood Limited, Chichester, 1989, 245–266. 19, 20

Rot1976. Rothstein, M.: *Aspects of Symbolic Integration and Simplification of Exponential and Primitive Functions.* Dissertation, Univ. of Wiscinson, Madison, 1976. 343

Roth2006. Roth, R.: *Introduction to Coding Theory.* Cambridge University Press, Cambridge, 2006. 110

RSA1978. Rivest, R., Shamir, A. Adleman, L.: A method for obtaining digital signatures and public key cryptosystems. *Comm. ACM* **21**, 1978, 120–126. 106

Sal1990. Salomaa, A.: *Public-Key Cryptography.* Springer, Berlin, 1990. 110

Sch2003. Schulz, R. H.: *Codierungstheorie.* Vieweg, Braunschweig/Wiesbaden, 2nd edition, 2003. First edition 1991. 97

SK2000. Schmersau, D., Koepf, W.: *Die reellen Zahlen als Fundament und Baustein der Analysis.* Oldenbourg, München, 2000. 62, 64

Spr2004. Sprenger, T.: Algorithmen für mehrfache Summen. Diploma thesis, 2004, Universität Kassel. www.mathematik.uni-kassel.de/~koepf/Diplome. 325

SS1971. Schönhage, V., Strassen, A.: Schnelle Multiplikation großer Zahlen. *Computing*, 1971, 281–292. 148

Tef1999. Tefera, A.: A multiple integral evaluation inspired by the multi-WZ method. *Electronic J. Combinatorics* **6(1)**, 1999, 1–7. 325

Tef2002. Tefera, A.: MultInt, a MAPLE package for multiple integration by the WZ method. *J. Symbolic Computation* **34**, 2002, 329–353. 325

Teg2020. Teguia Tabuguia, Bertrand: Power series representations of hypergeometric type and non-holonomic functions in Computer Algebra. PhD dissertation at the University of Kassel, 2020: www.mathematik.uni-kassel.de/~koepf/Diplome. 221, 244, 268, 281

Tra1976. Trager, B.: Algebraic factoring and rational function integration. *Proc. SYMSAC 1976*, ACM Press, New York, 1976, 219–226. 343

Tro2004a. Trott, M.: *The Mathematica Guidebook for Graphics.* Springer, New York, 2004. 34

Tro2004b. Trott, M.: *The Mathematica Guidebook for Programming.* Springer, New York, 2004. 34

Tro2005. Trott, M.: *The Mathematica GuideBook for Symbolics.* Springer, New York, 2005. 34

Ver1976. Verbaeten, P.: *Rekursiebetrekkingen voor lineaire hypergeometrische funkties.* Proefschrift voor het doctoraat in de toegepaste wetenschapen. Katholieke Universiteit te Leuven, Heverlee, Belgien, 1976. 293

Wal2000. Walter, W.: *Gewöhnliche Differentialgleichungen.* Springer, Berlin/Heidelberg, 7th edition, 2000. First edition 1972. 255

Wan1978. Wang, P. S.: An improved multivariate polynomial factorization algorithm. *Math. Comp.* **32**, 1978, 1215–1231. 217

Weg1997. Wegschaider, K.: *Computer Generated Proofs of Binomial Multi-Sum Identities.* Diploma thesis, RISC Linz, 1997. 293, 325

Wei2005a. Weisstein, E. W.: RSA Number. *MathWorld–A Wolfram Web Resource.* mathworld.wolfram.com/RSANumber.html. 60

Wei2005b. Weisstein, E. W.: RSA-200 Factored. *MathWorld–A Wolfram Web Resource.* mathworld.wolfram.com/news/2005-05-10/rsa-200. 60, 107

Wei2005c. Weisstein, E. W.: RSA-640 Factored. *MathWorld–A Wolfram Web Resource.* mathworld.wolfram.com/news/2005-11-08/rsa-640. 60

Wes1999. Wester, M. (Ed.): *Computer Algebra Systems.* Wiley, Chichester, 1999. 18

Wie1999. Wiesenbauer, J.: Public-Key-Kryptosysteme in Theorie und Programmierung. *Schriftenreihe zur Didaktik der ÖMG* **30**, 1999, 144–159. 112

Wie2000. Wiesenbauer, J.: Using DERIVE to explore the Mathematics behind the RSA cryptosystem. In: Etchells, T., Leinbach, C., Pountney, D. (Eds.): Proceedings of the 4th International DERIVE & TI-89/92 Conference *Computer Algebra in Mathematics Education* (Liverpool 2000), CD-ROM, bk-teachware, 2000. 112

Wiki2016. Babylonian Mathematics. www.wikipedia.org/wiki/Babylonian_mathematics. 61

Wol2003. Wolfram, St.: *The Mathematica Book.* Wolfram Media, 5th edition, 2003. First edition 1988. 34

WZ1992. Wilf, H. S., Zeilberger, D.: An algorithmic proof theory for hypergeometric (ordinary and "*q*") multisum/integral identities. *Invent. Math.* **108**, 1992, 575–633. 320, 325

Zas1969. Zassenhaus, H. On Hensel factorization I. *J. Number Theory* **1**, 1969, 291–311. 206, 210

Zei1990a. Zeilberger, D.: A fast algorithm for proving terminating hypergeometric identities. *Discrete Math.* **80**, 1990, 207–211. 315, 324

Zei1990b. Zeilberger, D.: A holonomic systems approach to special functions identities. *J. Comput. Appl. Math.* **32**, 1990, 321–368. 290, 316

List of Symbols

Mathematica List of Keywords

© Springer Nature Switzerland AG 2021

W. Koepf, *Computer Algebra*, Springer Undergraduate Texts in Mathematics and Technology,

https://doi.org/10.1007/978-3-030-78017-3

Index

© Springer Nature Switzerland AG 2021
W. Koepf, *Computer Algebra*, Springer Undergraduate Texts in Mathematics and Technology,
https://doi.org/10.1007/978-3-030-78017-3

Printed in the United States
by Baker & Taylor Publisher Services